Computer Modelling of Fluids Polymers and Solids

NATO ASI Series

Advanced Science Institutes Series

A Series presenting the results of activities sponsored by the NATO Science Committee, which aims at the dissemination of advanced scientific and technological knowledge, with a view to strengthening links between scientific communities.

The Series is published by an international board of publishers in conjunction with the NATO Scientific Affairs Division

A Life Sciences **B Physics**	Plenum Publishing Corporation London and New York
C Mathematical and Physical Sciences **D Behavioural and Social Sciences** **E Applied Sciences**	Kluwer Academic Publishers Dordrecht, Boston and London
F Computer and Systems Sciences **G Ecological Sciences** **H Cell Biology**	Springer-Verlag Berlin, Heidelberg, New York, London, Paris and Tokyo

Series C: Mathematical and Physical Sciences - Vol. 293

Computer Modelling of Fluids Polymers and Solids

edited by

C.R.A. Catlow

Davy Faraday Research Laboratory,
The Royal Institution,
London, United Kingdom

S.C. Parker

Department of Chemistry,
University of Bath,
Bath, United Kingdom

and

M.P. Allen

H.H. Wills Physics Laboratory,
University of Bristol,
Bristol, United Kingdom

Kluwer Academic Publishers

Dordrecht / Boston / London

Published in cooperation with NATO Scientific Affairs Division

Proceedings of the NATO Advanced Study Institute on
Computer Modelling of Fluids Polymers and Solids
Bath, United Kingdom
September 4–17, 1988

Library of Congress Cataloging in Publication Data

NATO Advanced Study Institute on Computer Modelling of Fluids Polymers and Solids (1988 : University of Bath, U.K.)
Computer modelling of fluids polymers and solids : proceedings of the NATO Advanced Study Institute on Computer Modelling of Fluids Polymers and Solids, held at the University of Bath, U.K., Sept. 4-17th,1988 / edited by C.R.A. Catlow, S.C.Parker, M.P. Allen.
p. cm. -- (NATO ASI series. Series C, Mathetical and physical sciences ; vol. 293)

1. Condensed matter--Mathematical models--Congresses.
2. Condensed matter--Computer simulation--Congresses. 3. Polymers--Congresses. 4. Amorphous substances--Congresses. I. Catlow, C. R. A. (Charles Richard Arthur), 1947- . II. Parker, S.C. III. Allen, M.P. IV. Title. V. Series: NATO ASI series. Series C, Mathematical and physical sciences ; no. 293.
QC173.4.C65N374 1988
530.4'1--dc20 89-28175

ISBN-13: 978-94-010-7621-0 e-ISBN-13: 978-94-009-2484-0
DOI: 10.1007/978-94-009-2484-0

Published by Kluwer Academic Publishers,
P.O. Box 17, 3300 AA Dordrecht, The Netherlands.

Kluwer Academic Publishers incorporates the publishing programmes of
D. Reidel, Martinus Nijhoff, Dr W. Junk and MTP Press.

Sold and distributed in the U.S.A. and Canada
by Kluwer Academic Publishers,
101 Philip Drive, Norwell, MA 02061, U.S.A.

In all other countries, sold and distributed
by Kluwer Academic Publishers Group,
P.O. Box 322, 3300 AH Dordrecht, The Netherlands.

Printed on acid-free paper

TABLE OF CONTENTS v

PREFACE

Computer Modelling techniques have developed very rapidly during the last decade, and interact with many contemporary scientific disciplines. One of the areas of greatest activity has concerned the modelling of condensed phases, including liquids solids and amorphous systems, where simulations have been used to provide insight into basic physical processes and in more recent years to make reliable predictions of the properties of the systems simulated. Indeed the predictive role of simulations is increasingly recognised both in academic and industrial contexts. Current active areas of application include topics as diverse as the viscosity of liquids, the conformation of proteins, the behaviour of hydrogen in metals, the diffusion of molecules in porous catalysts and the properties of micelles.

This book, which is based on a NATO ASI held at the University of Bath, UK, from September 5th-17th, 1988, aims to give a general survey of this field, with detailed discussions both of methodologies and of applications. The earlier chapters of the book are devoted mainly to techniques and the later ones to recent simulation studies of fluids, polymers (including biological molecules) and solids. Special attention is paid to the role of interatomic potentials which are the fundamental physical input to simulations. In addition, developments in computer hardware are considered in depth, owing to the crucial role which such developments are playing in the expansion of the horizons of computer modelling studies.

An important feature of this book is the exercises and problems in the Appendix. These proved to be one of the most successful aspects of the ASI, and they provide an introduction to and illustrations of most of the current techniques in the field.

The ASI was made possible by a generous grant from the NATO Scientific Affairs Division. We are also grateful for the additional support that was provided by the SERC Collaborative Computer Project CCP5 and by Chemical Design Ltd. We would further like to acknowledge the enormous contribution made to the success of the ASI by the organising committee, including Maurice Leslie, Bill Smith, David Fincham and David Heyes, by the University of Bath Computing Service and by graduate students from both Bristol and Bath.

The success of the ASI was also enhanced by the loan of 16 Inmos T800 transputers, and an Active Memory Technology Distributed Array Processor (DAP). Thanks are due to Andy Jackson, Tony Hey, Dave Nicolaides and John Alcock.

Finally, we would like to thank Mrs. H. Hitchen for her invaluable help in the organisation of the meeting and in the preparation of the proceedings.

C. R. A. Catlow, S. C. Parker, M. P. Allen

Lecturers

Dr. C. L. Brooks, Department of Chemistry, Carnegie-Mellon University, Pittsburgh, PA 15213, U.S.A.

Prof. C. R. A. Catlow, Department of Chemistry, University of Keele, Keele, Staffordshire. ST5 5BG, U.K.

Dr. J. Clarke, Department of Chemistry, UMIST, Sackville Street, Manchester, M60 1QD, U.K.

Dr. D. Evans, Research School of Chemistry, Australian National University, P.O.Box 4, Canberra, ACT 2600, Australia.

Dr. D. Fincham, Computer Centre, University of Keele, Keele, Staffs ST5 5BG, U.K.

Dr. D. Frenkel, Fysisck Laboratorium, Rijksuniversiteit, Sorbonnelaan 4, Utrecht, Netherlands.

Dr. M. J. Gillan, Department of Physics, University of Keele, Keele Staffs. ST5 5BG, U.K.

Dr. K. Heinzinger, 6500 Mainz, Mainz Saarstrasse 23, Postfach 3060, West Germany.

Dr. R. A. Jackson, Department of Chemistry, University of Keele, Keele, Staffs. ST5 5BG., U.K.

Dr. A.J.C.Ladd, Lawrence Livermore National Laboratory, University of California, P.O.Box 808, Livermore, California 94550 U.S.A.

Dr. Guilia de Lorenzi, Consiglio Nazionale delle Richerche, Centro di Fisica Stati Aggregati ed Impianto Ionico, 38050 Povo,Trento Italia.

Dr. M. Meyer, Laboratoire de Physique des Materiaux, Centre National de la Recherche Scientifique, 1 Place Aristide-Briand, Bellevue, 92195 Meudon Principal Cedex, France.

Dr. S. C. Parker, Department of Chemistry, University of Bath, Claverton Down, Bath. BA2 7AY, U.K.

Dr. S. Price, University of Cambridge, University Chemical Laboratory Lensfield Rd, Cambridge, CB2 1EW, U.K.

Dr. J.P. Ryckaert, Pool de Physique, Faculte de Science, Universite Libre de Bruxelles, C.P. 223, Bruxelles B 1050 Belgium.

AN INTRODUCTION TO COMPUTER MODELLING OF CONDENSED MATTER

C. R. A. CATLOW

Department of Chemistry, University of Keele, Keele, Staffs. ST5 5BG.

1. INTRODUCTION

This book is concerned with the computer simulation of condensed matter at the atomic and molecular levels. Indeed, we can define this area of simulation as the attempt to model and predict the structural and dynamical properties of matter using interatomic force models; the latter clearly play a central role in the field which is reflected by their extensive coverage in this book.

There are two broad philosophies in contemporary simulation studies. First, simulations may be used to provide insight and to illuminate the range and limitations of analytical theories. Much of the earlier work in this field, especially that concerned with the modelling of hard sphere systems, is in this category. And there have been impressive achievements notably the discovery of the long-time tail in the velocity auto-correlation function in dense fluids, a detailed discussion of which is given by Ladd in Chapter (3). The second approach uses simulation as a technique to predict the properties of real systems. One of the best examples here is the work of Parker and Price (summarised in Chapter (16)) concerning the mantle mineral Mg_2SiO_4 for which there have been successful predictive simulations of the behaviour of the material at high temperatures and pressures. This type of application makes high demands on the quality of the interatomic potential used.

The principle techniques used in the simulation field are energy minimisation, molecular dynamics and Monte-Carlo methods, all of which are reviewed in detail in this book. The great majority of calculations are based on a classical description of the system, but we should note that the incorporation of quantum effects into simulations is now possible; and in Chapter (6) Gillan reviews this important development. Hybrid methods which combine simulation with electronic structure techniques (for example, the recent work of Car and Parrinello[(1)]) are also of growing importance. In addition, in solid state studies the embedding of quantum mechanical cluster calculations by a simulated surrounding structure is becoming increasingly common, as in the recent studies of Harding et al[(2)] and Vail et al[(3)].

A brief introduction to the main features of each simulation technique is given later in this Chapter; and in the final section we give a short review of the applications of energy minimisation

C.R.A. Catlow et al. (eds.), Computer Modelling of Fluids Polymers and Solids, 1–28.

techniques, the use of which has been one of the most productive areas in the simulation field. However, to demonstrate the scope and extent of the field, we first present a general summary of the more important areas of application of simulations, which include the following:

(i) Structure and dynamics of molecular liquids and solids, where, for example, in recent studies[4] of diatomic liquids (e.g. Cl_2), impressive agreement between theoretical and experimental properties - both structural and dynamical - has been achieved. In addition, several successful studies are reported on phase transitions and dynamical properties of molecular solids.

(ii) Aqueous solutions and electrolytes, for which, as discussed in Chapter (14), simulations can now yield adequate models for the structure of water and have given considerable insight into the structures of hydrated ions.

(iii) Simulation of micelles and colloids where valuable qualitative insight has been gained into the behaviour of these complex systems.

(iv) Simulation of the structures, mechanical properties and dynamics of polymers - a very active field in recent years in which simulations using supercomputers have allowed phenomena such as polymer reptation to be modelled.

(v) Simulation of complex crystal structures, where energy minimisation methods can now make very detailed predictions of the structures and properties of crystals with very large unit cells, e.g. the microporous zeolites discussed in Chapter (15).

(vi) Defect structures and energies in solids, for which very detailed predictions are now available for a wide variety of materials as discussed later in this Chapter.

(vii) Sorption in porous media - an area where there is currently rapid progress in topics ranging from capillary action to the location by simulation of reactive molecules in zeolite pores.

(vii) Properties of surfaces, surface defects and impurities and of surface layers, where calculations have made realistic predictions of surface structural properties[5], and of the segregation of impurities and defects to surfaces[6]. In addition, elegant dynamical simulation studies of the behaviour of sorbed layers have also been performed[7]. Simulation studies of grain boundaries and interfaces is also a field of growing importance.

(ix) Structural properties of metal hydrides where work discussed by Gillan in Chapter (6), has shown the valuable role of quantum simulation techniques.

(x) Studies of liquid crystals where simulations have improved our understanding of the phase diagrams of these systems and of the nature of order-disorder transitions.

(xi) Structure and dynamics of glasses, for which simulation studies have been performed on both oxide and halide materials yielding structural models in good agreement with experiment.

(xii) Studies of viscosity and shear thinning where there have been several successful studies of the atomic processes responsible for these macroscopic phenomena.

(xiii) Investigation of protein dynamics, in which there has been an explosion of work over the past five years which is discussed in detail in Chapter (12).

(xiv) Modelling of pharmaceuticals, where energy minimisation procedures are now used routinely in many industrial applications.

From above brief summary (which is far from comprehensive) it is clear that computer simulation methods range in their application from solid state physics through physical and inorganic chemistry and materials science to biological sciences. Almost all these applications are discussed later in this book. Our discussion in this Chapter continues with a summary of basic considerations relating to techniques, potentials and computer hardware.

2. BASICS OF COMPUTER SIMULATIONS

Before discussing the features of the three principle types of simulation methods, it is necessary to consider two matters relating first to the types of ensembles and secondly to the use of periodic boundary conditions. All simulation methods rest on the specification of a finite number of particles. We need therefore to consider the statistical mechanical implications of the various techniques, and the ways in which our finite collection of particles can be made to mimic an infinite system.

(2.1) Ensembles

For Molecular Dynamics (MD) and Monte Carlo (MC) techniques that are discussed in greater detail below, there are a variety of statistical ensembles that may be employed. Simulations have been reported using the four following types:

(i) The Microcanonical ensemble in which the ensemble contains a constant number of particles (N), constant volume (V) and constant internal energy (E); hence the alternative denotation as the NVE ensemble.

(ii) The Canonical ensemble, where N, V and temperature (T) are constant - hence the NVT ensemble.

(iii) The Isothermal-Isobaric (or NPT) ensemble where pressure P is constant, in addition to N and T.

(iv) The Grand Canonical (or μVT) ensemble in which the number of particles is not constant but may vary in order to achieve constant chemical potential, μ.

M.D. simulations are most easily carried out in the microcanonical ensemble, while MC is naturally suited to the canonical ensemble. However, much modern MD work is undertaken using constant pressure (NPT) ensembles, while MC simulations using the Grand Canonical Ensemble have been extensively studied. Several illustrations of the use of all four types of ensemble will follow in later Chapters.

(2.2) Periodic Boundary Conditions (PBCs)

As we have noted, simulations necessarily concern finite numbers of particles which are contained in a 'simulation box'. However, by application of periodic boundary conditions, an infinite system may be simulated. This is achieved by generating an infinite number of images of the basic simulation box as shown in fig. (1). The resulting infinite system, of course, has no surfaces.

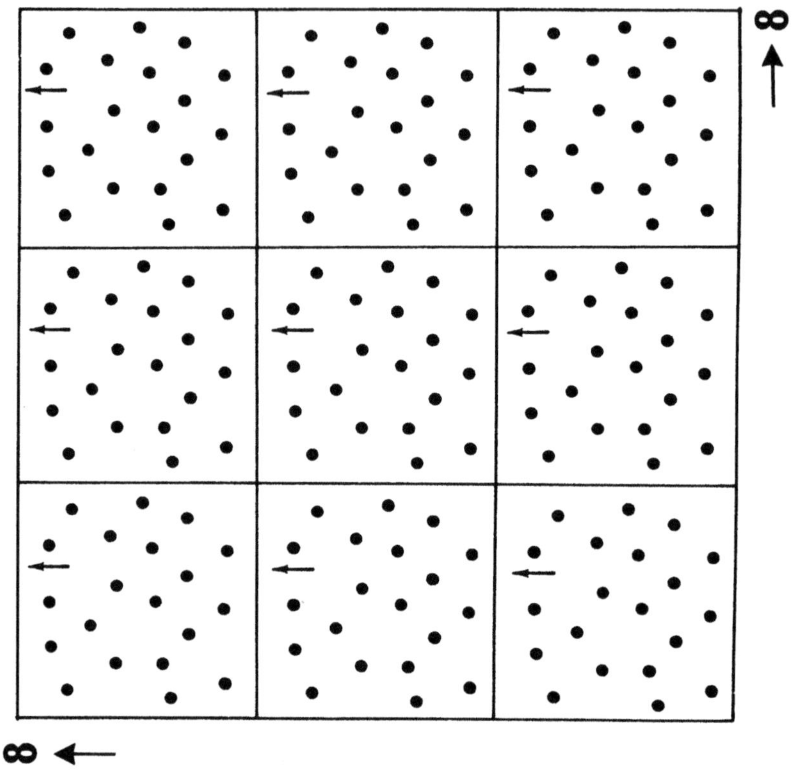

Fig. (1) Illustration of periodically repeated ensemble of particles. The arrows below one of the particles indicate it leaving the box, with its image in an adjacent box, re-entering.

In carrying out simulations with PBCs it is necessary to ensure that when a particle leaves the box on one side, its image from a neighbouring box re-enters on the opposite side, as shown in fig. (1). Care must also be taken with summations which will extend into the neighbouring boxes and will be discussed in greater detail in subsequent chapters.

The use of PBCs may correspond to physical reality as in simulations of crystalline materials, or be artificial as in work on liquids or amorphous systems. In the latter case the imposition of the artificial periodicity is rarely serious except where very long wave length properties (or very small simulation boxes) are considered.

There are, of course, cases where PBCs are not needed, as in modelling of droplets, small clusters and in some work on large macro-molecules. But in the vast majority of work on solids, liquids and amorphous materials the use of PBCs is standard practice with the number of particles in the basic simulation box ranging from a few hundred to several thousand.

These fundamental factors pertain to all simulation techniques; we now continue by discussing in further detail the three basic types of simulation.

(2.3) Energy Minimisation (EM)

EM methods are restricted to the prediction of static structures and of those properties which can be described within an harmonic (or quasi-harmonic)dynamical approximation; there is no explicit inclusion of atomic motions. Despite these limitations, the methods have proved to be powerful and remarkably flexible in their range of applications. The basis of the method is simple: the energy $E(\underline{x})$ is calculated, using knowledge of interatomic potentials, as a function of all the structural variables, $\underline{x}$, (e.g. atomic coordinates or bond lengths and angles); an initial configuration is specified and the variables are adjusted, using an iterative computational method, until the minimum energy configuration is obtained, i.e. the system runs 'down-hill' as shown diagramatically in fig.(2a). The method may be extended if vibrational properties of the energy minimum are calculated using the harmonic approximation; thus for a molecule, normal coordinate analysis may be used, while for a solid, standard lattice dynamical methods are employed (as discussed by Parker in Chapter (16)). This allows entropies in addition to enthalpies to be calculated and hence 'free-energy' minimisation may be performed.

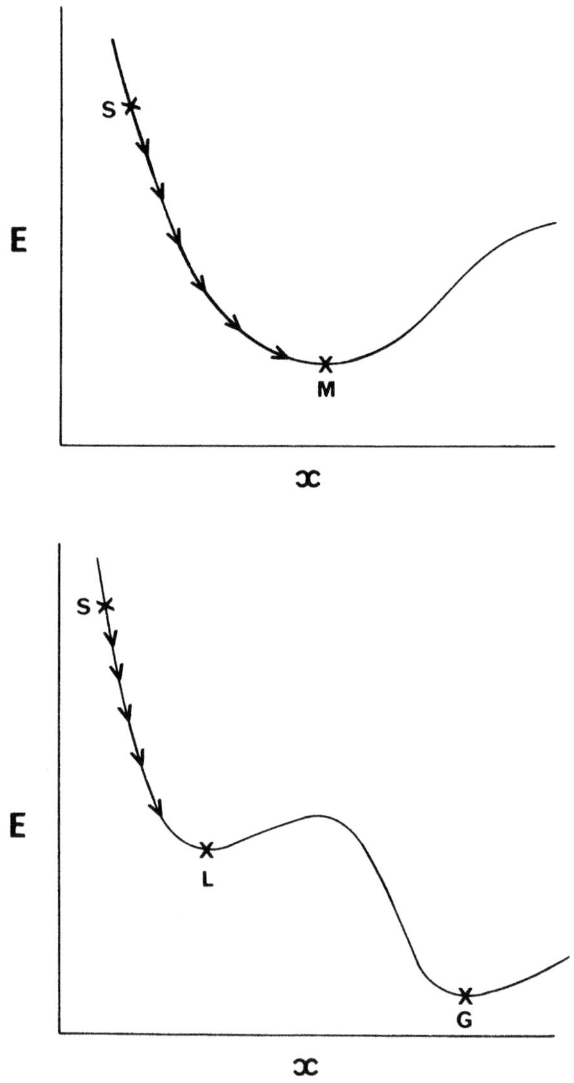

Fig (2). (a) shows energy (E) minimisation with respect to some structural variable (x). The system runs down from the starting point S to the minimum M. (b) illustrates the local minimum problem with the system running from S to the local mimimum L, despite the presence of lower global minimum G.

The most important technical features of energy minimisation methods concern first the type of summation procedures used in evaluating the total interaction energy; this problem is, however, common to all atomistic simulations and is discussed elsewhere in the book by Jackson (Chapter 15) and Brooks (Chapter 12). Secondly there is the choice of the computational minimisation method which is now considered in further detail.

Minimisation algorithms may be classified according to the type of derivative that is used in choosing the search direction. The simplest methods employ the energy function alone and search over configuration space until the minimum is located. While such methods may be suitable for very simple problems with few variables, they are unacceptably inefficient in almost all contemporary studies. Much greater efficiency is obtained using gradient techniques in which the first derivatives $\partial E/\partial x_i$ with respect to all the structural variables x_i are calculated. These then guide the direction of minimisation. The following two iterative gradient methods are widely used:

(i) Steepest descent, in which the minimisation 'follows' the gradient, i.e. the values of $x_i^{(k+1)}$ in the $(k+1)^{th}$ iteration are related to those in the kth by:

$$\underline{x}^{(k+1)} = \underline{x}^{(k)} + \alpha^{(k)}\underline{s}^{(k)} \qquad (1)$$

where $\underline{s}^{(k)} = -\underline{g}^{(k)}$ with $g_i^{(k)} = (\partial E/\partial_x)_i^{(k)}$: $\alpha^{(k)}$ is a numerical constant chosen each iteration in order to optimise the efficiency of the procedure.

(ii) Conjugate gradients. In this method the displacement vector $s^{(k)}$ uses information on the previous values of the gradients which speeds up convergence. Thus for $s^{(k)}$ we write

$$\underline{s}^{(k)} = - \underline{g}^{(k-1)} + \beta^{(k)}\underline{s}^{(k-1)}, \qquad (2)$$

$$\text{where } \beta^{(k)} = \frac{\underline{g}^{T(k-1)} \cdot \underline{g}^{(k-1)}}{\underline{g}^{T(k-2)} \cdot \underline{g}^{(k-2)}}, \qquad (3)$$

where the $g^{(k)}$ are vectors whose components are the derivatives with respect to individual coordinates and where the superscript, T, indicates the transpose of the vector.

Greater details of these methods will be found in reference (8). Their efficiency is greatly improved over search methods, but several hundred iterations are normally required even if the 'starting point' of the minimisation is relatively close to the final minimum.

Much more rapid convergence can be achieved when knowledge of second derivatives is used to guide the minimisation direction, as in Newton methods where the iterative minimisation proceeds according to the expression:

$$\underline{x}^{(k+1)} = \underline{x}^{(k)} - \underline{\underline{H}}^{(k)} \cdot \underline{g}^{(k)} \qquad (4)$$

where the matrix $\underline{\underline{H}} = \underline{\underline{W}}^{-1}$, in which the elements W_{ij} are the second derivatives $\left(\frac{\partial^2 E}{\partial x_i \partial x_j}\right)$. It can readily be shown that such methods must reach the minimum within one iteration if we are in a region of configurational space in which the energy is harmonic with respect to the minimum. This, of course, does not apply generally. The method is, however, far more rapidly convergent than gradient procedures.

The advantages of the improved convergences would, however, easily be lost in the extra computational effort required in calculating and inverting the second derivative matrix each iteration. It is fortunate therefore there are algorithms which enable the inverse second derivation matrix, $\underline{\underline{H}}$, to be updated each iteration without recalculation and inversion. The most widely used of these is the Davidon-Fletcher-Powell algorithm in which the matrix $\underline{\underline{H}}$ is updated each iteration according to the formula:

$$\underline{\underline{H}}^{(k+1)} = \underline{\underline{H}}^{(k)} - \frac{(\underline{\underline{H}}^{(k)} \cdot \underline{\Delta g}^{(k)}) \cdot (\underline{\Delta g}^{T(k)} \cdot \underline{\underline{H}}^{(k)})}{(\underline{\Delta g}^{T(k)} \cdot \underline{\underline{H}}^{(k)} \cdot \underline{\Delta g}^{(k)})} - \frac{(\underline{\Delta_x}^{(k)} \cdot \underline{\Delta_x}^{T(k)})}{(\underline{\Delta x}^{T(k)} \cdot \underline{\Delta_g}^{(k)})} \qquad (5)$$

where

$$\underline{\Delta g}^{(k)} = g^{(k+1)} - g^{(k)} \qquad (6)$$

and

$$\underline{\Delta x} = \underline{x}^{(k+1)} - \underline{x}^{(k)} \qquad (7)$$

and in which the superscript 'T' indicates the transpose of the vector.

Such algorithms are, of course, approximate, and it is necessary to recalculate $\underline{H}$ typically every 20–30 interations. However, with the use of update procedures, Newton methods converge much more rapidly and are far less computationally expensive than gradient techniques. There remains, however, one major computational problem in the need to store the inverse of the second derivative matrix. In systems with large numbers of variables, c.p.u. requirements soon become formidable. For example, if we are applying minimisation methods to model the crystal structures of zeolites - a problem discussed by Jackson in Chapter (15), then unit cells with 300 atoms are common. Since each Cartesian coordinate of each atom is a variable, a 900 x 900 matrix will be stored requiring ~ 1 Megaword of memory; c.p.u. memory must be used, otherwise unrealistic amounts of time will be spent paging the matrix into and out of the c.p.u. Clearly such memory requirements will prevent the use of Newton methods in large scale minimisation problems on machines without large c.p.u. memories; and even with modern supercomputers very large problems may not be feasible. When this occurs, recourse must be made to the gradient techniques which, although requiring more c.p.u. time, have far lower memory requirements as only the gradients of the energy need to be stored.

Some of the most successful applications of Newton minimisation techniques are in solid state studies, especially of defects. These will be considered in section (5) of this chapter, and in Chapter (15).

E.M. techniques clearly have the advantages of simplicity and versatility which has led to them being widely applied to e.g. crystal structure modelling (of both organic and inorganic materials), to studies of the conformation of molecules, including biological macromolecules (note that in these fields, E.M. is often referred to by the term '<u>molecular mechanics</u>') and to modelling of defects in solids. Compared with many other computer simulation techniques E.M. requires little c.p.u. time, and this factor allows the use of more complex and sophisticated potentials. Nevertheless E.M. methods are severely limited; they inherently omit any representation of atomic motions and time dependent phenomena. Moreover, even given the usefulness of the static approximation, there is a major additional difficulty in that E.M. techniques can only be guaranteed to locate the nearest local minimum to the starting point of the calculation as shown diagramatically in fig. (2b). The <u>local minimum problem</u> may be very severe as in studies of protein conformations, although less difficulties are encountered in solid state applications. There is no general solution to the problem. The use of several different starting points in a calculation is obviously advisable. In addition, energy minimised configurations may be input into dynamical simulations (using the techniques summarised below) which may allow energy barriers to be surmounted. There remains, however, no guarantee that the lowest energy or <u>global</u> minimum has been located.

E.M. remains, however, a widely used technique, which is of considerable value provided its limitations are borne in mind. It is undoubtedly most appropriate as a 'refinement technique' for improving structural models based on approximate knowledge from experiment and from other sources. Illustrations are given later in this Chapter and in Chapter (15).

(2.4) <u>Molecular Dynamics</u> (M.D.)

Unlike the energy minimisation techniques discussed above, molecular dynamics includes atomic kinetic energy explicitly. It does so in a simple and direct manner by assigning all particles in the simulation box a position and velocity. With knowledge of the interatomic potentials, the forces acting on the particle may be calculated. The simulation then proceeds by solving Newton's equation of motion for the ensemble by allowing it to evolve through a succession of time steps, each of Δt. In the limit of an infinitely small value of Δt, we can write for the coordinates x_i and velocities v_i of the ith particle before and after Δt:

$$x_i(t + \Delta t) = x_i(t) + v_i(t)\Delta t, \quad (8a)$$

$$v_i(t + \Delta t) = v_i(t) + \frac{f_i(t)\Delta t}{m_i}, \quad (8b)$$

where f_i is the force acting on the particle and m_i its mass. In practice a finite value of Δt is, of course used (typically in the range 10^{-15} - 10^{-14} sec) and more sophisticated updating algorithms are employed involving higher powers of Δt. The nature of the algorithms used together with the special strategies employed when simulating ensembles of hard spheres are described in later chapters.

M.D. simulations normally consist of the following steps:

(i) An initial set-up procedure in which the positions and velocities are assigned to particles in the simulation box, the velocities being chosen in line with a target temperature for the simulation.

(ii) An equilibration period in which the ensemble attains equipartition between potential and kinetic energy and a thermalised distribution of velocities. During this period, velocities will frequently be scaled to bring them in line with the target temperature. The extent of the period will depend on the temperature and on the degree of anharmonicity of the potential surface sampled by the particles in the ensemble: a high degree of an anharmonicity will promote the rapid redistribution of energy. Several thousand time steps are normally needed for complete equilibration.

(iii) The 'production run' then follows in which the equilibrated ensemble is allowed to evolve in time - normally for several thousand time steps. Coordinates and velocities for each time step are stored on disk or tape for subsequent analysis. This analysis will include the calculation of radial distribution functions, diffusion coefficients and a range of correlation functions including the velocity auto-correlation function (v.a.f.) and the van Hove correlation function. Further discussion of these quantities, their importance and the methods used in their calculation are given in Chapter (3), and in the excellent monograph of Allen and Tildesley[9]. Calculation of the diffusion coefficient is particularly simple; it relies on the result of random walk theory, which gives:

$$\langle r_\alpha^2 \rangle = 6 D_\alpha t + B_\alpha \quad (9)$$

where $\langle r_\alpha^2 \rangle$ is the mean square displacement of particles of type α in time t. D_α is the diffusion coefficient and B_α is related to the mean amplitude of the particle vibrational motion. Diffusion coefficients can therefore be obtained simply by plotting $\langle r^2 \rangle$ vs t, and measuring the slope of the plot which will show a linear increase with t if diffusion if occuring. Fig. (3) illustrates results obtained by Gillan and coworkers[10] for CaF_2 at 1200 K: Ca is not diffusing and there is no increase of $\langle r^2 \rangle$ with t; in contrast rapid diffusion of the F^- ions is clearly occuring - a feature of the simulations that accords well with experiment. Gillan's work also illustrates the value of M.D. simulation in yielding detailed mechanistic information concerning ion dynamics; thus by following the trajectories of the migrating ions, diffusion mechanisms may be deduced. Indeed M.D. techniques have made major contributions to our understanding of atomic diffusion mechanisms in solids.

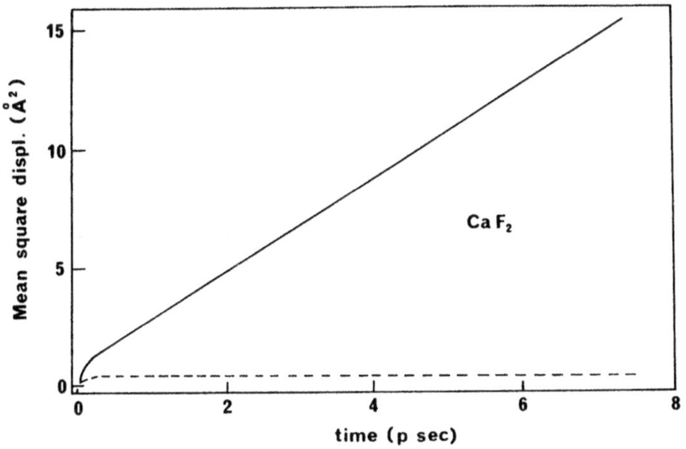

Fig.(3) Mean square displacements as a function of time for F(full line) and Ca (dotted line) in M.D. simulation of CaF_2 (see reference 10).

M.D. is probably the most powerful and widely used simulation method. Unlike other techniques, it yields detailed dynamical information and includes time as a parameter in the simulation. There are, however, a number of restrictions associated with M.D. the most important of which are as follows:

(i) The total amount of 'real time' available to the simulation is limited, generally to less than 100 ps. (although with increasing computer power, the horizons are constantly expanding). If the simulation is to be of value it is necessary that all processes of interest take place to a statistically significant extent within this period. Thus in studying, for example, diffusion in solids, it will not normally be possible to observe a sufficient number of atomic migration events within 100 psec., and M.D. will be of little value. However, the technique may be used for those solids which have exceptionally high atomic mobilities, e.g. the 'superionic conductors' which include materials such as $SrCl_2$, Li_3N, AgI and CaF_2 (referred to above).

(ii) The choice of interatomic potential is normally more restrictive than in E.M. methods. In particular it is difficult to include in M.D., effects of atomic polarisability without the expenditure of very large amounts of computer time, since the dipole moments on all atoms have to be calculated generally via an iterative procedure, each time step. Again, expansion in computer power is making such calculations increasingly feasible. But, to date, the vast majority of M.D. studies have omitted the effects of polarisability, which in many cases might reduce the reliability of the predictions of the simulations.

(iii) If periodic boundary conditions are used, then surface effects are excluded from the simulation. Of course, the use of periodic boundaries is not always necessary, and M.D. may be performed on a simple ensemble of particles without periodic images. In addition it is possible to do M.D. on infinite (and on finite) slabs and surfaces.

(iv) The method is computationally expensive. Simulations on systems with ~ 1000 particles for 'real-times' of ~ 50 p.sec. will normally take several hours on a modern supercomputer, e.g. the CRAY XMP. The continuing expansion in computer power is, however, reducing the problems associated with the computational demands of M.D.

Despite these limitations, M.D. is an increasingly flexible and widely used technique. M.D. may now be routinely performed in both NVE and NPT ensembles. Stochastic dynamical techniques have been developed in which the simulation box is in effect coupled to a thermal bath, which results in random, stochastic forces being applied to the particles during the course of the simulation. A variety of 'non-equilibrium' M.D. methods are available as discussed by Evans in Chapter (5). Perhaps the most exciting recent development has been the incorporation of quantum effects into M.D. via the path integral formalism discussed by Gillan in Chapter (6).

(2.5) Monte-Carlo Techniques

M.C. is a technique of computational statistical mechanics ideally suited for calculating ensemble averages in the canononical (NVT) ensemble. The simulation proceeds via the generation of successive configurations of the ensemble by a series of random moves each of which normally involves the displacement of only one particle. Once a sufficient number (normally several thousand) configurations have been generated, ensemble averages are straightforwardly calculated.

Possibly the most crucial technical feature of an M.C. simulation concerns the 'acceptance' procedure, i.e. the criteria used to decide whether a configuration generated after a move should be included in the final set of configurations which are stored and used in calculating ensemble averages. The most widely used approach is based on the Metropolis method, and is discussed in Chapter (4). The method in effect weights the probability of acceptance of a new configuration by its Boltzmann factor. In grand-canonical (μVT) M.C. a move may involve the inclusion of an additional particle in the ensemble; Chapter (4) presents a detailed account of the acceptance criteria used in such simulations.

M.C. has similar computational requirements to M.D. and like M.D., M.C. calculations normally have an equilibration period followed by a production run. But unlike M.D., the successive configurations in the simulation have no relationship in time. M.C. is therefore inherently more restricted than M.D. as time dependent phenomena cannot be directly investigated. However, the method is the simplest and most direct way of undertaking simulations in the canonical and grand-canonical ensembles. Moreover, it continues to have considerable vitality with important fundamental developments, as in the recent studies of phase equilibria which are discussed in Chapter (4), and with exciting applications such as the work of Cheetham and coworkers[11] on the behaviour of sorbed molecules in zeolite catalysts.

(2.6) Free Energy Calculations

We conclude this section of the Chapter by commenting on one of the most important topics in contemporary simulation studies, i.e. the determination of free energies. This is a difficult problem except for one type of system, i.e. those crystalline solids for which the harmonic approximation is acceptable. In this latter case, standard lattice dynamical techniques can be used to calculate internal energies and entropies and hence free energies. Indeed, such methods are proving to be of considerable value in simulation studies of phase transitions in solids as discussed by Parker in Chapter (16).

For liquids and disordered systems, the central problem in free energy calculations is the generation of a set of configurations for the ensemble which sample configurational space sufficiently well to permit the reliable calculation of a partition function. Both M.D. and M.C. (with Boltzmann sampling via e.g. the use of the Metropolis method) weight the sampling close to the energy minimum - a procedure which is generally acceptable for calculating ensemble averages, but not for the partition function to which appreciable contributions are made from configurations which may be remote from the energy minimum. Non-Boltzmann sampling techniques are available (as discussed in the book of Allen and Tildesley[9]). Greater success has, however, been enjoyed in calculating free energy differences, for which perturbation methods can be employed, as discussed by Brooks in chapter (12). Indeed several later chapters will return to this important theme.

This completes our brief introduction to the techniques of computer simulation. As we have already emphasised, the reliability of simulation techniques is largely dependent on the quality of the interatomic potentials used, the general features of which are discussed in the next section.

3. POTENTIALS

The interatomic potential V for a system of n particles describes the variation of the total potential energy of the system as a function of the nuclear coordinates $r_i \ldots r_n$. i.e.

$$V = V(r_i \ldots\ldots r_n) \tag{10}$$

In practice, V is generally broken down into a series of summations

$$V = \sum_{ij}{}' V(r_i, r_j) + \sum_{ijk}{}' V(r_i, r_j, r_k) + \sum_{ijkl}{}' V(r_i, r_j, r_k, r_l) \tag{11}$$

where the first term refers to a sum over all pairs of atoms, the second over all triplets, and the third over all quartets, with the summation continuing in principle up to 'N-body' terms. The primes on the summations indicate that the multiple counting of equivalent terms (e.g. ij and ji) is avoided.

The majority of simulations approximate V simply by the pair potential terms, which in turn are commonly decomposed as follows:-

$$V(r_i r_j) = \frac{q_i q_j}{r_{ij}} + \Phi(r_{ij}), \tag{12}$$

with the first term being the Coulomb potential between a pair of atoms with charges q_i and q_j and separation r_{ij}. $\Phi(r_{ij})$ is the 'short-range' potential acting between the atoms, which includes contributions from many terms, including covalence, non-bonded repulsion (itself a complex quantity comprising internuclear repulsion and electron ... electron Coulomb and exchange energies), and dispersion. Several types of analytical functions are used to model $\Phi(r_{ij})$ as will be discussed below.

The inclusion of many-body terms in a simulation will normally considerably increase the computational demands. However, the importance of including them in reliable simulations is increasingly recognised. The types of function employed will be discussed below.

(3.1) Potential Functions and Parameters

It is, of course, possible to use numerical potentials i.e. tabulations of Φ as a function of r; and indeed functions of this type have been widely and successfully used by Mackrodt and coworkers(12). The bulk of simulations have, however, employed analytical functions. In the case of the Coulomb term, the r^{-1} function is of course exact. But for other terms the functions are approximate. The following are in common use for the different classes of interaction:

A Two-body

(i) Bonded interactions. The simplest and most widely used function applied to a bonding pair of atoms is the bond harmonic function, i.e.

$$\Phi(r) = \tfrac{1}{2}K\,(r-r_o)^2, \tag{13}$$

where r_o is the equilibrium bond distance and K is the bond force constant. Functions of this type are quite adequate for values of r_{ij} close to r_o. Greater reliability over a wider range of separations can be achieved by use of the Morse function, which has the form:

$$\Phi(r) = D\,\{1 - \exp[-\beta(r-r_o)]\}^2, \tag{14}$$

where D is the dissociation energy of the bond, r_o is the equilibrium bond length and β is a variable parameter, which can, however, be determined from spectroscopic data.

(ii) Non-Bonded Interactions

Again, several functions are available, with possibly the most widely used being the Lennard-Jones potential, which takes the form:

$$V(r) = 4\,\epsilon\left[\left(\frac{\sigma}{r}\right)^{12} - \left(\frac{\sigma}{r}\right)^{6}\right], \tag{15}$$

with a steeply repulsive r^{-12} function describing the non-bonded repulsion and an attractive r^{-6} term modelling the dispersive interaction (the leading term of which shows exactly r^{-6} variation with distance). ϵ is the minimum energy of the function (with respect to the infinitely separated atoms) and σ can be interpreted as the approximate radius of the atom. Many simulations have been reported on model, Lennard-Jones fluids, and when suitably parameterised the potentials are well suited to modelling rare gas fluids, and indeed non-bonded interactions in molecular fluids and solids.

An alternative function that has been particularly popular in modelling solids is known as the Buckingham potential, in which the r^{-12} term is replaced by an exponential repulsive term, thus giving:

$$V(r) = Ae^{-r/\rho} - Cr^{-6}. \qquad (16)$$

There is some evidence from quantum mechanical studies that the exponential function is suitable for modelling the short range repulsion between closed shell species.

In concluding this section, we note that functions of the type discussed above are all non-directional atom ... atom potentials, i.e. they are simply functions of the internuclear distances between pairs of atoms. There is increasingly evidence that such models have definite shortcomings and that anisotopic terms must be included; a detailed discussion is given in Chapter (2) of this book.

B Many Body

(i) Bond-bending functions

These are the simplest many-body term, and are applied about trios of atoms in which the central atom subtends an angle Θ, with the bond-bending energy V(Θ), given by:

$$V(\Theta) = \tfrac{1}{2}K_B\ (\Theta - \Theta_o)^2, \qquad (17)$$

where K_B is the bond-bending force constant and Θ_o the equilibrium bond angle. Such functions are clearly most appropriate in covalently bonded systems, and they have been used successfully in modelling quartz and silicates where they are applied around 0-Si-0 bonds with Θ_o being the tetrahedral angle. They are also widely used in force fields for covalently bonded molecules and macromolecules.

(ii) Triple Dipole Terms

Application of third-order perturbation theory to dispersive interactions between atom triplets ijk yields a repulsive term of the following type:

$$V_{ijk} = \frac{K_T(1 + 3\cos\Theta_1\cos\Theta_2\cos\Theta_3)}{r_{ij}^3\ r_{jk}^3\ r_{ik}^3}, \qquad (18)$$

in which the angles and distances are as indicated in fig.(4a). K_T is the triple dipole force constant. The importance of such terms is well established for rare gas fluids and solids (although it is equally well established that such terms do not include all the many body contributions). In addition, triple dipole terms have recently been shown to be of value in modelling silver halide crystals(13) (AgCl and AgBr) whose properties manifest clear deviations from the predictions of pair potential models.

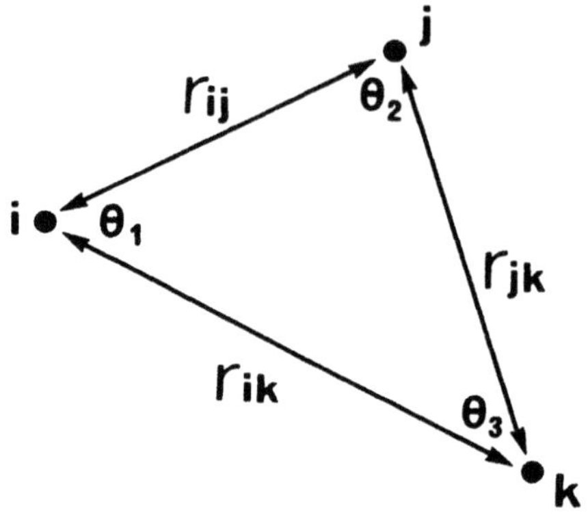

Fig. (4a): Triplet of atoms ijk with bond lengths and bond angles marked.

(iii) Torsional terms

In modelling molecules (especially macromolecules) it is commonly necessary to include a type of 4 body potential that depends on a torsional angle Φ. Thus in a system of atoms, 1,2, 3 and 4, the torsional angle, Φ, is that between the planes defined by atoms 1,2 and 3 and by 2, 3 and 4 as shown diagramatically in fig.(4b). Several functions may be used, for example:

$$V(\Phi) = K[1 \pm \cos(n\Phi)] \tag{19}$$

Further discussion of torsional terms is given in the Chapters (7) and (8) on polymer modelling and in the discussion of biological molecule simulations in chapter (12).

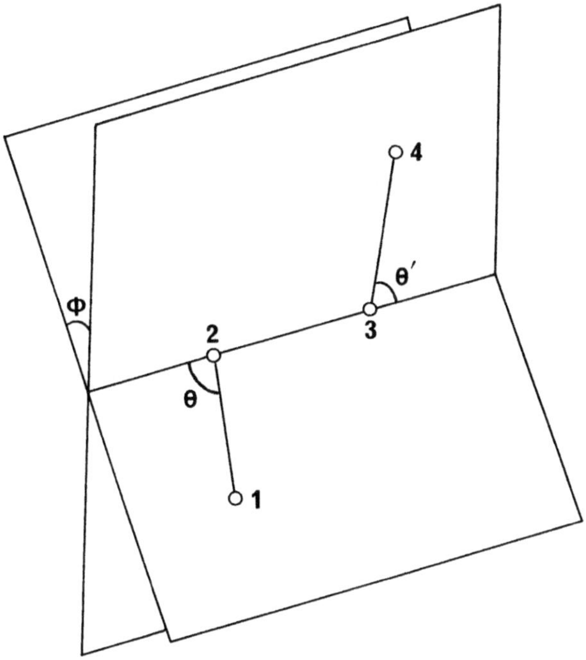

Fig. (4b) Quartet of atoms illustrating torsional angle Φ (after Allen and Tildesley[9]).

C Modelling of Polarisability

It is increasingly recognised that it is necessary to include in simulations the effects of the electronic polarisability of atoms; and indeed it has long been known that such effects are vital for accurate modelling of lattice dynamical, dielectric and defect properties of ionic solids. The simplest approach models polarisability in terms of a point dipole whose magnitude, μ, is proportional to the effective field, E, acting on the atoms, i.e.

$$\mu = \alpha E, \tag{20}$$

where α is the atomic (or ionic) polarisability. Point polarisable ion (or PPI) models are generally acceptable for modelling simple molecular systems. They do, however, fail badly in describing ionic solids. In these systems the coupling between short-range repulsion and polarisation is strong (since polarisation, which involves displacement of valence shell electrons, modifies the short range interactions between ions), and the omission of this coupling leads to a poor description of dielectric and defect properties. This coupling can, however, be well described by using the shell model, which models polarisation in terms of the displacement of a massless 'shell' (which is a point entity) relative to a core in which all the mass of the atom is concentrated, the core and shell being connected by an harmonic spring. The magnitude of the dipole model is determined by the magnitude of the core-shell separation. The present author has discussed in detail elsewhere(14)(15) the merits of the shell model description of polarisability in ionic solids. We consider that the model should be used increasingly in simulating other systems.

(3.2) Parameterisation

The choice of potential model is the first important step in setting up a simulation. The next is to fix the variable parameters in the model, for which there are two broad strategies, i.e. empirical parameterisation and the use of theoretical methods.

Empirical techniques

These methods are particularly simple. They involve the adjustment, usually via a least squares fitting routine, of all or some of the variable parameters in a model, until the best agreement is obtained between calculated and experimental properties (including structural, vibrational, elastic and dielectric properties) of one or several molecules or materials. The method has been very widely used in deriving potentials for molecules, macromolecules and solids (see e.g. references (14) and (16)). And it is the only generally reliable procedure for obtaining polarisation parameters (although there has been notable progress in recent years in calculating polarisabilities). Empirical methods are, however, inherently limited as they only yield information on potentials at spacings close to those in the model compounds, and, of course, because they require model compounds to be available.

Theoretical Methods

Both intra- and inter-molecular potentials may be calculated using a variety of theoretical methods ranging from electron gas techniques, which have been widely and successfully used by e.g. Gordon and Kim(17) and by Mackrodt and coworkers(18) in calculating non-bonded potentials, to ab-initio Hartree Fock methods which are being increasingly used in calculating parameters for both non-bonded and bonding interactions in molecules and solids. Given the continuing growth in computer power the techniques of quantum chemistry will unquestionably be increasingly used in this field.

4. HARDWARE ISSUES

We have already commented on the way in which the horizons of computer simulation studies are being greatly expanded by the growth in computer power. Progress in the field depends to a large extent upon our ability to exploit the special features of the available hardware. Chapters (10) and (11) in this volume will look critically and in detail at the varieties of hardware that are currently available and their adaptation to particular types of simulation. Here we wish to draw attention to the three following issues which are of prime importance:

(i) Parallelism in which different processors carry out operations concurrently is being increasingly exploited for high performance computing. Parallel architecture, which is discussed in detail by Fincham in Chapter (11) can be particularly suitable for simulations, and low cost machines based on e.g. transputer systems will play an increasingly important role in simulation studies.

(ii) Vectorisation

Many of the most powerful 'super-computer' systems rely on vector processing in which operations are carried on blocks of variables rather than successively on single variables (i.e. scalar processing). For efficient use of such machines, programs must be written to exploit the vector processing facilities, a detailed discussion of which is given in chapter (10).

(iii) Matching of Problems and Machines

This general point is of increasing importance with the growing diversity of computers. Simulation problems should be carefully matched to the power and architecture of available hardware. The largest, most powerful supercomputer is not necessarily the best system for a given problem; and dedicated smaller machines may be more effective than multi-purpose large machines.

5. ENERGY MINIMISATION : SOME RECENT APPLICATIONS

In this final section we aim to give a flavour of the types of problem, that are currently being investigated using simulations by presenting some of our recent studies using the simplest technique, energy minimisation. We will show how this technique has proved to be of value in studying structures, properties and defects in materials and molecules.

(5.1) Modelling of Structures

Our first illustration concerns the modelling of crystal structures which is considered in greater detail in Chapters (15) and (16). Energy minimisation techniques may now be used routinely and efficiently to model highly complex inorganic crystal structures. A good example is provided by recent work of Collins[19] on the layer structured mineral muscovite ($KAl_2AlSi_3O_{10}(OH)_2$). The energy minimised crystal structure for this compound is in good agreement with experiment as demonstrated by the comparison in Table (1) between calculated and experimental cell dimensions. We note that the free energy minimised structure at 300K shows improved agreement as regards the C axis lattice parameter. Chapters (15) and (16) show that similar success can be obtained for a wide variety of mineral systems.

TABLE 1 Experimental and calculated structural parameters of Muscovite (All lengths in Å and angles in °) (after ref.19).

CELL DIMENSION	EXPT[20]	MUSCOVITE EXPT[21]	0 K SIM*	300 K SIM¤
a	5.192	5.204	5.246	5.254
b	9.0153	9.018	9.179	9.195
c	20.046	20.073	19.783	20.009
β	95.73	95.82	96.53	96.53
THICKNESS				
TETRAHEDRAL SHEET	2.245	2.243	2.234	2.238
OCTAHEDRAL SHEET	2.089	2.106	2.109	2.113
INTERLAYER SEPARATION	3.393	3.393	3.250	3.359
PARAMETER**				
TETRAHEDRAL SHEET				
mean T-0	1.644	1.644	1.620	1.622
τ	110.9	111.0	107.5	107.6
α	11.3	10.8	7.6	7.5
ΔZ	0.21	0.22	0.33	0.32
OCTAHEDRAL SHEET				
mean M2-0,OH	1.930	1.934	1.932	1.934
Ψ	57.2	57.0	56.9	56.9
O-H	-	0.920	0.975	0.975
INTERLAYER SEPARATION				
$K\text{-}0_{outer}$		3.353	3.331	3.354
$K\text{-}0_{inner}$		2.872	2.980	3.020
Δ		0.481	0.351	0.334

* Energy minimised structure

¤ Free energy minimised structure

** The parameters which characterise the detailed structure of the octahedral and tetraderal sheets and interlayer separation are discussed in greater detail in reference (19).

Our second illustration concerns a small peptide molecule, apamin. This 18-residue poly peptide is a component of bee venom and possesses powerful neurotoxic properties owing to its ability to block calcium dependent potassium fluxes. It has not been possible to crystallise the molecule, although structural information has been obtained from circular dichroism and NMR studies. Freeman et al[22] carried out a detailed energy minimisation study using a number of previously proposed models. The most stable model is illustrated diagramatically in fig. (5); it includes both reverse turn and alpha helical structure; the dihedral angles are reported in table (2). The results are in good agreement with models based on circular dichroism studies and secondary structure prediction[23].

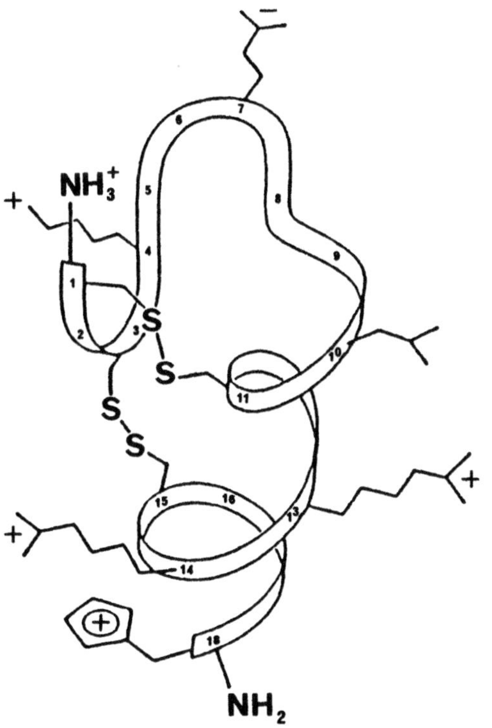

Fig.(5): Energy minimised conformation for the 18 residue peptide, apamin.

TABLE 2 Dihedral angles of proposed apamin models (in degrees) after energy minimisation (see ref. 22).

	Φ	Ψ	ω	χ1	χ2	χ3	χ4	χ5
Cys	36.5	38.4	160.2	-69.6				
Asn	-79.8	158.2	-178.7	166.8	-110.7			
Cys	-54.0	-34.8	179.3	-63.7				
Lys	-116.9	24.9	175.2	-170.8	-174.8	177.6	-178.2	179.9
Ala	-165.0	70.4	-174.2					
Pro	-86.1	22.2	-175.4					
Glu	-81.8	66.5	172.1	-59.9	67.3	27.2		
Thr	-144.6	168.8	-173.0	-43.6	-24.7			
Ala	-147.0	28.4	171.2					
Leu	-121.5	22.0	-156.1	-61.7	174.2			
Cys	-93.5	71.1	173.2	-45.3				
Ala	-53.2	-32.3	160.5					
Arg	-134.5	160.7	174.2	-176.2	175.2	175.9	-90.8	
Arg	-124.4	167.5	-178.1	-63.9	174.5	175.2	-116.9	
Cys	-24.1	-66.6	-177.0	165.4				
Gln	-152.9	145.9	178.9	179.8	50.2	-104.1	1.5	
Gln	-90.3	-30.2	175.2	-68.8	58.3	67.8	1.7	
His	70.6	37.1	-179.9	-47.0	-91.2			

(5.2) Calculation of Properties

For crystalline solids, following energy minimisation, it is possible from calculated first and second derivatives of the lattice energy with respect to atomic coordinates, to obtain a wide range of crystal properties, including elastic, dielectric, piezoelectric and lattice dynamical properties. A good illustration is again provided by Collins'[19] recent study of muscovite, where as shown in table (3), there is good agreement between calculated and experimental elastic constants. Further discussion of this type of calculation is given in Chapter (15). It is, however, now clear that modelling methods may be used to predict this type of property.

TABLE 3 Experimental and calculated Elastic Constants for Muscovite (after Collins[19])

Elastic Stiffness Constants	Experimental (Vaughan & Guggenheim[30]) (10^{10}NM^{-2})	Calculated OK simulation	Calculated 300K simulation
C_{11}	18.43	22.87	22.85
C_{22}	17.84	23.27	23.32
C_{33}	5.91	6.56	5.72
C_{44}	1.60	1.63	1.03
C_{55}	1.76	1.73	1.28
C_{66}	7.24	7.62	7.65
C_{23}	2.17	2.24	1.62
C_{13}	2.38	2.50	1.98
C_{12}	4.83	9.84	9.51
C_{15}	-0.20	-0.19	-0.17
C_{25}	0.39	0.69	0.44
C_{35}	0.12	0.17	0.07
C_{46}	0.05	0.48	0.28

(5.3) Simulation of Defects

Modelling of the structures and energies of defects in solids has been one of the most successful areas of application of simulation techniques in recent years. Several recent reviews are available[15][24][25] and the account here is therefore brief; further discussion is given in Chapter (15). Static lattice methods have been successfully applied to calculate the energies and entropies of formation, interaction and migration of defects. The techniques rest on the pioneering work 50 years ago of Mott and Littleton in which the defect structure and energy is evaluated by performing an energy minimisation operation on the defect and an immediately surrounding region of lattice containing typically 100-300 atoms. (Strictly speaking, these are 'force-balance calculations' as the coordinates of the atoms in the region are adjusted to zero-force, rather than the total energy of the defective lattice being minimised; for greater details see reference (14)). The response of more distant regions of the lattice is based on pseudo-continuum models.

TABLE 4 Defect Energy Calculations

Crystal	Process	Calculated energy (eV)
NaCl	Schottky pair formation	2.4-2.7 (2.3-2.7)
NaCl	Cation vacancy migration	0.66 (0.7-0.8)
CaF_2	Anion Frenkel pair formation	2.6-2.7(2.6-2.7)
CaF_2	Anion vacancy activation	0.35(0.38-0.47)
CaF_2	Anion interstitial formation	0.91(0.77-0.79)
MgO	Schottky pair formation	7.5-7.7(5.7)
MgO	Cation vacancy activation	1.8-2.2 (2.0-2.3)
NiO	Schottky pair formation	6-7
NiO	Cation vacancy activation	1.86(1.5)

The experimental values are given in parentheses. Detailed discussion of calculated and experimental results are given on p.356 of ref.25.

This two-region strategy, which is illustrated diagramatically in fig.(6) has proved to be highly successful in yielding defect parameters in good quantitative agreement with experiment. Particular success has been enjoyed in studies of ionic and semi-ionic solids as shown by the selection of results collected in table (4). The techniques have also proved valuable in more qualitative studies of defect phenomena . For example, a relevant recent study concerned the widely investigated rare-earth doped alkaline earth fluorides. Controversy has surrounded the nature of the aggregates formed by the substitutional rare-earth dopants and their charge compensating interstititals in heavily doped crystals (i.e. crystals containing > 5 mole % rare-earth). Calculations of Bendall et al[26] were of value in suggesting that there is a change in cluster structure on going from larger rare earths (e.g. La^{3+} and Nd^{3+}) to smaller cations, e.g. Er^{3+}. For the former, relatively small clusters comprising two dopant ions and two or three interstitials (see Figure 7a,b) were calculated to have the greatest stability; we note that these clusters are stabilised by a coupled lattice-interstitial relaxation mode that was identified in an early theoretical study of Catlow[27]. In contrast, for the smaller dopant ions, the beautiful, symmetrical cubo-octahedral cluster shown in figure (7c) has the greatest stability. The cluster consists of 6 rare-earth ions grouped around a central interstitial site. The eight F^- lattice sites of the cube are vacant, and $12F^-$ ions are each situated above one of the cube edges. Greatest stability is achieved when the central cube contains a pair of F^- interstitials orientated along the <111> direction.

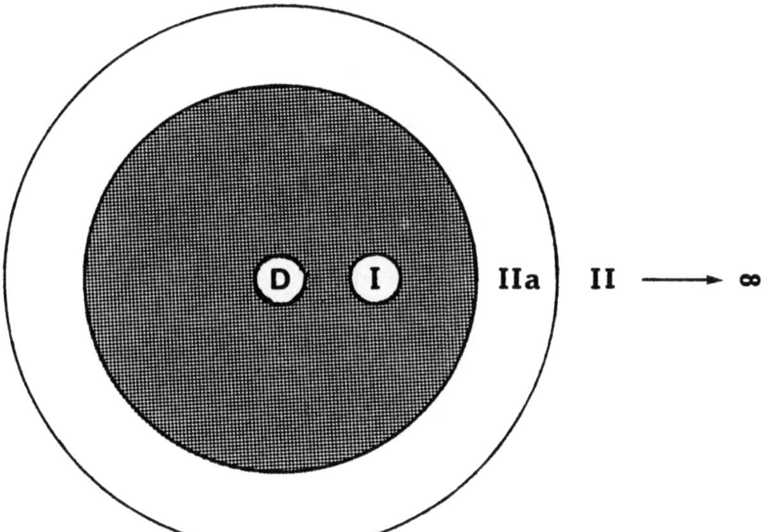

Fig.(6) Two region strategy used in defect calculations. The defect (D) is embedded in region I. Region II extends to infinity. Note that an interface region IIa is necessary.

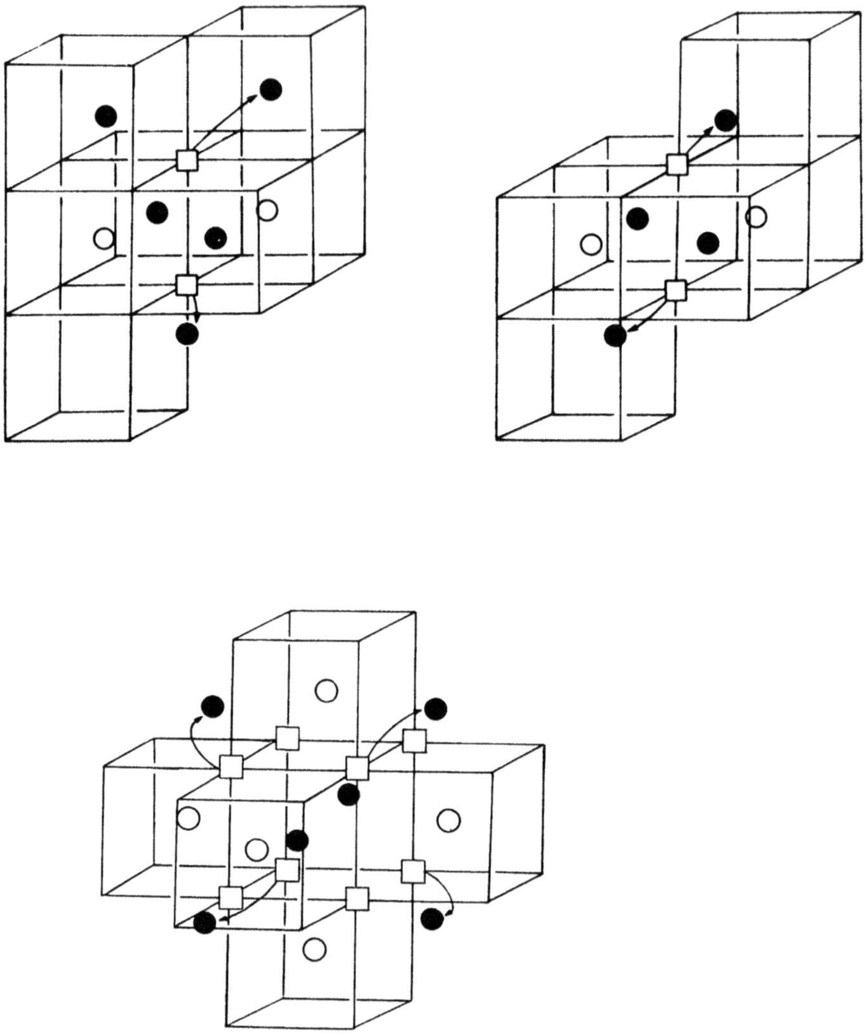

Fig. (7) Cluster in doped CaF_2. Open circle indicates rare earth dopant; filled circle is interstitial. Arrow indicates lattice ion relaxing to interstitial site from vacancy (open square). Figs. (7a and b) represent smaller dopant dimers; fig. (7c) is the large dopant hexamer.

The predictions of the calculations have recently been strongly supported by an EXAFS study of the local environment of the rare-earth cations in CaF_2. The EXAFS technique (see e.g. Hayes and Boyce[28], for a good review) allows us to probe the local structure of particular atomic species. EXAFS spectra were collected for a range of dopants in 10 mole % doped CaF_2, the data being collected using the synchrotron radiation source (SRS) at the SERC Daresbury Laboratory, U.K. The spectra for the larger rare-earth ions (e.g. Nd^{3+}) could be fitted accurately assuming the formation of the type of cluster shown in fig. (7a); whereas the spectra of the smaller ions (e.g. Er^{3+} indicated the presence of the cubo-octahedral clusters. The work (details of which are available in Catlow et al[29]) is a good illustration of the way in which simulations and experiments may be used in a concerted way to investigate complex problems in defect physics and chemistry.

6. CONCLUSIONS

We hope that the examples presented here show the extent to which simulations interact with and illuminate experiment. The subsequent chapters in this book will amplify all the technical topics we have discussed, and will present a wide range of applications, illustrating the major role that is now played by simulations in condensed matter sciences.

REFERENCES

1. R. Car and M. Parrinello, Phys. Rev. Lett., 55, 2471 (1985).
2. J. H. Harding, A. H. Harker, P. B. Keegstra, R. Pandey, J. M. Vail and C. Woodward, Physica, 131B, 152 (1985).
3. J. M. Vail, A. H. Harker, J. H. Harding and P. Saul, J. Phys. C. 17, 3401 (1984).
4. P. M. Rodger, A. J. Stone and D. J. Tildesley, Molec. Phys. 63, 173 (1988).
5. P. W. Tasker, Phil. Mag. A39, 119, (1979).
6. W. C. Mackrodt, Advances in Ceramics 23, 293 (1988).
7. J. T. Talbot, D. J. Tildesley, and W. A. Steele, Faraday Discuss. Chem. Soc., 80, 91 (1985).
8. P. R. Adby and M.A.H.Dempster, "Introduction to Optimisation Methods" (Chapman & Hall, London) (1974).
9. M. P. Allen and D. J. Tildesley "Computer Simulation of Liquids", Oxford University Press, (1987).
10. M. J. Gillan, Physica 131B, 157 (1985).
11. S. Yashonath, J. M. Thomas, A. K. Nowak, and A. K. Cheetham, Nature, 331, 601, (1988).
12. W. C. Mackrodt, E. A. Colbourn and J. Kendrick, Report No.CL-R/81/1637/A., ICI Corporate Laboratory (1981).
13. R. Baezold, C.R.A.Catlow, J. Corish, F. Healy, P.W.M.Jacobs and Y. Tan, J. Phys.Chem.Solids - in press.
14. C. R. A. Catlow and W. C. Mackrodt (eds), Lecture Notes in Physics, Vol. 166, Springer-Berlin, (1982)
15. C. R. A. Catlow, Ann. Rev. Mater. Sci., 16, 517 (1986).
16. M. Rigby, E. B. Smith, W. A. Wakeham and G. C. Maitland, "The Forces between Molecules" (Oxford University Press), (1986).

17. R. G. Gordon and Y. S. Kim, J. Chem. Phys. 56, 3122 (1972).
18. W. C. Mackrodt and R. F. Stewart, J. Phys.C. 12, 431 (1979).
19. D. R. Collins and C. R. A. Catlow - to be published.
20. R. Rothbauer, Neuse Jahrb. Mineral. Monatshefte, 143 (1971).
21. R. A. Knurr and S. W. Bailey, Clays and Clay Minerals, 34, 7, (1986).
22. C. M. Freeman, C. R. A. Catlow, A. M. Hemmings and R. C. Hider FEBS Lett. 197, 289 (1986).
23. R. C. Hider and U. Ragnarsson, FEBS Lett. 111, 189 (1980).
24. W. C. Mackrodt in "Transport in Non-Stoichiometric Compounds" (eds. G. Petot-Ervas, Hj. Matzke and C. Monty), North Holland, (1984).
25. F. Agullo-Lopez, C.R.A.Catlow and P. D. Townsend, "Point Defects in Materials", Academic Press (1988).
26. Bendall, P. J., C.R.A.Catlow, J. Corish and P.W.M.Jacobs, J. Solid State Chem. 51, 159 (1984).
27. C. R. A. Catlow, J. Phys. C. 6, L64 (1973).
28. T. L. Hayes and R. Boyce, Solids State Physics, 37, 173 (1983).
29. C.R.A.Catlow, A. V. Chadwick, G. N. Greaves and L. M. Moroney, Nature, 312, 601 (1984).
30. M. T. Vaughan and S. Guggenheim, J. of Geophysical Research, 91, 4657 (1986).

TOWARDS REALISTIC MODEL INTERMOLECULAR POTENTIALS

SARAH L. PRICE
University Chemical Laboratory,
Lensfield Road,
Cambridge CB2 1EW,
England

ABSTRACT: The realism of a computer simulation is usually limited by the accuracy of the fundamental scientific input into the calculation: the model intermolecular potential. We examine the problems in establishing accurate model potentials, by considering the physical origins of intermolecular forces, highlighting the approximations which are usually made in the potentials used in simulations, and discussing the problems in quantifying intermolecular potentials by ab initio methods and by fitting to experimental data. This emphasises the importance of choosing a realistic functional form for the potential. The isotropic atom-atom model potential, which is usually used for modelling polyatomic molecules, is contrasted with the recently developed anisotropic site-site approach to designing model potentials. The electrostatic interaction can be represented very accurately within the anisotropic site-site formalism, by the use of an ab initio based distributed multipole model. We show how empirical anisotropic site-site potentials have been used to great effect in a Molecular Dynamics simulation of liquid chlorine and Monte Carlo simulations of three condensed phases of benzene. Thus we can expect that the use of such model potentials will lead to more realistic simulations in the future.

1. Overview

Chapter (1) has shown how computer simulation can be an extremely powerful tool for developing our understanding of molecular systems at the atomic level. However, there are two major limitations which are preventing computer simulations from fulfilling their true potential for providing scientific insights and aiding the industrial development of new drugs and materials. The first problem is that simulations require considerable computing resources; this limitation is rapidly being overcome by the software and hardware developments. The second is the accuracy of the fundamental scientific input into the simulations. For atomic level simulations of molecular systems, this scientific input is the model intermolecular potential, which quantifies the forces between the molecules.

Since the fundamental law of computer modelling is that the quality of the results depends on the quality of the input, or more colloquially 'rubbish in gives rubbish out', it is

C.R.A. Catlow et al. (eds.), Computer Modelling of Fluids Polymers and Solids, 29–54.

important to be able to assess the likely accuracy of a model potential. The aim of this contribution is to provide some background in the theory of intermolecular forces, in order to assess critically the model potentials which are currently used in simulations. Intermolecular potentials are only known with very high accuracy for the rare gases. Much current work on deriving accurate model potentials is concentrating on small rigid polyatomic molecules, therefore these systems will dominate the examples used in this paper. However, the the non-bonded interactions of biologically important molecules, and the forces in ionic materials are essentially the same, so much of this paper is also relevant to such systems, although the additional problems which arise in modelling these species will be considered in the chapters of Parker, Jackson and Brooks. On the other hand, since model intermolecular potentials seek to describe the energy of a configuration of nuclei, averaged over electron positions, they are not appropriate for systems which have to be modelled at the level of the electrons (i.e. quantum mechanically), such as metals or chemically reacting systems.

Early simulation work was aimed at understanding general features of, for example, liquid behaviour, and so idealised model potentials were appropriate. Nowdays, many simulations are undertaken in order to model real systems. The simulations seek to produce results which are in agreement with experiment, which gives credibility to the predictions and insights which are also derived from the study. The first stage in such a computer simulation is to find a model for the intermolecular interactions in the chosen system, which is sufficiently realistic to give worthwhile results. We will see why there are very few molecular systems where the intermolecular potential has been established with sufficient accuracy that one can be confident of a realistic simulation of any phase. However, for many molecules or ions, there are either several proposed model potentials in the literature, with significant differences, or none at all. There are also difficulties in that there are no generally reliable procedures for developing intermolecular potentials, and one cannot often confidently recommend the 'best' model for a particular system, as when the models are simple, the 'best ' model will be very dependent on the nature of the intended simulation. Indeed, the choice of model potential is commonly the most difficult and frequently the least satisfactory feature of a simulation. In order to improve this situation, we need to develop more accurate intermolecular potentials, which will be more reliably transferable from study to study. This can only be done by going back to the theory of intermolecular forces, to develop more accurate models for the various contributions to the potential. This paper will outline some of the theory of intermolecular forces and illustrate our current knowledge by describing some recent work on the development of anisotropic site-site potentials. The use of more sophisticated models, which are more firmly based in theory, should make the development of intermolecular potentials a more rigorous procedure in the future.

2. Definitions and the Pairwise Additive Approximation

Let us start from the basic definition of an intermolecular pair potential $U(R,\Omega)$ as the energy of interaction of a pair of molecules as a function of their relative separation R and orientation Ω. This assumes that the molecule is rigid, which is usually a good approximation for small molecules, though the potential has to be made a function of the intramolecular bondlengths and angles for studying the transfer of energy between translational and vibrational motion. Organic molecules are not usually rigid, so it is usual to model their intermolecular forces by approximating the molecule as a set of fragments, usually atoms, and assuming that the contribution from each fragment does not depend on the molecular conformation. (This assumption will only be valid if the charge density associated with each fragment does not change with the conformation of the molecule.) We also assume that the molecules are in a non-degenerate electronic groundstate, and that interaction is weak compared with chemical bond strengths so it does not change the vibrational or electronic states of the molecule. If this is not so, then additional effects arise and the potential surface will be a function of many more variables, such as the electronic and vibrational states.

It is this intermolecular pair potential which determines the motion of two colliding molecules, and so is the pair potential which is measured in molecular beam scattering experiments, and determines the structure and spectroscopy of the dimer, and other dilute gas properties [1].

However, we can also define an *effective* intermolecular pair potential that can be used to calculate the interaction energy of a system of many (N) molecules,

$$U \approx \sum_{i<j}^{N} U_{ij}(R_{ij}, \Omega_{ij}), \tag{1}$$

by makin̈g the *pairwise additive approximation*. This is indeed an approximation because the exact expression the energy of just three molecules, i, j and k is $U_{ij} + U_{jk} + U_{ik} + U_{ijk}$, where U_{ij} is the true intermolecular pair potential describing the interaction of i and j in isolation, and U_{ijk} is a three body energy which reflects the error in the pairwise additive approximation. U_{ijk} will be highly dependent on the relative orientation of the three molecules. Thus, the exact energy of an ensemble of N molecules would involve the sum over the true pair potential, plus a sum over the three-body potential, plus the four-body terms, and so on up to the N-body terms. Many-body terms certainly exist, but their importance will depend on the system. An extreme example is that one ion close to an argon atom will polarize it, inducing a dipole which will interact with the charge to lower the energy. When a second charge is placed symmetrically on the other side of the argon atom, this does not double the dipole, but cancels it out, and there is only an induced quadrupole on the argon atom, leading to a much smaller induction energy. Thus non-additive induction effects are very important for ions, and models for such effects will be described elsewhere. The usual approach when modelling neutral molecules is to hope that all the many body effects are relatively small, so that an *effective* pair potential can be obtained, which includes the non-additive effects in some ill-defined averaged way. This hope is often ill-founded, and the pairwise additive approximation is one important source of error in most simulations. The existence of many-body effects implies that we cannot expect to derive a *pair*-potential which will be accurate for both the gas phase properties and condensed phase properties. Indeed we should question whether an effective pair potential can be successfully transferred between condensed phases with very different arrangements of the molecules. Although some work has been done on the theory and quantification of the many-body terms for some simple systems, (for example, the three-body terms contribute about 10% to the lattice energy of argon [1], and the three-body Axilrod-Teller dispersion term [2] alone is positive and equal to 6% of the experimental lattice energy of nitrogen [3]), much more work is needed before we can assess the errors inherent in the pairwise additive approximation for larger molecules.

3. Contributions to the Intermolecular Potential

The significant intermolecular forces all have an electrical origin, and are fundamentally the same as the forces involved in chemical bonding; although magnetic and gravitational effects do exist, they can normally be neglected [4]. When the molecules are well separated, so that there is negligible overlap of the molecular charge distributions, the presence of another molecule does not significantly change the wavefunction of each molecule, and the interaction energy U is very small compared with the sum of the total electronic energies $E_0^A + E_0^B$ of the two isolated molecules. In this situation, we can regard the electrons as being definitely assigned to one molecule or the other, and apply quantum mechanical perturbation theory, to give an expansion of the change in energy U which arises from the Coulombic interaction between the charges in the two molecules A and B. This interaction defines the perturbation operator $V = \sum_{ij} e_i^A e_j^B / r_{ij}$. Rayleigh-Schrödinger perturbation theory provides an expression for the long range potential in terms of integrals over the groundstate (0^A and 0^B) and excited state (n^A and n^B) wavefunctions of the isolated molecules. To second order in the perturbation theory expansion [4-6], the quantum mechanical expression for the long range pair potential is

$$\begin{aligned} U = \; & \langle 0^A 0^B | V | 0^A 0^B \rangle \\ & - \sum_{n^B \neq 0^B} \frac{|\langle 0^A 0^B | V | 0^A n^B \rangle|^2}{E_n^B - E_0^B} - \sum_{n^A \neq 0^A} \frac{|\langle 0^A 0^B | V | n^A 0^B \rangle|^2}{E_n^A - E_0^A} \\ & - \sum_{\substack{n^A \neq 0^A \\ n^B \neq 0^B}} \frac{|\langle 0^A 0^B | V | n^A n^B \rangle|^2}{E_n^A - E_0^A + E_n^B - E_0^B} \end{aligned} \tag{2}$$

which may be identified as

$$U = U_{estatic} + U_{induct}^B + U_{induct}^A + U_{disp}. \tag{3}$$

The first order term $U_{estatic}$ can be identified as the electrostatic energy, which is the classical energy of interaction of the undistorted molecular charge distributions $\rho(\mathbf{r})$. It can be evaluated exactly by integration over the ab initio charge distributions

$$U_{estatic} = \int_{all\ space} \frac{\rho^A(\mathbf{r}_1)\rho^B(\mathbf{r}_2)}{|\mathbf{r}_1 - \mathbf{r}_2|} d^3\mathbf{r}_1 d^3\mathbf{r}_2. \tag{4}$$

This is an extremely important contribution to the intermolecular potential, even for molecules like nitrogen, which are neutral and do not have a permanent dipole, because it is very dependent on orientation and can be either attractive or repulsive. It is strictly pairwise additive.

The traditional method of approximating $U_{estatic}$ for small molecules is based on the central multipole expansion, which uses an expansion of r_{ij}^{-1} in terms of the centre of mass separation of the molecules R_{AB}, and the distances r_i^A and r_j^B from each electron to the centre of the molecule to which it belongs [5,6]. Inserting this expansion for V in the perturbation expansion expression for $U_{estatic}$ in eqn. (2) leads to the following general expansion for the electrostatic energy of two molecules of any symmetry:

$$U_{estatic} = \sum_{l_1,l_2,k_1,k_2} \binom{l_1+l_2}{l_1} Q^A_{l_1k_1} Q^B_{l_2k_2} \overline{S}^{k_1k_2}_{l_1,l_2,l_1+l_2}(\Omega) R_{AB}^{-l_1-l_2-1}. \tag{5}$$

Here the integers l_1, k_1, l_2, k_2 define the different terms in the expansion in inverse powers of the intermolecular separation R_{AB}, associated with the different multipole moments $Q_{l_1k_1}$ of the isolated molecules, with the appropriate orientation dependence given by the function $\overline{S}^{k_1k_2}_{l_1,l_2,l_1+l_2}(\Omega)$. It is not necessary to be able to derive these $\overline{S}$ functions as the formulae for all terms in the multipole expansion for two interaction sites with no symmetry (i.e. all possible multipoles), up to R^{-5}, have been given in a simple, explicit form which is suitable for use in model potentials [8], along with the derivation of the associated forces and torques. The permanent multipole moments of the isolated molecules, which represent the molecular charge distributions in eqn. (5) are defined by the expectation values calculated from the molecular wavefunction by

$$Q_{lk} = \langle 0|(4\pi/(2l+1))^{\frac{1}{2}} \sum_i e_i r_i^l Y_{lk}(\theta,\phi)|0\rangle, \tag{6}$$

where Y_{lk} is a spherical harmonic function. (An equivalent multipole expansion of the long range potential in Cartesian tensors is possible [4]). If the molecule has any symmetry, many of the multipole moments will be zero. For example, a linear molecule only has non-zero multipole moments for $k = 0$. This considerably simplifies eqn. (5). For example, for two neutral linear molecules, with axis directions defined by unit vectors $\mathbf{z_A}$ and $\mathbf{z_B}$, and having a dipole ($\mu \equiv \mu_z = Q_{10}$), quadrupole ($\Theta \equiv \Theta_{zz} = Q_{20}$), octupole ($\Omega \equiv \Omega_{zzz} = Q_{30}$) moment etc., in a relative orientation defined by intermolecular separation R_{AB} in the direction of the unit vector $\hat{\mathbf{R}}$ from A to B, have an electrostatic energy

$$\begin{aligned}
&U_{estatic} = \\
&\mu^A\mu^B(\mathbf{z}_A.\mathbf{z}_B - 3\mathbf{z}_A.\hat{\mathbf{R}}\,\mathbf{z}_B.\hat{\mathbf{R}})R_{AB}^{-3} \\
&+ \Theta^A\mu^B\tfrac{3}{2}[\mathbf{z}_B.\hat{\mathbf{R}} + 2\mathbf{z}_A.\hat{\mathbf{R}}\,\mathbf{z}_A.\mathbf{z}_B - 5\mathbf{z}_A.\hat{\mathbf{R}}^2\,\mathbf{z}_B.\hat{\mathbf{R}}]R_{AB}^{-4} \\
&+ \Theta^B\mu^A\tfrac{3}{2}[-\mathbf{z}_A.\hat{\mathbf{R}} - 2\mathbf{z}_B.\hat{\mathbf{R}}\,\mathbf{z}_A.\mathbf{z}_B + 5\mathbf{z}_B.\hat{\mathbf{R}}^2\,\mathbf{z}_A.\hat{\mathbf{R}}]R_{AB}^{-4} \\
&+ \Theta^A\Theta^B\tfrac{3}{4}[35\mathbf{z}_A.\hat{\mathbf{R}}^2\,\mathbf{z}_B.\hat{\mathbf{R}}^2 - 5\mathbf{z}_A.\hat{\mathbf{R}}^2 - 5\mathbf{z}_B.\hat{\mathbf{R}}^2 - 20\mathbf{z}_A.\hat{\mathbf{R}}\,\mathbf{z}_B.\hat{\mathbf{R}}\,\mathbf{z}_A.\mathbf{z}_B + 2\mathbf{z}_A.\mathbf{z}_B{}^2 + 1]R_{AB}^{-5} \\
&+ \Omega^A\mu^B\tfrac{1}{2}[-35\mathbf{z}_A.\hat{\mathbf{R}}^3\,\mathbf{z}_B.\hat{\mathbf{R}} + 15\mathbf{z}_A.\hat{\mathbf{R}}^2\,\mathbf{z}_A.\mathbf{z}_B + 15\mathbf{z}_A.\hat{\mathbf{R}}\,\mathbf{z}_B.\hat{\mathbf{R}} - 3\mathbf{z}_A.\mathbf{z}_B]R_{AB}^{-5} \\
&+ \Omega^B\mu^A\tfrac{1}{2}[-35\mathbf{z}_B.\hat{\mathbf{R}}^3\,\mathbf{z}_A.\hat{\mathbf{R}} + 15\mathbf{z}_B.\hat{\mathbf{R}}^2\,\mathbf{z}_A.\mathbf{z}_B + 15\mathbf{z}_B.\hat{\mathbf{R}}\,\mathbf{z}_A.\hat{\mathbf{R}} - 3\mathbf{z}_A.\mathbf{z}_B]R_{AB}^{-5}\cdots
\end{aligned} \tag{7}$$

These expression may be familiar, as they have been widely used to model the electrostatic energy in intermolecular potentials for small linear molecules. Since most experimental techniques will only give the first non-zero multipole moment of a molecule, for example the dipole for HF, the quadrupole for N_2 and the octupole for CH_4, this multipole expansion seems an obvious way of encapsulating the best available information on the charge distributions of the molecules. It is only recently that reliable values of the higher multipole moments have become available from ab initio calculations, as high quality wavefunctions

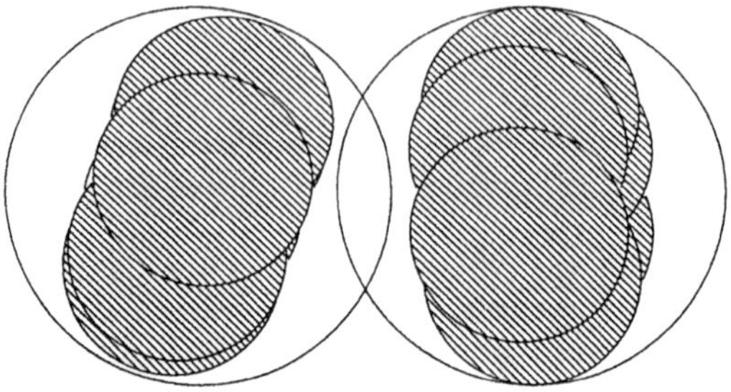

Figure 1. Two benzene molecules in an orientation where there is a negligible overlap of the molecular charge distribution, as defined by the shaded van der Waals surfaces. The large circles show the convergence spheres for the central multipole expansion, which have to contain all the charge distribution, and so such an expansion is invalid for this orientation. In contrast, an atomic site distributed multipole model would have a convergence sphere around each atom at approximately the van der Waals radius, and so would give a valid and convergent approximation to the electrostatic interaction energy.

are required because the multipole moments are very sensitive to the charge distribution at the edges of the molecule. There is an excellent compilation of experimental and ab initio values for the multipole moments of small molecules in a recent book by Gray and Gubbins [9]. However, such central multipole expansions are completely unsuitable for modelling the electrostatic interactions of molecules in condensed phases. This arises from the assumed multipole expansion of r_{ij}^{-1}, which is only valid for orientations where there is no intersection of the spheres around each molecule which contain (essentially) all the charge distribution. However, as Figure 1 shows, this condition will not be obeyed for markedly non-spherical molecules for some of the relative orientations which are sampled in condensed phases. (If the penetration effect, the error in the multipole expansion due to the overlap of the charge distributions, is modelled separately, then the convergence spheres only have to contain all the nucleii [7]. In this case the expansion is more likely to be valid, but the series will nevertheless converge poorly.) The central multipole expansion was derived for separations which are very much greater than the molecular dimensions [4], and so will not be an efficient approach to modelling the electrostatic energy for even fairly spherical molecules in condensed phases. However multipole expansion methods can be adapted to give distributed multipole models (§4), which are far more appropriate for modelling the electrostatic interactions in molecular simulations.

The second two terms in the perturbation expansion (eqn. (2)) of the long-range potential describe the induction or polarization energy, where for U^{A}_{induct} the permanent charge distribution of B polarizes the charge distribution of A, the distortions of A's wavefunction being mathematically described in terms of adding in contributions from the excited state wavefunctions. The induction energy is the extra energy which comes from the interaction of the induced multipole moments on A with the permanent charge distribution of B, and it is always attractive. Further corrections arise at higher orders of perturbation theory. This term can also be approximated by using the central multipolar expansion of V [4,5], in terms of the permanent multipoles and polarizabilities of the molecules. The polarizabilities of a molecule measure how easily the molecular charge cloud is distorted and are defined (c.f. eqns (2) and (6)),

$$\alpha(l_1 l_2 k_1 k_2) = \sum_{n \neq 0} \frac{\langle 0|Q_{l_1 k_1}|n\rangle\langle n|Q_{l_2 k_2}|0\rangle + \langle 0|Q_{l_2 k_2}|n\rangle\langle n|Q_{l_1 k_1}|0\rangle}{E_n - E_0}. \tag{8}$$

This approach to calculating the induction energy suffers from the same problems as the central multipole expansion of the electrostatic energy, and so requires the use of distributed polarizabilities [10]. Since the induction energy is highly non-additive, and difficult to include in simulations, it is usually ignored as one of the approximations in forming a model potential for uncharged molecules.

The last term in eqn (2), U_{disp}, is a purely quantum mechanical effect, and represents the stabilisation which results from the correlation of the charge fluctuations in the molecular charge distributions. It is always present, and is the only long range interaction between two inert-gas atoms. It is additive to second order in perturbation theory, but non-additive terms arise at higher orders, such as the Axilrod-Teller three-body dispersion term [2]. The multipolar expansion of V [4,5], gives the dispersion energy between two spherical atoms as

$$U_{disp} = -C_6/R_{AB}^6 - C_8/R_{AB}^8 - C_{10}/R_{AB}^{10} \cdots \tag{9}$$

where the C_n coefficients can be expressed as integrals over polarizabilities at imaginary frequency of the isolated molecules. The C_n coefficients are also functions of orientation for polyatomic molecules, and terms in R^{-7} etc can arise for certain symmetries. Values of C_6 can be derived from experimental spectral data, or from elaborate ab initio calculations, but these are generally only available for the smallest molecules. Although a multi-site model for the dispersion is required to avoid the problems of the central multipole expansion, a theoretically rigorous model is complicated as it has to describe the flow of charge from one-site to another [11].

At the intermolecular distances that occur in condensed phases, it is no longer possible to assume that the molecular charge densities do not overlap. The dominant effect of the overlap is a repulsive force resulting from the classical repulsion of the electrons, and the Pauli exclusion principle causing the electrons density between the nuclei to decrease, thereby increasing the nuclear repulsion. These effects are usually taken together in a repulsion model potential, which varies with separation [12] as $f(R)exp(-\alpha R)$, where $f(R)$ is a slowly varying function of R. The repulsion force provides a steep repulsive wall around the molecule, and effectively defines the shape of the molecule. It is difficult to treat theoretically, and so there is no rigorous analytical theory for the orientation dependence of the repulsive potential and various models have to be evaluated empirically (§5).

Another effect which can arise when there is molecular overlap is the transfer of charge from one molecule to another. Although this is a genuine physical effect, it is becoming clear that the importance of this term in ab initio studies is very dependent on the basis set used, making it difficult to evaluate its importance [7]. It is also extremely non-additive, so it is usually ignored in designing model potentials. The onset of overlap also modifies the electrostatic, dispersion and induction effects, so that the perturbation theory expressions are no longer valid.

Thus we can see that model intermolecular potentials for atoms which contain an exponential repulsion term and an R^{-6} dispersion term are using the simplest theoretically justified approximations to describe just the dominant contributions to the potential. Although the perturbation theory expressions allow us to develop models for the long range terms in the potential, which can be quantified using the properties of the charge distribution of the isolated molecules, we have no simple method of quantifying the repulsion potential at the onset of molecular overlap.

4. Determination of Quantitative Model Intermolecular Potentials

Conceptually, the simplest method of generating a quantitative intermolecular potential surface for two molecules would be to do an ab initio calculation on the dimer or supermolecule, with the positions of the nuclei of the two molecules fixed to correspond to a specified relative orientation, and then subtract the energies of the two isolated molecules. This procedure, if repeated at many different relative orientations of the two molecules, would give sufficient points on the potential surface to enable a model potential to be determined by fitting. There are several problems with this approach. Firstly, the intermolecular interaction energy is several orders of magnitude smaller than the total energy of the molecule, so high accuracy is required. Secondly, the potential needs to be calculated at a large number of points to determine both the parameters of the model potential, and check that the functional form of the model potential is able to represent the surface adequately. These problems are particularly severe because the dispersion energy is not included at all in a SCF (self-consistent-field) calculation, because it arises from the correlation in the motions of the electrons. Thus the supermolecule calculation requires a very large basis set and a good treatment of the electron correlation. The expense of such a calculation rises rapidly with the number of electrons, and so this approach is only feasible for the smallest systems, such as He_2 and $(H_2)_2$, though it has also been applied to $(N_2)_2$ [13] with an empirical adjustment of the dispersion term in the model potential after the fitting procedure.

An obvious way around the latter difficulty is to obtain the first order contributions, i.e. the repulsion, electrostatic and induction terms, from an SCF supermolecule calculation, and then add on the dispersion terms using the perturbation theory expression and the C_n coefficients derived from either spectroscopic analysis or ab initio calculations of the properties of the monomers. The problem with this approach is the modification of the dispersion energy with the onset of overlap. This is usually represented in the model potential by applying a damping function, which comes into effect at short range, to the long-range terms, for example [14]

$$U_{disp} = -\sum_{n\geq 3}^{\infty} \frac{C_{2n}}{R^{2n}} \left(1 - \left[\sum_{k=0}^{2n} \frac{(bR)^k}{k!}\right] \exp(-bR)\right). \qquad (10)$$

This approach is well developed for model potentials for spherical systems, though work is still in progress to improve on the rather ad-hoc damping functions which are used

[15], but it is only just being applied to the simplest polyatomic systems, such as $X \cdots Y_2$. Unfortunately the nature of the damping function is most important in the intermediate well region, which is the region which is critical for condensed phase simulations. Hence it is not surprising that a recent attempt to derive an accurate intermolecular potential for N_2, based on this approach, found that the ability to reproduce various properties of solid nitrogen was extremely sensitive to the damping function used [16].

Such damping functions do not appear in most model potentials, since when the model repulsive potential is fitted to some experimental property, it is assumed that the empirical fitting will absorb the errors in the representation of the dispersion and electrostatic energies in the intermediate and short range region. Indeed the problem of damping the long range forces has only come to the fore in the past decade, as the increase in computer power means we can contemplate developing intermolecular potentials completely ab initio. Most of the history of intermolecular forces has been concerned with developing theories for the macroscopic properties of molecular and ionic solids, liquids and gases in terms of the intermolecular potential. Estimates of the well-depth and minimum energy separation were obtained by the analysis of diverse experimental properties, using very simple two parameter model potentials. Indeed, the intellectual momentum that Lennard-Jones and others developed in the 1930s was so great, that it has hampered the acceptance of the inadequacy of the Lennard-Jones model potential for predicting modern, more accurate experimental data. The theories linking various properties of the gas, solid and liquid, to the potential are given in a recent encylopaedic book [1]; there is also an abbreviated student version [17], which is a useful introduction to the subject. Since most model potentials used in simulations are derived by fitting to experimental data, it is worth looking critically at the procedure.

To determine a model intermolecular potential from experimental data requires (i) some experimental data for some property of the molecular solid, liquid or gas,which is sensitive to the intermolecular potential, (ii) a quantitative theory for calculating that property from an intermolecular potential, and (iii) an assumed functional form for the intermolecular potential which contains some adjustable parameters. Let us consider, for the sake of argument, an inert gas system, and a very simple, early model potential, the Lennard-Jones potential

$$U(R) = \epsilon \left(\left[\frac{\rho}{R}\right]^{12} - 2\left[\frac{\rho}{R}\right]^{6} \right) . \tag{11}$$

This model potential has only two parameters, the minimum energy separation ρ and the well depth ϵ. It has a reasonable functional form at long range, as the leading term in the dispersion energy does go as R^{-6}, but the R^{-12} form for the repulsive wall was chosen purely for mathematical convenience in the pre-computer era, and an exponential form would be more realistic. This model potential is then fitted to the experimental data by first guessing initial values for ϵ and ρ, calculating the property from the guessed potential, comparing with the experimental values, and thereby deriving an improved estimate of ϵ and ρ, using some fitting criteria, such as a least squares minimisation. The set of parameters which gives the 'best' agreement with experiment defines the model potential, which can then be used in simulations.

The reliability of the derived potential depends on the accuracy and suitability of all three inputs. If there is insufficient data, or the property being analysed is not particularly sensitive to the intermolecular potential, or it has large error bars, it may be possible to get exact agreement between experiment and predicted values for a range of values of the parameters, or different model potentials, which correspond to rather different potential curves. If the theory includes approximations, such as the pairwise additive approximation, there may come a point at which you are improving the agreement with experiment, but getting a worse approximation to the true potential. Many people underestimate the importance of (iii), the choice of the model potential, particularly if acceptable agreement with the experimental data has been achieved. The resulting potential can be no better than the closest approximation to the accurate potential which is possible within the assumed functional form, and it will reproduce best the aspects of the potential to which the data are sensitive. For example the second virial coefficient $B(T)$, which measures the deviation in the PVT behaviour of a dilute gas from the ideal, is given by

$$B(T) = -2\pi N_A \int_o^{\infty} [exp\{-U(R)/kT\} - 1]R^2 dR, \tag{12}$$

and so is determined by a temperature weighted average over the potential. It can be shown [1] that at low temperatures, when the molecular collisions only sample the attractive region of the potential, $B(T)$ is approximately determined by the area within the well. Thus such data could be well reproduced by a potential with completely the wrong well shape! If a Lennard-Jones potential was used, the R^{-12} model for the repulsive wall would be a poor approximation to the true, approximately exponential form, and so high temperature, or high pressure properties would be predicted poorly.

This suggests that we need to fit the potential to a wide variety of experimental data, and this approach has been applied successfully to the argon potential [18]. Condensed phases properties were included in the fitting procedure by using the Axilrod-Teller three-body dispersion term [2] in the simulations: the assumption that this is the only significant many-body term was justified by the success of the fitting procedure and may be fortuitous. Since the state of the art potentials [1] can account for a very wide range of experimental data, simulations of the physical properties of argon can enjoy almost unique confidence in the model potential, as the errors in the model for the forces on the atoms which can be used in simulations are probably at the level of the many-body terms. However, the form of the Barker-Pompe model potential which can represent the argon intermolecular potential,

$$U(R)/\epsilon = \exp[\alpha(1 - R/\rho)] \sum_{i=0}^{5} A_i(R/\rho - 1)^i + \sum_{j=0}^{2} C_{2j+6}/(\delta + (R/\rho)^{2j+6}), \tag{13}$$

contains 11 parameters in addition to the well depth ϵ and minimum energy separation ρ, indicating that a large number of potential parameters are required to give the model potential sufficient flexibility to describe the actual pair potential. Hence, simpler forms of model intermolecular potential, which contain fewer parameters and have been fitted to only a few experimental properties, are unlikely to be accurate over a wide range of separations.

Progress in determining accurate intermolecular potentials for polyatomic molecules is far behind that for the inert gases. The potentials are intrinsically more complicated as they are now a function of the relative orientation of the two molecules. The model potential is generally a function of the five angular variables needed to define the orientation, as well as the centre of mass separation. Alternatively, if the orientation dependence is described implicitly by the use of an atom-atom type of model potential, there will usually be several types of atom in the system, requiring different parameters. The intermolecular interactions are also intrinsically more complicated, as there are now electrostatic and induction contributions which may need to be included explicitly in the model. Hence, it seems likely that when potentials for polyatomics reach the same accuracy as the argon potential, they will contain dozens of parameters, and have quite complex functional forms.

A further problem in determining accurate and reliable potentials for polyatomics is that extracting information about the potential from experimental data is more difficult than for monatomics. Some data, such as the second virial coefficients, depend on some orientational average of the potential, and so could be reproduced well by a potential with completely the wrong orientation dependence. Other experimental properties which depend on the dynamics of molecular collisions, such as transport data, are much more difficult to calculate for polyatomics than for atoms because of the possibility of inelastic collisions. Approximate calculations of transport properties can be used to test a proposed potential, but the calculations are too expensive to be used in an iterative fitting procedure. In contrast, the crystal structures of polyatomics have less symmetry than those of the rare gases, and so contain more information which can be used in deriving an effective potential. However, solid state properties, such as the structure, elastic properties and lattice vibration (phonon) frequencies, only sample the potential, and its derivatives, around a very limited set of relative orientations of the molecules, and so a potential derived from such properties may extrapolate poorly to other orientations unless the model potential has an appropriate functional form. This is a particular weakness with transferable isotropic atom-atom parameter sets, as the set of crystal structures used to derive the parameters may not sample any close contacts of a particular type, e.g. no C···N separated by approximately the sum of the van der Waals radii, and so the C···N repulsive potential parameters will not be tested, and so the potential may give very poor results for the crystal structure of a molecule which does contain close C···N contacts.

In conclusion, current model potentials for polyatomics are not very accurate, and even for diatomics, there are very few potentials which can successfully predict the properties of more than one phase [16,19]. This may well be due to the use of too simple a functional form for the model potential, but if more complex forms are used, then some justification for the chosen form, and sufficient data to determine the parameters, are required. Nevertheless, simulations with approximate model potentials are still worth doing, as feedback from the successes and disappointments of simulations provides important pieces of the jigsaw which will eventually form the picture of the accurate potential. The key question about any simulation is always whether the model potential is good enough for the purposes of the simulation.

Simple potentials will be good enough for many purposes. An isotropic atom-atom Lennard-Jones potential is not much more than a softened version of a hard-sphere model of the molecular shape, with a dispersion term which will tend to optimise the closeness of the intermolecular packing. This can give very useful results, as every undergraduate chemist has learnt a great deal from far cruder, physical space-filling molecular models which aid thinking in three dimensions.

However, in planning a simulation of a more subtle property, or seeking quantitative agreement with experimental data, the first hurdle is either the choice between rival model potentials, or the problem of developing one for the project. The choice between rival model potentials should not be determined solely by the claims of the developers, as those most experienced in the art of potential development generally have lower expectations of the reliability of their models in other applications! Indeed, the preceding discussion suggests that all polyatomic model potentials are probably rather crude approximations to the true potential for many orientations. More appropriate criteria for choosing between rival model potentials are firstly whether the potential was fitted to properties which are sensitive to the same features of the potential as those being studied in the simulation, and secondly whether the form of the model potential is likely to give a reasonable extrapolation to other orientations which might be sampled.

Carbon disulphide provides a salutary example of the problems which can be encountered in transferring potentials between studies. There were two rival isotropic atom-atom Lennard-Jones potentials for CS_2, with rather different parameters. One had been used in liquid simulations [20], the second in calculations on the low temperature crystal structure [21]. Both potentials gave reasonable predictions for the crystal phonon frequencies [22] in harmonic lattice dynamics calculations. However, the differences between the potentials became apparent when they were used to predict the crystal structure at about 80K [23], using the New Molecular Dynamics method [24], which allows the shape and volume of the simulation cell to change to balance the spontaneously generated internal stresses and the applied external pressure. The first potential gave a qualitatively correct prediction of the observed *Cmca* structure, but failed to reproduce the anomalous negative thermal expansion coefficient along one crystal axis. The simulation using the second potential immediately transformed from the initial experimental *Cmca* lattice to a radically different structure. Thus, although both simple potentials had been satisfactory for some calculations of solid state properties, they were both inadequate in a more realistic simulation of the solid. It has been proposed that a more realistic potential for CS_2 must include anisotropic electrostatic and repulsive forces [25]. The general conclusion must be that it is unsatisfactory to work with such crude models that it may be fortuitous if they prove satisfactory for a given simulation. Therefore, we must seek to develop model potentials on a firmer theoretical basis, even if they prove to be more complicated and so the simulations require more computer time.

5. The Isotropic Atom-Atom Model Potential

Most modelling of polyatomic molecules, from triatomics to bio-organic molecules is based on the assumption that the intermolecular potential between a pair of molecules can be modelled as a sum of interactions between every intermolecular pair of atoms, and that these interactions depend only on the separation of the atoms. A typical model is

$$U = \sum_{kl} B_{\kappa\lambda} \exp(-C_{\kappa\lambda} R_{kl}) - A_{\kappa\lambda}/R_{kl}^6 + q_\kappa q_\lambda / R_{kl}, \tag{14}$$

where atom k, of atomic type κ is at a distance R_{kl} from atom l of type λ in the other molecule. This model has an exponential model for the repulsion, a R^{-6} dispersion term, and a point charge electrostatic model, which are quantified by the derived parameters $B_{\kappa\lambda}$ and $C_{\kappa\lambda}$, $A_{\kappa\lambda}$, and charges q_κ and q_λ respectively. Other potentials of this general type may have a Lennard-Jones repulsion-dispersion model, or the atom-atom potential could be a more complex function of R_{kl} [16]. The total intermolecular potential has an orientation dependence which is implicit in the use of an isotropic inert-gas-like potential between every pair of atoms. The model is based on the assumption that molecules interact with each other as if they were a superposition of spherical charge distributions. The same assumption is also used for modelling the non-bonded interactions between atoms in flexible molecules in most force-fields designed for predicting molecular conformations [26]. These force-fields contain additional potential terms to describe the energy changes associated with bond-stretching, angle-bending and torsional rotations within the molecule, which compete in determining the molecular conformation with the steric repulsion, the attractive dispersion and the electrostatic interactions, between all the different pairs of atoms in the molecule which are separated by three or more bonds.

The isotropic atom-atom potential model is clearly a good zeroth order approximation, which recognises that the relative positions of the atoms within the molecule is the major factor in defining the molecular shape. Hence the model will give a fair approximation to the excluded volume of the molecule, which is vital for even the crudest type of simulation, and also gives a first approximation to the dispersion and electrostatic energies. However, the implicit orientation dependence may not be correct, for example, the model would predict that any homonuclear diatomic A_2 molecule would form an X-shaped dimer, with all four intermolecular atom-atom distances equal to the minimum energy separation of the atom-atom potential. (The basic isotropic atom-atom model cannot include any electrostatic interaction). However the chlorine dimer is polar [27], and probably has a floppy 'L' shaped structure.

The isotropic atom-atom approach is particularly convenient for organic molecules, as by assuming transferability of the potential parameters for a particular atom in different molecules, a limited set of parameters can be used to define a model potential for a whole series of organic molecules. This approach has been widely used, and is described in detail in a recent book by Pertsin and Kitaigorodsky [28]. This book is an excellent source of references to sets of potential parameters for many molecules, as it contains tables which give different sets of parameters, often for model potentials with exactly the same functional form. Many people are surprised that the same parameter can vary by an order of magnitude between different sets of parameters, and that occasionally some of the 'dispersion' coefficients (A in eqn. (14)) are negative. This reflects the different experimental data (often a collection of crystal structures and a few heats of sublimation) used in deriving the parameters, and the different weighting given to the residual errors in the fitting process. The problems in finding acceptable fits to the data, and the diversity of the resulting potentials are more severe for heteroatoms than for the basic hydrocarbon parameters. Thus the model is a crude approximation, and different errors are absorbed into the parameters according to which parts of the potential surface are sampled in the fitting. There is considerable cancellation of errors between different components of the model potentials, so taking the repulsion-dispersion potentials from one potential scheme, and combining them with the electrostatic terms from another source, is unlikely to give a useable model. Hence, it is not possible to state that a given parameter set is best, except in the sense that it has been fitted and tested against the widest range of data. The only way of establishing which potential will be best for the purposes of a given simulation is by trying a variety of models. The potential which has been derived using properties which are similar to those being studied in the simulation is the most likely to be the best, but it is often worth performing a simulation with a few different potentials.

Let us consider the basic approximation in the isotropic atom-atom model, that it is the sum of interactions between spherical atoms. This completely neglects the rearrangement of the valence electrons which occurs on bonding. Chemical theories of bonding not only invoke a shifting of electron density into the bonds, but also, for some atoms and types of bonding, invoke the formation of π and lone pair orbitals. These markedly non-spherical features in the electron density distribution can be clearly seen on electron density difference maps, obtained from high quality X-ray data [29,30], and are often invoked in theories of organic reactivity and hydrogen bonding. Recently we have been working on the development of model potentials which can represent the anisotropic effect of these non-spherical features on the intermolecular potential, and the rest of this chapter is devoted to describing the progress that has been made towards more accurate model potentials for polyatomic molecules by moving to anisotropic atom-atom potentials.

6. Modelling the Electrostatic Energy

The electrostatic energy is a very important component of the intermolecular potential, not only for ions and strongly polar molecules, but for all molecules, with the possible exception of the saturated hydrocarbons. For example, a model potential for benzene, which only included repulsion and dispersion terms, would predict that the dimer would adopt a parallel-plate structure, since this maximises the dispersion energy by packing as closely as the repulsive walls of the molecules allow. However, the structure of the benzene dimer, although the subject of controversy for both experimentalists and theoreticians [31], is definitely more open than a parallel-plate structure, because of the electrostatic repulsion of the π electrons.

In regions where there is negligible overlap of the molecular charge distributions, the electrostatic contribution to the potential can be considered as arising solely from the rearrangement of the valence electrons from spherical atoms, because a superposition of neutral spherical charge distributions has no long range electrostatic potential. Hence an atom-based electrostatic model is intuitively reasonable. (It was the early reliance on experimental data to quantify the electrostatic interaction which lead to the emphasis on the central multipole expansion for small molecules.)

Ab initio calculations are providing increasingly accurate electron density distributions, which can be analysed to obtain either the higher order multipole moments, or alternative models for the electrostatic interactions. An ab initio electron density difference map for Cl_2 is shown in Figure 2, where the subtraction of spherical atomic charge densities clearly reveals the build up of the lone pair density at the sides of the atoms. Beside this figure are representations of the electron distribution corresponding to the first few pure multipole moments. A qualitative comparison shows that the total molecular charge density of Cl_2 will be poorly represented as a neutral sphere plus a quadrupole, hexadecapole, and the other symmetry allowed central multipoles, as the lone pair density will not be represented accurately by the expansion unless extremely high order central multipoles are included. In contrast, the electron density around each atom is fairly well modelled by an atomic quadrupole, which gives a rough approximation to the lone pair density, plus a dipole and octupole to refine the description of the bonding region and the lone pair density. Thus the

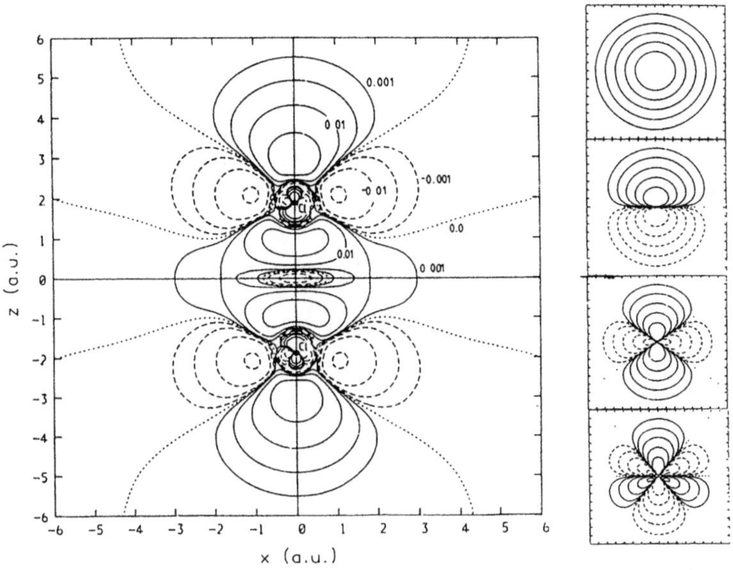

Figure 2. [32] Chlorine SCF density difference, defined by subtracting spherical atomic charge densities from the molecular charge density calculated using a $(7s6p3d)$ basis set. The solid contours are positive, dotted zero and the dashed negative (build up of electron density). The non-zero contours are at 10^{-3}, 3×10^{-3}, 10^{-2}, 3×10^{-2}, and 10^{-1} atomic units. The figures at the side show pure multipolar distributions corresponding to a charge Q_{00}, dipole Q_{10}, quadrupole Q_{20} and octupole Q_{30}, in descending order.

multipole expansion at each atom would give quite an accurate model for the molecular charge density with relatively few terms.

A quantitative distributed multipole model requires a recipe for dividing up the molecular charge distribution into the contributions to be represented at each expansion site. There are several methods available [33-39], but the general approach can be illustrated by considering one method. The distributed multipole analysis of Stone (DMA) [33] exploits the fact that modern wavefunctions are expanded in terms of Gaussian functions (ϕ). Thus the charge density can be expressed as a sum of contributions from every pair of Gaussian primitives in the basis set.

$$\rho(\mathbf{r}) = \sum_{tu} \rho_{tu}\phi_t(\mathbf{r})\phi_u(\mathbf{r}). \tag{15}$$

Each product of Gaussian functions can be represented by another Gaussian at a site whose position is determined by their centres and exponents. Thus the contribution to the electrostatic potential, outside the charge distribution, from a density element $\rho_{tu}\phi_t\phi_u$ involving two s-orbital primitives is equivalent to a point charge at the centre of the product Gaussian, and an s and a p orbital generate a point dipole and charge at that site, and two p orbitals generate a point quadrupole, dipole and charge, and etc.

Thus the charge density calculated from an s and p basis set can be exactly represented by a charge, dipole and quadrupole at each atom, summed from the pairs of orbitals which are both centred on that atom, plus a very large number of point charges, dipoles and quadrupoles at sites between the nuclei, representing the charge density contributions from pairs of orbitals on different atoms. Each of the latter contributions can be represented by an infinite multipole expansion at another centre, which is taken to be the nearest of the chosen expansion sites. The formula for this 'shifting' process, in the special case of a linear molecule, is

$$Q_{L0}(z_0) = \sum_{l=0}^{L} \binom{L}{l} (z_1 - z_0)^{L-l} Q_{l0}, \tag{16}$$

giving the multipole moments Q_{L0} generated at the chosen point z_0 on the molecular axis which represent the set of multipoles Q_{l0} at the overlap centre z_1. The qualitative argument above suggests that every atomic centre is a good choice of expansion sites for a distributed multipole model, but this can be supplemented by extra sites in the centre of bonds for high accuracy, or sites on hydrogen atoms in CH_n groups can be omitted for computational efficiency. Provided that there are sufficient expansion sites that every contribution to the charge density is represented at a sufficiently adjacent site, the multipoles above quadrupole which appear in the final distributed multipole analysis will be small.

We can use the sets of distributed multipoles at each site to evaluate the long range electrostatic energy of two (or more) molecules, by summing all the terms in the multipole expansion of the electrostatic energy up to R^{-5} [8] (to include the quadrupole-quadrupole contribution), between every intermolecular pair of sites. Thus a distributed multipole analysis of an ab initio wavefunction can be used to generate an anisotropic atom-atom electrostatic model which will represent the effects of features such as equatorial lone pairs or π electrons by a significant atomic quadrupole, or an axial lone pair by a significant atomic dipole, without requiring the input of any assumptions about the bonding. A program for performing a DMA is included in the CADPAC [40] suite of ab initio programs, and it has been used to obtain the DMAs of aromatic hydrocarbons [41], azabenzenes [42], an undecapeptide [43], and various peptide fragments [44], as well as smaller molecules.

Thus the use of a distributed multipole model for the electrostatics is theoretically desirable, but is it worth using in simulations, where some weight is always given to keeping the model potential simple to save computer time? The answer, which will depend on the molecule concerned and the aim of the simulation, is suggested by looking at the results of molecular modelling work which has employed distributed multipole models.

One highly desirable property of a model potential is that it should predict the correct dimer structure. Buckingham and Fowler [45] have shown that the electrostatic interaction is very important in accounting for the variety of structures adopted by the van der Waals complexes of small polyatomics. This was done by using a distributed multipole model to determine the minimum in the electrostatic energy of the complexes, subject to the constraint that the molecules could not approach closer than allowed by the Pauling, van der Waals radii of the non-hydrogenic atoms. This simple model correctly predicted the relative orientation of the two molecules within 29 van der Waals complexes, most of which were hydrogen bonded. The success of this simple model has been analysed using ab initio perturbation theory [46], which showed that for many van der Waals complexes, the electrostatic energy dominates the orientation dependence of the potential, as the angular variations in the exchange-repulsion and charge transfer terms tend to cancel.

The structures of these complexes can be qualitatively rationalised by the Legon-Millen rule [47], which states that hydrogen bonds will tend to form to regions of high electron density in the proton acceptor, such as lone pairs and π bonds. Buckingham and Fowler's results show that this rule of thumb can be accounted for by the electrostatic attraction to these regions of high electron density, but this is only apparent when the electrostatic potential around these features is modelled accurately. For example, HCN$\cdots$HF is linear because the axial lone pair on nitrogen gives it a significant atomic dipole, whereas the acetylene HF complex is T-shaped as the DMA represents the π electron density. Similarly the HF dimer is bent because the equatorial lone pair density on F produces a significant atomic quadrupole. A point charge atomic model would predict that this dimer was linear (H^+-$F^-\cdots H^+$-F^-).

Thus an electrostatic model which represents the effects of the non-spherical features in the molecular charge distribution is necessary to describe the orientation dependence of strong electrostatic interactions such as hydrogen bonds. However, a simplification of a distributed multipole model may be adequate for some purposes. Indeed, there would be little point in using the full DMA from a good ab initio wavefunction in conjunction with very crude models for the other intermolecular effects, such as a Lennard-Jones atom-atom potential. The DMA, or some intuition as to the nature of the bonding in the molecule, may be very useful in devising a suitable electrostatic model. This requires the use of the 'shifting' formula (eqn. (16) for linear molecules). Table I compares various simple electrostatic models for nitrogen and chlorine, which correspond to the same total quadrupole moment, with their DMAs. In the case of chlorine, the electron density difference (Figure 2) looks very much like a quadrupolar distribution around each atom, and this is reflected in the DMA. Thus we would expect that a simple electrostatic model of a quadrupole on each atom would be a fairly good approximate model, certainly better than a charge model, which requires a substantial negative charge at the bond centre. A simple charge model is even less realistic for nitrogen, where representing the negative quadrupole requires negative charges on the atoms and a positive charge in the middle of the triple bond! The substantial charge in the multiple bond means that the nitrogen charge distribution has a substantial negative charge and quadrupole moment on the central site, if one is included in the DMA, and the axial lone pairs give rise to significant dipole moments on the atoms. This complex charge distribution is more difficult to simplify than that of chlorine, and the higher moments have to be used to derive a more complex model, such as the charge and dipole model in Table I, which is the only one of the simplified models which is chemically reasonable.

Most simulations of larger polyatomic molecules use a point-charge model for the electrostatic interaction. A set of Mulliken analysis charges, which is often used for bio-organic molecules, is essentially a brutally truncated distributed multipole analysis, and thus the generally perceived inadequacies of such point charge models can be directly attributed to their neglect of the electrostatic contribution from the atomic dipoles, quadrupoles etc. These errors can be partially absorbed by using point charges which have been fitted to the electrostatic potential outside the molecule [48,49]. However, various schemes for potential-derived point charge models have found, either as an empirical necessity, as in the case of the azabenzenes [50], or from a theoretical analysis of the problem [51], that extra sites are needed to describe the potential arising from bonding features such as lone pairs. The use of extra point charge sites has also sometimes been found necessary to provide satisfactory models for use in simulations. For example, it was necessary to use a five-point charge model to give a satisfactory account of the lattice frequencies of nitrogen [52], and a three

Table I
Electrostatic models corresponding to the same total quadrupole moment
(All quantities in atomic units)

(a) For Nitrogen		N1	Centre	N2
z		1.037	0	−1.037
One centre	Q_{20}	0	−1.12	0
Two centre dipole	Q_{10}	−0.27	0	0.27
Two centre quadrupole	Q_{20}	−0.56	0	−0.56
Three centre charge	Q_{00}	−0.52	1.04	−0.52
Three centre model	Q_{00}	1.64	−3.28	1.64
also fitted to Q_{40}	Q_{10}	−1.12	0	1.12
Ab initio DMA [33]	Q_{00}	0.43	−0.86	0.43
	Q_{10}	−0.73	0	0.73
	Q_{20}	−0.12	1.21	−0.12

(b) For Chlorine		Cl1	Centre	Cl2
z		1.878	0	−1.878
One centre	Q_{20}	0	2.52	0
Two centre quadrupole	Q_{20}	1.26	0	1.26
Three centre charge	Q_{00}	0.36	−0.72	0.36
Ab initio DMA [32]	Q_{10}	−0.16	0	0.16
	Q_{20}	1.86	0	1.86
Ab initio DMA [32]	Q_{00}	0.09	−0.17	0.09
	Q_{10}	−0.34	0	0.34
	Q_{20}	2.09	0.28	2.09

point charge model (derived from a DMA) was needed to model the lone pair density on the sulphur atom in thiourea [53], in order to obtain a model potential which was capable of predicting the hydrogen bonding networks found in the crystal. Thus the accuracy of a point charge model will depend on the bonding in the molecule, but the use of point charges at only atomic sites will generally be a rather crude model. The increase in the cost of simulation with the number of sites means that the addition of extra charge sites is a rather inefficient method of improving the model, and so the use of distributed multipole moments is probably the way forward.

7. Repulsion Potential

The deviation of a molecular charge distribution from a superposition of spherical atomic charge densities will not only determine the electrostatic interactions but will also affect the other contributions to the potential. Indeed, it is inconsistent to use a distributed multipole model for the electrostatics, acknowledging that the charge distribution around each atom is non-spherical, and combine it with an isotropic atom-atom potential. However, although it is generally acknowledged that the atom-atom repulsion-dispersion terms are not strictly spherical, it is generally hoped that this effect is so minor that it can be neglected in model potentials. However, there is empirical evidence that the repulsive wall around some atoms is not spherical from the variation in the effective van der Waals radius of atoms with orientation. Nyburg and Faerman [54] analysed the X···X (X=N, O, F, S, Cl, Se, Br and I) van der Waals diameter in hundreds of close intermolecular contacts of the form C-X···X-C from the Cambridge Structural Database, as a function of the angles between the X···X vector and the C-X bond vectors. This showed that two iodine atoms can approach each other 0.7 Å closer if they are end on than side on. The corresponding difference for Cl is 0.4 Å. These differences are consistent with the equatorial lone pair density pushing out the repulsive wall at the sides of the atoms.

Thus it will be necessary to develop models for the effects of the anisotropy in the charge distribution on the repulsion and dispersion contributions to the intermolecular potential. The discussion in §1 suggests that this will be more difficult than the development of the ab initio based anisotropic site-site electrostatic models described above. However, the evidence is mounting that such anisotropic site-site potential models will considerably improve the accuracy of model potentials [55]. For example, it is necessary to introduce anisotropy into atom-atom potential models to achieve acceptable fits to ab initio potential surfaces for $(N_2)_2$ [56] and $(HF)_2$ [57]. The remainder of this paper will describe two studies using anisotropic site-site potentials which have lead to realistic simulations of condensed phases.

One of the earliest studies of the importance of anisotropy in the repulsion potential was an investigation of the crystal structure of Cl_2 [58]. It had been established that an isotropic atom-atom potential, plus any reasonable electrostatic model, would predict that Cl_2 would adopt the *Pa*3 structure rather than the observed *Cmca* packing. This had been interpreted as demonstrating that intermolecular bonding was sufficiently strong in Cl_2 to stabilise the observed structure [59], though it was later pointed out that a reduction in the repulsion in the same direction as the proposed increased attraction would have the same effect [60]. This was demonstrated by the development of an anisotropic atom-atom repulsive potential [58], which when added to electrostatic terms calculated from a DMA and an isotropic atom-atom dispersion model, gave an excellent quantitative prediction of the experimental crystal structure, and its stability relative to *Pa*3. The ability to predict the crystal structure required a fairly specific form of anisotropy in the repulsive potential, which can be interpreted as arising from the lone pair electron density. Since this lone pair density appears on the experimental X-ray deformation density map of the solid [29] in a position which packs well in the observed structure (Figure 3), but there is no detectable peak corresponding to intermolecular bonding, it seems probable that the anisotropy in the repulsion and electrostatic potential arising from the lone pair density, plays a major role in stabilising the *Cmca* structure of Cl_2. It has since been shown that quite small anisotropies in the atom-atom repulsive potential, or changes between an atomic dipole or atomic quadrupole description of the electrostatic interactions, can change the type of crystal packing predicted for X_2 molecules [61].

This raised the question as to whether the anisotropy in the chlorine atom-atom potential would explain the unusual asymmetry in the first peak in the experimental structure factor S(k) of the liquid. The original anisotropic atom-atom potential is not completely satisfactory for the liquid [62], which is not unexpected, as the potential has a minimum in a relative orientation which was not sampled in the solid [58]. (This illustrates the problem of transferring model potentials which may extrapolate incorrectly to unsampled regions of the potential surface.) A superior anisotropic atom-atom potential, with anisotropy in

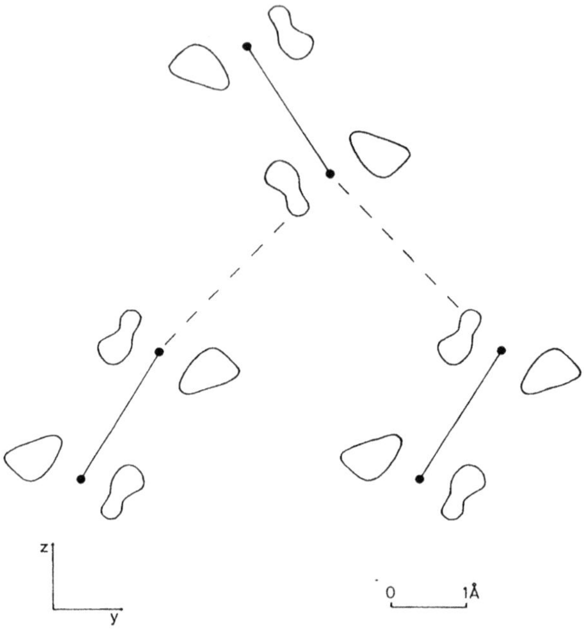

Figure 3. Schematic representation of the packing of the lone pairs in the $x = 0$ plane of solid Cl_2. The approximate position of the lone pairs is taken from the experimental electron deformation density of solid Cl_2 obtained by Stevens [29].

the dispersion, has been developed [19] using lower temperature crystal structure data, and the constraints that no alternative crystal structure had a lower energy, that the potential extrapolated reasonably to unsampled orientations, and that the pressure in the liquid at one state point was reasonable. The potential had the following form:

$$U = \sum_{i,j=1}^{2} K \exp[-\alpha(R_{ij} - \sigma_0 - \sigma(\Omega_{ij}))] - C(R_{ij} + \eta_0 - \sigma(\Omega_{ij}))^{-6} + U_{estatic}$$

where

$$\sigma(\Omega) = \sigma_1 \mathbf{z}_A.\mathbf{z}_B + \sigma_2 \tfrac{1}{2}[3\mathbf{z}_A.\hat{\mathbf{R}}^2 + 3\mathbf{z}_B.\hat{\mathbf{R}}^2 - 2] + \sigma_3 \tfrac{1}{2}[5\mathbf{z}_A.\hat{\mathbf{R}}^3 - 3\mathbf{z}_A.\hat{\mathbf{R}} - 5\mathbf{z}_B.\hat{\mathbf{R}}^3 + 3\mathbf{z}_B.\hat{\mathbf{R}}], \quad (17)$$

shifts the position of the repulsive wall around each atom in and out according to the relative orientation of the two atoms, and $U_{estatic}$ is the electrostatic interaction arising from a point dipole and quadrupole on each atom, as given within eqn. (7). Here the unit vectors $\mathbf{z}_A$ and $\mathbf{z}_B$ are along the molecular bond, pointing outwards at the atom in question, and $\hat{\mathbf{R}}$ is a unit intermolecular atom-atom vector. This potential can also predict the lattice frequencies, liquid structure factor, and a diverse range of thermodynamic properties of the liquid satisfactorily. The gaseous structure factor and the second virial coefficients are also reasonable for an effective pair potential. The paper even has a note added in proof comparing new experimental data on $S(k)$ at 290 K with the curve predicted by the potential. The agreement is even better than with the experimental data available when the paper was written, which is the sort of achievement that justifies the development of simulations as a scientific tool. Since most other proposed potentials for chlorine have had difficulty in simultaneously describing even two of these properties, the success of this model shows that the use of anisotropic atom-atom potentials represents an important advance in our ability to model real polyatomic systems. Incidentally, the liquid simulations with this potential only required $2\frac{1}{2}$ times the computer time need for a 2-centre Lennard-Jones potential, which is a small price to pay to make contact with experimental results!

The anisotropic atom-atom potential above has the symmetry appropriate for an atom in a linear molecule $C_{\infty h}$, but atoms in less symmetrical bonding environments, and atoms with different shapes, will require different functional forms in a model potential. Suitable, symmetry adapted functions can be taken from the complete set of expansion functions for any scalar property of the relative orientation of two molecules of arbitrary symmetry, as defined by Stone [63]. Since these $\overline{S}$ functions are defined using spherical tensor theory, it is straightforward to extract the functions which are appropriate to the symmetry of the molecular fragment [64]. Indeed, a subset of these $\overline{S}$ functions describes the orientational dependence of the perturbation theory multipole expansion of the long range electrostatic, induction and dispersion energy [5,8] c.f. eqn. (5)). The forces and torques corresponding to any anisotropic potential expressed in terms of $\overline{S}$ functions can be derived analytically, using spherical tensor theory [8,63]. The formalism can be used with any set of coordinates which define the relative orientation of the two molecular sites, but we have found the use of the unit vectors which define the local axis system and the direction of the inter-site vector particularly convenient, as exemplified in eqns (5) and (17).

The $\overline{S}$ functions appropriate to $C_{\infty h}$ and D_{3h} symmetry were used in developing a six-site anisotropic potential scheme for the azabenzene molecules [65], which are derived by substituting N atoms for CH groups in a benzene ring. This substitution has a major effect on the intermolecular potential, and the six molecules adopt very different crystal structures. The DMAs of the azabenzenes [42] were simplified to provide the electrostatic model, which comprised point charges to model the short range inductive effect in the CN bond, except for benzene where the dominant electrostatic effect was represented by distributed quadrupole moments. Anisotropic CH$\cdots$CH, CH$\cdots$N and N$\cdots$N repulsion-dispersion potentials were developed by fitting various functional forms to the crystal structures of benzene, pyrazine, pyrimidine and s-tetrazine, using the static crystal structure analysis program WMIN [66]. The form of anisotropy in the repulsive potentials which was required to model these crystal structures satisfactorily could be rationalised as representing the effects of the CH bond and nitrogen lone pair density. These potentials also gave satisfactory predictions of the crystal structures of s-triazine and pyridine, showing that the potential scheme was transferable.

A parallel study of the azabenzene interactions, using an isotropic site-site potential [50], found that it was necessary to include extra sites to model the effects of the lone pair density, resulting in a 12-site model. Thus, the lone pair density plays an important role in determining the diverse crystal structures of the azabenzenes, and so must be modelled, either by extra isotropic sites or anisotropic functions, for even such simple applications. Since the crystal structures sample the potential at relatively few orientations, these potentials have only passed a preliminary test, so it is noteworthy that the six-site potential has more flexibility for improvement using more testing experimental data.

This six-site potential for benzene has recently been improved [67], as it was found inadequate for the modelling the high pressure monoclinic phase. Monte-Carlo simulations on the orthorhombic phase suggested that the original potential was too attractive in the plane parallel sandwich geometry. This fault was remedied by increasing the quadrupole moment on each carbon to a sixth of the total quadrupole moment, thus absorbing the effects of the omitted contributions in the DMA. The potential was reparameterised by fitting to both the high pressure monoclinic and the low temperature orthorhombic crystal structures. The resulting potential was then used in constant pressure and temperature Monte-Carlo simulations of the orthorhombic, monoclinic and liquid phases of benzene. In such Monte-Carlo simulations the size and shape of the simulation cell is allowed to vary (c.f. the New Molecular Dynamics method), and so this provides a stringent test of the potential.

The new six-site model gave a satisfactory account of the orthorhombic phase at 218K and 0.7kbar, the monoclinic phase at 294K and 25kbar and the liquid at 300K and 1 kbar. The simulation showed the experimentally observed reorientations of the molecules in the orthorhombic phase. In contrast, Williams 12-site isotropic 'atom-atom' model [68] gave a somewhat less satisfactory simulation of these phases, and did not predict any reorientation, suggesting that the model may overestimate the anisotropy of the potential around the hydrogen atoms. This would be consistent with the empirical observation that the effective van der Waals shape of a hydrogen protrudes less when bonded to an sp^2 carbon than when bonded to an sp^3 carbon [69]. Further improvements in the benzene potential will require more work on the anisotropy of the CH group, particularly the anisotropy of the dispersion. The anisotropic carbon representation of the CH group is not only flexible for further improvement, but is also very efficient in computer time. The simulations using the six-site model took less than half the computer time required for the twelve site isotropic potential.

The use of anisotropic site-site potentials for organic molecules is attractive because the anisotropic functions can be used to describe important non-spherical features such as lone pairs and π electrons effectively, and also the potential can be made computationally efficient by representing methyl groups etc. [70] by a single site. Most model potentials for organic molecules are made up of isotropic atom-atom potentials and it is generally assumed that these potentials will be transferable to the same atomic species in other molecules. The systematic nature of organic chemistry shows that the molecular charge distributions are made up of recognisable, transferable functional groups and the intermolecular potential arising from a particular atom or functional group will be as transferable as the associated charge distribution. Hence, the transferable potential approach to modelling organic molecules has a reasonable foundation. The use of isotropic atom-atom potentials as the basic building block is more questionable, as if an atom is markedly non-spherical, an isotropic atom-atom potential will not be transferable between studies which sample the atom in different orientations. This has been demonstrated by an analysis of the three polymorphic forms of p-dichlorobenzene [71], which clearly showed the need for anisotropy in the Cl interaction potential. Thus, by using anisotropic site-site potentials, we hope to derive transferable potentials for bio-organic molecules which reflect the bonding environment of the atom and so are as transferable as the 'atomic' charge density. We have started by deriving transferable DMAs for modelling the electrostatic interactions of amides and some polypeptides [44].

Much more work needs to be done before we can hope to derive model potentials for polyatomics which are as accurate as those available for the inert gases, and therefore be able to simulate molecular systems with considerable confidence in the resulting predictions.

However, moving from the isotropic atom-atom model to anisotropic site-site model potentials is a step in the right direction. Unfortunately, the problem of being able to determine the potential parameters remains, and so to make progress we need to be able to derive quantitative anisotropic site-site models for the repulsion and dispersion interactions from the properties of the ab initio charge distribution. Such work, based on a distributed polarizability analysis [10] for the dispersion and the overlap of the molecular charge distributions for the repulsion, is in progress. Until the day when we do know more intermolecular potentials accurately, the problem of whether a model potential is 'good enough' for a simulation study will remain. In cases where an isotropic atom-atom potential proves unsatisfactory, consideration of the non-spherical features in the molecular charge distribution may suggest the means of improving the model.

Acknowledgements I wish to thank The Royal Society for support in the form of a 1983 University Research Fellowship.

References

[1] G. C. Maitland, M. Rigby, E. B. Smith, and W. A. Wakeham, *Intermolecular Forces*, Clarendon press, Oxford., 1981.

[2] B. M. Axilrod and E. Teller,' Interaction of van der Waals type between three atoms', *J. Chem. Phys.*, **11**, 299 (1943).

[3] P. A. Monson and M. Rigby,' Non-additive three body contribution to the lattice energies of nitrogen and carbon dioxide', *Molec. Phys.*, **39**, 1163 (1980).

[4] A. D. Buckingham,' Permanent and induced molecular moments and long range intermolecular forces ', *Adv. Chem. Phys.*, **12**, 107 (1967).

[5] A. J. Stone and R. J. A. Tough,' Spherical tensor theory of long range intermolecular forces ', *Chem. Phys. Letts.*, **110**, 123 (1984).

[6] A.J. Stone, 'Intermolecular Forces' in *Molecular Liquids Dynamics and Interactions*, A.J. Barnes, W.J. Orville-Thomas and J. Yarwood, eds., NATO ASI Series C 135, 1984.

[7] A. J. Stone and S. L. Price,'Some new ideas in the theory of intermolecular forces: anisotropic atom-atom potentials', *J. Phys. Chem*, **92**, 3325 (1988).

[8] S. L. Price, A. J. Stone and M. Alderton,' Explicit formulae for the electrostatic energy, forces and torques between a pair of molecules of arbitrary symmetry ', *Molec. Phys.*, **52**, 987 (1984).

[9] Gray C.G. and Gubbins K.E., *Theory of Molecular Fluids Vol. 1*, Oxford, 1984.

[10] A. J. Stone,' Distributed polarizabilities', *Molec. Phys.*, **56**, 1065 (1985).

[11] C.S. Tong, 'Anisotropy in repulsion and dispersion forces between atoms in molecules', Ph.D. Thesis, University of Cambridge (1988).

[12] E.A. Mason and L. Monchick,'Methods for the determination of intermolecular forces', *Adv. Chem. Phys.*, **12**, 329 (1967).

[13] H. J. Böhm and R. Ahlrichs,' The N_2-N_2 interaction A theoretical investigation', *Molec. Phys.*, **55**, 1159 (1985).

[14] K. T. Tang and J. P. Toennies,'An improved model for the van der Waals potential based on universal damping functions for the dispersion coefficients', *J. Chem. Phys.*, **80**, 3726 (1984).

[15] P. J. Knowles and W. J. Meath,‘ A separable method for the calculation of dispersion and induction energy damping functions with applications to the dimers arising from He, Ne and HF’, *Molec. Phys.*, **60**, 1143 (1987).

[16] M. S. H. Ling and M. Rigby,‘ Towards an intermolecular potential for nitrogen’, *Molec. Phys.*, **51**, 855 (1984).

[17] M. Rigby, E.B. Smith, W.A. Wakeham and G.C. Maitland, *The forces between molecules*, Clarendon press, Oxford., 1986.

[18] J.A. Barker, R.A. Fisher and R.O. Watts,‘Liquid argon: Monte-Carlo and Molecular Dynamics calculations’, *Molec. Phys.*, **21**, 657 (1971).

[19] P. M. Rodger, A. J. Stone and D. J. Tildesley,‘ The intermolecular potential of chlorine: a three phase study’, *Molec. Phys.*, **63**, 173 (1988).

[20] D. J. Tildesley and P. A. Madden,‘ Time correlation functions for a model of liquid carbon disulphide’, *Molec. Phys.*, **48**, 129 (1983).

[21] P. J. Grout and Leech J. W.,‘ Lattice dynamics of crystalline carbon disulphide revisited’, *Molec. Phys.*, **45**, 51 (1982).

[22] P. J. Grout and Leech J. W.,‘ Intermolecular modes of solid carbon disulfide’, *J. Phys. C*, **15**, L1083 (1982).

[23] R. W. Impey and M. L. Klein,‘ Intermolecular force models and the crystal structure of carbon disulphide’, *Chem. Phys. Letts.*, **103**, 143 (1983).

[24] S. Nose and M. L. Klein,‘ Constant pressure molecular dynamics for molecular systems’, *Molec. Phys.*, **50**, 1055 (1983).

[25] E. Burgos and R. Righini,‘ The effects of anisotropic atom-atom interactions on the crystal structure and lattice dynamics of solid CS_2’, *Chem. Phys. Letts.*, **96**, 584 (1983).

[26] U. Burkert and N.L. Allinger, *Molecular Mechanics*, ACS Monograph 177, 1982.

[27] S. J. Harris, S. E. Novick, J. S. Winn and W. Klemperer,‘ $(Cl_2)_2$: A polar molecule’, *J. Chem. Phys.*, **61**, 3866 (1974).

[28] A. J. Pertsin and A. I. Kitaigorodsky, *The Atom-Atom Potential Method*, Springer Series in Chemical Physics, vol 43, 1987.

[29] E. D. Stevens,‘ Experimental electron density distribution of molecular chlorine’, *Molec. Phys.*, **37**, 27 (1979).

[30] G. Moss and D. Feil,‘ Electrostatic molecular interactions from X-ray diffraction data. I Development of method: test on pyrazine’, *Acta Cryst. A*, **37**, 414 (1981).

[31] S. L. Price and A. J. Stone,‘ The electrostatic interactions in van der Waals complexes involving aromatic molecules’, *J. Chem. Phys.*, **86**, 2859 (1987).

[32] M. J. Alderton, ‘Distributed Multipole Analysis’, Ph.D. Thesis, University of Cambridge, (1983).

[33] A. J. Stone and M. Alderton,‘ Distributed multipole analysis Methods and applications’, *Molec. Phys.*, **56**, 1047 (1985).

[34] W.A. Sokalski and A. Sawaryn,‘Correlated molecular and cumulative atomic multipole moments’, *J. Chem. Phys.*, **87**, 526 (1987).

[35] F. Vigné-Maeder and P. Claverie,‘The exact multicenter multipolar part of a charge distribution and its simplified representation’, *J. Chem. Phys.*, **88**, 4934 (1988).

[36] J.F. Rico, J.R. Alvárez-Collado and M. Paniagua,‘1.1.1. electrostatic description of molecular systems ’, *Molec. Phys.*, **56**, 1145 (1985).

[37] A. Pullman and D. Perahia,‘Hydration scheme of uracil and cytosine’, *Theor. Chim. Acta*, **48**, 29 (1978).

[38] D. L. Cooper and N. C. J. Stutchbury,' Distributed multipole analysis from charge partitioning by zero-flux surfaces: the structure of HF complexes', *Chem. Phys. Letts.*, **120**, 167 (1985).
[39] Z. Berkovitch-Yellin and L. Leiserowitz,' The role of Coulomb forces in the crystal packing of amides. A study based on experimental electron densities', *J. Amer. Chem. Soc.*, **102**, 7677 (1980).
[40] R. D. Amos, *CADPAC: The Cambridge Analytical Derivatives Package*, publication CCP1/84/4, S.E.R.C. Daresbury Laboratory, Daresbury, Warrington WA4 4AD, England, 1984.
[41] S. L. Price,'A distributed multipole analysis of the charge densities of some aromatic hydrocarbons', *Chem. Phys. Letts.*, **114**, 359 (1985).
[42] S. L. Price and A. J. Stone,' A distributed multipole analysis of the charge densities of the azabenzene molecules', *Chem. Phys. Letts.*, **98**, 419 (1983).
[43] S. L. Price, R. J. Harrison and M. F. Guest, 'An ab initio distributed multipole study of the electrostatic potential around a undecapeptide cyclosporin derivative and a comparison with point charge electrostatic models', J. Comput. Chem., in press.
[44] C. H. Faerman and S. L. Price, manuscript in preparation.
[45] A. D. Buckingham and P. W. Fowler,' A model for the geometries of van der Waals complexes', *Canad. J. Chem.*, **63**, 2018 (1985).
[46] G. J. B. Hurst, P. W. Fowler, A. J. Stone and A. D. Buckingham,' Intermolecular forces in van der Waals dimers', *Int. J. Quant. Chem.*, **29**, 1223 (1986).
[47] A. C. Legon and D. J. Millen,'Directional character, strength and nature of the hydrogen bond in gas-phase dimers', *Acc. Chem. Res.*, **20**, 39 (1987).
[48] U. C. Singh and P. A. Kollman,' An approach to computing electrostatic charges for molecules', *J. Comput. Chem.*, **5**, 129 (1984).
[49] S. R. Cox and D. E. Williams,' Representation of the molecular electrostatic potential by a net atomic charge model', *J. Comput. Chem.*, **2**, 304 (1981).
[50] D. E. Williams and R. R. Weller,' Lone-pair electronic effects on the calculated ab initio SCF-MO electric potential and the crystal structures of azabenzenes', *J. Amer. Chem. Soc.*, **105**, 4143 (1983).
[51] R. Bonaccorsi, E. Scrocco and J. Tomasi,' An approximate expression for the electrostatic molecular potential in terms of completely transferable group contributions', *J. Amer. Chem. Soc.*, **99**, 4546 (1977).
[52] C. S. Murthy, S. F. O'Shea and I. R. McDonald,' Electrostatic interactions in molecular crystals Lattice dynamics of solid nitrogen and carbon dioxide', *Molec. Phys.*, **50**, 531 (1983).
[53] M. T. Dove and R. M. Lynden-Bell,' A model of the paraelectric phase of thiourea', *Philosophical Mag. B*, **54**, 443 (1986).
[54] S. C. Nyburg and C. H. Faerman,' A revision of van der Waals atomic radii for molecular crystals: N, O, F, S, Cl, Se, Br and I bonded to carbon', *Acta Cryst. B*, **41**, 274 (1985).
[55] S. L. Price,'Is the isotropic atom-atom model potential adequate?', *Molec. Simulation*, **1**, 135 (1988).
[56] S. L. Price,' The limitations of isotropic site-site potentials to describe a N_2-N_2 intermolecular potential surface', *Molec. Phys.*, **58**, 651 (1986).
[57] J. T. Brobjer and J. N. Murrell,' The intermolecular potential of HF', *Molec. Phys.*, **50**, 885 (1983).

[58] S. L. Price and A. J. Stone,‘ The anisotropy of the Cl_2-Cl_2 pair potential as shown by the crystal structure Evidence for intermolecular bonding or lone pair effects? ’, *Molec. Phys.*, **47**, 1457 (1982).

[59] L.-Y. H. Hsu and D. E. Williams, ‘Potential energy models for nonbonding and bonding interactions in solid chlorine’, *Inorganic Chem.*, **18**, 79 (1979); errata **19**, 2200 (1980).

[60] S. C. Nyburg and W. Wong-Ng,‘ Anisotropic atom-atom forces and the space group of solid chlorine’, *Proc. Roy. Soc. A*, **367**, 29 (1979).

[61] S. L. Price,‘The structure of the homonuclear diatomic solids revisited -a distorted atom approach to the intermolecular potential ’, *Molec. Phys.*, **62**, 45 (1987).

[62] F. P. Ricci, D. Rocca and R. Vallauri,‘ A Monte Carlo simulation study of liquid chlorine’, *Molec. Phys.*, **60**, 1245 (1987).

[63] A. J. Stone,‘ The description of bimolecular potentials, forces and torques: the S and V function expansions’, *Molec. Phys.*, **36**, 241 (1978).

[64] A. J. Stone, ‘Intermolecular forces’ in *The molecular physics of liquid crystals*, G. R. Luckhurst and G. W. Gray, eds., Academic press, 1979, ch. 2.

[65] S. L. Price and A. J. Stone,‘ A six-site intermolecular potential scheme for the azabenzene molecules, derived by crystal structure analysis’, *Molec. Phys.*, **51**, 569 (1984).

[66] Busing W.R., *WMIN, a computer program to model molecules and crystals in terms of potential energy functions*, Oak Ridge National Laboratory Report ORNL-5747, 1981.

[67] S. Yashoneth, S.L. Price and I. R. McDonald,‘ A six-site anisotropic atom-atom potential model for the condensed phases of benzene’, *Molec. Phys.*, **64**, 361 (1988).

[68] D. E. Williams and S. R. Cox,‘ Nonbonded potentials for azahydrocarbons: the importance of the Coulombic interaction’, *Acta Cryst. B*, **40**, 404 (1984).

[69] S. C. Nyburg, C. H. Faerman and L. Prasad,‘ A revision of van der Waals atomic radii for molecular crystals II: hydrogen bonded to carbon’, *Acta Cryst. B*, **43**, 106 (1987).

[70] S. L. Price,‘ Model anisotropic intermolecular potentials for saturated hydrocarbons’, *Acta Cryst. B*, **42**, 388 (1986).

[71] M. G. Munowitz, G. L. Wheeler and S. D. Colson,‘ A critical evaluation of isotropic potential functions for chlorine Calculations on the three phases of p-dichlorobenzene at 100K’, *Molec. Phys.*, **34**, 1727 (1977).

MOLECULAR DYNAMICS

Anthony J.C. Ladd

Lawrence Livermore National Laboratory, Livermore, California 94550

The basic methodology of equilibrium molecular dynamics is described. Examples from the literature are used to illustrate how molecular dynamics has been used to resolve theoretical controversies, provide data to test theories, and occasionally to discover new phenomena. The emphasis is on the application of molecular dynamics to an understanding of the microscopic physics underlying the transport properties of simple fluids.

1. Introduction

Molecular dynamics is concerned with the time evolution, under the laws of classical mechanics, of systems of particles interacting via conservative forces. From a technical point of view it is simply a numerical solution of the classical N-body problem, but its main purpose is in elucidating the equilibrium and transport properties of condensed matter. Since solids are well treated, at least to a first approximation, by the phonon theory (lattice dynamics), and gases likewise by the kinetic theory of Maxwell and Boltzmann, the major contributions from molecular dynamics have been in the area of dense gases and liquids. In these lectures I will give a brief outline of the molecular dynamics method, primarily as applied to dense fluids, and describe some of the important results obtained since the beginning of the subject in the 1950's. Although for many years there was a dearth of general material on molecular dynamics, there have been several very useful books published recently; the revision of Hansen and McDonald's book [1], the proceedings of the 1985 Varenna summer school [2], Hoover's Vienna lecture notes [3] and the reprint collection edited by Ciccotti, Frenkel and McDonald [4] have all appeared within the last two years. More detailed discussions of much of the material contained in these lectures can be found in these references.

These lectures focus on the basic methodology of equilibrium molecular dynamics, pioneered by Alder and Wainwright, Rahman, and Verlet. At this point in time it is impossible, and perhaps not even desirable, to survey the vast literature of molecular dynamics. Therefore I have selected illustrative examples that, by and large, have had a lasting impact in the field of liquid state theory, and which exemplify a style of simulation that focuses on solving specific problems or answering specific questions, rather than simply reporting the output of a computer program. The examples are limited to simple force laws, hard spheres, hard disks, and Lennard-Jones interactions; they focus on time correlation functions because it is only in the calculation of dynamic properties that molecular dynamics offers any advantage over the computationally simpler and in some ways more flexible Monte-Carlo methods.

2. Intermolecular Forces

The first element in a study of the statistical mechanics of liquids is a knowledge of the intermolecular forces. This is the subject of another series of lectures, so I will confine myself to a few general remarks. Intermolecular potentials fall roughly into three classes; model potentials, empirical potentials, and quantum-mechanical potentials. Model potentials, specifically the hard-core repulsion,

C.R.A. Catlow et al. (eds.), Computer Modelling of Fluids Polymers and Solids, 55–82.

dominated the early literature of molecular dynamics [5,6] and to a lesser extent Monte-Carlo [7,8]. This was partly because of their computational simplicity, but there was also the specific intention of showing that the structure of simple liquids was more or less universal and could be well approximated by the interaction of rigid spherical particles. Of course this idea was not new in itself; it was present in the work of van der Waals and also in the hard-sphere glasses built from ball bearings by Bernal [9]. However the key insight was that, with the aid of the digital computer, a model for the liquid state, akin to the harmonic model of a solid and the kinetic theory of gases, was possible. The most striking example of the new numerical statistical mechanics was Alder and Wainwright's demonstration of the solid-fluid phase transition in a system of hard elastic disks, accompanied by a rather spectacular picture of the two-phase coexistence region [10]. Subsequent calculations of the free energy of the coexisting phases confirmed that there was indeed a first-order solid-fluid phase transition in both hard-disk and hard-sphere systems [11]. These results demonstrated convincingly that systems of only a few hundred particles could model complex and ill-understood macroscopic phenomena, which was at that time a very surprising result to most theoreticians who were focusing on apparently simpler theories of liquids based on integral equation approximations to the radial distribution function. The realization that the computer could be used to solve problems previously considered intractable, such as the solid-fluid phase transition, helped to change the thrust of liquid-state theory away from increasingly fruitless efforts to solve approximate integral equations towards the much more profitable perturbation theories developed in the 1970's (for an extensive review of this work see ref. [12]). The age of the model potential is not yet over. In a recent series of papers, Frenkel and coworkers have applied the same approach to liquid crystals. The interactions are again hard repulsive collisions, but the particles are elongated to approximate the shape of molecules in a liquid crystal. Simulations have been used to demonstrate the relationship between the geometric packing of ellipsoidal and cylindrical shapes and various simple liquid crystalline phases [13,14]. Again this work will shift the emphasis of the theory of liquid crystals away from the mean-field theories towards perturbation methods.

As time went on and computers became more powerful, more realistic potentials could be used to approximate the intermolecular forces [15,16]. The focus of such simulations shifts away from the general properties of liquids to the more specific, for instance the relationship between intermolecular forces and the structure of molecular liquids and solids. Such a procedure is much more ambitious than the hard-core molecular dynamics pioneered by Alder and Wainwright, and even today it is impossible, for instance, to model the intermolecular forces with sufficient precision to reliably predict the phase diagrams of solids from first principles [17]. This is due in part to the many-body interactions that exist in condensed phases, which are usually only accounted for in an average way by effective pair potentials. These effective potentials are unsatisfactory inasmuch as they do not represent the true interactions between isolated pairs of particles, and moreover depend on the thermodynamic state of the system. They are usually determined by fitting appropriate functional forms to experimental thermodynamic, transport, and spectroscopic data. A further problem introduced by the search for more realistic potentials is that real intermolecular forces are long-ranged, with at least an R^{-6} attractive potential due to the van der Waals interactions. If the system is homogeneous then the long-range terms can be treated as a uniform background potential, which does not affect the forces but only the thermodynamic properties in ways that can be easily accounted for, as described long ago by Verlet [18]. The situation is somewhat worse in inhomogeneous systems, for instance two-phase systems or in hydrodynamic flow problems, and in such cases very short range potentials are usually used. Of course the thermodynamics of such systems is not exactly the same as for potential functions that include the van der Waals interactions. The difficulties involved in solving many-body quantum mechanical problems, particularly for fermion systems such as electrons, means that the goal of a first principles theory of condensed phases, based on the Coulomb interaction remains elusive.

Nevertheless, some progress has been made recently towards incorporating "in situ" quantum mechanical calculations into Monte-Carlo and molecular dynamics simulations. If successful, such calculations will make simulations of real liquids both more realistic and much simpler than current methods, since complicated empirical potentials can be replaced by Schrödinger's equation for the

Coulomb interactions. At present such a scheme can only be carried out within the framework of an approximate density functional [19], although attempts to simulate low-Z liquid metals, incorporating a classical Monte-Carlo treatment of the nuclei with a quantum Monte-Carlo simulation of the electronic ground state of each nuclear configuration, are underway at Livermore [20].

3. Boundary Conditions

Once the intermolecular forces are specified, only the boundary conditions and the initial conditions are required to completely specify the problem to be solved. Periodic boundary conditions are to be preferred in most instances since they preserve a macroscopic homogeneity or translational invariance. As a result, thermodynamic and transport properties deviate only by terms of order N^{-1} from the thermodynamic limit. On the other hand, if an external potential is used to confine the system to a fixed volume or to a fixed pressure convergence is much slower, as $N^{-1/3}$ in three dimensions. Useful exceptions to periodic boundary conditions occur only when the imposition of macroscopic inhomogeneities is an essential part of the problem, for instance when modeling hydrodynamic flows such as Rayleigh-Bernard convection [21,22].

A simple way of understanding periodic boundary conditions is as a large array of replicas of a basic unit cell. Thus the system is analogous to a crystal made up of identical unit cells, with the difference that the unit cell contains a large number of atoms or molecules. When a particle leaves a unit cell, its image enters at the same time from the opposite side thereby conserving the overall mass and momentum in the cell. It is computationally simpler, though not essential, to keep track of the particles in one particular unit cell, rather than follow the motion of specific particles, since in this case the coordinates are bounded by the dimensions of a single cell. It should be emphasized that the implementation of periodic boundary conditions in no way implies the existence of a physical boundary. The location of the unit cells is quite arbitrary, and the system is translationally invariant, i.e. unaffected by a change in origin.

In principle, the potential energy (and forces) should be calculated by summing over all the interactions in the infinite periodic system [15]. In practice, the long range van der Waals interactions, proportional to R^{-6} are usually truncated at a distance of less than one-half the box length, so that for each pair interaction only the nearest image need be considered in calculating the potential energy and forces [18]. This approximation breaks down if the potential falls off as R^{-3} or slower (i.e. dipolar or Coulombic forces) and in this case all the interactions, including all periodic images, must be included [23]. Moreover the sums are only conditionally convergent, which means they depend on the macroscopic boundary conditions imposed on the whole sample [24]. A direct summation of the interactions over a large number of images is representative of a macroscopic sample in a vacuum; a more useful result is obtained by introducing an artificial screening length, summing the now convergent interactions to any desired accuracy and then taking the limit that the screening length become infinite [25]. In the latter case there are no external boundaries; the system is always larger than the range of the interactions. This boundary condition is the one used by Ewald in determining the electrostatic energy of ionic crystals, and is also the one most suitable for computer simulation [26]. The key point is that the constitutive properties of the medium, such as the dielectric constant, depend only on local short-range correlations; otherwise they would have no meaning as an intrinsic property of the medium. On the other hand, the macroscopic response of the system, say the polarization of a dipolar fluid by an electric field, does depend on the boundary conditions as is well known. The boundary dependent relationships between polarization, dielectric constant, and electric field are now well understood [24,25], and accurate dielectric properties can be calculated from systems of as few as 108 dipolar particles [25].

A further advantage of the Ewald method is that it can be applied directly to a wide variety of systems in which long-range interactions are present. For instance an analogous situation to the dielectric problem occurs in the simulation of colloidal suspensions. In this case the interactions between the solid particles are primarily hydrodynamic, that is they are transmitted by the motion of the fluid. This results in complicated many-body forces, but at long range the interaction between a pair of particles falls off as R^{-3} just as for dipolar molecules [27]. The stress and strain-rate fields in a suspension under shear (analogous to the polarization and local electric field in the dipolar case) are boundary-condition dependent, but the viscosity (analogous to the dielectric constant) is not. Surprisingly the number dependence of the viscosity of such suspensions is very small when Ewald-type boundaries are used; the viscosity of a 32 particle system was recently found to be indistinguishable from the viscosity of 54 or 108 particle systems, and quite similar to the viscosity of a 16 particle system [28]. A slightly different application occurs in atomistic simulations of defects in crystalline solids, since although the intermolecular forces are usually short range, stress and strain fields around such defects are in general long-range. As an alternative to the somewhat clumsy procedure of introducing macroscopically derived displacement boundary conditions, periodic boundaries can be used. Again the long-range interactions between a defect, such as a dislocation, and all its periodic images can be exactly accounted for by an Ewald-type sum of the interaction energies [29]. Such a procedure leads to a very small number dependence in the defect energies. For instance, an accurate calculation of the energy of dislocations in two-dimensional crystals requires of the order of 100 atoms [30].

From a numerical point of view the implementation of Ewald-type boundary conditions is quite straightforward. A long-range potential $\phi(r)$ is multiplied by a screening function $f(r/\lambda)$ where λ is the screening length and $f(x)$ vanishes quickly for $x > 1$. The limiting procedure described above would involve calculating the potential energy for several different values of λ and extrapolating to the $\lambda \rightarrow \infty$ limit. This is of course impractical, though it can be used in test calculations to detect gross errors in programming the Ewald sum. Rather, we make use of the identity that for a system periodic on a length L,

$$\sum_{n=-\infty}^{\infty} g(x+nL) = L^{-1} \sum_{k=-\infty}^{\infty} e^{ikx} g(k), \tag{1}$$

where $k = 2\pi m/L$ and $g(k)$ is an infinite Fourier transform, i.e.

$$g(k) = \int_{-\infty}^{\infty} e^{-ikx} g(x) dx. \tag{2}$$

Then we sum the short-range function $\phi(r)\, f(r/\lambda)$ directly in real space, and the long-range function $\phi(r)\,[1 - f(r/\lambda)]$ in Fourier space. This decomposition allows for rapidly converging sums in both real space and Fourier space. With a suitable choice of screening function and screening length, the Fourier sum typically requires about the same amount of computer time as the direct sum.

4. Numerical Solution of the Equations of Motion.

Having specified the intermolecular forces and the boundary conditions the next step is to solve Newton's equations of motion

$$m\ddot{r} = F(r^{3N}), \tag{3}$$

subject to an appropriate set of initial conditions. Initial coordinates can be generated from a crystal lattice, usually face-centered or body-centered cubic, at the required density. For small systems at high fluid densities it is important that the periodic system be commensurate with the crystalline structure (generally face-centered cubic for monatomic solids), so that at least the correct short-range order can be realized. Velocities are usually assigned randomly and scaled so that the system is at the desired total energy. The solution of Newton's equations of motion is somewhat different for hard-core and continuous potentials and will be discussed separately.

4.1 Hard-Core Potentials

For a hard-core potential, the particles move at constant speed along straight line trajectories, punctuated by binary, elastic collisions. The collision times t_c can be found from the contact condition for every pair of particles $r_{12}(t_c) = \sigma$, where σ is the collision diameter. The smallest positive real root, from all the $N(N-1)/2$ quadratic equations for t_c, is the next collision time. The post collision velocities are determined by the constraints of conservation of momentum, angular momentum, and energy of the colliding pair of particles, and can be computed by reversing the normal (to the contact surface) component of the relative momentum, i.e. for equal mass spheres

$$\mathbf{v}'_1 = \mathbf{v}_1 - \Delta\mathbf{v}\ , \quad \mathbf{v}'_2 = \mathbf{v}_2 + \Delta\mathbf{v}, \tag{4}$$

where $\Delta\mathbf{v}$ is the projection of the relative velocity $\mathbf{v}_{12} = \mathbf{v}_1 - \mathbf{v}_2$ along the line of centers defined by the unit vector $\hat{\mathbf{r}}_{12}$,

$$\Delta\mathbf{v} = (\hat{\mathbf{r}}_{12}\cdot\mathbf{v}_{12})\hat{\mathbf{r}}_{12}. \tag{5}$$

Somewhat more complex hard-core systems can be studied in a similar way. In the first place the collisions may be inelastic, that is some energy is absorbed on collision, so that

$$\Delta\mathbf{v} = \varepsilon(\hat{\mathbf{r}}_{12}\cdot\mathbf{v}_{12})\hat{\mathbf{r}}_{12}, \tag{6}$$

with $\varepsilon < 1$. Because the collisions absorb rather than conserve energy the equilibrium state is one with all particles at rest unless energy, from say a gravitational potential or from moving boundary walls, is supplied. Such models are simple approximations to a wide class of macroscopic granular materials, such as powders, sand grains, coal dust, and oil shale [31]. More realistic models of these materials incorporate frictional forces between the contact surfaces thus allowing for the exchange of translational and angular momentum [32]. The transport of such materials is of great importance in many industrial processes, and is amenable to statistical mechanical methods. Moreover continuum approximations are in many cases inadequate since the dimensions of the macroscopic system are often only a few particle diameters and discrete particle effects are therefore of crucial importance.

The simplest collision searching algorithm, outlined above, requires of the order of N^2 operations to find each collision. The computer time can be reduced to order N by keeping lists of collision sequences and only updating those parts of the list affected by each collision [33]. A further reduction to order 1 can be achieved by constructing tables of near neighbors, though careful programming is then required for large systems to eliminate all order N operations. Neighbor lists can be usefully applied to any system with short-range forces, either hard core or continuous, in which only a small fraction of the particles lie within the radius of the intermolecular potential r_c. It is advantageous in such cases to restrict the volume searched, thereby reducing the time taken to find the interactions of a single particle from order N to order n, where n is the number of particles within the range of the intermolecular forces. There are two commonly used techniques; the simplest, due to Verlet [18], just keeps track of all pairs of particles separated by a distance less than $r_c + \Delta$, where the "skin" thickness Δ allows for the diffusion of particles between updates of the table. Although an order N^2 operation is periodically required to update the table, the computation of the forces at each time step is reduced to order Nn. The choice of skin thickness Δ is related to the frequency with which the table is to be updated and the optimum values of these parameters can be determined by trial and error. A more reliable method, devised by Hockney [34], is to partition the system into cubic cells, each at least of length r_c. Then all the interactions of a particle *i* are calculated by summing over the particles in the cube containing *i* and the 26 adjacent cubes. The primary virtue of this approach is its robustness and simplicity of use; there are no adjustable parameters. Moreover, the storage requirements can be elegantly reduced from Nn to about N by using link lists. However, now that even personal computers have large core memories this is largely of historic significance. The primary disadvantage of Hockney's method is that it involves a larger number of neighbors, about 3 times as many at high density as Verlet's method, though this is partially compensated for by the fact that no order N^2 updates of the table are required. At lower densities, when in Verlet's method either the skin thickness or the update frequency must be increased to cope with the more rapid diffusion, and possibly also for very large systems, Hockney's method is preferable.

4.2 Continuous Potential

The equations of motion for a system of hard-core particles can be solved exactly, or more strictly speaking to within the precision of the computer, typically 64 bits or about 14 decimal digits with the usual floating point arithmetic. On the other hand, for continuous potentials a further approximation is introduced by a finite difference approximation to the differential equations of motion, of which the simplest is the centered-difference approximation popularized by Verlet [18]. It can be easily derived by a Taylor expansion of the coordinates forwards and backwards in time to obtain the result

$$r(t+\delta t) - 2r(t) + r(t-\delta t) = \frac{F}{m}(\delta t)^2 \tag{7}$$

with an error of order $(\delta t)^4$. The Verlet algorithm is simple, robust and fairly accurate. It is not self starting, that is it requires information from the previous time step not just the current one. The Runge-Kutta method is slightly more complex than Verlet's algorithm but it is more accurate and is self starting. It requires recasting the 3N Newtonian equations into 6N first-order differential equations, i.e.

$$\dot{r} = v\ , \quad \dot{v} = F/m. \tag{8}$$

The Runge-Kutta method is often disparaged on the grounds that it requires 4 function evaluations per time step, but in fact numerical tests show that over a wide range of time steps it is more efficient than Verlet's algorithm, where the efficiency is measured in terms of the number of force evaluations required for a particular accuracy of the trajectory [35]. Even better than Runge-Kutta are the Gear predictor-corrector algorithms. They also solve first-order differential equations, proceeding in two stages. In the predictor cycle information from previous time intervals is used to estimate the coordinates and velocities at the next time step. Then the derivatives v and F/m are used to correct these estimates. In principle several corrector cycles can be used until self consistency is achieved, but in practice one is usually sufficient. These methods are not self starting but are often initialized by a short Runge-Kutta trajectory. Explicit formulae for the various Runge-Kutta and Gear algorithms are available in standard reference works on numerical methods, and also in the paper given at the Varenna summer school by Berendsen and van Gunsteren [36].

The above discussion of algorithmic accuracy made the implicit assumption that the equations of motion were integrable; that is by some limiting process, say successively reducing the time step, we are able to converge on an "exact" trajectory starting from some reasonably well specified initial state.

Such is the expectation of classical Newtonian mechanics as evidenced by the regular motion of the planets. However such motion is now known to be the exception rather than the rule, and only applies to harmonic or weakly anharmonic systems. In reality the equations of motion of a molecular fluid contain a Lyapunov instability, which causes neighboring trajectories to diverge exponentially with time. Thus although it is legitimate to expect to converge on a trajectory from a well specified starting point for some *finite* time, this dynamical coherence cannot be maintained for very long. As a simple example of the exponential Lyapunov instability consider a hard sphere in a gravitational potential, vertically above another fixed sphere [3]. If unlimited precision were available the sphere would bounce up and down indefinitely. If, on the other hand, the sphere is slightly displaced from the vertical axis by an amount δ, then after collision the tangential velocity will no longer be zero but equal to $v\delta/\sigma$ were v is the vertical velocity at collision. This means that the next collision will be even further off center, say 2δ. Again the tangential velocity is increased, to about $2v\delta/\sigma$, so that the next impact point is 4δ off the center axis, doubling on each collision to give an exponential divergence of the trajectory. From the initial conditions in Fig. 1 the sphere bounces only 9 times of the surface of the fixed particle. The effect of the Lyapunov divergence on a many-body hard-sphere system can be more accurately estimated from the diffusion of a single particle in a fixed array of scatterers, the Lorentz model. In this case for 64 bit arithmetic, corresponding to about 14 decimal digits, the average time of dynamical coherence, measured by the time at which a collision sequence is no longer reversible, is about 12 collisions [37]. A mean collision time is about 10^{-13} sec. in the case of dense liquid argon so that the coherence time in simulations of liquid argon can be no more than 1 psec.

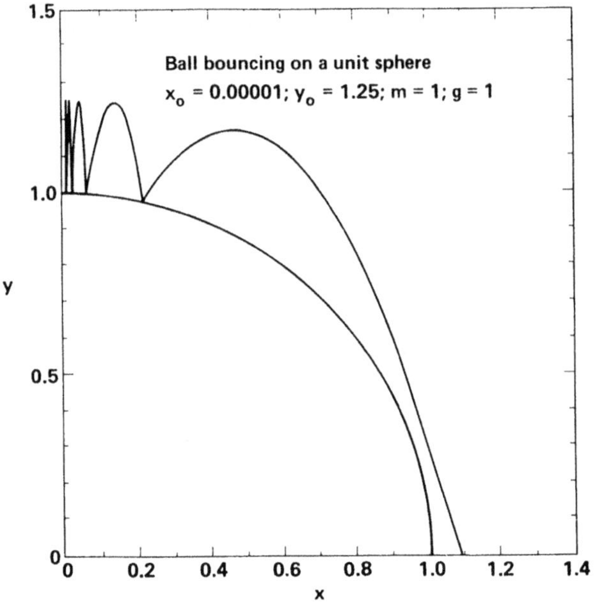

Figure 1. Lyapunov instability for hard sphere collisions (from ref. [3]). A single sphere falling under a gravitational potential undergoes elastic collisions with the surface of a fixed sphere. A small tangential displacement grows exponentially with each collision. In this case the moving sphere falls off the fixed particle surface after 9 collisions.

5. The Road to Equilibrium

In molecular dynamics experiments, macroscopic information is obtained from averages along a phase-space trajectory, i.e.

$$<A> = \lim_{t_M \to \infty} \frac{1}{t_M} \int_0^{t_M} A[r^{3N}(t), v^{3N}(t)]dt, \tag{9}$$

where t_M indicates a macroscopic time. Such averages are only consistent with thermodynamics if the system is ergodic, that is if a single sufficiently long trajectory visits all regions of phase space. A proof of ergodicity is extremely difficult and does not exist in general [38]. Therefore we must rely on numerical evidence and the first such test was not auspicious. In a famous simulation of a linear chain of particles coupled by weakly anharmonic forces, Fermi, Pasta, and Ulam [39] investigated the question of whether or not this simple model of a crystalline solid would come to equilibrium from any initial condition, as predicted by the second law of thermodynamics. The results were negative, much to the surprise of the investigators. At sufficiently low temperatures the system was found to return periodically to a state very close to the initial one. Later investigations [40] showed that even closer repetitions of the initial state occurred over periods about 15 times longer than the repeat period found by Fermi et. al.. On the other hand numerical experiments on hard-sphere systems showed that equilibrium could be achieved rapidly [5]. As an initial condition all the particles were assigned the same speed but with a random direction for the velocity, so that the initial energy of each particle was the same. After one collision the distribution of energy ranges from zero to twice the mean, after two collisions from zero to 4 times the mean and so on. Alder and Wainwright found that after about 3 collisions per particle, an equilibrium Maxwell-Boltzmann distribution of velocity was achieved. Moreover the entropy, as measured by the logarithm of the one-particle velocity distribution function, increased monotonically as predicted by Boltzmann's H-theorem. The apparent contradiction between these results can be understood from the fact that the hard-sphere system contains the Lyapunov instability, whereas the Fermi-Pasta-Ulam model does not. A simple model of a system that does *not* contain a Lyapunov instability is a hard-sphere in a gravitational field, bouncing on an elastic plate [3]. In this case, as in the Fermi-Pasta-Ulam model, the divergence of neighboring trajectories is linear not exponential. Thus the Lyapunov instability, although destroying the dynamical coherence, has the valuable property of ensuring a rapid and efficient exploration of phase space. It should be emphasized that the Lyapunov instability is *not* simply a numerical or algorithmic problem. Rather, the complete unpredictability of long-time trajectories (chaos) is fundamental to the majority of dynamical systems, in contrast to the predictability of periodic and quasi-periodic (integrable) systems. The recent book by Gleick provides an excellent and very readable overview of the development of the theory of chaotic phenomena [41].

Newton's equations of motion are reversible; that is a system run forwards or backwards in time satisfies the same differential equations. In more detail, reversibility means that if the system is at the phase point $[r^{3N},v^{3N}]$ at time $t = 0$ and moves along a trajectory to the phase point $[r'^{3N},v'^{3N}]$ at time t, then if the system is started at time $t = 0$ at the phase point $[r'^{3N},-v'^{3N}]$ it will follow the reversed trajectory to the phase point $[r^{3N},-v^{3N}]$ at time t. This implies that a system can evolve in time with decreasing entropy, in violation of the second law of thermodynamics. This apparent contradiction between the microscopic and macroscopic views, known as Loschmidt's paradox, was rejected by Boltzmann, who intuitively understood the essentially chaotic nature of the microscopic dynamics [42]. It can be seen that the Lyapunov instability provides a mechanism for overcoming the apparent reversibility of the microscopic equations of motion; even a perturbation on the level of quantum-mechanical uncertainty will prevent the reversal of the trajectory for more than a few mean collision times, so that macroscopically, the second law is never violated. Moreover, recent work by Hoover et. al. [43,44]

indicates that the phase space of a non-equilibrium system is fractal so that the dimensionality of accessible phase-space is less than the number of degrees of freedom. Thus trajectories that may violate the second law, in addition to being dynamically unstable, occupy a volume of zero-measure in the phase space and initial conditions for such trajectories cannot be generated by sampling techniques, but only as a result of a non-equilibrium experiment.

6. Thermodynamic Properties

Since the object of molecular dynamics is to make a connection between the microscopic world described by Newton's equations of motion and the macroscopic world of thermodynamics and hydrodynamics, we need to find dynamical variables, functions of the coordinates and velocities of the particles, whose time or ensemble averages correspond to macroscopically observable quantities. The simplest of these quantities, corresponding to the equilibrium temperature and pressure, will be derived here. More complicated quantities, corresponding to the fluxes observed in non-equilibrium systems, will be discussed in the section on transport coefficients. Since the entropy of the system is dependent on the volume of accessible phase space, the free energy cannot be determined from a phase-space average of any dynamical variable. However, the free energy difference between two states can be calculated from the reversible thermodynamic work required to transform the system from one state to the other [45].

Dynamical variables, corresponding on average to thermodynamic quantities, can be derived from the canonical distribution function. For example, the classical thermodynamic temperature is related to the canonical-ensemble average of the kinetic energy. Thus the dynamical variable for temperature is

$$T = \frac{2}{3Nk_B}\sum_i \frac{1}{2} m v_i^2, \tag{10}$$

where it should be remembered that the equivalence between T and the thermodynamic temperature is only on average. In the microcanonical ensemble usually used in molecular dynamics simulations, this expression is only exact in the large system limit. For finite systems, there are corrections arising from the conservation of total momentum and energy [46]. Momentum conservation just reduces the number of degrees of freedom, and requires replacing N by N − 1 in the above equation. However, the energy conservation correction is more complex and involves a temperature derivative of the heat capacity.

An expression for the thermodynamic pressure can also be obtained from the canonical partition function, but we will use an alternative method based on the virial theorem which can equally well be applied to non-equilibrium fluxes, as will be seen later. We define the dynamical variable W(t) as

$$W(t) = \sum_i m\mathbf{r}_i \cdot \mathbf{v}_i, \tag{11}$$

where the sum extends over all particles in a *fixed* unit cell. Thus W is not a continuous function of time but has discontinuous jumps when a particle leaves the unit cell and its image enters from the other side. Moreover, since the maximum value of any coordinate is fixed by the dimensions of the unit cell and the maximum velocity is fixed by the constant energy constraint, W is a bounded function. Consequently the time average of $\dot{W}$ vanishes, so that

$$\lim_{t_M \to \infty} \frac{1}{t_M}\int_0^{t_M} \dot{W}(t)dt = \lim_{t_M \to \infty} \frac{1}{t_M}\int_0^{t_M} \sum_i (m v_i^2 + \mathbf{r}_i \cdot \mathbf{F}_i)dt = 0, \tag{12}$$

where $\mathbf{F}_i$ is the force on particle i. The second term in the sum, $\mathbf{r}_i \cdot \mathbf{F}_i$, contains a discontinuity $-L\mathbf{n}\cdot\mathbf{F}_i$ whenever a particle leaves the unit cell and its image enters from the opposite face, where $\mathbf{n}$ is a unit vector outward normal to the face. The time integral of this quantity is just the total momentum crossing each of the three pairs of faces bounding the periodic unit cube multiplied by the box length, and macroscopically is given by $-3PVt_M$, where P is the pressure. We note that this is the same momentum transfer that would occur if the unit cube were instead bounded by rigid walls or by walls with short range (compared to the range of the intermolecular potential) forces. In this case the walls would exert a force $-\mathbf{n}P\,dA$ on each area element of the surface bounding the fluid. Thus we write the potential part of the virial as a continuous local contribution from the intermolecular forces and associate the discontinuous contributions with the external pressure, so that for pairwise-additive forces

$$\sum_i \mathbf{r}_i \cdot \mathbf{F}_i = \sum_{i>j} (\mathbf{r}_i - \mathbf{r}_j) \cdot \mathbf{F}_{ij} = \sum_{i>j} \mathbf{r}_{ij} \cdot \mathbf{F}_{ij} - 3PV. \tag{13}$$

It should be noted that the quantity $\mathbf{r}_i - \mathbf{r}_j$ in Eq. (13) is calculated *without* using the minimum-image convention, but simply using the coordinates of the particles in the fixed unit cell. On the other hand $\mathbf{r}_{ij}$ *is* evaluated using the minimum-image convention; the difference between the resulting averages is just -3PV. The force $\mathbf{F}_i$ involves a sum over all periodic images in principle, or at least over the nearest images; thus $\mathbf{F}_{ij} \equiv \sum_{\mathbf{n}} \mathbf{F}(\mathbf{r}_{ij} + \mathbf{n}L)$ [15], and is invarient under translations of particle coordinates by multiples of the box length L. From the above considerations, we can see that the dynamical variable corresponding to the thermodynamic pressure is

$$P = \frac{1}{3V} \left\{ \sum_i m v_i^2 + \sum_{i>j} \mathbf{r}_{ij} \cdot \mathbf{F}_{ij} \right\}, \tag{14}$$

where we note that the quantity $\mathbf{r}_{ij} \cdot \mathbf{F}_{ij}$ is continuous when either particle crosses the periodic boundaries.

Other expressions, equivalent macroscopically though not microscopically, can be derived for the pressure. One such expression, valid for crystalline solids in the absence of diffusion, can be obtained from the virial

$$W(t) = \sum_i m \mathbf{r}_i^0 \cdot \mathbf{v}_i \tag{15}$$

were $\mathbf{r}_i^0$ is the coordinate of the lattice site [47]. The resulting expression for the pressure is somewhat simpler than Eq. (14), but completely equivalent thermodynamically,

$$P = \frac{1}{3V} \sum_{i>j} \mathbf{r}_{ij}^0 \cdot \mathbf{F}_{ij}. \tag{16}$$

The difference between the average values of $\mathbf{r}_{ij}^0 \cdot \mathbf{F}_{ij}$ and $\mathbf{r}_{ij} \cdot \mathbf{F}_{ij}$ exactly compensates for the missing kinetic contribution. In a similar vein, slightly different expressions for the pressure of molecular liquids can be obtained depending on whether the virial is summed over atoms or molecules. In either case expressions similar to those above are obtained, except that the quantities m, $\mathbf{v}$, $\mathbf{r}$, and $\mathbf{F}$ refer either to atoms or to molecules. Again, these expressions are thermodynamically equivalent.

For hard-core systems the expression for the pressure can be somewhat simplified. First, the kinetic contribution $\sum_i m v_i^2$ is just twice the kinetic energy, and is a constant of the motion. Second, the potential contributions to the virial occur only during collisions, so that the time average can be written as [Eqs. (4) and (14)],

$$\frac{1}{t_M}\sum_{t_c=0}^{t_M}(-m\mathbf{r}_{ij}\cdot\mathbf{v}_{ij}),$$

where ij refers to the pair colliding at t_c. The relative velocity of the colliding pair before collision $\mathbf{v}_{ij}$ can be averaged separately from the coordinates, and independently of density [33]

$$\overline{(-m\mathbf{r}_{ij}\cdot\mathbf{v}_{ij})} = (m\pi k_B T)^{1/2}\sigma, \tag{17}$$

so that the pressure depends only on the collision frequency (per particle) τ^{-1},

$$PV/Nk_BT - 1 = \frac{\sqrt{\pi}}{6}\tau^{-1}(m\sigma^2/k_BT)^{1/2}, \tag{18}$$

where the mean collision time τ is related to the total collision rate by

$$\tau^{-1} = \frac{2}{Nt_M}\sum_{t_c=0}^{t_M}1. \tag{19}$$

Averages of dynamical variables corresponding to first-order thermodynamic properties, temperature, pressure, etc. are ensemble independent in the large system limit. Second order properties, heat capacity, compressibility, etc. can be calculated from fluctuations in dynamical variables, but the relationship between these fluctuations and thermodynamic quantities depends on the ensemble, even in the large system limit. For instance, in the canonical ensemble we find that the constant volume heat capacity $C_V = (\partial E/\partial T)_V$ is related to fluctuations in the total energy,

$$C_V/Nk_B = \frac{1}{N(k_BT)^2}(\langle E^2\rangle - \langle E\rangle^2). \tag{20}$$

On the other hand, in a microcanonical ensemble the total energy is a constant of the motion so that this formula cannot be correct. In fact, the heat capacity is related to fluctuations in kinetic energy K [46]

$$\frac{3}{2}\left[1 - \frac{3Nk}{2C_V}\right] = \frac{1}{N(k_BT)^2}(\langle K^2\rangle - \langle K\rangle^2). \tag{21}$$

Many other such relationships can be derived. One simple and useful rule is that only fluctuations about a non-zero mean have ensemble dependent averages [46].

7. Numerical Errors

Of course any numerical calculation contains errors of various kinds, and it is important to sort out what these errors may be and make estimates of them. Not only does this set bounds on the precision of the final results, but it is also a useful guide to the most efficient use of the available computer time. At this point we will just consider errors involved in calculating equilibrium averages, additional errors involved in calculating time-dependent properties will be discussed later. There are three major sources of error; number dependence, errors arising from truncation of the potential, and statistical errors. Programming errors can be largely eliminated by the powerful checks of momentum and energy conservation, reversibility for short times, and in the case of hard-core systems a periodic check for overlapping particles. This is an advantage of molecular dynamics over the Monte-Carlo methods which have few such constraints. Other general checks involve comparison with published results or, for a static lattice, with analytic calculations.

The number dependence of thermodynamic properties when periodic boundary conditions are used is almost invariably small, with corrections of order 1/N to the temperature and pressure, so that results with about 100 particles are sufficient for 1% accuracy. This is true even for systems of particles interacting via long-range forces, if the interactions are summed by the Ewald method, and if the effects of the long-range interactions on the observed properties are taken into account explicitly [25,48]. For very small systems other corrections to the thermodynamic properties can occur from percolation effects, when a cluster of interacting particles extends across the whole system [49]. In practice the number dependence is often the smallest error and in many simulations unnecessarily large systems are used, with an adverse effect on the statistical errors. It is usually more efficient to do several runs on small systems, say N = 32, N = 54, and N = 108 and extrapolate, rather than perform one large calculation on an N = 256 particle system. The total computer time for a given statistical accuracy would be somewhat less in the first project than in the second. One important exception is in measuring wavelength dependent fluctuations. In this case one is restricted to k-vectors commensurate with the periodic boundary conditions, so that larger systems can be useful in permitting a wider range of k-vectors to be studied.

More serious errors can often arise from truncation of the intermolecular forces. If the potential is simply truncated at the cut-off radius then large errors in energy conservation are observed, because of the discontinuous jumps in potential energy that occur when pairs of particles move in and out of the cut-off sphere. To avoid this problem the potential is often shifted by a constant so that it vanishes at the cut-off radius. This does not affect the forces and hence causes no change in the pressure, but the potential energy is affected and a correction must be applied. Moreover when a high-order predictor corrector algorithm, such as the Gear method, is used to solve the equations of motion the force should also be a continuous function. This can be achieved by taking the original potential and fitting a quadratic spline so that near the cut-off radius both the potential energy and force vanish smoothly. This smoothing affects both the thermodynamic energy and pressure. For pairwise additive forces, the effects of truncation errors or potential smoothing can be estimated from the radial distribution function g(r) [18]. For instance, the pressure can be written as

$$PV/Nk_BT = 1 - \frac{N}{6Vk_BT}\int_0^\infty r\frac{d\phi}{dr}g(r)dr, \qquad (22)$$

and if we assume that the truncation distance of the potential r_c is sufficiently large that g(r) can be set equal to unity, then a simple form for the pressure correction is

$$\Delta PV/Nk_BT = -\frac{2\pi N}{3Vk_BT}\int_{r_c}^\infty r^3\frac{d\phi}{dr}dr. \qquad (23)$$

A somewhat more accurate but more complicated expression can be used by including the known asymptotic expression for the pair distribution function $g(r) = \exp(-\phi(r)/k_BT)$. These corrections are not small; for instance at the triple point of the Lennard-Jones potential with a cut-off $r_c = 2.5\sigma$, the long-range corrections result in a pressure correction corresponding to about a 5% change in volume of the liquid phase. Thus such corrections are essential for a quantitative calculation of the equation of state. Obtaining a large cut-off radius for the potential, within the minimum image approximation, is one reason for using larger system sizes.

The most problematic and often the largest source of error in computer simulations are the statistical errors that arise from insufficient sampling. Successive averages over sufficiently large blocks of data or sufficiently long times are expected to be uncorrelated with each other and thus these block averages should have a Gaussian distribution about the mean. Typically, there might be 10-20 such blocks in the total length of the run, and it is useful to check if these block averages do indeed have a Gaussian distribution so that one can be sure they are uncorrelated. Assuming that the the block averages are independent, then the standard deviation σ of the set of n observations X_i,

$$\sigma^2 = \langle X^2 \rangle - \langle X \rangle^2 \tag{24}$$

is a meaningful measure of the reliability of the observed mean value $\langle X \rangle$. The expected confidence in $\langle X \rangle$ can be expressed in terms of the standard deviation divided by the square root of the number of samples. There is about a 95% chance that the true mean lies between $\langle X \rangle - 2\sigma/\sqrt{n}$ and $\langle X \rangle + 2\sigma/\sqrt{n}$. Note that for small n, a more accurate error estimate will result from replacing n by $n-1$. The main difficulty in reducing the statistical errors is that they are proportional to the inverse of the square of the run time, so that if a particular accuracy can be obtained in one day of running, then more than three months will be required to get one extra significant figure. In practice this sets rather firm bounds on the accuracy that can be achieved from a particular simulation.

8. Self Diffusion and the Velocity Autocorrelation Function

About the simplest time-dependent property of a fluid is the motion of a tagged particle that is physically identical to the other particles in the system. The phenomena of self diffusion has been the subject of extensive investigations by kinetic theory [50], hydrodynamics [51], and molecular dynamics [52] and is a prime example of a fruitful collaboration between computer experiments and theory. It is well known that the mean-square displacement of a particle from its initial position grows linearly in time after some initial relaxation, with a slope proportional to the self-diffusion coefficient D,

$$\langle | r_1(t) - r_1(0) |^2 \rangle = 6Dt + C. \tag{25}$$

The average can be calculated either over an ensemble of initial conditions

$$\int f(r^{3N}, v^{3N}) \, | r_1(t) - r_1 |^2 dr^{3N} dv^{3N},$$

or as a time average

$$\frac{1}{t_M} \int_0^{t_M} | r_1(t+s) - r_1(s) |^2 ds$$

which is somewhat simpler. Combining results from all N particles in the system reduces the statistical uncertainty by a factor of $\sqrt{N}$. In practice, the diffusion constant is usually calculated from the equivalent expression involving a time integral of the velocity autocorrelation function (VAF),

$$D = \frac{1}{3}\int_0^\infty <\mathbf{v}_1(t)\cdot\mathbf{v}_1(0)>dt. \tag{26}$$

The first molecular dynamics calculations of the VAF were reported at the 1956 IUPAP meeting by Alder and Wainwright [5]. They focused on a comparison between molecular dynamics results and the theoretical predictions of Boltzmann and Enskog [53], with the aim of assessing the density range over which these approximations were valid. Enskog improved on the low-density Boltzmann equation by including a more accurate (for hard spheres exact) calculation of the collision frequency, but the essential approximation of both theories is that successive collisions are uncorrelated with each other (the molecular chaos assumption) and so can be computed stochastically. Thus we expect an exponential decay of the VAF with a relaxation time of the order of one mean collision time, after which the velocity of the tagged particle should be randomized. The Enskog result is actually

$$\frac{1}{3}<\mathbf{v}_1(t)\cdot\mathbf{v}_1(0)> = \frac{k_BT}{m}e^{-2t/3\tau}, \tag{27}$$

where τ is the mean collision time for a single particle. These early molecular dynamics runs, within a very limited statistical accuracy, appeared to validate the Enskog approximation except for a negative region in the high density VAF which was attributed to collisions with the surrounding cage of particles [5]. Deviations from Enskog theory at high density were not surprising, and it was not for about another 10 years that much more accurate calculations showed that there were also positive deviations from Enskog theory even at low density [54], the famous long-time tails [52].

In between these calculations a study of the VAF in a high density liquid was reported in Rahman's pioneering work on liquid argon [16]. As was emphasized recently by Verlet [55], the innovation in Rahman's work was not primarily the solving of technical problems associated with using continuous forces, but rather the bold step towards simulating real as opposed to model liquids. Rahman's main motivation was to examine theoretical models for the dynamical correlations in simple liquids, particularly those probed by neutron scattering experiments. The most striking feature of Rahman's VAF is the strong negative region followed by a smaller negative plateau, both features being in qualitative disagreement with the exponential behavior predicted by the Enskog theory or by the Langevin equation with which Rahman compared his results. The Langevin equation involves a decomposition of the force on the tagged particle into a systematic drag force, proportional to the tagged particle velocity, and a random force arising from collisions between the tagged particle and other "bath" particles,

$$m\dot{\mathbf{v}} = -\zeta\mathbf{v} + \mathbf{R}(t), \tag{28}$$

where ζ is the friction coefficient. The key approximation in the Langevin equation is that the random force is independent of the tagged particle velocity, so that

$$\frac{d}{dt}<\mathbf{v}(t)\cdot\mathbf{v}(0)> = -\frac{\zeta}{m}<\mathbf{v}(t)\cdot\mathbf{v}(0)>, \tag{29}$$

leading to an exponential decay of the VAF with a relaxation time m/ζ and a diffusion constant

$D = k_BT/\zeta$, the Stokes-Einstein relation. If the tagged particle is of macroscopic size, then the friction coefficient can be calculated from hydrodynamics. Einstein showed that experimental results for the diffusion of colloidal particles were well represented by stick boundary conditions for which $\zeta = 6\pi\eta a$, where η is the viscosity of the fluid and a is the radius of the colloidal particle. On the other hand, on a microscopic scale, Zwanzig and Bixon showed that Rahman's diffusion constant was consistent with a slip boundary condition, i.e. $\zeta = 4\pi\eta a$ [56]. It turns out however, that this empirical expression is only valid when the tagged particle and bath particle are of similar size and mass [37,57]. Moreover the association of the particle radius with the hydrodynamic radius is suspect when the tagged particle and bath particles are of comparable size. A more valid expression would use the radius of the excluded volume sphere, that is $2a$, in which case the microscopic extension of Stokes' law no longer holds. The size dependence of the friction coefficient was eventually explained more convincingly by a generalized wavelength-dependent hydrodynamics [57].

A later paper by Rahman analyzed the decay of the VAF in liquid argon in terms of a "sliding" motion in the direction of diffusion, combined with a faster "rattling" motion perpendicular to this direction [58]. He found that the sliding motion, which dominates the diffusion, is strongly correlated with structural deformations in the surrounding cage. The relaxation of these elongated structures sets the time scale for the minimum in the VAF. In order to explain the minimum in Rahman's VAF quantitatively Zwanzig and Bixon used a generalized Langevin equation, with a frequency dependent friction coefficient obtained from a viscoelastic model of the fluid. However, the Zwanzig-Bixon VAF returns to zero too quickly at subsequent times, although the agreement with Rahman's molecular dynamics results can be improved by a better description of the frequency-dependent shear viscosity [59]. The hard-sphere VAF in this time regime, between 10-30 mean collision times (one mean collision time is about 10^{-13}s for liquid argon), has recently been explained quantitatively by an extended, wavelength-dependent mode-coupling theory [60].

9. Long-Time Tails

By 1967 it was known that the VAF did not decay exponentially as predicted by the Boltzmann and Enskog theories, even at low densities [54]. In addition to the Enskog-like exponential decay at short times, there was in addition a non-exponential long-time tail, indicating the presence of correlated collision sequences at all densities. It was established numerically that this long-time tail decays algebraically, as $\alpha t^{-d/2}$ where d is the dimensionality of the system. The search for an explanation of the observed magnitude of α led to a direct numerical comparison of the velocity field around a slowly moving disk, calculated both by molecular dynamics and numerical hydrodynamics [52]. The surprising result was that after only 10 mean collision times, the two velocity fields were in quantitative agreement (Fig. 2), indicating that hydrodynamics could be applied to molecular length scales and time scales. Moreover, the velocity field indicates the physical mechanism causing the persistence in the VAF, namely the recirculation of some of the initial momentum or vortex flow. In two dimensions, the hydrodynamic model predicts that the long time tail decays as $m/8\pi\eta t$, but actually the diffusion of the tagged particle away from the center of the vortex reduces the strength of the tail by a factor $1+D/\nu$, where $\nu = \eta/\rho$ is the kinematic viscosity. The final result in two dimensions, $\alpha = m/[8\pi\rho(\nu+D)]$, also predicted by kinetic theory and by hydrodynamic mode-coupling theory is in quantitative agreement at long times with the hard-disk data at several densities. An important consequence of the discovery of long-time tails was a resurgence of interest in the kinetic theory of gases, as theoreticians sought a more fundamental explanation in terms of correlated collision sequences [61]. A direct experimental verification of the long-time tails was not achieved until more than 10 years later [62].

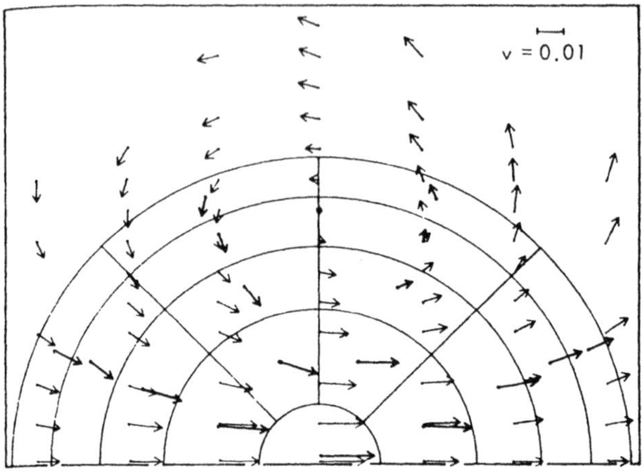

Figure 2. Velocity field around a moving disk calculated by hydrodynamics and by molecular dynamics (from ref. [52]). After 10 mean collision times the flow fields calculated by hydrodynamics (light arrows) and molecular dynamics (heavy arrows) are in quantitative agreement. The vortex flow pattern illustrates the recirculation of momentum that is responsible for the t^{-1} *long-time tail. The magnitude of the velocity, relative to the initial velocity, is indicated by the horizontal bar.*

There has been some debate as to whether the actual transport coefficients or their Enskog approximations should be used to evaluate α. Hydrodynamics implies that the real transport coefficients should be used [51,52], kinetic theory prefers the Enskog coefficients [50,61]. This dilemma is particularly acute in two-dimensional systems where the long-time t^{-1} tail implies that the hydrodynamic transport coefficients contain a logarithmic divergence, although a "self-consistent" scheme to overcome this problem has been proposed [63]. Numerically the evidence is unclear, but in general the use of Enskog coefficients seems to be favored [64,65].

10. Numerical Errors in Time Correlation Functions

The validity of the long-time correlation functions calculated by molecular dynamics has been questioned on the grounds that in a typical 14 digit precision simulation the trajectories are only reversible for about 10 mean collision times, beyond which dynamical coherence is lost [66]. However numerical comparisons of single precision (14 digit) and double precision (28 digit) hard-disk simulations indicate that valid statistical averages of the VAF can be obtained even when the individual trajectories are inaccurate [64]. There is at present no theoretical justification for this empirical observation, but few people doubt its validity, partly because of the good agreement at long times between theory and computer simulation.

A more serious practical difficulty arises because of a time-dependent finite-size effect, which can be understood qualitatively from the hydrodynamic model described in the previous section. When the volume element corresponding to the tagged particle is given a push, some of the initial momentum is carried away by sound waves. In an infinite system this part of the initial momentum is dissipated exponentially in time, but for a finite system it recurs via the periodic boundaries to give the particle an additional forward momentum at a time given roughly by L/c, where L is the box length and c is the sound speed. For a 500 particle hard sphere system at high density the sound traversal time is only about 25 mean collision times, whereas statistically the VAF can be measured for several hundred mean collisions. An elegant solution was used by Erpenbeck and Wood, who have reported extensive calculations of the VAF for hard disks [64] and hard spheres [65]. They compare the molecular dynamics results with a hydrodynamic (mode-coupling) calculation [51] which has been modified for periodic boundary conditions. This requires replacing the integrals over wave vector (modes) by sums over all wave vectors commensurate with the periodic boundary conditions. The agreement between molecular dynamics and "periodic hydrodynamics" is excellent and provides much stronger evidence of the asymptotic $t^{-3/2}$ decay of the three-dimensional VAF than was achieved by Alder and Wainwright [52]. This is another example in which the most important finite-size effects can be calculated analytically for periodic systems.

11. Hydrodynamics and Transport Coefficients

The macroscopic behavior of fluids is described by the equations of hydrodynamics [67], which relate the transport of mass, momentum, and energy to gradients of the fluxes,

$$\frac{\partial \rho}{\partial t} + \nabla\cdot(\rho\mathbf{u}) = 0, \tag{30}$$

$$\frac{\partial}{\partial t}(\rho\mathbf{u}) + \nabla\cdot(\rho\mathbf{u}\mathbf{u}) = \nabla\cdot\boldsymbol{\sigma}, \tag{31}$$

$$\frac{\partial e}{\partial t} + \nabla\cdot(e\mathbf{u}) = -\nabla\cdot\mathbf{q} + \boldsymbol{\sigma}:\nabla\mathbf{u}. \tag{32}$$

In these equations ρ, $\rho\mathbf{u}$, and e are the densities of mass, momentum and internal energy (excluding the kinetic energy of the fluid motion), and $\boldsymbol{\sigma}$ and $\mathbf{q}$ are the stress tensor and heat flux vector. These equations can all be derived from the principle of continuity, which simply points out that in the absence of internal sources and sinks, any change within a fixed volume element dV must take place by flow through the boundary surface, i.e.

$$\frac{\partial}{\partial t}\int \rho dV = -\int \rho\mathbf{u}\cdot\mathbf{n}dS, \tag{33}$$

where $\mathbf{n}$ is an outward normal at the surface point dS. Using Green's theorem to transform the surface integral into a volume integral leads directly to the equation of continuity or mass conservation. In the momentum equation, the total momentum flux $\rho\mathbf{u}\mathbf{u} - \boldsymbol{\sigma}$ has been split into a convective part $\rho\mathbf{u}\mathbf{u}$ due to the macroscopic motion of the fluid, and a part due to pressure forces and viscous forces that arise from thermal fluctuations and intermolecular forces. Similarly, in the energy conservation equation, the total energy flux is split into convective (e**u**) and conductive (**q**) parts. However there is a further dissipative *source* of energy due to the work done within a volume element by the viscous forces. The term $\boldsymbol{\sigma}:\nabla\mathbf{u}$ is the tensorial analogue, for a general deformation, of the adiabatic ($-P\Delta V$) work done during a small volume change.

The hydrodynamic equations can also be derived from Newtonian mechanics by localizing the mass, momentum, and energy of a particle at its center, as described in the classic paper by Irving and Kirkwood [68]. They emphasized that the potential parts of the energy and stress tensor cannot be localized precisely, so that there are many possible microscopic expressions for the fluxes, all of which are equivalent on length scales larger than the range of the intermolecular forces. Here we point out that the Irving-Kirkwood expressions for the microscopic fluxes can also be derived from generalized virial theorems. A microscopic expression for the stress tensor can be obtained from a tensorial virial

$$\mathbf{W}(t) = \sum_i \mathbf{r}_i \mathbf{v}_i, \tag{34}$$

with an analogous result to that found earlier for the thermodynamic pressure [Eq. (14)]

$$\rho\mathbf{uu} - \sigma = \frac{1}{V}\left\{\sum_i m\mathbf{v}_i\mathbf{v}_i + \sum_{i>j}\mathbf{r}_{ij}\mathbf{F}_{ij}\right\}. \tag{35}$$

In this derivation we have obtained the total momentum flux, rather than just the stress tensor, because the virial $\mathbf{W}$ was defined in a space-fixed frame. If the momentum of a particle $\mathbf{p}_i$ is defined relative to the local stream velocity, i.e. $\mathbf{p}_i = m(\mathbf{v}_i - \mathbf{u})$ then

$$\sigma = -\frac{1}{V}\left\{\sum_i \frac{\mathbf{p}_i\mathbf{p}_i}{m} + \sum_{i>j}\mathbf{r}_{ij}\mathbf{F}_{ij}\right\}. \tag{36}$$

Microscopically the stream velocity $\mathbf{u}$ is just a local average of the particle velocities, so that the average value of thermal velocity $\mathbf{p}_i/m$ is zero. An expression for the heat flux can be derived in the same fashion from an "energy virial",

$$\mathbf{W}(t) = \sum_i \mathbf{r}_i e_i, \tag{37}$$

with e_i the energy associated with particle i. The final result is

$$\mathbf{q} = \frac{1}{V}\left\{\sum_i (\frac{\mathbf{p}_i}{m} e_i + \sum_{i>j}\mathbf{r}_{ij}\mathbf{F}_{ij}\cdot\frac{(\mathbf{p}_i+\mathbf{p}_j)}{2m}\right\}. \tag{38}$$

Macroscopically, Navier-Stokes hydrodynamics postulates that the stress tensor and heat flux are proportional to the linear gradients of velocity and temperature respectively [67],

$$\sigma = -P + \eta_V \nabla\cdot\mathbf{u} + \eta[(\nabla\mathbf{u}) + (\nabla\mathbf{u})^t], \tag{39}$$

$$\mathbf{q} = -\lambda\nabla T, \tag{40}$$

which are the Newtonian model of a viscous fluid, and Fourier's law of heat conduction. Thus the fluxes can be described by three transport coefficients, the bulk viscosity η_V describing the added non-equilibrium resistance to changes in volume, the shear viscosity η describing the resistance to changes in shape, and the thermal conductivity λ. In most instances, these simple linear constitutive laws are valid because of the large difference between macroscopic and molecular scales (length and time). Turbulence occurs because of the coupling between macroscopic length scales, which is made possible by the non-linear advection term $\nabla\cdot(\rho\mathbf{uu})$ in the momentum equation. On a molecular scale these non-linear terms are negligible, so that long-wavelength thermal fluctuations can be described by linearized hydrodynamic equations, which nevertheless are governed by the same transport coefficients as the non-linear fluid equations.

12. Transport Coefficients by Molecular Dynamics

The connection between microscopic molecular mechanics and macroscopic hydrodynamics is completed by the Green-Kubo relations, which relate the transport coefficients to the decay of the corresponding fluxes [69],

$$\eta = \frac{V}{k_BT}\int_0^\infty <\sigma_{xy}(t)\sigma_{xy}(0)>dt, \tag{41}$$

$$\eta_V = \frac{V}{k_BT}\int_0^\infty <\delta P(t)\delta P(0)>dt, \tag{42}$$

$$\lambda = \frac{V}{k_BT^2}\int_0^\infty <q_x(t)q_x(0)>dt. \tag{43}$$

The Green-Kubo relation for the bulk viscosity is a bit more complicated since the pressure fluctuates about a non-zero mean, which is a time average *not* an ensemble average. In the microcanonical ensemble the time averaged pressure along a single trajectory is identical to the ensemble averaged pressure, but in other ensembles this equality does not hold because a single Newtonian trajectory cannot explore all the available phase space, due to the constraints of constant total energy and momentum. To evaluate transport coefficients in a canonical ensemble for instance, the procedure would be to select initial phase points from a canonical distribution and for each trajectory use the pressure corresponding to the microcanonical (time) average at that energy [69]. This unnecessarily complex procedure has never been used in practice.

One of the simplest and to my mind one of the most convincing derivations of the Green-Kubo relations follows directly from the linearized hydrodynamic equations [1,70]. This method is particularly simple in the case of the shear viscosity, because the transverse momentum current fluctuates independently of the other modes. Thus we can write down an equation of motion for a transverse momentum current, say $j_x(y)$ ($\mathbf{j} = \rho\mathbf{u}$)

$$\frac{\partial j}{\partial t} = \nu\nabla^2 j, \tag{44}$$

where $\nu = \eta/\rho$ is the kinematic viscosity, and we have used the result that to linear order in the fluctuations $\rho\nabla\mathbf{u} = \nabla(\rho\mathbf{u}) = \nabla\mathbf{j}$. By taking a Fourier transform in space and a Laplace transform in time of the equation of motion, we obtain an expression relating the decay of ensemble-average fluctuations to the shear viscosity, namely

$$<j(k;s)j(-k;0)> = \frac{<j(k)j(-k)>}{s+\nu k^2}, \tag{45}$$

where j and k point in orthogonal directions. This expression, derived from hydrodynamics, is only valid at long wavelengths and long times, and predicts that the transverse momentum fluctuations decay exponentially in time with a slope νk^2. The validity of this expression can be extended to molecular length scales and time scales by allowing the viscosity itself to be wavelength and frequency dependent. This generalized viscosity can then be used to explain phenomena in which molecular scale fluctuations are important [57,70]. The hydrodynamic viscosity follows by taking the long-wavelength limit first, followed by the low frequency limit. From the relation

$$\lim_{k \to 0} \frac{d}{dt} j(k;t) = ik\sigma_{xy}(t)V, \tag{46}$$

and the result $\langle j(k)j(-k)\rangle = Nmk_BT$ we find, after some elementary manipulations, the expression given above for the shear viscosity. Derivations of the other transport coefficients are somewhat more complicated, because of the coupling between fluctuations in density, longitudinal momentum, and energy, but follow a similar route.

Numerical studies of transport coefficients began in the 1970's using the Green-Kubo fluctuation formulae, first for hard spheres [71] and then for the Lennard-Jones potential [59]. The hard-sphere calculations were carried out at many state points for small systems, primarily 108 particles supplemented by some 500 particle runs. The Lennard-Jones simulation was a single run for 864 particles at the triple point. Soon after that extensive results for shear viscosity and thermal conductivity of the Lennard-Jones and inverse 12th-power potentials were obtained by the direct non-equilibrium molecular dynamics studies of Ashurst and Hoover [72]. Results for the thermal conductivity from these early simulations were found to be in accord with Enskog's theory of dense fluid transport. The results for the bulk viscosity seem a little more complex. For hard spheres the simulation results were in agreement with Enskog theory, within the rather large statistical errors, except that the bulk viscosity at the freezing point was somewhat smaller than the Enskog prediction. On the other hand, for the inverse 12th power potential more recent NEMD simulations indicate that Enskog theory overestimates the bulk viscosity by about a factor of 7 over a wide range of densities [73].

The shear viscosity for both hard and soft potentials was found to increase much faster with density in the region around the freezing transition than predicted by Enskog theory. In fact the shear viscosity apparently diverges at densities somewhat higher than the thermodynamic freezing point, and this divergence, or equivalently (via the Stokes-Einstein relation) the vanishing of the diffusion constant, is often associated with a transition to a kinetically frozen "glassy" phase [74]. These early results also confirmed that the qualitative similarities between the thermodynamic properties of hard spheres and more realistic potentials extended to transport properties also. The corrections to Enskog theory, necessary to fit the hard-sphere shear viscosity data applied quite well to the Lennard-Jones and inverse-power potentials as well [59,72].

13. "Molasses" Tail

The enhanced shear viscosity was found to be due to a long-time tail in the potential part of the Green-Kubo integrand [59,71], which has been called the "molasses" tail to emphasize that it is suspected to be of structural rather than hydrodynamic origin [75]. To clarify what is meant by kinetic and potential contributions, we note that if we take the kinetic and potential (collisional) contributions to the shear stress separately,

$$\sigma^K = -\frac{1}{V}\sum_i m\mathbf{v}_i\mathbf{v}_i \quad ; \quad \sigma^P = -\frac{1}{V}\sum_{i>j}\mathbf{r}_{ij}\mathbf{F}_{ij}, \tag{47}$$

then the Green-Kubo integrand can be split into kinetic $\langle\sigma^K_{xy}(t)\sigma^K_{xy}(0)\rangle$, potential $\langle\sigma^P_{xy}(t)\sigma^P_{xy}(0)\rangle$, and cross $2\langle\sigma^K_{xy}(t)\sigma^P_{xy}(0)\rangle$, contributions. In a non-equilibrium simulation in the linear response regime, the cross contribution is divided equally into the kinetic and potential parts of the shear stress [76]. The slow relaxation of the shear stress is the most important high density deviation from Enskog theory and has been the subject of extensive investigations, by both equilibrium and non-equilibrium molecular dynamics. Substantial disagreement (30%) existed for some time between equilibrium Green-Kubo [59] and non-equilibrium molecular dynamics [72,76] results for the shear viscosity of the Lennard-Jones system at the triple point. It is now known that the original Green-Kubo result [59] was anomalous, for reasons that are still unclear. A recent suggestion [77] is that the statistical errors may in fact be larger than reported in ref. [59]. Other Green-Kubo calculations, particularly for smaller systems sizes, are consistent with non-equilibrium molecular dynamics results [77-79]. Speculation that the Green-Kubo method contains a larger number dependence than direct non-equilibrium molecular dynamics has not been justified either theoretically or numerically; in the limit of small gradients non-equilibrium molecular dynamics should be identical to a Green-Kubo calculation carried out under the same conditions [80]. The good agreement between the Lennard-Jones viscosity and that of liquid argon at the triple point may be fortuitous. Recent work suggests that the viscosity is very sensitive to the tail of the intermolecular potential [77], so that the truncation of the intermolecular forces may cause large discrepancies if compared with an untruncated Lennard-Jones potential.

Mode-coupling [51] and hydrodynamic theory [52] predict a long-time $t^{-d/2}$ tail in the kinetic correlation function *only*. Unfortunately, this tail is so small in amplitude that extensive molecular dynamics simulations have failed to either confirm or deny its existence [81]. However, evidence of a weak logarithmic divergence in the viscosity of a two-dimensional lattice gas (in which there are only kinetic fluxes) has been discovered recently [82]. In three dimensional systems the kinetic tail is much too small to have a noticeable effect on the viscosity.

Evans has suggested that the potential and cross contributions to the shear stress autocorrelation function (SSAF) also decay algebraically as $t^{-d/2}$ at long times, but with coefficients about two orders of magnitude larger than the kinetic tail [83]. This conclusion was based on frequency-dependent non-equilibrium molecular dynamics simulations, which showed a cusp-like behavior in the viscosity at low frequencies. Moreover, Green-Kubo results showed that the potential and cross correlations decay in a very similar fashion to the kinetic correlation function, although the data cannot be unambiguously fitted to a $t^{-3/2}$ dependence [84]. The arguments against the need for a fundamental revision of the asymptotic mode-coupling theory are basically twofold. Firstly, the time or frequency range over which the SSAF can be measured is limited statistically to times prior to the advent of the long-time tail in the VAF. Thus it seems unlikely that computer simulations are measuring the asymptotic decay in the SSAF. Secondly, the magnitude of the potential and cross tails are very strongly density dependent, and cannot be detected at densities less than one half close packing. By contrast, the kinetic tails are present at all densities, with about the same amplitude. Thus the mechanism leading to the molasses tail is still elusive, but the fact that it is restricted to high densities implies that it is linked to the persistence of solid-like ordering in the dense fluid phase [75].

To investigate the possibility of a structural relaxation mechanism in dense hard sphere fluids, which may also be related to a dynamic theory of the glass transition, we have examined the decay of angular xy correlations in hard sphere fluids [85]. The form of the potential contribution to the shear stress in hard-sphere systems, namely

$$J^{P}_{xy} = \sum_{i>j}(-m\mathbf{r}_{ij}\cdot\mathbf{v}_{ij})\hat{x}_{ij}\hat{y}_{ij}\delta(t-t_c), \tag{48}$$

$(\mathbf{J} = -\boldsymbol{\sigma}V)$ suggests a factorization into an average collisional part $\overline{(-m\mathbf{r}_{ij}\cdot\mathbf{v}_{ij})} = (\pi m k_B T)^{1/2}\sigma$ and an angle dependent part $K_{xy} = \sum_{i>j}\hat{x}_{ij}\hat{y}_{ij}\delta(t-t_c)$. Since the pressure fluctuations are known to relax quickly for hard-sphere systems, we expected that at long-times the decay of the shear stress fluctuations would be quantitatively related to the decay of the angular fluctuations, i.e.

$$\langle J^{P}_{xy}(t)J^{P}_{xy}(0)\rangle = \pi m k_B T\sigma^2\langle K_{xy}(t)K_{xy}(0)\rangle. \tag{49}$$

Numerical experiments showed that this factorization was valid beyond about 10 mean collision times

(Fig. 3), confirming that the long-time tail in the SSAF is due to a persistence in orientation of the colliding pairs of particles, rather than a persistence in the collisional momentum transfer that might be associated with a hydrodynamic mechanism.

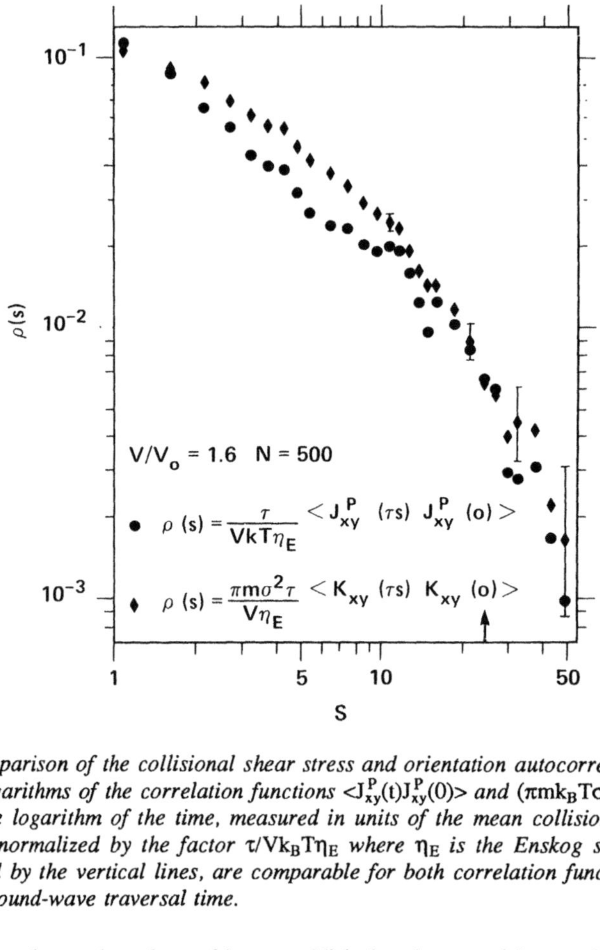

Figure 3. Comparison of the collisional shear stress and orientation autocorrelation functions (from ref. [85]). The logarithms of the correlation functions $<J^P_{xy}(t)J^P_{xy}(0)>$ *and* $(\pi m k_B T\sigma^2)<K_{xy}(t)K_{xy}(0)>$ *are plotted against the logarithm of the time, measured in units of the mean collision time* τ. *The correlation functions are normalized by the factor* $\tau/Vk_BT\eta_E$ *where* η_E *is the Enskog shear viscosity. The error bars, indicated by the vertical lines, are comparable for both correlation functions. The vertical arrow indicates the sound-wave traversal time.*

More recently, we have been able to establish that the potential part of the SSAF does *not* decay algebraically as $t^{-3/2}$. This could only be done at the highest fluid density $V/V_0 = 1.5$ (V_0 is the close packed volume) where the molasses tail can be measured out to nearly 100 mean collision times. The results for N = 500 (10^8 collisions) and N = 4000 (5×10^8 collisions) are of similar statistical accuracy and do not exhibit a significant N-dependence, except in the region of the sound traversal time, about 25 and 50 mean collision times respectively for N = 500 and N = 4000. The correlation function shown in Fig. 4 is clearly inconsistent with an algebraic $t^{-3/2}$ decay, but is consistent with a slow exponential relaxation as predicted by the recent wavelength-dependent extended mode coupling theories [86,87]. However, the amplitude of the tail predicted by these theories is somewhat uncertain.

Moreover, we cannot confirm that the relaxation is exponential. Our data is also consistent with a "stretched" exponential relaxation, $\exp[-(t/\tau)^{\beta}]$ with $\beta \approx 0.7$, which is similar to experimental observations of viscous and dielectric relaxation [88]. In either case, the tail corrections to the viscosity are now expected to be quite small, as compared with the corrections predicted from an algebraic decay.

To try to elucidate further the underlying mechanism we have calculated the stress field correlation function, analogous to the velocity field correlation function discussed earlier. The potential part of the shear stress in a small cube (side length about σ) at $t = 0$ is correlated with that in another cube at time t. From these results we can estimate the range of the angular correlations in various directions, and also calculate the corresponding relaxation times. The results (for $V/V_0 = 1.6$, $N = 4000$) show that in the direction perpendicular to the plane of shear, the correlations are short range, dying off in 1-2 particle diameters. In the plane of the shear, the correlations are longer range, out to about 5σ. The relaxation times do not vary much with either distance or orientation and are of the same order of magnitude as that for the complete SSAF, i.e. about 20 mean collision times. It is hoped that these investigations will eventually provide sufficient clues for a quantitative theory of the molasses tail.

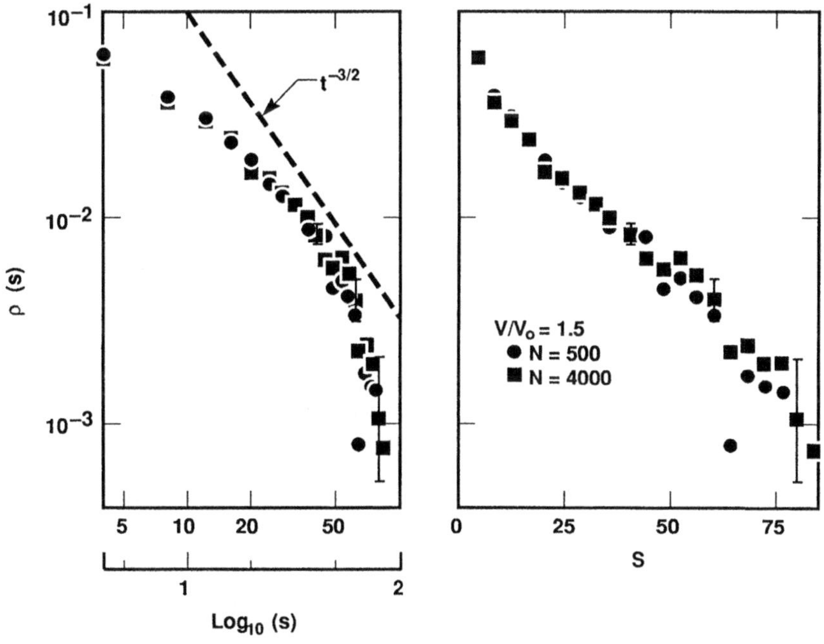

Figure 4. Relaxation of the collisional contribution to the shear-stress autocorrelation function at $V/V_0 = 1.5$. *The logarithm of the collisional shear stress correlation function* $\rho(s)$, *normalized as in Fig. 3, is plotted vs the logarithm of the time and vs the time itself, measured in mean collision times* $(s = t/\tau)$. *The* $t^{-3/2}$ *line is plotted with an arbitrary amplitude. The error bars are comparable for both* $N = 500$ *and* $N = 4000$.

14. Other Ensembles

Until recently, molecular dynamics simulations were almost invariably used to generate microcanonical averages by integrating along a single trajectory. On the other hand Monte-Carlo simulations were quickly extended from the canonical ensemble to the isobaric (N,p,T) ensemble [89] and to the grand canonical ensemble [90]. Time correlation functions can be obtained by integrating Newton's equations of motion over many different trajectories with initial conditions sampled from, say, a canonical distribution [65,84], but there is no indication that this procedure leads to statistically more accurate results than a single microcanonical trajectory of the same total length. In the early 1980's, significant technical innovations in molecular dynamics came from the realization that other dynamics, besides Newtonian mechanics, could be used to generate thermodynamic and transport properties. These innovations have both extended the range of problems that can be studied by molecular dynamics, and have helped clarify theoretical issues related to the validity of non-equilibrium molecular dynamics methods. Here I will briefly summarize the effect of these innovations on equilibrium molecular dynamics; Denis Evans will describe the non-equilibrium aspects in his lectures.

In 1980, Andersen showed that by allowing the volume of the system to fluctuate a constant pressure (N,p,H) ensemble (H is the enthalpy) could be generated by an average along a single trajectory [91]. At the same time, Hoover et. al. showed that similar equations of motion could be used with externally driven volume fluctuations to simulate non-equilibrium cycles of expansion and compression, and thus measure the bulk viscosity [73]. The equations of motion used by Andersen and by Hoover et. al. contain a macroscopic deformation, described by a strain rate $\dot{\varepsilon} = \frac{1}{3}\dot{V}/V$, together with microscopic thermal fluctuations,

$$\dot{r} = \frac{p}{m} + \dot{\varepsilon} r \quad ; \quad \dot{p} = F - \dot{\varepsilon} p. \tag{50}$$

The velocity in a space-fixed frame $\dot{r}$ contains a thermal contribution p/m and a macroscopic deformation $\dot{\varepsilon} r$. The momentum p is identical for all periodic images of a particle; it is related to the microscopic temperature [Eq. (10)] and kinetic pressure [Eq. (14)] as described earlier, with the local velocity p/m substituted for v. The term $\dot{\varepsilon} r$ describes the motion of a fixed point in a macroscopic continuum under a homogeneous expansion or compression. Thus for different periodic images $\dot{r}$ differs by multiples of $\dot{\varepsilon} L$ (L is the box length). The equation of motion for $\dot{p}$ contains both interparticle forces and a scaling of the momentum that corresponds to work done against the kinetic pressure. For instance in an ideal gas (F = 0) it leads to the adiabatic equation of state, $P \propto V^{-5/3}$. More generally we see that these equations satisfy an exact microscopic analogue of the first law of thermodynamics,

$$\dot{E} = \sum_i \frac{p_i \cdot \dot{p}_i}{m} - \sum_{i>j} \dot{\mathbf{r}}_{ij} \cdot \mathbf{F}_{ij} = -3PV\dot{\varepsilon} = -P\dot{V}. \tag{51}$$

These equations of motion with a prescribed time dependent strain rate $\dot{\varepsilon}(t) = \varepsilon_0 \omega \cos\omega t$ were used in non-equilibrium simulations of the bulk viscosity [73], and also to demonstrate a link between non-equilibrium molecular dynamics and equilibrium Green-Kubo fluctuation formulae [76]. By contrast, in Andersen's method the volume fluctuations are driven by imbalances between the internal pressure, measured by the instantaneous value of the microscopic dynamical variable, and the required external pressure,

$$M\ddot{V} = -P_{ext} + P \tag{52}$$

where P is the microscopic pressure [Eq. (14) with **v** replaced by **p**/m]. The volume does not respond instantaneously to these pressure differences, but has a finite relaxation time, as if it were connected to an external piston of mass M. The equilibrium properties are independent of the mass of the piston, but from a practical point of view it cannot be too large, since in that case the system will not equilibrate in a reasonable amount of computer time. On the other hand if the mass is too small then the dynamics will be noticeably non-Newtonian. In the extreme case that the volume is adjusted so that the microscopic pressure is a constant of the motion a different ensemble, with different fluctuations, results. A "realistic" relaxation time for the volume fluctuations is L/c, i.e. the sound traversal time, in which case Newtonian dynamics is recovered in the large system limit.

Andersen's constant pressure method was extended to general fluctuations in shape and size by Rahman and Parrinello [92]. The analogous non-equilibrium methods for general deformations were being developed at about the same time [76,93]. An important application of the Rahman-Parrinello technique lies in the study of the often complex phase diagrams of molecular solids. The fluctuations in size and shape of the unit cell permit the system to move readily from one solid phase to another of lower free energy, whereas in a small periodic system with fixed size and shape this is usually impossible. Thus the phase diagram, or strictly speaking the boundaries of mechanical stability, can be mapped out by simply observing what the equilibrium state of the system is at a particular pressure and energy (or temperature). However, the Rahman-Parrinello method does not necessarily locate an accurate coexistence boundary, nor even correctly predict the thermodynamically stable phase in all cases. Particularly in the region of complex phase boundaries, it may be necessary to calculate the free energy of each phase to be sure that the thermodynamically stable phase is correctly identified [17].

Another significant extension of the traditional microcanonical molecular dynamics was the introduction of a deterministic thermostat by Nosé [94]. Earlier work had led to a stochastic thermostat that produced a canonical distribution [91] and a deterministic thermostat that produced an unusual hybrid distribution, microcanonical in the kinetic energy and canonical in the potential energy [95,96]. Nosé showed that by introducing an extra degree of freedom, corresponding to a scaling of the time, a microcanonical (time) average in the extended 6N + 1 ensemble produced a canonical distribution in the physical 6N ensemble. Later, Hoover derived a significantly simpler set of equations of motion that did not incorporate a time scaling, but instead introduced a friction coefficient that responded to differences between the microscopic temperature, as measured by the kinetic energy, and the required temperature [97],

$$\dot{r} = v \quad ; \quad \dot{v} = \frac{F}{m} - \zeta v \quad ; \quad Q\dot{\zeta} = \sum_i m v_i^2 - 3(N-1)k_B T_{ext}. \tag{53}$$

Again there is an arbitrary parameter Q which sets the relaxation time of the microscopic temperature. A physically reasonable relaxation time is the time for thermal diffusion across the system L^2/D_T where D_T is the thermal diffusivity.

It is also possible to calculate time correlation functions in constant pressure and constant temperature ensembles. If the relaxation times for the pressure and temperature are long compared with the relaxation times being probed then it is reasonable to assume that the time correlation functions will be similar to ones generated from Newtonian trajectories but with an isobaric and/or isothermal initial distribution. Even if there is coupling between the "external" degrees of freedom and the internal dynamical processes, Evans and Holian have shown that discrepancies in the time correlation functions are only expected to be of order 1/N [98].

15. Concluding Remarks

In these lectures I have given a brief overview of the basic methodology of equilibrium molecular dynamics. As is clear from a cursory inspection of the literature the problem of setting up a molecular dynamics experiment on the computer is quite straightforward, in most cases trivial in fact. The extraction of worthwhile information from the mass of data produced by the computer is much more difficult. It has been my intention that the examples selected illustrate some of the ways to accomplish this. The choice of material has been quite subjective, though more in terms what was left out rather than what was included. A more complete list of landmarks in the field of computer simulation, both molecular dynamics and Monte-Carlo, is contained in the collection of reprints edited by Ciccotti et. al. [4].

Acknowledgments

I would like to thank Bill Hoover (Lawrence Livermore National Laboratory) and Giovanni Ciccotti (Università di Roma) who read and commented on earlier drafts of this manuscript. This work was supported by the U.S. Department of Energy and Lawrence Livermore National Laboratory under Contract No. W-7405-Eng-48.

References

[1] J.P. Hansen and I.R. McDonald, *Theory of Simple Liquids*, (Academic Press, London, 1986).
[2] G. Ciccotti and W.G. Hoover, editors, *Molecular Dynamics Simulation of Statistical-Mechanical Systems*, (North-Holland, Amsterdam, 1986).
[3] W.G. Hoover, *Molecular Dynamics*, (Springer-Verlag, Berlin, 1986).
[4] G. Ciccotti, D. Frenkel and I.R. McDonald, *Simulation of Liquids and Solids*, (North-Holland, Amsterdam, 1987).
[5] B.J. Alder and T.E. Wainwright, in *Transport Processes in Statistical Mechanics*, edited by I. Prigogine, (Interscience, New York, 1958).
[6] B.J. Alder and T.E. Wainwright, *J. Chem. Phys.*, **27**, 1208 (1957).
[7] N. Metropolis, A.W. Rosenbluth, M.N. Rosenbluth, A.H. Teller and E. Teller, *J. Chem. Phys.*, **21**, 1087 (1953).
[8] W.W. Wood and J.D. Jacobson, *J. Chem. Phys.*, **27**, 1207 (1957).
[9] J.D. Bernal, *Proc. Roy. Soc.*, **A280**, 299 (1964).
[10] B.J. Alder and T.E. Wainwright, *Phys. Rev.*, **127**, 359 (1962).
[11] W.G. Hoover and F.H. Ree, *J. Chem. Phys.*, **49**, 3609 (1968).
[12] J.A. Barker and D. Henderson, *Rev. Mod. Phys.*, **48**, 587 (1976).
[13] D. Frenkel and B.M. Mulder, *Mol. Phys.*, **55**, 1171 (1985).
[14] A. Stroobants, H.N.W. Lekkerkerker, and D. Frenkel, *Phys. Rev. A*, **36**, 2929 (1987).
[15] W.W. Wood and F.R. Parker, *J. Chem. Phys.*, **27**, 720 (1957).
[16] A. Rahman, *Phys. Rev.*, *136*, **A405**, (1964).
[17] D. Frenkel, *Phys. Rev. Lett.*, **56**, 858, (1986).
[18] L. Verlet, *Phys. Rev.*, **159**, 98, (1967).
[19] R. Car and M. Parrinello, *Phys. Rev. Lett.*, **55**, 2471 (1985).
[20] G. Sugiyama, M.E. Colvin and B.J. Alder, (work in progress).
[21] M. Mareschal and E. Kestemont, *J. Stat. Phys.*, **48**, 1187 (1987).
[22] D.C. Rapaport, *Phys. Rev. Lett.*, **60**, 2480 (1988).
[23] S.G. Brush, H.L. Sahlin and E. Teller, *J. Chem. Phys.*, **45**, 2102 (1966).
[24] S.W. DeLeeuw, J.W. Perram, and E.R. Smith, *Proc. Roy. Soc.*, **A373**, 27 (1980).
[25] E.L. Pollock and B.J. Alder, *Physica*, **102A**, 1 (1980).

[26] J.W. Perram and E.R. Smith, *J. Stat. Phys.*, **46**, 179 (1987).
[27] J. Happel and H. Brenner, *Low-Reynolds Number Hydrodynamics*, (Martinus Nijhoff, Dordrecht, 1986).
[28] A.J.C. Ladd, UCRL-98907 (1988), *J. Chem. Phys.*, (In Press).
[29] D.S. Fisher, B.I. Halperin and R. Morf, *Phys. Rev. B*, **20**, 4692 (1979).
[30] A.J.C. Ladd and W.G. Hoover, *Phys. Rev. B*, **26**, 5469 (1982).
[31] O.R. Walton and R.L. Braun, *Acta Mechanica*, **63**, 73 (1986).
[32] O.R. Walton and R.L. Braun, *J. Rheology*, **30**, 949 (1986).
[33] B.J. Alder and T.E. Wainwright, *J. Chem. Phys.*, **31**, 459 (1959).
[34] R.W. Hockney and J.W. Eastwood, *Computer Simulations Using Particles*, (McGraw-Hill, New York, 1981).
[35] G.D. Venneri and W.G. Hoover, *J. Comput. Phys.*, **73**, 468 (1987).
[36] H.J.C. Berendsen and W.F. van Gunsteren in *Molecular Dynamics Simulation of Statistical-Mechanical Systems*, edited by G. Ciccotti and W.G. Hoover, (North-Holland, Amsterdam, 1986).
[37] W.E. Alley, *Ph.D. Thesis*, University of California-Livermore UCRL-52815 (1979).
[38] Ya. G. Sinai, in *The Boltzmann Equation*, edited by E.G.D. Cohen and W.Thirring, (Springer-Verlag, Vienna, 1973).
[39] E. Fermi, J.G. Pasta, S.M. Ulam, reprinted in *Collected Works of Enrico Fermi, Vol II*, edited by E. Amaldi, H.L. Anderson, E. Persico, F. Rasetti, C.S. Smith, A. Wattenburg and E. Segré, (University of Chicago Press, 1965).
[40] J.L. Tuck and M.T. Menzel, *Adv. in Math.*, **9**, 399 (1972).
[41] J. Gleick, *Chaos: Making a New Science*, (Viking, New York, 1987).
[42] M.J. Klein, in *The Boltzmann Equation*, edited by E.G.D. Cohen and W.Thirring, (Springer-Verlag, Vienna, 1973).
[43] B.L. Holian, W.G. Hoover and H.A. Posch, *Phys. Rev. Lett.*, **59**, 10 (1987).
[44] W.G. Hoover, *Phys. Rev. A*, **37**, 252 (1988).
[45] D. Frenkel in *Molecular Dynamics Simulation of Statistical-Mechanical Systems*, edited by G. Ciccotti and W.G. Hoover, (North-Holland, Amsterdam, 1986).
[46] J.L. Lebowitz, J.K. Percus, and L. Verlet, *Phys. Rev.*, **153**, 250 (1967).
[47] W.G. Hoover and Z.W. Salzburg, *J. Chem. Phys.*, **54**, 2129 (1971).
[48] J.-P. Hansen in *Molecular Dynamics Simulation of Statistical-Mechanical Systems*, edited by G. Ciccotti and W.G. Hoover, (North-Holland, Amsterdam, 1986).
[49] J.L. Lebowitz and J.K. Percus, *Phys. Rev.*, **124**, 1673 (1961).
[50] J.R. Dorfman and E.G.D. Cohen, *Phys. Rev. Lett.*, **25**, 1257, (1970).
[51] M.H. Ernst, E.H. Hauge and J.M.J. van Leeuwen, *Phys. Rev. Lett.*, **25**, 1254, (1970).
[52] B.J. Alder and T.E. Wainwright, *Phys. Rev. A*, **1**, 18, (1970).
[53] S. Chapman and T.G. Cowling, *The Mathematical Theory of Non-Uniform Gases*, (Cambridge University Press, 1960)
[54] B.J. Alder and T.E. Wainwright, *Phys. Rev. Lett.*, **18**, 988, (1967).
[55] L. Verlet, in *Aneesur Rahman Memorial Meeting*, edited by H.J.C. Berendsen, (CECAM, Orsay, 1987).
[56] R. Zwanzig and M. Bixon, *Phys. Rev. A*, **2**, 2005, (1970).
[57] W.E. Alley and B.J. Alder, in *Molecular Structure and Dynamics*, edited by M. Balaban, (International Science Services, Philadelphia, 1980).
[58] A. Rahman, *J. Chem. Phys.*, **45**, 2585 (1966).
[59] D. Levesque, L. Verlet, and J. Kürkijarvi, *Phys. Rev. A*, **7**, 1690 (1973).
[60] T.R. Kirkpatrick and J.C. Nieuwoudt, *Phys. Rev. A*, **33**, 2658 (1986).
[61] J.R. Dorfman and E.G.D. Cohen, *Phys. Rev. A*, **6**, 776, (1972), ibid., **12**, 292 (1975).
[62] G.L. Paul and P.N. Pusey, *J. Phys. A*, **14**, 3301 (1981).
[63] T.E. Wainwright, B.J. Alder and D.M. Gass, *Phys. Rev. A*, **4**, 233, (1971).
[64] J.J. Erpenbeck and W.W. Wood, *Phys. Rev. A*, **26**, 1648 (1982).

[65] J.J. Erpenbeck and W.W. Wood, *Phys. Rev. A*, **32**, 412 (1985).
[66] R.F. Fox, *Phys. Rev. A*, **27**, 3216 (1983).
[67] L.D. Landau and E.M. Lifshitz, *Fluid Mechanics*, (Addison-Wesley, London, 1959).
[68] J.H. Irving and J.G. Kirkwood, *J. Chem. Phys.*, **18**, 817 (1950).
[69] R.W. Zwanzig, *Ann. Rev. Phys. Chem.*, **16**, 67 (1965).
[70] W.E. Alley and B.J. Alder, *Phys. Rev. A*, **27**, 3158 (1983).
[71] B.J. Alder, D.M. Gass, and T.E. Wainwright, *J. Chem. Phys.*, **53**, 3813, (1970).
[72] W.T. Ashurst and W.G. Hoover, *Phys. Rev. A*, **11**, 658 (1975).
[73] W.G. Hoover, A.J.C. Ladd, R.B. Hickman, and B.L. Holian, *Phys. Rev. A*, **21**, 1756 (1980).
[74] T.R. Kirkpatrick, *Phys. Rev. A*, **31**, 939 (1985).
[75] B.J. Alder in *Molecular Dynamics Simulation of Statistical-Mechanical Systems*, edited by G. Ciccotti and W.G. Hoover, (North-Holland, Amsterdam, 1986).
[76] W.G. Hoover, D.J. Evans, R.B. Hickman, A.J.C. Ladd, W.T. Ashurst, and B. Moran, *Phys. Rev. A*, **22**, 1690 (1980).
[77] J.J. Erpenbeck, LA-UR-88-2180 (1988), *Phys. Rev. A.* (submitted).
[78] M. Schoen and C. Hoheisel, *Mol. Phys.*, **56**, 653, (1985).
[79] D. Levesque and L. Verlet, *Mol. Phys.*, **61**, 143, (1987).
[80] J.-P. Ryckaert, A. Bellemans, G. Ciccotti, and G.V. Paolini, *Phys. Rev. Lett.*, **60**, 128 (1988).
[81] B.J. Alder and T.E. Wainwright, *Phys. Rev. A*, **4**, 233 (1971).
[82] L. P. Kadanoff, G.R. McNamara, and G. Zanetti, *Phys. Rev. A*, (to be published).
[83] D.J. Evans, *J. Stat. Phys.*, **22**, 81 (1980).
[84] J.J. Erpenbeck and W.W. Wood, *J. Stat. Phys.*, **24**, 455 (1981).
[85] A.J.C. Ladd, W.E. Alley, and B.J. Alder, *J. Stat. Phys.*, **48**, 1147 (1987).
[86] T.R. Kirkpatrick and J.C. Nieuwoudt, *Phys. Rev. A*, **33**, 2651 (1986).
[87] I.M. de Schepper, A.F.E.M. Haffmans, and H. van Beijeren, *Phys. Rev. Lett.*, **57**, 1715 (1986).
[88] W. Götze, in *Liquid and Amorphous Metals Vol 1*, edited by W. Gläser, F. Hensel, and E. Lüscher (R. Oldenbourg Verlag, Munich, (1988).
[89] W.W. Wood, *J. Chem. Phys.*, **48**, 415 (1968).
[90] D.J. Adams, *Mol. Phys.*, **28**, 1241 (1974).
[91] H.C. Andersen, *J. Chem. Phys.*, **72**, 2384, (1980).
[92] M. Parrinello and A. Rahman, *J. Appl. Phys.*, **52**, 7182 (1981).
[93] A.J.C. Ladd, *Mol. Phys.*, **53**, 459, (1984).
[94] S. Nosé, *J. Chem. Phys.*, **81**, 511 (1984).
[95] W.G. Hoover, A.J.C. Ladd, and B. Moran, *Phys. Rev. Lett.*, **48**, 1818 (1982).
[96] D.J. Evans, *J. Chem. Phys.*, **78**, 3297 (1983).
[97] W.G. Hoover, *Phys. Rev. A*, **31**, 1695 (1985).
[98] D.J. Evans and B.L. Holian, *J. Chem. Phys.*, **83**, 4069 (1985).

MONTE CARLO SIMULATIONS

D. Frenkel
FOM Institute for Atomic and Molecular Physics
PO Box 41883
1009 DB Amsterdam
The Netherlands

February 28, 1989

Abstract

Monte Carlo simulation techniques are discussed, with special emphasis on those technical aspects that are important for the simulation of dense liquids and solids. In these notes, the Metropolis sampling scheme is introduced as a special case of importance sampling. The choice of the optimum trial move is discussed, as are the problems encountered when sampling orientational and internal degrees of freedom. After introducing the MC method as a technique to measure averages in the canonical ensemble, we briefly look at simulations in other ensembles, in particular: microcanonical, isobaric-isothermal and grand-canonical. The latter ensemble forms the starting point for a discussion of particle insertion techniques to measure the chemical potential and more general acceptance-ratio and overlapping distribution schemes. We thereupon discuss the application of computer simulation techniques to the study of phase transitions. Although the emphasis is on first-order transitions, a few remarks are made about the technical problems posed by continuous phase transitions. We end with a discussion of the relative merits of Monte Carlo and Molecular Dynamics simulations.

1 Introduction

Monte Carlo sampling is a numerical technique that can be used to evaluate multidimensional integrals. Such integrals play an important role in many branches of physics and chemistry. In particular in the field covered by the present school, i.e. the numerical study of dense molecular systems. While preparing these notes, I realized that, in this school, there are no lectures on the statistical mechanics of (classical) many-body systems. I suppose that this implies that the audience is very familiar

C.R.A. Catlow et al. (eds.), Computer Modelling of Fluids Polymers and Solids, 83–123.

with the latter subject. However that may be, it is useful to recall, for future reference, the essence of statistical mechanics. This is easy to do, because I can simply quote from the opening lines of Feynman's book on statistical mechanics [1] :

> The key principle of statistical mechanics is as follows:
>
> If a system in equilibrium can be in one of N states, then the probability of the system having energy E_n is $(1/Q)\exp(-E_n/kT)$, where
>
> $$Q = \sum_{n=1}^{N} \exp(-E_n/kT) \;, \tag{1}$$
>
> k = Boltzmann's constant, and T is the temperature. Q is called the *partition function.*
>
> If we take $|i>$ as a state with energy E_i and A as a quantum-mechanical operator for a physical observable, then the expected value of the observable is
>
> $$< A >= \frac{1}{Q} \sum_{|i>} \exp(-E_i/kT) < i|A|i > \;. \tag{2}$$
>
> This fundamental law is the summit of statistical mechanics, and the entire subject is either a slide-down from this summit, as the principle is applied to various cases, or the climb-up to where the fundamental law is derived and the concepts of thermal equilibrium and temperature T clarified.

I shall restrict myself to the *'slide-down'* and I refer the reader who is interested in the more fundamental aspects of statistical mechanics to any one of the many excellent textbooks on the subject, for instance refs [1] and [2]. The only additional relation that we need in order to establish the link with thermodynamics is the equation that relates the partition function Q with the Helmholtz free energy F :

$$F = -kT \log Q \;. \tag{3}$$

Equations 1 - 3 are the starting point for the whole of equilibrium statistical mechanics.

After having thus introduced statistical mechanics in a rather axiomatic way, our next aim is to indicate where the Monte Carlo method comes in. This is done most conveniently in *classical* statistical mechanics. As shall be discussed in the lectures by Gillan, it straightforward (but not computationally cheap) to extend Monte Carlo simulation techniques from classical many-body systems to their quantum-mechanical counterparts. In the present lectures we focus on classical statistical mechanics. The classical expression for the partition function Q is:

$$Q = a \int d\mathbf{p}^N d\mathbf{q}^N \exp(-H(\mathbf{q}^N, \mathbf{p}^N)/kT) \,, \tag{4}$$

where $\mathbf{q}^N$ stands for the coordinates of all N particles, and $\mathbf{p}^N$ for the corresponding momenta. The function $H(\mathbf{q}^N, \mathbf{p}^N)$ is the Hamiltonian of the system. It expresses the total energy of an isolated system as a function of the coordinates and momenta of the constituent particles: $H = K + U$, where K is the kinetic energy of the system and U is the potential energy. Finally, a is a constant of proportionality, chosen such that the sum over quantum states in eqn. 1 approaches the classical partition function in the limit $\hbar \to 0$. For instance, for a system of N identical atoms, $a = 1/(h^{3N} N!)$. The classical equation corresponding to equation 2 is:

$$< A >= \frac{\int d\mathbf{p}^N d\mathbf{q}^N A(\mathbf{p}^N, \mathbf{q}^N) \exp(-\beta H(\mathbf{p}^N, \mathbf{q}^N))}{\int d\mathbf{p}^N d\mathbf{q}^N \exp(-\beta H(\mathbf{p}^N, \mathbf{q}^N))} \,, \tag{5}$$

where $\beta = 1/kT$. In the above equation, the observable A has been expressed as a function of coordinates and momenta. As K is a quadratic function of the momenta the integration over momenta can be carried out analytically. Hence, averages of functions that depend on momenta only are usually easy to evaluate. The difficult problem is the computation of averages of functions $A(\mathbf{q}^N)$. Only in a few exceptional cases can the multi-dimensional integral over particle coordinates be computed analytically, in all other cases numerical techniques must be used.

Having thus defined the nature of the numerical problem that we must solve, let us next look at possible solutions. It might appear that the most straightforward approach would be to evaluate $< A >$ in eqn. 5 by numerical quadrature, for instance using Simpson's rule. It is easy to see, however, that such a method is completely useless even if the number of independent coordinates DN (D is the dimensionality of the system) is still very small ($O(10)$). Suppose that we plan to carry out the quadrature by evaluating the integrand at on a mesh of points in the DN-dimensional configuration space. Let us assume that we take m equidistant along each coordinate axis. The total number of points at which the integrand must be evaluated is then equal to m^{DN}. For all but the smallest systems this is number becomes astronomically large, even for small values of m. For instance, if we take 100 particles in three dimensions, and $m = 5$, then we would have to evaluate the integrand at 10^{210} points! Computations of such magnitude cannot be performed in the known universe. And this is fortunate, because the answer that would be obtained would have been subject to a large statistical error. After all, numerical quadratures work best on functions that are 'smooth' over distances corresponding to the mesh-size. But for most intermolecular potentials, the Boltzmann-factor in eqn. 5 is a rapidly varying function of the particle coordinates. Hence an accurate quadrature requires a small mesh-spacing (i.e. a large value of m). Moreover, when evaluating the integrand for a dense liquid (say), we would find that for the overwhelming majority of points this Boltzmann factor is vanishingly small. For instance, for a fluid of 100 hard spheres at the freezing point, the Boltzmann factor would be non-zero for 1 out of every 10^{260} configurations!

The above example clearly demonstrates that better numerical techniques are needed to compute thermal averages. One such a technique is the Monte Carlo method or, more precisely, the Monte Carlo importance-sampling scheme introduced 1954 by Metropolis, Rosenbluth, Rosenbluth, Teller and Teller [3]. The application of this method to the numerical simulation of dense molecular systems is the subject of the present set of lectures.

2 Importance Sampling

Before discussing importance sampling, let us first look at the simplest Monte Carlo technique, i.e. random sampling. Suppose we wish to evaluate numerically a one dimensional integral I :

$$I = \int_a^b f(x)\, dx. \tag{6}$$

Instead of using a 'conventional' quadrature where the integrand is evaluated at predetermined values of the abscissa, we could do something else. Note that eqn. 6 can be rewritten as:

$$I = (b-a) < f(x) > , \tag{7}$$

where $< f(x) >$ denotes the unweighted average of $f(x)$ over the interval $[a, b]$. In 'brute force' Monte Carlo, this average is determined by evaluating $f(x)$ at a large number (say, L) of x-values randomly distributed over the interval $[a, b]$. It is clear that as $L \to \infty$, this procedure should yield the correct value for I. However, as with the conventional quadrature procedure, this method is of little use to evaluate averages such as in eqn. 5 because most of the computing is spent on points where the Boltzmann factor is negligible. Clearly it would be much preferable to sample many points in the region where the Botzmann factor is large, and few elsewhere. This is the basic idea behind importance sampling.

How should we distribute our sampling through configuration space? To see this, let us first consider a simple, one-dimensional example. Suppose we wish to compute the definite integral in eqn. 6 by Monte Carlo sampling, but with the sampling points distributed non-uniformly over the interval $[a, b]$ (for convenience we assume $a = 0$ and $b = 1$), according to some non-negative probability-density $w(x)$. Clearly, we can rewrite eqn. 6 as:

$$I = \int_0^1 w(x)\, \frac{f(x)}{w(x)}\, dx. \tag{8}$$

Let us assume that we know that $w(x)$ is the derivative of another (non-negative, non-decreasing) function $u(x)$, with $u(0) = 0$ and $u(1) = 1$. Then I can be written as :

$$I = \int_0^1 du \, \frac{f(x(u))}{w(x(u))}. \tag{9}$$

In equation 9 we have written $x(u)$ to indicate that if we consider u as the integration variable, then x must be expressed as a function of u. The next step is to generate L random values of u uniformly distributed in the interval $[0,1]$. We then obtain the following estimate for I :

$$I \approx \frac{1}{L} \sum_{i=1}^{L} \frac{f(x(u_i))}{w(x(u_i))} \, . \tag{10}$$

What have we gained by rewriting I in this way? The answer depends crucially on our choice for $w(x)$. To see this, let us estimate σ_L^2, the variance in I_L, where I_L denotes the estimate for I obtained from equation 10 with L random sample points:

$$\sigma_L^2 = \frac{1}{L^2} \sum_{i=1}^{L} \sum_{j=1}^{L} \langle \left(\frac{f(x(u_i))}{w(x(u_i))} - < \frac{f}{w} > \right) \left(\frac{f(x(u_j))}{w(x(u_j))} - < \frac{f}{w} > \right) \rangle \, , \tag{11}$$

where the angular brackets denote the 'true' average, i.e. the one that would be obtained in the limit $l \to \infty$. As different samples i and j are assumed to be totally independent, all cross terms in eqn. 11 vanish, and we are left with:

$$\begin{aligned} \sigma_L^2 &= \frac{1}{L^2} \sum_{i=1}^{L} \langle \left(\frac{f(x(u_i))}{w(x(u_i))} - < \frac{f}{w} > \right)^2 \rangle \\ &= \frac{1}{L} [< (\frac{f}{w})^2 > - < \frac{f}{w} >^2] \, . \end{aligned} \tag{12}$$

Eqn. 12 shows that the variance in I still goes as $\frac{1}{L}$, but the magnitude of this variance can be reduced greatly by choosing $w(x)$ such that $(f(x)/w(x))$ is a smooth function of x. Ideally we should have $(f(x)/w(x))$ constant, in which case the variance would vanish altogether. In contrast, if $w(x)$ is constant, as is the case for the 'brute force' Monte Carlo sampling, then the relative error in I can become very large. For instance, if we are sampling in a (multi-dimensional) configuration space of volume Ω, of which only a small fraction f is accessible (for instance, $f = 10^{-260}$ (section 1)), then the relative error that results in a brute-force MC sampling will be of order $1/(Lf)$. As the integrand in eqn. 5 is only non-zero for those configurations where the Boltzmann factor is non-zero, it would clearly be advisable to carry out a non-uniform Monte Carlo sampling of configuration space, such that the weight-function w is approximately proportional to the Boltzmann factor. Unfortunately, the simple importance sampling scheme described above cannot be used to sample multidimensional integrals over configuration space, such as eqn. 5. The reason is simply that we do not know how to construct a transformation such as the one from eqn. 8 to eqn. 9 that would enable us to generate points in configuration-space with a probability-density proportional to the Boltzmann factor. In fact, a necessary (but not nearly sufficient) condition for the solution of the latter problem is that we must be able to compute analytically the partition function of the system under study. If we could do that for the systems that are of interest to us, there would hardly be any need for computer simulation.

3 The Metropolis Method

The closing lines of the previous section suggest that it is in general not possible to evaluate an integral, such as $\int d\mathbf{q}^N \exp(-\beta U(\mathbf{q}^N))$ by direct Monte Carlo sampling. However,in many cases we are not interested in the configurational part of the partition function itself, but in averages of the type:

$$< A >= \frac{\int d\mathbf{q}^N \exp(-\beta U(\mathbf{q}^N)) A(\mathbf{q}^N)}{\int d\mathbf{q}^N \exp(-\beta U(\mathbf{q}^N))} . \tag{13}$$

Hence, we wish to know the *ratio* of two integrals. What Metropolis et al. [3] showed is that it is possible to devise an efficient Monte Carlo scheme to sample such a ratio. In order to explain the Metropolis scheme, let us first look more closely at the structure of eqn. 13. In what follows we denote the configurational part of the partition function by Q_c:

$$Q_c \equiv \int d\mathbf{q}^N \exp(-\beta U(\mathbf{q}^N)) . \tag{14}$$

Note that the ratio $\exp(-\beta U)/Q_c$ in eqn. 13 is the probability density to find the system in a configuration around $\mathbf{q}^N$. Let us denote this probability density by $P(\mathbf{q}^N)$. Clearly, $P(\mathbf{q}^N)$ is non-negative. Suppose now that we are somehow able to randomly generate points in configuration-space according to this probability distribution $P(\mathbf{q}^N)$. This means that, on average, the number of points n_i generated per unit volume around a point $\mathbf{q}_i^N$ is equal to $L\, P(\mathbf{q}_i^N)$, where L is the total number o f points that we have generated. In other words:

$$< A >\approx \frac{1}{L} \sum_{i=1}^{L} n_i \, A(\mathbf{q}_i^N) . \tag{15}$$

By now the reader is almost certainly confused about the difference, if any, between eqn. 15 and eqn. 10 of section 2. The difference is that in the case of eqn. 10 we know *a priori* the probability of sampling a point in a (hyper)volume $d\mathbf{q}^N$ around $\mathbf{q}_i^N$. In other words we know both $\exp(-\beta U(\mathbf{q}_i^N))$ *and* Q_c. In contrast, in eqn. 15 we know only $\exp(-\beta U(\mathbf{q}_i^N))$, i.e. we know only the relative, but not the absolute probability of visiting different points in configuration space. This may sound rather abstract: let me therefore try to clarify the difference with the help of an example: suppose we wish to measure the average methane concentration in a coal mine. The essence of importance sampling is that we only attempt to sample inside the coal mine and not in the rock around it. In order to use eqn. 10 we must know the total volume V of the coal mine, and then we randomly generate measuring points with a probability per unit of accessible volume equal to $1/V$. In contrast, to use eqn. 15 we need not know V. Rather, we randomly generate measuring points in the coal mine (for instance, by carrying out a random walk through the mine). If we sample enough points, both 'experiments' will yield the average methane concentration, but the latter method achieves this without any prior information about the size or shape of the volume to be sampled. This, then, is the essence of the Metropolis method.

Let us next consider how to generate points in configuration space with a relative probability proportional to the Boltzmann factor. The general approach is first to prepare the system under study in a configuration $\mathbf{q}_i^N$ that has a non-vanishing Boltzmann factor $\rho(\mathbf{q}_i^N)$. This configuration may, for example, correspond to a regular crystalline lattice with no hard-core overlaps. Next, we generate a new trial configuration $\mathbf{q}_j^N$ by adding a small random displacement $\delta\mathbf{q}^N$ to $\mathbf{q_i}^N$. The Boltzmann factor of this trial configuration is $\rho(\mathbf{q}_j^N)$. We must now decide whether we will accept or reject the trial configuration. There are many possible rules for making this decision, that satisfy the constraint that on average the probability of finding the system in a configuration $\mathbf{q}_n^N$ is proportional to $w(\mathbf{q}_n^N)$. Here we discuss only the Metropolis scheme, because it is simple and generally applicable. Moreover, in almost all situations of practical interest it is no less efficient than its competitors (for a discussion, see the book by Allen and Tildesley [4]).

Let us now 'derive' the Metropolis scheme to determine the transition probability π_{ij} to go from configuration $\mathbf{q}_i^N$ to $\mathbf{q}_j^N$. It is convenient to start with a thought experiment (actually a thought simulation). We carry out a very large number (say M) Monte Carlo simulations in parallel, where M is much larger than the total number of accessible configurations. We denote the number of points in any configuration $\mathbf{q}_i^N$ by m_i. We wish that, on average, m_i is proportional to $\rho(\mathbf{q}_i^N)$. There is one obvious condition that the matrix elements π_{ij} must satisfy: namely that they do not destroy such an equilibrium distribution once it is reached. This means that, in equilibrium, the average number of accepted trial moves that result in the system leaving state i must be exactly equal to the number of accepted trial moves from all other states j to state i. It is convenient to impose a much stronger condition, namely that in equilibrium the average number of accepted moves from i to any one state j is exactly canceled by the number of reverse moves. This 'detailed balance' condition implies the following:

$$\rho_i \pi_{ij} = \rho_j \pi_{ji} \, . \tag{16}$$

There are many possible forms of the transition matrix π_{ij} that satisfy eqn. 16. Let us look how π_{ij} is constructed in practice. We recall that a Monte Carlo move consists of two stages: first we perform a trial move from state i to state j. We denote the transition matrix that determines the probability to perform a trial move from i to j by α_{ij}. $\boldsymbol{\alpha}$ is usually referred to as the underlying matrix of the Markov chain. The next stage is the decision to either accept or reject this trial move. Let us denote the probability of accepting a trial move from i to j by P_{ij}. Clearly, $\pi_{ij} = \alpha_{ij}P_{ij}$. We now *choose* $\boldsymbol{\alpha}$ to be a symmetric matrix. With this choice for $\boldsymbol{\alpha}$, we can rewrite eqn. 16 as in terms of the P_{ij} :

$$\rho_i P_{ij} = \rho_j P_{ji} \ . \tag{17}$$

From eqn. 17 follows:

$$P_{ij}/P_{ji} = \rho_j/\rho_i = \exp(-\beta(U_j - U_i)) \ . \tag{18}$$

Again, there are many possible choices for P_{ij} that satisfy this condition (and the obvious condition that the probability P_{ij} cannot exceed 1). The choice of Metropolis et al. is:

$$\begin{aligned} P_{ij} &= \rho_j/\rho_i \quad \text{if } \rho_j < \rho_i \\ &= 1 \qquad\quad \text{if } \rho_j \geq \rho_i \ . \end{aligned} \tag{19}$$

Other choices for P_{ij} are possible (for a discussion, see for instance [4]), but the original choice of Metropolis et al. appears to result in a more efficient sampling of configuration space than most other strategies that have been proposed.

In summary, then, in the Metropolis scheme, the transition probability for going from state i to state j is given by:

$$\begin{aligned} \pi_{ij} &= \alpha_{ij} && (\rho_j \geq \rho_i) \\ &= \alpha_{ij}(\rho_j/\rho_i) && (\rho_j < \rho_i) \\ \pi_{ii} &= 1 - \textstyle\sum_{j \neq i} \pi_{ij} && . \end{aligned} \tag{20}$$

Note that we still have not specified the matrix $\boldsymbol{\alpha}$, but for the fact that it must be symmetric. This reflects the fact that there is a considerable freedom in the choice of our trial moves. We will come back to this point in the next section. One thing that we have not yet explained is how to decide whether a trial move is to be accepted or rejected. The usual procedure is as follows. Suppose that we have generated a trial move from state i to state j, with $U_j > U_i$. According to eqn. 18 this trial move should be accepted with a probability $P_{ij} = \exp(-\beta(U_j - U_i)) < 1$. In order to decide whether to accept or reject the trial move, we generate a random number r from a uniform distribution in the interval $[0, 1]$. Clearly, the probability that r is less than P_{ij} is equal to P_{ij}. We now accept the trial move if $r < P_{ij}$, and reject it otherwise. This rule guarantees that the probability to accept a trial move from i to j is indeed equal to P_{ij}. Obviously it is very important that our random number generator does indeed generate numbers uniformly in the interval $[0, 1]$. Otherwise the Monte Carlo sampling will be biased. The quality of random number generators should never be taken for granted. A good discussion of random number generators can be found in the book 'Numerical recipes' [5].

3.1 Exercise

In order to familiarize yourself with the Metropolis scheme, consider the following simple problem: We wish to study the properties of a simple two-level system at temperature T by Monte Carlo sampling. The lower level (1) has an energy $E_1 = 0$, while the upper level has an energy $E_2 = E$. We construct the symmetric matrix $\boldsymbol{\alpha}$ as follows:

$$\boldsymbol{\alpha} = \begin{pmatrix} 0 & 1 \\ 1 & 0 \end{pmatrix}$$

Next, evaluate the matrix P_{ij} according to the Metropolis prescription. Construct the matrix π_{ij} and determine its eigenvalues. Verify that the eigenvector belonging to the eigenvalue 1 corresponds to the equilibrium distribution. Repeated application of the matrix π_{ij} to a non-equilibrium distribution should lead to equilibrium. Estimate how many iterations are needed to let any deviation from equilibrium decay to $1/e$ of its original value. Now repeat the same calculations for the so-called 'symmetric' sampling scheme which has sometimes been used as an alternative to the Metropolis scheme. In the symmetric sampling scheme, $P_{ij} = \rho_j/(\rho_i + \rho_j)$. Which scheme leads to the fastest convergence to equilibrium?

Note that the optimal choice of the sampling scheme depends on the nature of the model. Usually, a numerical test is needed to find out which scheme is best in any particular case.

4 Trial moves

Now that we have specified the general structure of the Metropolis algorithm, we should consider its implementation. As the problem of selecting intermolecular potentials and setting up a system with periodic boundary conditions is amply discussed in other lectures of this school, I will simply assume that we have an atomic or molecular model system in a suitable starting configuration, and that we have specified all intermolecular interactions. We must now set up the underlying Markov chain, i.e. the matrix $\boldsymbol{\alpha}$. In more down to earth terms: we must decide how we are going to generate trial moves. We should distinguish between trial moves that involve only the molecular centers-of-mass and those that change the orientation or possibly even the conformation of a molecule.

4.1 Translational moves

We start our discussion with trial moves of the molecular centers-of-mass. A perfectly acceptable method to create a trial displacement is to add random numbers between $-\Delta$ and $+\Delta$ to the x, y and z coordinates of the molecular center-of-mass:

$$x_i \quad \rightarrow \quad x_i + \Delta\,(r_1 - 0.5)$$

$$\begin{aligned} y_i &\rightarrow y_i + \Delta\,(r_2 - 0.5) \\ z_i &\rightarrow z_i + \Delta\,(r_3 - 0.5)\,. \end{aligned} \tag{21}$$

where r_i are random numbers uniformly distributed between 0 and 1. Clearly, the reverse trial move is equally probable (hence α is symmetric) [6]. However, we are faced with two questions, namely: how large should we choose Δ ? And: should we attempt to move all particles simultaneously, or rather one at a time? In the latter case we should pick the molecule that is to be moved at random, to ensure that the underlying Markov chain remains symmetric. All other things being equal, we should choose the most efficient sampling procedure. However, we must first define what we mean by efficient sampling. In very vague terms, sampling is efficient if it gives you good value for money. 'Good value' in a simulation corresponds to high statistical accuracy and 'money' is simply *money*: the money that buys your computer time and even your own time. Let us assume that you are very poorly paid. In that case we only have to worry about your computer budget. Then we could use the following definition of an optimal sampling scheme: *A Monte Carlo sampling scheme can be considered optimal, if it yields the lowest statistical error in the quantity to be computed for a given expenditure of 'computing budget'.* Usually, 'computing budget' is equivalent to CPU time.

From this 'definition' it is clear that, in principle, a sampling scheme may be optimal for one quantity but not for another. Actually, the above definition is all but useless in practice (as are most definitions). It is just not worth the effort to measure the error estimate in, for instance, the pressure, for a number of different Monte Carlo sampling schemes in a series of runs of fixed length. However, it is reasonable to assume that the mean-square error in the 'observables' is inversely proportional to the number of 'uncorrelated' configurations visited in a given amount of CPU time. And the number of independent configurations visited is a measure for the distance covered in configuration space. This suggests a more manageable criterion to estimate the efficiency of a Monte Carlo sampling scheme: namely the sum of the squares of all accepted trial displacements divided by computing time. This quantity should be distinguished from the mean-square displacement per unit of computing time, because the latter quantity goes to zero in the absence of diffusion (e.g. in a solid or a glass), whereas the former does not.

Using this criterion it is easy to show that for simulations of condensed phases it is usually advisable to move particles one at a time. To see this, let us consider a system of N spherical particles, interacting through a potential energy function $U(\mathbf{r}^N)$. Typically, we expect that a trial move will be rejected if the potential energy of the system changes by much more than kT. At the same time, we try to make the MC trial steps as large as is possible without having a very low acceptance. A displacement that would, on average, give rise to an increase of the potential energy by kT would still have a reasonable acceptance. In the case of a single-particle trial move we then have:

$$\begin{aligned} < \Delta U > &= \langle \frac{\partial U}{\partial r_i^\alpha} \rangle \overline{\Delta r_i^\alpha} + \frac{1}{2} \langle \frac{\partial^2 U}{\partial r_i^\alpha \partial r_i^\beta} \rangle \overline{\Delta r_i^\alpha \Delta r_i^\beta} + \cdots \\ &= 0 + f(u) \overline{\Delta r_i^2} + \mathcal{O}(\Delta^4) \end{aligned} \tag{22}$$

where the angular brackets denote averaging over the ensemble, and the horizontal bar denotes averaging over random trial moves. The second derivative of U has been absorbed into the function $f(U)$, the precise form of which does not concern us here. If we now equate $< \Delta U >$ on the right-hand side of eqn. 22 to kT, we find the following expression for $\overline{\Delta r_i^2}$:

$$\overline{\Delta r_i^2} \approx kT/f(U) \tag{23}$$

If we attempt to move N particles, one at a time, most of the computation involved is spent on the evaluation of the change in potential energy. Assuming that we use a neighbour list or a similar time-saving device, the total time spent on evaluation the potential energy change is proportional to nN, where n is the average number of interaction partners per molecule. The sum of the mean-square displacements will be proportional to $N\overline{\Delta r^2} \sim NkT/f(U)$. And hence the mean-square displacement per unit of CPU-time will be proportional to: $kT/(n\,f(U)$. Now suppose that we try to move all particles at once. The cost in CPU time will still be proportional to nN. But, using the same reasoning as in equations 22-23, we estimate that the sum of the mean-square displacements is smaller by a factor $1/N$. Hence the total efficiency will be down by this same factor. This simple argument explains why most simulators use single-particle, rather than collective trial moves [7].

Next, consider the choice of the parameter Δ which determines the size of the trial move. How large should Δ be? If it is very large, it is likely that the resulting configuration will have a high energy and the trial move will probably be rejected. If it is very small, the change in potential energy is probably small and most moves will be accepted. In the literature one often finds the mysterious statement that an acceptance of approximately 50 % should be optimal. This statement is not necessarily true. The optimum acceptance ratio is the one that leads to the most efficient sampling of configuration space. If we express efficiency as mean-square displacement per CPU-time, it is easy to see that different Monte Carlo codes will have different optimal acceptance ratio's. The reason is that it makes a crucial difference if the amount of computing required to test whether a trial move is accepted depends on the magnitude of the move. For continuous potentials where *all* interactions have to be computed before a move can be accepted of rejected, the amount of computation does not depend on the size of a trial move. But the situation is very different for simulations on molecules with a hard repulsive cores. Here a move can be rejected as soon as overlap with *any* neighbor is detected. In this case, a rejected move is cheaper than an accepted one, and hence the average computing time per trial move goes down as the step-size is increased. As a result, the optimal acceptance ratio for hard-core systems is appreciably lower than for systems with continuous interactions. Exactly how much depends on the nature of the program, in particular on whether it is a scalar or a vector code (in the latter case hard-core systems are treated much like continuous systems), how the information about neighbor-lists is stored, and even on the computational 'cost' of random numbers and exponentiation. The consensus seems to be that for hard-core systems the optimum acceptance ratio is closer to 20 % than to 50 %, but this is just another rule-of-thumb. When in doubt, try it out.

4.2 Orientational moves

If we are simulating molecules rather than atoms we must also generate trial moves that change the molecular orientation. As we suggested in section 4.1 and in note [6], it almost requires an effort to generate translational trial moves with a distribution that does not satisfy the symmetry requirement of the underlying Markov chain. For rotational moves, the situation is very different. It is only too easy to introduce a systematic bias in the orientational distribution function of the molecules by using a non-symmetrical orientational sampling scheme. Several different strategies to generate rotational displacements are discussed in ref. [4]. Here I will only mention one possible approach.

4.2.1 Rigid, linear molecules

Consider a system consisting of N linear molecules. We specify the orientation of the i-th molecule by a unit vector $\mathbf{u}_i$. One possible procedure to change $\mathbf{u}_i$ by a small, random amount is the following. First, we generate a unit vector $\mathbf{v}$ with a random orientation. This is quite easy to achieve (and is left as an exercise to the reader; if you wish to check your result, see ref. [4]). Next we multiply this random unit vector $\mathbf{v}$ by a scale factor γ. The magnitude of γ determines the magnitude of the trial rotation. Next we add $\gamma\mathbf{v}$ to $\mathbf{u}_i$. Let us denote the resulting sum vector by $\mathbf{t}$: $\mathbf{t} = \gamma\mathbf{v} + \mathbf{u}_i$. Note that $\mathbf{t}$ is not a un it vector. Finally, we normalize $\mathbf{t}$, and the result is our trial orientation vector $\mathbf{u}'_i$. We still have to fix γ, which determines the acceptance probability for the orientational trial move. The optimum value of γ is determined by essentially the same criteria as for translational moves. We have not yet indicated whether or not the translational and orientational trial moves should be performed simultaneously. Both procedures are acceptable. However, if rotation and translation correspond to separate moves, then the selection of the type of move should be probabilistic rather than deterministic.

4.2.2 Rigid, non-linear molecules

Only slightly more complex is the case of a non-linear rigid molecule. It is conventional to describe the orientation of non-linear molecules in terms of the Eulerian angles (ϕ, θ, ψ). However, for most simulations, use of these angles is less convenient because all rotation operations should then be expressed in terms of trigonometric functions, and these are computationally 'expensive'. It is usually better to express the orientation of such a molecule in terms of quaternion parameters (for a discussion of quaternions in the context of computer simulation, see ref. [4]). The rotation of a rigid body can be specified by a quaternion of unit norm **Q**. Such a quaternion may be thought of as a unit vector in four-dimensional space:

$$\mathbf{Q} \equiv (q_0, q_1, q_2, q_3) \quad \text{with } q_0^2 + q_1^2 + q_2^2 + q_3^2 = 1 \tag{24}$$

There is a one-to-one correspondence between the quaternion components q_α and the Eulerian angles:

$$\begin{aligned} q_0 &= \cos\frac{\theta}{2}\cos(\frac{\phi+\psi}{2}) \\ q_1 &= \sin\frac{\theta}{2}\cos(\frac{\phi-\psi}{2}) \\ q_2 &= \sin\frac{\theta}{2}\sin(\frac{\phi-\psi}{2}) \\ q_3 &= \cos\frac{\theta}{2}\sin(\frac{\phi+\psi}{2}), \end{aligned} \tag{25}$$

and the rotation matrix **R**, which describes which describes the rotation of the molecule-fixed vector in the laboratory frame, is given by (see e.g. [8]):

$$\mathbf{R} = \begin{pmatrix} q_0^2 + q_1^2 - q_2^2 - q_3^2 & 2(q_1q_2 - q_0q_3) & 2(q_1q_3 + q_0q_2) \\ 2(q_1q_2 + q_0q_3) & q_0^2 - q_1^2 + q_2^2 - q_3^2 & 2(q_2q_3 - q_0q_1) \\ 2(q_1q_3 - q_0q_2) & 2(q_2q_3 + q_0q_1) & q_0^2 - q_1^2 - q_2^2 + q_3^2 \end{pmatrix} \tag{26}$$

In order to generate trial rotations of non-linear rigid bodies, we must rotate the vector (q_0, q_1, q_2, q_3) on the four-dimensional unit sphere. The procedure described above for the rotation of a $3D$ unit vector is easily generalized to $4D$. An efficient method to generate random vectors uniformly on the $4D$ unit sphere has been suggested by Vesely [8].

4.2.3 Non-rigid molecules

If the molecules under consideration are not rigid then we must also consider Monte Carlo trial moves that change the internal degrees of freedom of a molecule. In practice, it makes an important difference whether or not we have 'frozen out' some of the internal degrees of freedom of a molecule by imposing rigid constraints on, say, bond-lengths and possibly even some bond-angles. If not, the situation is relatively simple: we can carry out normal trial moves on the cartesian coordinates of the individual atoms in the molecule (in addition to center-of-mass moves). If some of the atoms are strongly bound, it is advisable to carry out small trial moves on those particles (there is no rule that forbids the use of trial moves of different size for different atoms, as long as the moves for one particular atom are always sampled from the same distribution).

However, when the bonds between different atoms become very stiff, the above procedure does not sample conformational changes of the molecule efficiently. In Molecular Dynamics simulations it is common practice to replace very stiff intramolecular interactions by rigid constraints, and this topic is discussed in the lectures of Ryckaert and Clarke. For Monte Carlo simulations this is also possible. In fact, elegant techniques have been developed for this purpose [9]. However, the corresponding MD techniques [10] are so much easier to use, in particular for large molecules, that I cannot recommend the use of the Monte Carlo technique for any but the smallest flexible molecules with internal constraints.

4.3 'Unphysical' moves

An important advantage of the Monte Carlo scheme is that it is possible carry out 'unphysical' trial moves that greatly enhance the speed with which configuration space is sampled. An example is provided by the simulation of mixtures. In a *Molecular Dynamics* simulation of, for instance, a binary $(A - B)$ mixture, the efficiency with which configuration space is sampled is greatly reduced by the fact that concentration fluctuations decay very slowly (typically the relaxation time $\tau \sim D_{AB}/\lambda^2$, where D_{AB} is the mutual diffusion coefficient and λ is the wavelength of the concentration fluctuation). This implies that very long runs are needed to ensure equilibration of the local composition of the mixture. In solids, equilibration may not take place at all. In contrast, in a *Monte Carlo* simulation, it is permissible to carry out trial moves that swap the identities of two particles of species A and B. Such moves, even if they have only a moderate rate of acceptance (a few percent will do), greatly speed up the sampling of concentration fluctuations.

Many more 'unphysical' trial moves are possible, not all of them useful. However, in the next section we shall describe a special class of such sampling methods that are of great practical importance.

5 Other Ensembles

In the previous sections we have introduced Monte Carlo simulation as a technique to compute the average value of an arbitrary function $A(\mathbf{q}^N)$ of the coordinates of a system of N particles at constant temperature T and volume V:

$$< A >_{NVT} = \frac{\int d\mathbf{q}^N \, A(\mathbf{q}^N) \exp(-\beta U(\mathbf{q}^N))}{\int d\mathbf{q}^N \, \exp(-\beta U(\mathbf{q}^N))} \,. \tag{27}$$

This average is usually referred to as the 'canonical' ensemble average. To be more precise, the true canonical ensemble average (eqn 5) includes the particle momenta. However, in all cases of practical interest, the integration over momenta can be carried out analytically, and therefore does not require numerical sampling. I should hasten to add, though, that this does not necessarily mean that one should never include sampling over momenta in a Monte Carlo run. In particular in the case of flexible molecules with constraints (section 4.2.3), such sampling might well be advantageous. An interesting suggestion in this direction was made by Cassandro et al. [11].

Most Monte Carlo calculations to date have been performed in the constant-NVT ensemble. However, it is sometimes advisable to perform a simulation where thermodynamic variables other than N,V or T are kept fixed. In the present section we discuss such simulations.

5.1 Microcanonical Monte Carlo

Most experimental observations are performed at constant N, P, T, sometimes at constant μ, V, T, and occasionally at constant N, V, T. Experiments at constant N, V, E are very rare, to say the least. Under what circumstances, then, would anyone wish to perform Monte Carlo simulations at constant N, V and E ? I suppose that, if you are interested in the simulation of dense liquids or solids, the answer would be: 'hardly ever'. Still there are situations where a microcanonical MC method, first suggested by Creutz [12], may be of use. In particular, if you are worried that a poor random-number generator may introduce a bias in the sampling of the Boltzmann distribution, or in the unlikely case that the exponentiation of the Boltzmann factor $\exp(-\beta(U_{new} - U_{old}))$ accounts for a non-negligible fraction of the computing time. The micro-canonical MC method uses no random numbers to determine the acceptance of a move. Rather, it uses the following procedure: We start with the system in a configuration $\mathbf{q}^N$. Denote the potential energy for this state by $U(\mathbf{q}^N)$. We now fix the total energy of the system at a value $E > U$. To this end, we introduce an additional degree of freedom (the 'demon'), that carries the remainder of the energy of the system: $E_D = E - U$. E_D must always be non-negative. Now we start our Monte Carlo run. After each trial move we compute the change in potential energy of the system ($\Delta U = U(\mathbf{q}'^N) - U(\mathbf{q}^N)$). If $\Delta U < 0$, we accept the move and increase the energy carried by the demon by $|\Delta U|$. If $\Delta U > 0$, we test if the demon carries enough energy to make up the difference. Otherwise, we reject the trial move. Note that no random numbers were used in this decision. Using elementary statistical mechanics it is easy to see that, after equilibration, the probability density to find the demon with an energy E_D is given by the Boltzmann distribution: $P(E_D) = (kT)^{-1} \exp(-E_D/kT)$. Hence the demon acts as a *thermometer*. Note that this method does not really simulate the microcanonical ensemble. What is kept (almost) constant is the total *potential* energy. We can, however, mimic the real N, V, E ensemble by introducing a demon for every quadratic term in the kinetic energy. We then apply the same rules as above, randomly selecting a demon to pay or accept the potential energy change for every trial move.

I know of no published application of the microcanonical MC method to the simulation of molecular systems.

5.2 Isobaric-isothermal ensemble

In contrast, the isobaric-isothermal (constant-NPT) ensemble is widely used in Monte Carlo simulations. This is not surprising because most real experiments are also carried out under conditions of controlled pressure and temperature. Moreover, constant-NPT simulations can be used to measure the equation of state of a model system even if the virial expression for the pressure cannot be readily evaluated. This may, for instance, be the case for certain models of non-spherical hard-core molecules. But also for the increasingly important class of models where the (non-pairwise additive) potential energy function is computed *numerically* for each new configuration. Finally, it is often convenient to use constant-NPT MC to simulate systems in the vicinity of a first-order phase transition, because at constant pressure the system is free (given enough time, of course) to transform completely into the state of lowest (Gibbs) free energy, whereas in a constant-NVT simulation the system may be kept at a density where it would like to phase-separate into two bulk phases of different density, but is prevented from doing so by finite size effects.

Constant-pressure Monte Carlo simulations were first described by Wood [13] in the context of a simulation-study of two-dimensional hard disks. Although method introduced by Wood is very elegant, it is not readily applied to systems with arbitrary continuous potentials. McDonald [14] was the first to apply constant-NPT simulations to a system with continuous intermolecular forces (a Lennard-Jones mixture), and it is the constant-pressure method of McDonald that is now being used almost universally, and that is discussed below.

I will derive the basic equations of constant-pressure Monte Carlo in a way that may appear unnecessarily complicated. However, this derivation has the advantage that the same framework can be used to introduce some of the other non-NVT Monte Carlo methods to be discussed below. For the sake of convenience we shall initially assume that we are dealing with a system of N identical atoms. The partition function for this system is given by:

$$Q(N,V,T) = \frac{1}{\Lambda^{3N}N!}\int_0^L \cdots \int_0^L d\mathbf{q}^N \exp(-\beta U(\mathbf{q}^N)) \, , \tag{28}$$

where $\Lambda = \sqrt{h^2/(2\pi mkT)}$ is the thermal de Broglie wavelength. It is convenient to rewrite eqn. 28 in a slightly different way. We have assumed that the system is contained in a cubic box with diameter $L = V^{\frac{1}{3}}$. We now defined scaled coordinates $\mathbf{s}^N$, by:

$$\mathbf{q}_i = L\mathbf{s}_i$$

for $i = 1, 2, \cdots, N$. If we now insert these scaled coordinates in eqn 28 we obtain:

$$Q(N,V,T) = \frac{V^N}{\Lambda^{3N}N!}\int_0^1 \cdots \int_0^1 d\mathbf{s}^N \exp(-\beta U(\mathbf{s}^N; L)) \, . \tag{29}$$

In equation 29, we have written $U(\mathrm{s}^N;L)$ to indicate that U depends on the *real* rather than the *scaled* distances between the particles. The expression for the Helmholtz free energy of the system is:

$$\begin{aligned} F(N,V,T) &= -kT\ln Q \\ &= -kT\ln(\frac{V^N}{\Lambda^{3N}N!}) - kT\ln(\int d\mathrm{s}^N \exp(-\beta U(\mathrm{s}^N;L))) \\ &= F_{id}(N,V,T) + F_{ex}(N,V,T)\,. \end{aligned} \tag{30}$$

In the last line of the above equation we have identified the two contributions to the Helmholtz free energy on the previous line as the ideal-gas expression plus an excess part. Let us now assume that the system is separated by a 'piston' [15] from an ideal gas reservoir. The total volume of the system plus reservoir is fixed at a value V_0. The total number of particles is M. Hence the volume accessible to the $m \equiv M - N$ ideal gas molecules is $V_0 - V$. The partition function of the total system is simply the product of the partition functions of the constituent sub-systems:

$$Q(N,m,V,V_0,T) = \frac{V^N(V_0-V)^m}{\Lambda^{3M}N!m!}\int d\mathrm{s}^m \int d\mathrm{s}^N \exp(-\beta U(\mathrm{s}^N;L))\,. \tag{31}$$

Note that the integral over the s^m scaled coordinates of the ideal gas yields simply 1. For the sake of compactness we have assumed that the thermal wavelength of the ideal gas molecules is also equal to Λ. The total free energy of this combined system is $F_{tot} = -kT\ln Q(N,m,V,V_0,T)$. Now let us assume that the 'piston' between the two subsystems is free to move, so that the volume V of the N-particle subsystem can fluctuate. Of course, the most probable value of V will be the one that minimizes the free energy of the combined system. The probability density $\mathcal{P}(V)$ that the N-particle subsystem has a volume V is given by:

$$\mathcal{P}(V) = \frac{V^N(V_0-V)^m \int d\mathrm{s}^N \exp(-\beta U(\mathrm{s}^N;L))}{\int_0^{V_0} dV'\, V'^N (V_0-V')^m \int d\mathrm{s}^N \exp(-\beta U(\mathrm{s}^N;L'))} \tag{32}$$

We now consider the limit that the size of the reservoir tends to infinity ($V_0 \to \infty, M \to \infty, (m/V_0 \to \rho)$. In that limit, a small volume change of the small system does not change the pressure P of the large system. In other words, the large system works as a manostat for the small system. In that case we can simplify eqns. 31 and 32. Note that in the limit $V/V_0 \to 0$ we can write:

$$(V_0 - V)^m = V_0^m(1-(V/V_0))^m \to V_0^m \exp(-mV/V_0).$$

Note that for $m \to \infty$, $\exp(-mV/V_0) \to \exp(-\rho V)$. But, as the reservoir contains an ideal gas, ρ can be written as βP. With these substitutions, eqn. 32 becomes:

$$\mathcal{P}(V) = \frac{V^N \exp(-\beta PV) \int d\mathrm{s}^N \exp(-\beta U(\mathrm{s}^N;L))}{\int_0^{V_0} dV'\, V'^N \exp(-\beta PV') \int d\mathrm{s}^N \exp(-\beta U(\mathrm{s}^N;L'))}\,. \tag{33}$$

In the same limit, the difference in free energy between the combined system and the ideal-gas system in the absence of the N-particle subsystem is the well-known Gibbs free energy G :

$$G(N,P,T) = -kT\ln \int dV \frac{V^N \exp(-\beta PV)}{\Lambda^{3N} N!} \int ds^N \exp(-\beta U(\mathbf{s}^N;L)) \qquad (34)$$

Equation 33 is the starting point for constant-NPT Monte Carlo simulations. The idea is that the probability-density to find the small system in a particular configuration of the N atoms (as specified by $\mathbf{s}^N$) at a given volume V is given by:

$$\mathcal{P}(V;\mathbf{s}^N) = \frac{V^N \exp(-\beta PV)\exp(-\beta U(\mathbf{s}^N;L))}{\int_0^{V_0} dV' \, V'^N \exp(-\beta PV') \int d\mathbf{s}^N \exp(-\beta U(\mathbf{s}^N;L'))}\,.$$

We can now carry out Metropolis sampling on the reduced coordinates $\mathbf{s}$ *and* the volume V, with a weight-function $\rho(\mathbf{s}^N,V)$ proportional to $\exp(-\beta\{U(\mathbf{s}^N,V)+PV-N\beta^{-1}\ln V\})$. In the constant-$N,P,T$ MC method, V is simply treated as an additional coordinate, and trial moves in V must satisfy the same rules as trial moves in $\mathbf{q}$ (in particular, we should maintain the symmetry of the underlying Markov chain). Let us assume that our trial moves consist of an attempted change of the volume from V to $V' = V + \Delta V$, where ΔV is a random number uniformly distributed between over the interval $[-\Delta V_{max}, +\Delta V_{max}]$. In the Metropolis scheme such a random, volume changing move will be accepted if:

$$\exp(-\beta[U(\mathbf{s}^N,V') - U(\mathbf{s}^N,V) + P(V'-V) - N\beta^{-1}\ln(V'/V)]) > \mathcal{R}\,, \qquad (35)$$

where $\mathcal{R}$ is a random number, uniformly distributed over the interval $[0,1]$. Instead of attempting random changes in the volume itself, one might construct trial moves in the box-length L [14], or in the logarithm of the volume [16]. Such trial moves are equally legitimate, as long as the symmetry of the underlying Markov chain is maintained. However, such alternative schemes result in a slightly different form for eqn. 35.

The frequency with which trial moves in the volume should be attempted is dependent on the efficiency with which 'volume-space' is sampled. If, as before, we use as our criterion of efficiency: *(sum of squares of accepted volume changes)/(CPU time)*, then it is obvious that if the frequency with which we attempt moves depends on their 'cost'. In general, a volume trial move will require that we recompute *all* intermolecular interactions. It is therefore comparable in cost to carrying out N trial moves on the molecular positions. In such cases it is common practice to perform one volume trial move for every cycle of positional trial moves. Note that, in order to guarantee the symmetry of the underlying Markov chain, volume moves should not be attempted periodically after a fixed number (say N) positional trial moves. Rather, at every step there should be a probability $1/N$ to attempt a volume move instead of a particle move. The criteria determining the optimal acceptance of volume moves are no different than those for particle moves.

There exists a class of potential energy functions for which volume trial moves are very cheap: namely, those for which the total interaction energy can be written as a sum of powers of the interatomic distances:

$$\begin{aligned} U_n &= \sum_{i<j} \epsilon(\sigma/r_{ij})^n \\ &= \sum_{i<j} \epsilon(\sigma/(Ls_{ij})^n , \end{aligned} \tag{36}$$

or, possibly, a linear combination of such sums (the famous Lennard-Jones potential is an example of the latter category). Note that U_n in equation 36 changes in a trivial way if the volume is modified such that the linear dimensions of the system change for L to L':

$$U_n(L') = \left(\frac{L}{L'}\right)^n U_n(L) . \tag{37}$$

Clearly, in this case, computing the probability of acceptance of a volume-changing trial move is extremely cheap. Hence such trial moves may be attempted with high frequency, e.g. as frequent as particle moves. One should be very careful when using the scaling property (eqn 37) if one uses at the same time a cutoff (say r_c) on the range of the potential. Use of eqn. 37 implicitly assumes that the cutoff-radius r_c scales with L, such that $r'_c = r_c(L'/L)$. The corresponding tail-correction to the potential (and the virial) should also be recomputed to take into account both the different cutoff-radius and the different density of the system. Finally, it is always useful to compute the virial pressure during a constant pressure simulation. On average, the virial pressure should always be equal to the applied pressure. This is easy to prove as follows. First of all, note that the virial pressure $P_v(V)$ of an N-particle system at volume V is equal to :

$$P_v(V) = -\left(\frac{\partial F}{\partial V}\right)_{NT} . \tag{38}$$

In an isothermal-isobaric ensemble, the probability-density $\mathcal{P}(V)$ of finding the system with volume V is equal to: $(\exp[-\beta(F(V) + PV)])/Z_{NPT}$, where $Z_{NPT} \equiv \int dV \exp[-\beta(F(V) + PV)]$. Let us now compute the average value of the virial pressure:

$$\begin{aligned} < P_v > &= \frac{\int dV - \left(\frac{\partial F(V)}{\partial V}\right) \exp[-\beta(F(V) + PV)]}{Z_{NPT}} \\ &= \frac{\int dV(\beta)^{-1} \left(\frac{\partial \exp(-\beta F(V))}{\partial V}\right) \exp[-\beta PV]}{Z_{NPT}} \\ &= \frac{\int dV P \exp[-\beta(F(V) + PV)]}{Z_{NPT}} \\ &= P . \end{aligned} \tag{39}$$

The third line in this equation follows from partial integration.

Thus far we have limited our discussion of constant-pressure Monte Carlo to pure, atomic systems. Extension of the technique to mixtures is straightforward. The method is also easily applicable to molecular systems. However, in the latter case, it is crucial to note that *only the center-of-mass positions of the molecules should be scaled in a volume move*, but never the relative positions of the constituent atoms in the molecule. This has one practical consequence, namely that the simple scaling relation (eqn. 37) can never be used in molecular systems with site-site interactions. The reason is that, even if the center-of-mass separations between molecules scale as a simple power of the system-size, the site-site separations do not.

5.3 Isotension-isothermal ensemble

The NPT-MC method is perfectly adequate for homogeneous fluids. However, for inhomogeneous systems, in particular crystalline solids, it may not be sufficient that the simulation box can change size. Often we are interested in the transformation of a crystal from one structure to another, or even in the change of the shape of the crystalline unit cell with temperature or with applied stress. In such cases it is essential that there is enough freedom in the shape of the simulation box to allow for such changes in crystal structure without creating grain-boundaries or other highly stressed configurations. This problem was first tackled by Parrinello and Rahman [17,18] who developed an extension of the constant-pressure Molecular Dynamics technique introduced by Andersen [19]. The extension of the Parrinello-Rahman method to Monte Carlo simulations is straightforward (actually, the method is quite a bit simpler in MC than in MD). To my knowledge, the first published account of constant 'stress' MC is a paper by Najafabadi and Yip [20]. At the core of the constant-stress MC method lies the transformation from the scaled coordinates $\mathbf{s}$ to the real coordinates $\mathbf{q}$. If the simulation box is not cubic and not orthorhombic, the transformation between $\mathbf{s}$ and $\mathbf{q}$ is given by a matrix $\mathbf{h}$: $q_\alpha = h_{\alpha\beta}s_\beta$. The volume of the simulation box V is equal to $|\det\mathbf{h}|$. Without loss of generality we can choose $\mathbf{h}$ to be a symmetric matrix. In the constant stress MC procedure certain moves consist of an attempted change of one or more of the elements of $\mathbf{h}$. Actually, it would be equally realistic (but not completely equivalent) to sample the elements of the metric tensor $\mathbf{G} = \tilde{\mathbf{h}}\mathbf{h}$, where $\tilde{\mathbf{h}}$ is the transpose of $\mathbf{h}$. If only hydrostatic external pressure is applied, the constant-stress MC method is almost equivalent to constant-pressure MC [21]. Under non-hydrostatic pressure (e.g. uni-axial stress), there is again some freedom of choice in deciding how to apply such deforming stresses. Probably the most elegant method (and the method that reflects most closely the statistical thermodynamics of deformed solids) is to express all external deforming stresses in terms of the so-called *thermodynamic tension* (see e.g. [22]). A discussion of the precise nature of the statistical-mechanical ensemble corresponding to such 'iso-tension' simulations is beyond the scope of this introductory text, and the reader is referred to ref. [23] for details.

5.4 Grand canonical ensemble

In section 3, we introduced the Metropolis sampling scheme as a method to compute thermal averages of functions $A(\mathbf{q}^N)$, that depend explicitly on the coordinates of the molecules in the N-body system under study. Examples of such 'mechanical' properties are the potential energy or the virial contribution to the pressure. However, the Metropolis method could not be used to determine the integral $\int d\mathbf{q}^N \exp(-\beta U(\mathbf{q}^N))$ itself. The latter quantity measures the effective volume in configuration space that is accessible to the system. Hence the original Metropolis scheme could not be used to determine those thermodynamic properties of a system that depend explicitly on the configurational integral. Examples of such 'thermal' properties are: the Helmholtz free energy F, the entropy S and the Gibbs free energy G. However, although the Metropolis method cannot be used to measure, for instance, free energies directly, it can be used to measure the *difference* in free energy between two possible states of an N-body system. This fact is exploited in the Grand-Canonical Monte Carlo method (GCMC) developed first implemented for classical fluids by Norman and Filinov [24], and later extended and improved by a number of other groups [25,26,27,28,29,30,31,32,33]. The basic idea of the GCMC method is explained below.

In order to understand the statistical mechanical basis for the GCMC technique, let us return to eqn. 31 of section 5.2. This equation gives the partition function of a combined system of N interacting particles in volume V and $m = M - N$ ideal gas molecules in volume $V_0 - V$:

$$Q(N, m, V, V_0, T) = \frac{V^N (V_0 - V)^m}{\Lambda^{3M} N! m!} \int ds^m \int ds^N \exp(-\beta U(\mathbf{s}^N)) \,. \quad (31)$$

Now, instead of allowing the two systems two exchange volume, let us see what happens if the systems can exchange particles. To be more precise, we assume that the molecules in the two sub-volumes are actually identical particles. The only difference is that when they find themselves in volume V they interact and when they are in volume $V_0 - V$, they do not. If we transfer a molecule i from a reduced coordinate $\mathbf{s}_i$ in the volume $V_0 - V$ to the same reduced coordinate in volume V, then the potential energy function U changes from $U(\mathbf{s}^N)$ to $U(\mathbf{s}^{N+1})$. The expression for the total partition function of the system, including all possible distributions of the M particles over the two sub-volumes is:

$$Q(M, V, V_0, T) = \sum_{N=0}^{M} \frac{V^N (V_0 - V)^{M-N}}{\Lambda^{3M} N! (M - N)!} \int ds^{M-N} \int ds^N \exp(-\beta U(\mathbf{s}^N)) \,. \qquad (40)$$

Following the approach of section 5.2, we now write the probability density to find a system with $M - N$ particles at reduced coordinates $\mathbf{s}^{M-N}$ in volume $V' \equiv V_0 - V$ and N particles at reduced coordinates $\mathbf{s}^N$ in volume V:

$$\mathcal{P}(\mathbf{s}^M; N) = \frac{V^N V'^{M-N}}{Q(M, V, V', T) \Lambda^{3M} N! (M - N)!} \exp(-\beta U(\mathbf{s}^N)) \,. \qquad (41)$$

Let us now consider a trial move in which a particle is transferred from V' to the same scaled coordinate in V. First we should make sure that we construct an underlying Markov that is symmetric. Symmetry, in this case, implies that the *a priori* probability to move a particle from V' to V should be equal to the *a priori* probability of the reverse move. The probability of acceptance of a trial move in which we move a particle *to* or *from* volume V is determined by the ratio of the corresponding probability densities (eqn. 41):

$$P_{N \to N+1} = \frac{V(M-N)}{V'(N+1)} \exp(-\beta(U(\mathrm{s}^{N+1}) - U(\mathrm{s}^N))) \tag{42}$$

$$P_{N+1 \to N} = \frac{V'(N+1)}{V(M-N)} \exp(-\beta(U(\mathrm{s}^{N}) - U(\mathrm{s}^{N+1}))) \,. \tag{43}$$

Now let us consider the limit that the ideal gas system is very much larger than the interacting system: $M \to \infty, V' \to \infty, (M/V') \to \rho$. Note that for an ideal gas the chemical potential μ is related to the particle-density ρ by: $\mu = kT \ln \Lambda^3 \rho$. Therefore, in the limit $(M/N) \to \infty$, eqns. 42- 43 become:

$$P_{N \to N+1} = \frac{V}{N+1} \Lambda^{-3} \exp(\beta\mu) \exp(-\beta(U(\mathrm{s}^{N+1}) - U(\mathrm{s}^N))) \tag{44}$$

$$P_{N+1 \to N} = \frac{(N+1)}{V} \Lambda^{3} \exp(-\beta\mu) \exp(-\beta(U(\mathrm{s}^{N}) - U(\mathrm{s}^{N+1}))) \,. \tag{45}$$

In the Metropolis scheme, an attempted particle addition to (removal from) volume V is accepted if $P_{N \to N+1} > \mathcal{R}$ ($P_{N+1 \to N} > \mathcal{R}$), where $\mathcal{R}$ is a random number uniformly distributed in the interval $[0,1]$.

Equations 44- 45 are the basic equations for Grand Canonical Monte Carlo simulations. Acceptable trial moves are: *i*) the random insertion of an additional particle at any point in volume V, *ii*) the random removal of any of the N particles in volume V. In addition, the particles in volume V can sample the accessible configuration-space by conventional Monte Carlo sampling. Note that in equations 44- 45 all explicit reference to the ideal-gas system has disappeared. The rules only refer to the system of volume V.

The most salient feature of the GCMC technique is that in such simulations the chemical potential μ is *imposed*, while the number of particles N is a fluctuating quantity. During the simulation we may measure other thermodynamic quantities, such as the pressure P, the average density $< \rho >$ or the internal energy $< U >$. As we know μ, we can derive all other 'thermal' properties, such as the Helmholtz free energy or the entropy. This may seem surprising. After all, in section 3 we stated that Metropolis sampling *cannot* be used to sample absolute free energies and related quantities. Yet, with Grand Canonical MC we appear to be doing precisely that. The answer is that, in fact, we do not. What we measure is not an absolute but a relative free energy. In GCMC, we are equating the chemical potential of a molecule in an ideal gas at density ρ (for the ideal-gas case we know how to compute μ) and the chemical potential of the same species in an interacting system at density ρ'.

Grand canonical MC works best if the acceptance of trial moves by which particles are added or removed is not too low. For atomic fluids this condition effectively limits the maximum density at which the method can be used to about twice the critical density. Special tricks are needed to extend GCMC to somewhat higher densities [31]. GCMC is easily implemented for mixtures and inhomogeneous systems, such as fluids near interfaces. In fact, some of the most useful applications of the GCMC method are precisely in these areas of research. Although GCMC can be applied to simple models of non-spherical molecules, it converges very poorly for all but the smallest polyatomic molecules. For more details on the Grand Canonical Monte Carlo method, the reader is referred to the book by Allen and Tildesley [4] and a review paper by Frenkel [34].

5.5 Gibbs ensemble

After struggling through the two previous sections, the reader may well wonder what makes simulations at constant P or constant μ so desirable. One important reason, though not the only one, is that the condition for coexistence of two or more phases $I, II, \cdots$ is that the pressure of all coexisting phases must be equal ($P_I = P_{II} = \cdots = P$), as must be the temperature ($T_I = T_{II} = \cdots = T$) and the chemical potentials of all species ($\mu_I^\alpha = \mu_{II}^\alpha = \cdots = \mu^\alpha$). Hence one might be inclined to think that the best ensemble to study would be the 'constant-μPT ensemble'. The quotation marks around the name of this 'ensemble' are intentional, because, strictly speaking, no such ensemble exists. The reason is simple: if we specify only intensive parameters, such as P, T and μ, the extensive variables (such as V) are unbounded. Another way to say the same thing is that the set P, T, μ is linearly dependent. In order to get a decent ensemble, we must fix at least one extensive variable. In the case of constant-pressure MC this variable is the number of particles N, while in Grand Canonical Monte Carlo the volume V of the system is fixed.

Recently, however, Panagiotopoulos [35] has developed a new Monte Carlo scheme, called the 'Gibbs method', that comes very close to achieving the impossible: namely simulating phase equilibria under conditions where the pressure, temperature and chemical potential(s) of the coexisting phases are equal. The reason that this method can work is that, although the *difference* between chemical potentials in different phases is fixed (namely, at $\Delta\mu = 0$), the absolute values are still free to change. Below we shall show how the Gibbs method can be derived using the description developed in the previous sections.

We start our discussion by recalling the expression for the partition function for a system of M particles distributed over two volumes V and $V' = V_0 - V$, i.e. eqn. 40

in section 5.4:

$$Q(M,V,V',T) = \sum_{N=0}^{M} \frac{V^N V'^{M-N}}{\Lambda^{3M} N!(M-N)!} \int d\mathbf{s}^{M-N} \int d\mathbf{s}^N \exp(-\beta U(\mathbf{s}^N)) .$$

In the previous section we assumed that the particles in volume V' behaved as ideal-gas molecules. Now we consider the case that the particles in both volumes are subject to the same intermolecular interactions. In that case, the partition function becomes:

$$Q(M,V,V',T) = \sum_{N=0}^{M} \frac{V^N V'^{M-N}}{\Lambda^{3M} N!(M-N)!} \int d\mathbf{s}^{M-N} \exp(-\beta U(\mathbf{s}^N)) \int d\mathbf{s}^N \exp(-\beta U(\mathbf{s}^N)) . \tag{46}$$

We now allow trial moves to transport particles from one sub-system to the other. Using the results of the previous section, it is easy to see that the probability ratio associated with a trial move that transfers one particle from volume V' to volume V is given by:

$$P_{N\to N+1} = \frac{V(M-N)}{V'(N+1)} \exp(-\beta[U(\mathbf{s}^{N+1};V) - U(\mathbf{s}^N;V) + U(\mathbf{s}^{M-N-1};V') - U(\mathbf{s}^{M-N};V')]) . \tag{47}$$

The corresponding expression for the reverse trial move follows by making the appropriate substitutions in eqn. 48. As before, the acceptance or rejection of this trial move can be decided by comparing $P_{N\to N+1}$ with a random number in the interval $[0,1]$. At this stage, we could set up a Monte Carlo simulation in which we allow trial moves that transfer particles from V to V' and *vice versa*. In the course of such a simulation, the distribution of particles over V and V' would become such that both subsystems are at the same chemical potential. This may correspond to a situation where both subsystems have the same density and pressure, but this need not be the case. For example, if the overall density of the total system (M/V_0), corresponds to a point in the two-phase region of a first-order phase transition, the two subsystems may end up in different phases (which is desirable), at different pressures (which is undesirable). In order to ensure that the two subsystems are always at the same pressure, the Gibbs method also allows for volume changing moves. These may either keep the total volume V_0 fixed (this was the formulation of the original paper of Panagiotopoulos [35]), or the two systems may both be in contact with a constant-pressure bath at pressure P (this generalization is described in a subsequent paper by Panagiotopoulos et al. [36]). Both methods follow directly from the discussion in section 5.2. For the sake of completeness we write down the expressions for the probability ratios associated with a volume-changing move in the constant-V_0

ensemble:

$$
\begin{aligned}
&P(V \rightarrow V+\Delta V, V' \rightarrow V'-\Delta V) = \\
&\quad \exp(-\beta[U_N(V+\Delta V) - U_N(V) + U_{M-N}(V'-\Delta V) - U_{M-N}(V')]). \\
&\quad \exp[N \ln((V+\Delta V)/V) + (M-N)\ln((V'-\Delta V)/V')] \, . \qquad (48)
\end{aligned}
$$

In the constant-P ensemble volume changing moves on the two subsystems can be carried out independently. The probability of acceptance of such trial moves is given directly by eqn. 35 of section 5.2. Of course, this constant-P method can only be applied to systems containing two or more components because in a one-component system, the two-phase region is a *line* in the P–T-plane. Hence, the probabibility that any specific choice of P and T will actually be *at* the phase transition, is vanishingly small. In contrast, for two-component systems, the two-phase region corresponds to a finite *area* in the $P-T$-plane.

Note that in either formulation of the Gibbs method, the total number of particles is fixed. The method can be extended to study inhomogeneous systems [37] and is particularly suited to study phase equilibria in multicomponent mixtures and equilibria across semi-permeable membranes [36]. The great advantage of the Gibbs method over the conventional techniques to study phase-coexistence is that in the Gibbs method the system spontaneously 'finds' the densities and compositions of the coexisting phases. In contrast, the conventional approach was, for instance, to compute the relevant chemical potentials as a function of pressure at a number of different compositions (this might require an appreciable number of simulations), and then *construct* the coexistence line. Although there still remain areas where the Gibbs method in its present form cannot be applied (e.g. phase equilibria involving solids), it seems clear that for the study of most fluid-fluid phase equilibria, the Gibbs method has effectively made all earlier methods obsolete.

6 Virtual moves

In the previous sections we have been discussing several Monte Carlo schemes that all aimed to sample thermal averages in one ensemble or another. The procedure by which this was achieved was by some kind of random walk through configuration-space, where the probability of acceptance of a trial move from one configuration to another was determined by the ratio of the probabilities of these configurations. In the present section we discuss a number of sampling schemes in which certain trial moves are attempted *but never accepted!*. I realize that such a scheme sounds very much like an exercise in futility. In fact, it is nothing of the sort, as I hope to demonstrate below.

6.1 Particle insertion method

Our first example of a Monte Carlo technique employing virtual moves concerns the so-called 'particle-insertion' method (often referred to as the Widom-method [38]) to measure the chemical potential μ of a species in a pure fluid or in a mixture. The statistical mechanics that is the basis for this method is quite simple. Consider the definition of the chemical potential μ_α of a species α. From thermodynamics we know that μ is defined as:

$$\begin{aligned}\mu &= \left(\frac{\partial G}{\partial N}\right)_{PT} \\ &= \left(\frac{\partial F}{\partial N}\right)_{VT} \\ &= -T\left(\frac{\partial S}{\partial N}\right)_{VE} .\end{aligned} \tag{49}$$

Where G, F and S are the Gibbs free energy, the Helmholtz free energy and the entropy, respectively. Here, and in the next few paragraphs we focus on a one-component system, and hence we drop the subscript α. Let us first consider the situation at constant NVT. If we express the Helmholtz free energy of an N-particle system in terms of the partition function Q_N (eqn. 30), then it is obvious from eqn. 49 that, for sufficiently large N the chemical potential is given by: $\mu = -kT\ln(Q_{N+1}/Q_N)$. If we use the explicit form (eqn. 30) for Q_N, we find:

$$\begin{aligned}\mu &= -kT\ln(Q_{N+1}/Q_N) \\ &= -kT\ln\left(\frac{V}{\Lambda^3(N+1)}\right) - kT\ln\left(\frac{\int ds^{N+1}\exp(-\beta U(s^{N+1})))}{\int ds^N \exp(-\beta U(s^N))}\right) \\ &= \mu_{id}(V) + \mu_{ex} .\end{aligned} \tag{50}$$

In the last line of eqn. 50, we have indicated the separation in the ideal-gas contribution to the chemical potential, and the excess part. As $\mu_{id}(V)$ can be evaluated analytically, we focus on μ_{ex}. We now separate the potential energy of the $N+1$-particle system into the potential energy function of the N-particle system, $U(s^N)$, and the interaction energy of the $N+1$-th particle with the rest: $\Delta U \equiv U(s^{N+1}) - U(s^N)$. Using this separation, we can write μ_{ex} as:

$$\mu_{ex} = -kT\ln < \int ds_{N+1}\exp(-\beta\Delta U) >_N , \tag{51}$$

where $< \cdots >_N$ denotes canonical ensemble averaging over the configuration space of the N-particle system. The important point to note is that equation 51 expresses μ_{ex} as an ensemble average that can be sampled by the conventional Metropolis scheme. There is only one aspect of this equation that makes it different form the averages that we considered before, namely the fact that we compute the average of an *integral* over the position of particle $N+1$. This last integral can be sampled by brute-force (unweighted) Monte Carlo sampling. In practice the procedure is as follows: we carry out a perfectly normal constant NVT Monte Carlo simulation on the system of N particles. At frequent intervals during this simulation (for instance, after every MC trial move) we randomly generate a coordinate $\mathbf{s}_{N+1}$, uniformly over the unit cube. With this value of $\mathbf{s}_{N+1}$, we then compute $\exp(-\beta\Delta U)$. By averaging the latter quantity over all generated trial positions, we obtain the average that appears in eqn. 51. So, in effect, we are computing the average of the Boltzmann factor associated with the random insertion of an additional particle in an N-particle system, *but we never accept any such trial insertions*, because then we would no longer be sampling the average needed in eqn. 51. The Widom method is a very powerful method to compute the chemical potential of (not too dense) atomic and simple molecular liquids. Its main advantage is its great simplicity, and the fact that it can be added on to an existing constant-NVT MC program, without any modifications to the original sampling scheme: we are simply computing one more thermal average. There is something else about the Widom method that makes it appealing, but that has nothing to do with computationally efficiency: it really provides an insight into the *meaning* of the chemical potential. An insight that is often hard to extract from most text-books on statistical thermodynamics. The extension of the Widom method to other ensembles, in particular NPT"and NVE is relatively straightforward. However, it would be incorrect simply to apply equation 51 to these other ensembles. As this point is not always fully appreciated in the literature, I shall briefly discuss the application of the Widom method to the NPT-ensemble (see refs. [39,40]) and to the NVE-ensemble (see ref. [34]). To derive the expression for the chemical potential in the NPT-ensemble, we start from eqn. 34:

$$G(N,P,T) = -kT\ln\int dV \frac{V^N\exp(-\beta PV)}{\Lambda^{3N}N!}\int d\mathbf{s}^N \exp(-\beta U(\mathbf{s}^N;V)) .$$

We must evaluate $\mu = (\partial G/\partial N)_{PT}$. Entirely analogously to the NVT case we find that $\mu = G(N+1,P,T) - G(N,P,T)$ equals:

$$\begin{aligned}
\mu &= -kT\ln < \frac{V}{\Lambda^3(N+1)}\int d\mathbf{s}_{N+1}\exp(-\beta\Delta U) > \\
&= -kT\ln(kT/(\Lambda^3 P)) - kT\ln < \frac{PV}{(N+1)kT}\int d\mathbf{s}_{N+1}\exp(-\beta\Delta U) > \\
&= \mu_{id}(P) + \mu_{ex}(P) \qquad (52)
\end{aligned}$$

Two points should be noted: first of all we now define the ideal gas reference state at the same pressure, rather than at the same average density as the system under study. And secondly, the fluctuating quantity that we are averaging is no longer $\exp(-\beta\Delta U)$, but $V\exp(-\beta\Delta U)$. In practice, one should only expect the fluctuating volume term in eqn. 52 to be important if large volume fluctuations are possible, for instance in the vicinity of phase transitions. But chemical potentials are often calculated precisely to locate such phase transitions.

In the constant-NVE ensemble, i.e. the one probed by conventional Molecular Dynamics simulations, we start from the relation: $\mu/T = -(\partial S/\partial N)_{VE}$. In the microcanonical ensemble, the entropy S is related to $\Omega(N,V,E)$, the total number of accessible states, by the famous relation $S = k\ln\Omega(N,V,E)$. The classical expression for $\Omega(N,V,E)$ is:

$$\Omega(N,V,E) = \frac{1}{h^{3N}N!}\int \delta(H(\mathbf{p}^N,\mathbf{q}^N) - E)d\mathbf{p}^N d\mathbf{q}^N \, . \tag{53}$$

Again the derivation proceeds much as before, but for the fact that we must now compute $\Omega(N+1,V,E)/\Omega(N,V,E)$. This is slightly more cumbersome (see [34]) and we only quote the final result:

$$\beta\mu_{ex} = -\ln\{< T >^{-3/2} < T^{3/2}\exp(-\Delta U/kT) >\} \, , \tag{54}$$

where T is the (fluctuating) temperature (as determined from the instantaneous kinetic energy of the particles). Such fluctuations tend to be large where the heat capacity of the system is large (see ref. [41]).

The particle insertion method can be modified to measure the *difference* in chemical potential between two species α and β in a mixture. In this case a trial move consists of an attempt to transform a particle of species α into species β (without, of course, ever accepting such trial moves). For more details, the reader is referred to refs. [39,40]. Let us finally mention that the particle insertion and swapping techniques are not limited to the measurement of chemical potentials. In fact, a wide class of partial molar quantities (such as, for example the partial molar enthalpy h_α or the partial molar volume v_α) can be measured in this way. For details, see refs. [40,42]

The particle insertion method fails under the same conditions where Grand Canonical Monte Carlo simulations become inefficient, and for the same reason: namely that the probability of 'accepting' a trial insertion becomes very small. Techniques exist to improve the acceptance of trial moves (see e.g. ref. [34]), but the disadvantage of most such schemes is that they tend to make the particle-insertion method more complicated to use.

6.2 Overlapping distribution method

The reader may wonder why, in the previous section, we have only been discussing trial move that attempt to *add* a particle to the system, and not the reverse move. After all, the excess chemical potential can also be written as:

$$\begin{aligned} \mu &= kT\ln(Q_N/Q_{N+1}) \\ &= \mu_{id} + kT\ln < \exp(+\beta\Delta U) >_{N+1} \, , \end{aligned} \tag{55}$$

where ΔU denotes the interaction energy of particle $N+1$ with the remaining N particles. It would seem that eqn. 55 can be sampled by straightforward Metropolis Monte Carlo. In general, however, this is not true. The reason is that the function $\exp(\beta\Delta U)$ is, in principle not bounded. It can become arbitrarily large, as ΔU grows. (Incidentally, this is not true for $\exp(-\beta\Delta U)$, because one of the conditions that a system must satisfy in order to be describable by classical statistical mechanics is that its the potential energy function has a lower bound). The problem with equation 55 is that very large values of the integrand coincide with very small values ($\mathcal{O}(\exp(-\beta\Delta U))$ of the Boltzmann factor (that determines how often a configuration is sampled during a Monte Carlo run). As a consequence, an appreciable contribution to the average in eqn. 55 comes from a part of configuration space that is hardly ever, or indeed never, sampled during a run. Hard spheres offer a good illustration. As the potential energy function of non-overlapping hard spheres is always zero, a simple Monte Carlo sampling of eqn. 55 for a dense fluid of hard spheres would always yield the nonsensical estimate $\mu_{ex} = 0$ (in fact, at freezing, $\mu_{ex}/kT \sim 15$). The correct way to obtain chemical potentials from simulations involving both particle insertions and particle removals has been indicated by Shing and Gubbins [43,44]. However, I find it convenient to discuss this problem in the context of a more general technique to measure free energy differences, first introduced by Bennett [45], called the overlapping distribution method.

Consider two N-particle systems, labeled 0 and 1 with partition functions Q_0 and Q_1. For convenience we assume that both systems have the same volume V, but this is not essential. From eqn 30 it follows that the free energy difference $\Delta F = F_1 - F_0$ can be written as:

$$\begin{aligned} \Delta F &= -kT\ln(Q_1/Q_0) \\ &= -kT\ln\left(\frac{\int ds^N \exp(-\beta U_1(s^N))}{\int ds^N \exp(-\beta U_0(s^N))}\right) . \end{aligned} \tag{56}$$

Suppose that we are carrying out a (Metropolis) sampling of the configuration space of system 1. For every configuration visited during this sampling of system 1 we can compute the potential energy of system-0 ($U_0(s^N)$) for the same configuration, and hence the potential energy difference $\Delta U = U_1(s^N) - U_0(s^N)$. We use this information to construct the a histogram that measures the probability density for the potential energy difference ΔU. Let us denote this probability density by $p_1(\Delta U)$. In the canonical ensemble, $p_1(\Delta U)$ is given by:

$$p_1(\Delta U) = \frac{\int ds^N \exp(-\beta U_1)\,\delta(U_1 - U_0 - \Delta U)}{q_1} , \tag{57}$$

where we have denoted the scaled, configurational part of the partition function by a lower-case q (e.g. $q_1 = \int ds^N \exp(-\beta U_1(s^N))$). The δ-function in eqn. 57 makes that we can substitute $U_0 + \Delta U$ for U_1 in the Boltzmann factor, hence:

$$\begin{aligned} p_1(\Delta U) &= \frac{\int ds^N \exp(-\beta(U_0 + \Delta U))\,\delta(U_1 - U_0)\Delta U)}{q_1} \\ &= \frac{q_0}{q_1}\exp(-\beta\Delta U)\frac{\int ds^N \exp(-\beta U_0)\,\delta(U_1 - U_0)\Delta U)}{q_0} \\ &= \frac{q_0}{q_1}\exp(-\beta\Delta U)p_0(\Delta U)\,, \end{aligned} \tag{58}$$

where $p_0(\Delta U)$ is the probability density to find a potential energy difference ΔU between systems 1 and 0, while Boltzmann-sampling the available configurations of system 0. As the free energy difference between systems 1 and 0 is simply $\Delta F = -kT\ln(q_1/q_0)$, we find from equation 58:

$$\ln p_1(\Delta U) = \beta(\Delta F - \Delta U) + \ln p_0(\Delta U)\,. \tag{59}$$

In order to obtain ΔF from eqn. 59 in practical cases, it is convenient to define two functions f_0 and f_1 by:

$$f_0(\Delta U) = \ln p_0(\Delta U) - \frac{\beta\Delta U}{2}$$

and

$$f_1(\Delta U) = \ln p_1(\Delta U) + \frac{\beta\Delta U}{2}$$

such that

$$f_1(\Delta U) = f_0(\Delta U) + \beta\Delta F\,.$$

Suppose that we have measured f_0 and f_1 in two separate simulations (one sampling system 0, the other system 1). We can then obtain ΔF by fitting the functions f_0 and f_1 to two polynomials in ΔU that are identical but for the constant term. The constant offset between the two polynomials yields our estimate for ΔF. Note that, in order to perform such a fit, it is not even necessary that there exists a range of ΔU where both f_0 and f_1 can be measured. However, in the absence of such a range of overlap, the statistical accuracy of the method is usually poor.

Now consider the particle insertion/removal problem. Let us assume that system 1 is a system with N interacting particles, while system 0 contains $N-1$ interacting particles and 1 ideal-gas particle. The difference in free energy between these two systems is obviously equal to μ_{ex}. Applying eqn. 59 to this particular case, we find:

$$\Delta\mu_{ex} = f_1(\Delta U) - f_0(\Delta U)\,. \tag{60}$$

Eqn. 60 is equivalent to the result obtained by Shing and Gubbins. Using the overlapping distribution method it is possible to combine the results of simulations with trial insertions and trial removals to arrive at a more accurate estimate for the chemical potential.

We end this section with the description a rather different application of virtual Monte Carlo moves, namely a method to measure the pressure of a system in a simulation at constant volume. Usually, of course, the pressure is determined by measuring the virial of the force (see [4]). However, for certain classes of intermolecular interactions (e.g. complicated non-pairwise additive potentials or non-spherical hard-core interactions) the Monte Carlo sampling of the virial may be problematic. In such cases it may be advantageous to measure the pressure, using a finite difference version of the relation $P = -(\partial F/\partial V)_{NT}$, i.e. $P \approx -(\Delta F/\Delta V)_{NT}$. To this end, we must compute the free energy difference between a system at volume V and the same system at a smaller volume $V' = V + \Delta V$. Using the extension of eqn. 56 to the situation where the volume of system 1 is V' and the volume of system 0 is V, we obtain:

$$
\begin{aligned}
-\frac{\Delta F}{\Delta V} &= \frac{kT}{\Delta V}\ln(Q_1/Q_0) \\
&= \frac{kT}{\Delta V}\ln\left(\frac{V'^N \int ds^N \exp(-\beta U(\mathrm{s}^N;V'))}{V^N \int ds^N \exp(-\beta U(\mathrm{s}^N;V))}\right) . \qquad (61)
\end{aligned}
$$

or:

$$
P = P_{id} - \frac{kT}{\Delta V}\ln < \exp(-\beta\Delta U(\mathrm{s}^N)) > , \qquad (62)
$$

where $\Delta U \equiv U(\mathrm{s}^N;V) - U(\mathrm{s}^N;V')$. Note that eqn. 62 resembles the particle-insertion method: the pressure is related to the probability of acceptance of a volume-changing move. For more details, see [16].

7 Phase transitions

The study of phase transitions by computer simulation has grown to become a subject in its own right to which entire books are devoted (see e.g. [46]). The reason is not so much that unconventional methods of computer simulation are involved in such studies, but rather that a numerical study of the precise location and character of a phase transition often requires a lot subtle analysis. In other words, it is not a matter of *how* to compute but *what*. Actually there is quite a difference between the tools used to study first-order phase transitions and those applied to analyse critical phenomena near continuous phase transitions. In this section we shall discuss first order phase transitions in some detail. In contrast, we shall devote only a few remarks to continuous transitions. This may seem inappropriate, because the technical problems involved in the numerical analysis of critical phenomena are much more complex (see, for instance the book by Binder [47] and ref. [46]). The reason why, in the present lectures, I devote relatively little attention to the fascinating subject is the following. In order to study critical phenomena by computer simulation it is necessary to study the growth in amplitude and extent of critical fluctuations in some quantity or other. To get meaningful results it is desirable to see the range of these fluctuations vary with temperature (say) by at least a decade or two. In very simple three-dimensional models systems (often models of spins on lattices) this is just (and sometimes, just not) feasible with the fastest of today's computers. Not surprisingly, the numerical study of critical phenomena in realistic models for molecular systems is simply not possible (at least, not yet). A possible exception are effectively two-dimensional systems, such as adsorbed (mono)layers. There is another reason why I do not wish to devote too much attention to the simulation of critical phenomena in molecular systems: namely the fact that critical phenomena can be grouped in universality classes. This means that the characteristic critical behaviour depends on the dimensionality of the system and on the mathematical properties of the 'order parameter' that distinguishes the two phases. In some cases it may make a difference whether or not the intermolecular interactions are long-ranged. But apart from that, the specific nature of the intermolecular interactions is believed to be totally irrelevant for the critical behaviour. Unless you have strong reasons to doubt this 'central dogma' of renormalization group theory, you would be ill-advised to spend prodigious amounts of computer time to measure critical exponents in a realistic molecular model for, say, a liquid crystal.

7.1 First order phase transitions

In sections 5.5 and 6.1 it was mentioned on several occasions that knowledge of 'thermal' quantities, such as μ, is usually necessary to locate the coexistence line for a first-order phase-transition. At first sight knowledge of μ may appear superfluous. After all, a computer simulation mimics the behaviour of a real solid or liquid. If the simulation is ergodic it should spontaneously transform to whatever phase is thermodynamically most stable, and then we would know all there is to know. Unfortunately, this approach does not work. At least, not for phase transitions involving three-dimensional solids. At a solid-solid or solid-liquid phase transition very strong hysteresis effects are usually observed in a simulation. In fact, it is very difficult to nucleate a crystal from a liquid during a simulation. Hence, to locate the point where two phases coexist, we must compute the chemical potential of the homogeneous phases at the same temperature and pressure and find the point where the two μ's are equal. For not too dense fluids, one might use the Gibbs method 5.5 to allow the computer to find this point. However, the latter method does not work for phase transitions involving solids. In such cases we really must compute μ. General methods to compute μ and related thermal quantities by computer simulation are discussed in the lectures of de Lorenzi and in ref. [34]. In the present section I wish to describe a simple, and usually reliable, method to compute the free energy F (and hence μ).

Let us first recall how free energies are measured in real experiments. In the real world free energies cannot be obtained from a single measurement either. What can be measured, however, is the derivative of the free energy with respect to volume V and temperature T:

$$\left(\frac{\partial F}{\partial V}\right)_{NT} = -P \tag{63}$$

and

$$\left(\frac{\partial F/T}{\partial 1/T}\right)_{VT} = E\,. \tag{64}$$

Here P is the pressure and E the energy of the system under consideration. The trick is now to find a reversible path that links the state under consideration to a state of known free energy. The change in F along that path can then simply be evaluated by integration of eqns. 63 and 64. There are only very few thermodynamic states where the free energy of a substance is known. One state is the ideal gas phase, the other is the perfectly ordered ground state at $T = 0K$.

In computer simulations, the situation is quite similar. In order to compute the free energy of a dense liquid, one may construct a reversible path to the very dilute gas phase. It is not really necessary to go all the way to the ideal gas. But at least one should reach a state that is sufficiently dilute that the free energy can be computed accurately, either from knowledge of the first few terms in the virial expansion of the compressibility factor $PV/(NkT)$, or that the chemical potential can be computed by other means (see section 6.1 and 6.2). For the solid, the ideal gas reference state is less useful (although techniques have been developed to construct a reversible path from a dense solid to a dilute (lattice-) gas [48]). The obvious reference state for solids is the harmonic lattice. Computing the absolute free energy of a harmonic solid is relatively straightforward, at least for atomic and simple molecular solids. However, not all solid phases can be reached by a reversible route from a harmonic reference state. For instance, in molecular systems it is quite common to find a strongly anharmonic plastic phase just below the melting line. This plastic phase is not (meta-) stable at low temperatures.

Fortunately, in computer simulations we do not have to rely on the presence of a 'natural' reversible path between the phase under study and a reference state of known free energy. If such a path does not exist, we can construct an artificial path. This is in fact a standard trick in statistical mechanics (see e.g. [49]). It works as follows: Consider a case where we need to know the free energy $F(V,T)$ of a system with a potential energy function U_1, where U_1 is such that no 'natural' reversible path exists to a state of known free energy. Suppose now that we can find another model system with a potential energy function U_0 for which the free energy *can* be computed exactly. Now let us define a generalized potential energy function $U(\lambda)$, such that $U(\lambda = 0) = U_0$ and $U(\lambda = 1) = U_1$. The free energy of a system with this generalized potential is denoted by $F(\lambda)$. Although $F(\lambda)$ itself cannot be measured directly in a simulation, we can measure its derivative with respect to λ:

$$\left(\frac{\partial F}{\partial \lambda}\right)_{NVT\lambda} = \langle \frac{\partial U(\lambda)}{\partial \lambda} \rangle_{NVT\lambda} \tag{65}$$

If the path from $\lambda = 0$ to $\lambda = 1$ is reversible, we can use eqn. 65 to compute the desired $F(V,T)$. We simply measure $< \partial U/\partial \lambda >$ for a number of values of λ between 0 and 1. Typically, 10 quadrature points will be sufficient to get the absolute free energy per particle accurate to within $0.01 k_B T$. It is however important to select a reasonable reference system. One of the safest approaches appears to be to choose as a reference system an Einstein crystal with the same structure as the phase under study [50]. This choice of reference system makes it extremely improbable that the path connecting $\lambda = 0$ and $\lambda = 1$ will cross an (irreversible) first order phase transition from the initial structure to another, only to go back to its original structure for still larger values of λ. Nevertheless, it is important that the parametrization of $U(\lambda)$ be chosen carefully. Usually, a linear parametrization (i.e. $U(\lambda) = \lambda U_1 + (1-\lambda) U_0$) is quite satisfactory. But occasionally such a parametrization may lead to weak (and relatively harmless) singularities in eqn. 65 for $\lambda \rightarrow 0$. More details about such free energy computations can be found in ref. [34].

Similar techniques can be used to locate first-order phase transitions involving phases with partial order (e.g. liquid crystals). For details, the reader is referred to refs [51,52,53]

7.2 Continuous phase transitions

As indicated in the introduction to section 7, our discussion of computer simulations of critical phenomena in realistic models of molecular systems is biased. At present, there seems to be little scientific justification for the tremendous expenditure of computer-time that would be required. This does not mean that you should stay away from continuous phase transitions, but simply that a study of the true critical regime is not advisable. Note, however, that this regime is usually quite narrow (certainly by MC/MD standards) and that there is a lot of interesting physics to be learned from simulations that study the behaviour of molecular systems near, but not very near, such phase transitions.

In the present section, I mention some of the problems that you may run into when the system that you study is approaching a continuous phase transition. Usually, continuous phase changes are symmetry-breaking transitions. The two phases on either side of such a transition are distinguished by the fact that in one phase there is some kind of ordering that is absent in the other. It is convenient to quantify the degree of ordering by an 'order parameter' O, i.e. a property of the system that vanishes on average in the disordered phase but is non-zero in the ordered phase. An example the net magnetization in a ferromagnetic material. The first manifestation of the vicinity of a critical point is that the fluctuations of O in the disordered phase grow in amplitude and, at the same time, persist longer. This makes it more difficult to obtain reliable estimates for those properties that depend on O, because such fluctuations affect the statistical accuracy of our MC results. In fact, for a simulation of a given length, the estimated variance in the Monte Carlo estimate of a quantity A goes as [54,55,4]:

$$\frac{<(\Delta A_n)^2>}{<A>^2} \approx (2n_A^c/n)\frac{<A^2>-<A>^2}{<A>^2}\ , \tag{66}$$

where $<(\Delta A_n)^2>$ is the expected variance in the MC estimate of A, n is a measure for the length of the run (expressed in trial moves per particle, say) and n_A^c is a measure for the 'decay time' (expressed in the same units) of spontaneous fluctuations in A. As both n_A^c and $<A^2>-<A>^2$ may diverge on approach of a continuous phase transition, very long runs may be needed to gather enough statistics on all quantities of interest.

But things are even worse. As the continuous phase transition is approached the fluctuations in the order parameter O tend to be correlated over an increasingly large distance, characterized by the 'correlation length' ξ. Clearly, in order to sample critical fluctuations correctly, the diameter of the simulation box must be considerably larger than ξ. So, close to a continuous phase transition, not only do we have to carry out longer runs, but at the same time we must go the larger systems. Finally, as if to add insult to injury, the critical fluctuations have one final card to play. If we go to larger system sizes, the persistence time of the critical fluctuations *increases.* Not too close to the transition this increase is proportional to the square of the box-size, but very close to the transition this divergence will even be worse. All these factors combined explain why the numerical study of critical phenomena in three dimensional systems require extraordinary care and considerable computer power, even for the very simplest Ising models.

Returning now to molecular systems, the question is: what to do if you approach something that looks like a continuous phase transition? The first thing we wish to know, of course, is if the transition is indeed continuous or not. Although powerful 'finite-size scaling' techniques have been developed specifically to answer such questions [46,47], such calculations are usually prohibitively expensive when applied to molecular systems (see however [56]). Hence, in general, we cannot even tell the order of the phase transition that we are approaching. It may be either continuous, or weakly first-order. However, there are many instances where the difference really does not matter: it really depends on what you are interested in. If it is the overall phase diagram that you are after, then not knowing if a given phase transition is continuous is like being a cartographer who makes a map of Spain without indicating if, at Torremolinos, the land ends in a beach or a cliff. Of course, the matter is not without interest to some, but you can have a good knowledge of the geography of Spain without knowing this.

So, if you run into something that may or may not be a continuous phase transition, then the first thing to do is to decide whether or not you really need to know the answer (at this stage). If not, proceed with simulations using the same system size that you used for the bulk phases. Even so, the (almost) critical fluctuations will make it difficult enough to get reasonable statistics on, say, the equation-of-state around the phase transition. Only after you have gained a good understanding of all the different phases that your model system may exhibit, comes the time to decide if you wish to know more about the phase transition itself.

8 Conclusions

At the end of these lecture notes on Monte Carlo techniques, the most important question remains to be asked: **Why Monte Carlo ?** After all, Molecular Dynamics simulations can be used to simulate molecular systems just as well, and often better. Indeed, all other things being equal, the Molecular Dynamics method is certainly preferable, because it yields information not just about the static equilibrium properties but also about the transport coefficients of the system under study (see the lecture notes by Ladd). Even if you are not interested in the dynamics, MD is preferable to MC when simulating systems subject to constraints (large, non-rigid molecules (see section 4.2.3) are just one example). In all other cases, the choice between MC and MD should be based on a comparison of the relative efficiency of the two techniques. As it turns out, there are quite a few problems where phase-space is sampled much more efficiently by Monte Carlo simulation than by Molecular Dynamics. This happens, for instance, if the system has certain modes of vibration that are weakly coupled to the remaining degrees of freedom. This is quite a common phenomenon in a low-temperature solid or glass, where long-wavelength phonons may have very long life-times. Another example is a high-frequency internal vibration of a molecule. Energy exchange between such a mode and the other degrees of freedom may be extremely slow in a MD simulation. In contrast, Monte Carlo does not suffer from such equilibration problems. Another situation where the 'unphysical' nature of Monte Carlo moves can be exploited is in (binary) mixtures, in particular when the interdiffusion of the two species is slow (see 4.3). However, Monte Carlo is often not the most efficient technique to sample phase-space. There are many cases where the route from one pocket in phase-space to another requires a collective rearrangement of the coordinates of many particles. Examples are: conformational changes in large molecules or structural phase-transitions in solids. In such cases, Molecular Dynamics often finds a 'natural' reaction-path from one state to the other, where random and uncorrelated Monte Carlo trial moves are much less successful. One possible solution to this problem would be to use 'smart' Monte Carlo [57,4], but only if you believe that straightforward MD will not do the job.

Finally, there may be a very mundane, though not altogether respectable reason for preferring Monte Carlo simulations. It may simply be that, as the potential energy functions that are used become more sophisticated, and the mathematical functions that describe these quantities grow more complex, the explicit evaluation of the forces and torques for a Molecular Dynamics program becomes quite cumbersome. This implies that there is a distinct risk of introducing human errors in the MD code, unless computer-algebra is used to derive the correct expressions. In such cases it is safest to start with a Monte Carlo simulation.

Suggested Literature

Almost everything written in these notes can be found in the (very extensive) literature on Monte Carlo simulations. In the text I have referred to only a small fraction of the papers that could have been mentioned. Below, I list a few books and review articles that may help the reader find his or her way in this field.

First, a few books about the numerical techniques: A good discussion of Monte Carlo sampling (with examples) can be found in Koonin: 'Computational Physics' [58]. The discussion of (quasi)random-number generators in 'Numerical Recipes' [5] reads like a novel. A discussion of Monte Carlo simulations with emphasis on techniques relevant for atomic and molecular systems may be found two articles by J. Valleau et al. in 'Modern Theoretical Chemistry' (volume 5) [59]. The books by Binder [47] and Mouritsen [46] emphasize the application of MC simulations to discrete system and discuss in detail the analysis of critical phenomena. Real textbooks that discuss the nuts and bolts of MC simulations are the book by Heermann [60] and the book by Allen and Tildesley [4]. The latter book gives more details, in particular on the simulation of realistic molecular models. Finally, quite a few of the articles referred to in the present paper can be found in an annotated reprint collection, edited by Ciccotti et al. [61].

ACKNOWLEDGMENT

The work of the FOM Institute is part of the research program of FOM and is supported by the 'Nederlandse Organisatie voor Wetenschappelijk Onderzoek'(NWO).

References

[1] R. P. Feynman, **Statistical Mechanics**, Benjamin, Reading (Mass.), 1972.

[2] L. D. Landau and E. M. Lifshitz, **Statistical Physics**, 3rd edition, Pergamon Press, London, 1980.

[3] N. Metropolis, A. W. Rosenbluth, M. N. Rosenbluth, A. H. Teller and E. Teller, J.Chem. Phys. **21**:1087 (1953).

[4] M. P. Allen and D. J. Tildesley, **Computer Simulation of Liquids**, Clarendon, Oxford, 1987.

[5] W. H. Press, B. P. Flannery, S. A. Teukolsky and W. T. Vetterling, **Numerical Recipes**, Cambridge U.P., Cambridge, 1986.

[6] Although almost all published MC simulations on atomic and molecular systems generate trial displacements in a cube centered around the original center-of-mass position, this is by no means the only possibility. Sometimes it is more convenient to generate trial moves in a spherical volume, and it is not even necessary that the distribution of trial moves in such a volume is uniform, as long as it has inversion symmetry. For an example of a case where another sampling scheme is preferable, see:
W. G. T. Kranendonk and D. Frenkel, Mol. Phys. **64**:403 (1988).

[7] The situation is of course different for trial moves that affect not all particles, but only a non-interacting subset. In that case the acceptance or rejection can be decided for each particle individually. Such a procedure is in fact very suited for Monte Carlo simulations on both vector and parallel computers.

[8] F. J. Vesely, J. Comp. Phys. **47**:291 (1982).

[9] M. Fixman, Proc. Nat. Acad. Sci. **71**:3050 (1974).

[10] G. Ciccotti and J. P. Ryckaert, Comp. Phys. Reports **4**:345 (1986).

[11] Unfortunately, the referees of this paper thought otherwise:
A. Cassandro, G. Ciccotti, V. Rosato and J. P. Ryckaert, unpublished.

[12] M. Creutz, Phys. Rev. Lett. **50**:1411 (1983).

[13] W. W. Wood, J. Chem. Phys. **48**:415 (1968).

[14] I. R. McDonald, Mol. Phys. **23**:41 (1972).

[15] Actually, there is no need to assume a *real* piston. The systems with volume V and $V_0 - V$ may both be isolated systems subject to their individual (periodic) boundary conditions. The only constraint that we impose is that the sum of the volumes of the two systems equals V_0.

[16] R. Eppenga and D. Frenkel, Mol. Phys. **52**:1303 (1984).

[17] M. Parrinello and A. Rahman, Phys. Rev. Lett. **45**:1196 (1980).

[18] M. Parrinello and A. Rahman, J. Appl. Phys. **52**:7182 (1981).

[19] H. C. Andersen, J. Chem. Phys. **72**:2384 (1980)

[20] R. Najafabadi and S. Yip, Scripta Metall. **17**:1199 (1983).

[21] Except that one should *never* use the constant stress method for uniform fluids, because the latter offer no resistance to the deformation of the unit box, and very strange (flat, elongated etc.) box-shapes may result. This may have serious consequences because simulations on systems that have shrunk considerably in any one dimension tend to exhibit appreciable finite-size effects.

[22] D. C. Wallace, in: **Solid State Physics**, H. Ehrenreich, F. Seitz and D. Turnbull, eds., Academic Press, New York, (1970), Volume **25**, p 301.

[23] J. R. Ray and A. Rahman, J. Chem. Phys. **80**:4423 (1984).

[24] G. E. Norman and V. S. Filinov, High Temp. Res. USSR **7**:216 (1969).

[25] D. J. Adams, Mol. Phys. **28**:1241 (1974).

[26] D. J. Adams, Mol. Phys. **29**:307 (1975).

[27] D. J. Adams, Mol. Phys. **32**:647 (1976).

[28] D. J. Adams, Mol. Phys. **37**:211 (1979).

[29] L. A. Rowley, D. Nicholson and N. G. Parsonage, J. Comp. Phys. **17**:401 (1975).

[30] J. Yao, R. A. Greenkorn and K. C. Chao, Mol. Phys. **46**:587 (1982).

[31] M. Mezei, Mol. Phys. **40**:901 (1980).

[32] J. P. Valleau and K. L. Cohen, J. Chem. Phys. **72**:3935 (1980).

[33] W. van Meegen and I. Snook, J. Chem. Phys. **73**:4656 (1980).

[34] D. Frenkel in: **Molecular Dynamics Simulations of Statistical Mechanical Systems**, Proceedings of the 97th International School of Physics 'Enrico Fermi', G. Ciccotti and W. G. Hoover, editors. North-Holland, Amsterdam, 1985, p.151.

[35] A. Z. Panagiotopoulos, Mol. Phys. **61**:813 (1987).

[36] A. Z. Panagiotopoulos, N. Quirke, M. Stapleton and D. J. Tildesley, Mol. Phys. **63**:527 (1988).

[37] A. Z. Panagiotopoulos, Mol. Phys. **62**:701 (1987).

[38] B. Widom, J. Chem. Phys. **39**:2808 (1963).

[39] K .S. Shing and S. T. Chung, J. Phys. Chem. **91**:1674 (1987).

[40] P. Sindzingre, G. Ciccotti, C. Massobrio and D. Frenkel, Chem. Phys. Lett. **136**:35 (1987).

[41] J. L. Lebowitz, J. K. Percus and L. Verlet, Phys. Rev. **153**:250 (1967).

[42] P. Sindzingre, C. Massobrio, G. Ciccotti and D. Frenkel, Chemical Physics (to appear).

[43] K. S. Shing and K. E. Gubbins, Mol. Phys. **46**, 1109 (1982).

[44] K. S. Shing and K. E. Gubbins, Mol. Phys. **49**, 1121 (1983).

[45] C. H. Bennett, J. Comput. Phys. **22**, 245 (1976).

[46] O. G. Mouritsen, **Computer Studies of Phase Transitions and Critical Phenomena**, Springer, Berlin, 1984.

[47] K. Binder, Applications of the **Monte Carlo Method** in Statistical Physics, Springer, Berlin, 1984.

[48] W. G. Hoover and F. H. Ree, J. Chem. Phys. **47**:4873 (1967).

[49] J. P. Hansen and I. R. McDonald, **Theory of Simple Liquids** , 2nd edition, Academic Press, London, 1986.

[50] D. Frenkel and A. J. C. Ladd, J. Chem. Phys. **81**:3188 (1984).

[51] D. Frenkel and B. M. Mulder, Mol. Phys. **55**:1171 (1985).

[52] A. Stroobants, H. N. W. Lekkerkerker and D. Frenkel, Phys. Rev. **A36**:2929 (1987).

[53] D. Frenkel, H. N. W. Lekkerkerker and A. Stroobants, Nature **332**:822 (1988).

[54] R. Zwanzig and N. K. Ailawadi, Phys. Rev. *182*:280 (1969).

[55] G. Jacucci and A. Rahman, Nuovo Cimento *D4*:341 (1984).

[56] D. Frenkel and R. Eppenga, Phys. Rev. **A31**:1776 (1985).

[57] P. J. Rossky, J. D. Doll and H. L. Friedman, J. Chem. Phys. **69**:4628 (1978).

[58] S. E. Koonin, **Computational Physics**, Benjamin/Cummings, Menlo Park, 1986.

[59] J. P. Valleau and S. G. Whittington, in **Statistical Mechanics, part A**, B. J. Berne (editor), Plenum, New York (1977), p. 137
J. P. Valleau and G. M. Torrie, *ibidem*, p 169.

[60] D. W. Heermann,**Computer Simulation Methods in Theoretical Physics**, Springer, Berlin, 1986.

[61] G. Ciccotti, D. Frenkel and I. R. McDonald, **Simulation of Liquids and Solids**, North-Holland, Amsterdam, 1987. This reprint collection contains, among others, references [3,14,18,19,32,41,45,48,50,54]

NON-EQUILIBRIUM STATISTICAL MECHANICS AND MOLECULAR DYNAMICS COMPUTATIONS

Denis J. Evans
Research School of Chemistry
Russell K. Standish
Research School of Physical Sciences
Australian National University
GPO Box 4,
Canberra, ACT. Australia.

ABSTRACT. In these lectures we review the relationship of Non-equilibrium Molecular Dynamics (NEMD) computer simulations to non-equilibrium statistical mechanics. We derive the Green-Kubo relations for the frequency and wavevector dependent shear viscosity coefficients that describe the phenomenon known as viscoelasticity. We introduce thermostatted equations of motion and use them to discuss the nonlinear steady state response of many body systems. We generalize the Green-Kubo relations for thermostatted systems far from equilibrium and show how these generalized response theory relations, the Transient Time Correlation Function and the Kawasaki formalisms, can be used to derive fluctuation relations for the derived properties of nonequilibrium steady states, such as the compressibilities and specific heats.

1. CONTINUUM MECHANICS

1.1 Conservation relations.

Consider an isolated system of N particles with coordinates $\mathbf{q}_1, \dots \mathbf{q}_N$ and momenta $\mathbf{p}_1, \dots \mathbf{p}_N$. These are interacting with each other, $\mathbf{F}_j$ being the force acting on particle j due to all the other particles. The equation of motion for this system can be derived from the Hamiltonian H. Thus

$$\dot{\mathbf{q}}_i = \frac{\partial H}{\partial \mathbf{p}_i} \tag{1.1}$$

$$\dot{\mathbf{p}}_i = -\frac{\partial H}{\partial \mathbf{q}_i}. \tag{1.2}$$

The system is completely specified by the initial conditions at time t = 0 and the Hamiltonian. In the following, we will take the system to be completely described by a potential Φ that depends only upon the coordinates $\mathbf{q}_j$. This leads to Newton's equations of motion:

C.R.A. Catlow et al. (eds.), Computer Modelling of Fluids Polymers and Solids, 125–154.

$$\dot{\mathbf{q}}_j = \frac{\mathbf{p}_j}{m} \tag{1.3}$$

$$\dot{\mathbf{p}}_j = \mathbf{F}_j = -\nabla_j \Phi \ . \tag{1.4}$$

If each of the particles has mass m, then the equations of motion define the time evolution of macroscopic properties such as the mass density $\rho(\mathbf{r},t)$ and velocity density $\mathbf{u}(\mathbf{r},t)$, which may be defined

$$\rho(\mathbf{r},t) = \sum_{i=1}^{N} m\delta(\mathbf{q}_i(t) - \mathbf{r}) \tag{1.3}$$

$$\rho(\mathbf{r},t)\mathbf{u}(\mathbf{r},t) = \sum_{j=1}^{N} \mathbf{p}_j(t)\ \delta(\mathbf{q}_j(t) - \mathbf{r}) , \tag{1.4}$$

The conservation laws of mechanics (conservation of total mass, total momentum) in isolated systems give rise to the hydrodynamic conservation equations (de Groot & Mazur 1962)

$$\frac{d\rho}{dt} = -\rho\ \nabla\cdot\mathbf{u} \tag{1.7}$$

$$\rho\frac{d\mathbf{u}}{dt} = -\ \nabla\cdot\mathbf{P} \tag{1.8}$$

where $\mathbf{P}$ is the pressure tensor. There is a similar equation expressing energy conservation, which involves the heat flux vector.

Consider a force $\mathbf{dF}$ exerted by the fluid across a surface element $\mathbf{dS}$. This force need not be aligned with the normal to the area element so the most general relationship is given as

$$\mathbf{dF} \equiv -\,\mathbf{dS}\cdot\mathbf{P} \tag{1.9}$$

This equation defines the pressure tensor $\mathbf{P}$.

At equilibrium in fluids, the pressure tensor is isotropic,

$$\mathbf{P} = \begin{pmatrix} p & 0 & 0 \\ 0 & p & 0 \\ 0 & 0 & p \end{pmatrix} \tag{1.10}$$

and the force is normal to the area element

$$\mathbf{dF} = -\,p\mathbf{1}\cdot\mathbf{dS} = -\,p\mathbf{dS} \quad . \tag{1.11}$$

In spite of the fact that the **dF** and **dS** are in general not parallel, we still refer to 1/3tr(**P**) as the hydrostatic pressure, p.

We will now give a microscopic derivation of the pressure tensor for a bulk fluid close to equilibrium. The spatial Fourier transform of the momentum current $\mathbf{J} \equiv \rho\mathbf{u}$ is microscopically given by

$$\bar{\mathbf{J}}\,(\mathbf{k},t) = \sum_{j=1}^{N} \mathbf{p}_j\, e^{i\mathbf{k}\cdot\mathbf{q}_j} \; .$$

The derivative with respect to time of this quantity can be related to the pressure by means of the Fourier transform of (1.7).

$$\dot{\bar{\mathbf{J}}}\,(\mathbf{k},t) = \sum_{j=1}^{N} \left(\mathbf{F}_j + \mathbf{p}_j \frac{i\mathbf{k}\cdot\mathbf{p}_j}{m} \right) e^{i\mathbf{k}\cdot\mathbf{q}_j}$$

$$= i\mathbf{k}\cdot\bar{\mathbf{P}}\,(\mathbf{k},t) \tag{1.12}$$

In the bulk limit for a homogenous fluid, the pressure tensor becomes $\mathbf{P}_0$, independent of position and can be found from the limit as $\mathbf{k} \longrightarrow 0$ of $\bar{\mathbf{P}}(\mathbf{k})$.

$$\lim_{\mathbf{k}\to 0} \bar{\mathbf{P}}\,(\mathbf{k},t) = \lim_{\mathbf{k}\to 0} \int_V \mathbf{P}_0\, e^{i\mathbf{k}\cdot\mathbf{r}}\, d\mathbf{r} = \mathbf{P}_0 V \tag{1.13}$$

If we expand the exponentials in (1.14), we find that

$$i\mathbf{k}\cdot\bar{\mathbf{P}}\,(\mathbf{k},t) = \sum_{j=1}^{N} \left[\mathbf{F}_j + i\mathbf{k}\cdot\left(\mathbf{F}_j\mathbf{q}_j + \frac{\mathbf{p}_j\mathbf{p}_j}{m} \right) + o(\mathbf{k}^2) \right] .$$

Since the system is isolated, $\Sigma_{j=0}^{N}\mathbf{F}_j = 0$, and so we find that

$$\mathbf{P}_0 V = \lim_{\mathbf{k}\to 0} \bar{\mathbf{P}}(\mathbf{k},t) = \sum_{j=1}^{N} \left(\mathbf{F}_j\,\mathbf{q}_j + \frac{\mathbf{p}_j\mathbf{p}_j}{m} \right) . \tag{1.14}$$

Now for fluids in which intermolecular forces are pair-wise additive, the force $\mathbf{F}_j$ is just the sum of all the forces $\mathbf{F}_{mj}$ due to molecule m acting on molecule j, and so we can reexpress (1.16) as a double sum

$$\mathbf{P}_0 V = \sum_{\ell,j=1}^{N} \mathbf{F}_{\ell j}\, \mathbf{q}_j + \sum_{j=1}^{N} \frac{\mathbf{p}_j \mathbf{p}_j}{m}$$

$$= \frac{1}{2} \sum_{\ell,j=1}^{N} \left(\mathbf{F}_{\ell j}\, \mathbf{q}_j + \mathbf{F}_{j\ell}\, \mathbf{q}_\ell \right) + \sum_{j=1}^{N} \frac{\mathbf{p}_j \mathbf{p}_j}{m} \ .$$

Now the force due to molecule m acting on molecule j is just the negative of the force due to molecule j acting on molecule m, so

$$\mathbf{P}_0 V = \frac{1}{2} \sum_{\ell,j=1}^{N} \mathbf{F}_{\ell j} \left(\mathbf{q}_j - \mathbf{q}_\ell \right) + \sum_{j=1}^{N} \frac{\mathbf{p}_j \mathbf{p}_j}{m} \tag{1.15}$$

1.2 Constitutive relations.

The conservation equations do not completely determine the state of the system, as there are still the unknown quantities of the pressure tensor and the heat flux vector. To have a closed system of equations, one must have relations, called constitutive relations, expressing the pressure tensor, and the heat flux vector, in terms of the density, velocity and energy distributions. For an introduction to the subject see de Groot & Mazur (1962). We shall give only the briefest of accounts here, using the pressure tensor as an example.

The non-equilibrium part of the pressure tensor $\mathbf{\Pi}$ is called the viscous pressure tensor.

$$\mathbf{\Pi} = \mathbf{P} - p_{eq}\, \mathbf{1} \tag{1.16}$$

Two centuries ago Newton realized that $\mathbf{\Pi}$ which is zero at equilibrium, can be driven by the strain rate tensor $\nabla\mathbf{u}$. For atomic fluids close to equilibrium, the most general linear relation between the viscous pressure tensor and the strain rate tensor is

$$\mathbf{\Pi} = - \mathbf{L}^{(4)} : \nabla\mathbf{u} \quad . \tag{1.17}$$

The fourth rank transport tensor $\mathbf{L}$ is a function of the thermodynamic state of the system (ρ,T). It is independent of the strain rate $\nabla\mathbf{u}$. In fluids $\mathbf{L}$ must be isotropic i.e. rotationally invariant.

There are three independent isotropic rank four tensors, (see Temple 1961) defined by the possible multilinear invariants of four vectors **u, v, w, x.**

$$\overset{(4)}{\mathbf{I}}_1 :: \mathbf{uvwx} = (\mathbf{u}\cdot\mathbf{v})\,(\mathbf{w}\cdot\mathbf{x})$$

$$\overset{(4)}{\mathbf{I}}_2 :: \mathbf{uvwx} = (\mathbf{u}\cdot\mathbf{w})\,(\mathbf{v}\cdot\mathbf{x})$$

$$\overset{(4)}{\mathbf{I}}_3 :: \mathbf{uvwx} = (\mathbf{u}\cdot\mathbf{x})\,(\mathbf{v}\cdot\mathbf{w})$$

In atomic fluids $\mathbf{q}_j - \mathbf{q}_\ell$ is parallel to the force $\mathbf{F}_{j\ell}$ acting between the atoms. From (1.15) we can see that $\mathbf{P}$ is symmetric and so (1.17) becomes

$$\boldsymbol{\Pi} = \left[-\eta\left(\overset{(4)}{\mathbf{I}}_2 + \overset{(4)}{\mathbf{I}}_3 - \frac{2}{3}\overset{(4)}{\mathbf{I}}_1 \right) - \frac{\eta_v}{3}\overset{(4)}{\mathbf{I}}_1 \right] : \nabla\mathbf{u} \tag{1.18}$$

$$= 2\eta\ \{\text{Traceless, symmetrized part of } \nabla\mathbf{u}\} - \frac{1}{3}\eta_v(\nabla\cdot\mathbf{u})\,\mathbf{1}$$

where: η = shear viscosity
η_v = bulk viscosity.

In molecular fluids, $\mathbf{q}_j - \mathbf{q}_\ell$ is not necessarily parallel to $\mathbf{F}_{j\ell}$ and the pressure is not necessarily symmetric, so a further viscosity term, vortex viscosity [de Groot & Mazur (1962)] must be introduced.

The conservation equation (1.7) gives an exact relationship between the velocity field and the pressure tensor. The Newtonian constitutive relation gives a relation between the pressure tensor and the velocity field which is exact for fluids close to equilibrium. By combining the conservation and Newtonian constitutive relations we obtain the Navier-Stokes equation of hydrodynamics.

$$\rho\frac{d\mathbf{u}}{dt} = -\nabla p + \left(\eta_v - \frac{\eta}{3}\right)\nabla\,(\nabla\cdot\mathbf{u}) + \eta\nabla^2\,\mathbf{u} \tag{1.19}$$

2. NON-MARKOVIAN CONSTITUTIVE RELATIONS: VISCOELASTICITY

Consider a fluid undergoing planar Couette flow. This flow is defined by a streaming velocity.

$$\mathbf{u}(\mathbf{r},t) = (u_x,u_y,u_z) = (\gamma\, y,\ 0,\ 0) \tag{2.1}$$

According to (1.18) the pressure tensor is,

$$\mathbf{P}(\mathbf{r},t) = \begin{bmatrix} p & -\eta\gamma & 0 \\ -\eta\gamma & p & 0 \\ 0 & 0 & p \end{bmatrix} \tag{2.2}$$

where η is the shear viscosity and γ is the strain rate. If the strain rate is time dependent then the shear stress is, $-P_{xy} = -P_{yx} = \eta\gamma(t)$. It is known that many fluids do not satisfy this relation regardless of how small the strain rate is. There must therefore be a linear but time dependent constitutive relation for shear flow which is more general than the Navier-Stokes constitutive relation.

Maxwell pointed out that there is a deep correspondence between the relationship for the shear stress induced by a strain rate in a fluid, and the shear stress induced by a strain in an elastic solid. The strain tensor is, $\nabla\boldsymbol{\epsilon}$ where $\boldsymbol{\epsilon}(\mathbf{r},t)$ gives the displacement of atoms at $\mathbf{r}$ from their equilibrium lattice sites. It is clear that,

$$\frac{d\boldsymbol{\epsilon}(\mathbf{r},t)}{dt} = \mathbf{u}(\mathbf{r},t) \quad . \tag{2.3}$$

Maxwell realized that if a displacement were applied to a liquid then for a short time the liquid must behave as if it were an elastic solid. After a Maxwell relaxation time the liquid would relax to equilibrium since a liquid by definition cannot support a strain.

It is easier to analyse this matter by transforming to the frequency domain. Maxwell said that at low frequencies the shear stress of a liquid is generated by the Navier-Stokes constitutive relation for a Newtonian fluid (2.2). In this frequency domain this states that,

$$\tilde{P}_{xy}(\omega) = -\eta\,\tilde{\gamma}(\omega) \tag{2.4}$$

where,

$$\tilde{A}(\omega) = \int_{-\infty}^{+\infty} dt\; e^{-i\omega t} A(t) \tag{2.5}$$

denotes the Fourier transform of A(t).

At very high frequencies we should have,

$$\tilde{P}_{xy}(\omega) = -G\,\frac{\partial\tilde{\epsilon}_x}{\partial y} \tag{2.6}$$

where G is the infinite frequency shear modulus. From (2.3) we can transform the terms involving the strain in (2.6) into terms involving the strain rate. At high frequencies therefore,

$$\tilde{P}_{xy}(\omega) = -\frac{G}{i\omega}\frac{\partial \tilde{u}_x}{\partial y} \quad . \tag{2.7}$$

The Maxwell model of viscoelasticity is obtained by simply summing the high and low frequency compliances, η^{-1} and $i\omega/G$ respectively, to obtain

$$\tilde{\gamma} = -\left(\frac{i\omega}{G} + \frac{1}{\eta}\right)\tilde{P}_{xy} \quad . \tag{2.8}$$

The expression for the frequency dependent Maxwell viscosity is,

$$\tilde{\eta}_M(\omega) = \frac{G}{i\omega + \frac{G}{\eta}} = \frac{\eta}{1 + i\omega\tau_M} \quad . \tag{2.9}$$

It is easily seen that this expression smoothly interpolates between the high and low frequency limits. The Maxwell relaxation time τ_M controls the transition frequency between low frequency viscous behaviour and high frequency elastic behaviour. The Maxwell model provides a rough approximation to the visoelastic behaviour of so-called viscoelastic fluids such as polymer melts or colloidal suspensions. It is important to remember that viscoelasticity is a linear phenomenon. The resulting shear stress is a linear function of the strain rate. It is also important to point out that Maxwell believed that all fluids are viscoelastic. The reason why polymer melts are observed to exhibit viscoelasticity is that their Maxwell relaxation times are macroscopic, of the order of seconds. On the other hand the Maxwell relaxation time for argon at its triple point is approximately 10^{-12} seconds! Using standard viscometric techniques viscoelastic effects are completely unobservable in argon.

If we rewrite the Maxwell constitutive relation in the time domain using an inverse Fourier-Laplace transform we see that,

$$P_{xy}(t) = -\int_0^t ds\, \eta_M(t-s)\,\gamma(s) \quad . \tag{2.10}$$

In this equation $\eta_M(t)$ is called the Maxwell memory function. It is called a memory function because (2.10) shows that the shear stress at a time t is not simply linearly proportional to the strain rate at the current time, time t, but to the entire strain rate history s, $0 \le s \le t$. Constitutive relations which are history dependent are called Non-Markovian. A Markovian process is one in which the present state of the system is all that is required to determine its future. The Maxwell model of viscoelasticity describes Non-Markovian behaviour. The Maxwell memory function is easily identified as an exponential,

$$\eta_M(t) = G\, e^{-t/\tau_M} \quad . \tag{2.11}$$

Although the Maxwell model of viscoelasticity is approximate the basic idea that liquids take a finite time to respond to changes in strain rate, and equivalently that liquids remember their strain rate histories, is correct. The most general linear relation between the strain rate and the shear stress for a homogeneous fluid can be written in the time domain as,

$$P_{xy}(t) = -\int_0^t ds\, \eta\,(t-s)\,\gamma\,(s) \ . \tag{2.12}$$

There is an even more general linear relation between stress and strain rate which is appropriate in fluids where the strain rate varies in space as well as in time.

$$P_{xy}(\mathbf{r},t) = -\int_0^t ds \int d\mathbf{r}'\ \eta(\mathbf{r}-\mathbf{r}',\, t-s)\,\gamma\,(\mathbf{r}',\, s) \tag{2.13}$$

We reiterate that the differences between these constitutive relations and the Newtonian constitutive relation, (1.18) are only observable if the strain rate varies significantly over either the time or length scales characteristic of the molecular relaxation for the fluid. The surprise is not so much that the validity of the Newtonian constitutive relation is limited. The more remarkable thing is, for example in argon, that strain rates can vary in time from essentially zero frequency to 10^{12}Hz, or in space from essentially zero wavevector to 10^{-9}m^{-1}, before Non-Newtonian effects are observable. It is clear from this discussion that analogous corrections will be needed for all the other Navier-Stokes transport coefficients if their corresponding thermodynamic fluxes vary on molecular time or distance scales.

3. GREEN - KUBO RELATIONS

For nearly a century, theoretical expressions for transport coefficients have been based on the Boltzmann equation. This equation is only valid for dilute gases. General expressions for transport coefficients at arbitrary density were found by Green and Kubo. These expressions are exact for **linear** transport coefficients for systems close to equilibrium. These relations express the transport coefficients in terms of integrals of equilibrium time correlation functions. For example, the relations for shear and bulk viscosity are

$$\eta = \frac{V}{k_B T}\int_0^\infty \left\langle P_{xy}(t)\, P_{xy}(0)\right\rangle dt \tag{3.1}$$

$$\eta_V = \frac{1}{V k_B T}\int_0^\infty \left\langle \left[p(t)V(t) - \langle pV\rangle\right]\left[p(0)V(0) - \langle pV\rangle\right]\right\rangle dt \tag{3.2}$$

where < .. > denotes an equilibrium ensemble average.

We will now give a derivation of an equilibrium fluctuation expression for the wavenumber and frequency dependent shear viscosity. It must be born in mind that this is still a linear transport theory, and generalizations to non-linear transport will be provided in the following sections.

Take a fluid with momentum current induced by a shear stress $\nabla\mathbf{u}$ at the time t = 0,

$$\bar{\mathbf{J}}(\mathbf{k},t) = \int \rho(\mathbf{r},t)\mathbf{u}(\mathbf{r},t)\, e^{i\mathbf{k}.\mathbf{r}}\, d\mathbf{r} \tag{3.3}$$

and consider the component of the momentum current transverse to **k**.

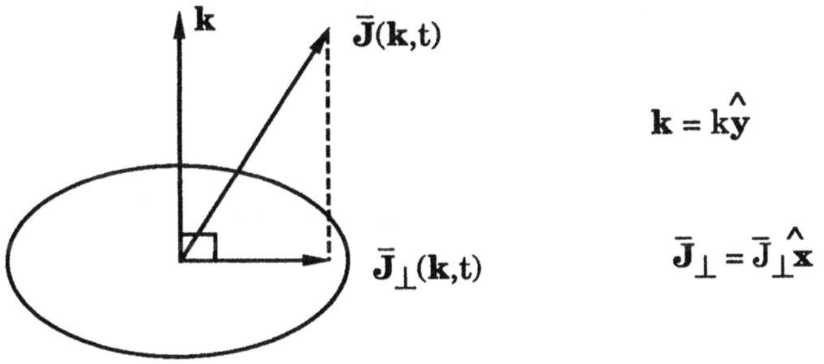

Fig. 1 The transverse component of the momentum current

For simplicity, define the co-ordinate system so that **k** is in the y direction and $\bar{\mathbf{J}}_\perp$ is in the x direction, as shown in figure 1. From the momentum conservation equation (1.15) we find

$$\dot{\bar{J}}_\perp (k,t) = ik\, \bar{P}_{yx}(k,t)\,. \tag{3.4}$$

Define the transverse momentum current autocorrelation function

$$C(k,t) = \left\langle \bar{J}_\perp(k,t)\, \bar{J}_\perp(-k,0) \right\rangle \quad . \tag{3.5}$$

From (2.13)

$$\bar{P}_{yx} = ik \int_0^t \bar{\eta}(k,t-s)\, \bar{u}_x (k,s)\, ds\ , \tag{3.6}$$

and substituting this into (3.4)

$$\dot{\bar{J}}_{\perp}(k,t) = -\frac{k^2}{\rho}\int_0^t \bar{\eta}(k,t-s)\,\bar{J}_{\perp}(k,s)\,ds \quad . \tag{3.7}$$

If we multiply by $\bar{J}_{\perp}(k,0)$ and take the ensemble average, we get

$$\dot{C}(k,t) = -\frac{k^2}{\rho}\int_0^t \eta(k,\, t-s)\, C(k,s)\,ds \quad . \tag{3.8}$$

Taking the Fourier-Laplace transform of C,

$$\tilde{C}(\omega) = \int_0^\infty e^{i\omega t}\, C(t)\,dt \,,$$

we have

$$-i\omega\tilde{C}(k,\omega) - C(k,0) = -\frac{k^2}{\rho}\tilde{\tilde{\eta}}(k,\omega)\,\tilde{C}(k,\omega)$$

or rearranging,

$$\tilde{C}(k,\omega) = \frac{C(k,0)}{-i\omega + \dfrac{k^2\tilde{\tilde{\eta}}(k,\omega)}{\rho}} \quad . \tag{3.11}$$

Now let us define the flux autocorrelation function

$$N(k,t) \equiv \frac{1}{Vk_BT}\left\langle \bar{P}_{yx}(k,t)\,\bar{P}_{yx}(-k,0)\right\rangle \tag{3.10}$$

Then from (3.3)

$$N(k,t) = \frac{1}{k^2Vk_BT}\left\langle \dot{\bar{J}}_{\perp}(k,t)\,\dot{\bar{J}}_{\perp}(-k,0)\right\rangle \tag{3.11}$$

Since correlation functions are invariant under time translations, i.e.

$$\langle A(t)\,B(0)\rangle = \langle A(0)\,B(-t)\rangle$$

we can express (3.11) in terms of the second derivative of C.

$$\ddot{C}(\mathbf{k},t) = \frac{\partial^2}{\partial t^2}\left\langle \overline{J}_\perp(\mathbf{k},t)\,\overline{J}_\perp(-\mathbf{k},0)\right\rangle$$

$$= \frac{\partial}{\partial t}\left\langle \dot{\overline{J}}_\perp(\mathbf{k},t)\,\overline{J}_\perp(-\mathbf{k},0)\right\rangle$$

$$= \frac{\partial}{\partial t}\left\langle \dot{\overline{J}}_\perp(\mathbf{k},0)\,\overline{J}_\perp(-\mathbf{k},-t)\right\rangle$$

$$= -\left\langle \dot{\overline{J}}_\perp(\mathbf{k},t)\,\dot{\overline{J}}_\perp(-\mathbf{k},0)\right\rangle$$

This implies

$$N(\mathbf{k},t) = -\frac{\ddot{C}(\mathbf{k},t)}{k^2 V k_B T}$$

Now

$$\tilde{\ddot{C}}(\mathbf{k},\omega) = -i\omega\,\tilde{\dot{C}}(\mathbf{k},\omega) - \dot{C}(\mathbf{k},0)$$

$$= -\omega^2\,\tilde{C}(\mathbf{k},\omega) - i\omega C(\mathbf{k},0)\ ,$$

so

$$\tilde{N}(\mathbf{k},\omega) = \frac{1}{k^2 V k_B T}\left\{\omega^2\,\tilde{C}(\mathbf{k},\omega) - i\omega C(\mathbf{k},0)\right\} \tag{3.12}$$

$C(\mathbf{k},0)$ can be evaluated by noting that equilibrium correlations are macroscopically small

$$\langle \mathbf{J}(\mathbf{r},0)\cdot\mathbf{J}(\mathbf{r}',0)\rangle = 3\rho k_B T\,\delta(\mathbf{r}-\mathbf{r}')$$

and so

$$C(\mathbf{k},0) = \left\langle \overline{J}_\perp(\mathbf{k},0)\,\overline{J}_\perp(-\mathbf{k},0)\right\rangle$$

$$= \frac{1}{3}\left\langle \overline{\mathbf{J}}(\mathbf{k},0)\cdot\overline{\mathbf{J}}(-\mathbf{k},0)\right\rangle$$

$$= \frac{1}{3}\int_V e^{i\mathbf{k}\cdot\mathbf{r}}\int_V \langle \mathbf{J}(\mathbf{r}',0)\cdot\mathbf{J}(\mathbf{r}-\mathbf{r}',0)\,d\mathbf{r}'\,d\mathbf{r}\rangle$$

$$= \rho V k_B T$$

If we substitute for C(**k**,0) in 3.12, we find

$$\tilde{N}(k,\omega) = \frac{\omega^2\tilde{C}(k,t)}{k^2 V k_B T} - \frac{i\omega\rho}{k^2} \tag{3.13}$$

Now we can solve (3.13) and (3.9) to find an explicit expression for the wavenumber and frequency dependent shear velocity [Evans (1981)].

$$\tilde{\tilde{\eta}}(k,\omega) = i\omega\rho \frac{\tilde{N}(k,\omega)}{i\omega\rho + k^2\tilde{N}(k,\omega)} \tag{3.14}$$

$$\tilde{\tilde{\eta}}(0,\omega) = \tilde{N}(0,\omega) \quad , \quad \tilde{N}(k,0) = 0$$

The shear viscosity in the bulk limit of a fluid near equilibrium is found from $\eta = \tilde{\tilde{\eta}}(0,0)$. This can be seen from (3.14), where

$$\tilde{\tilde{\eta}}(0,0) = \tilde{N}(0,0) = \int_0^\infty N(0,t)\, dt$$

$$= \frac{1}{Vk_BT} \int_0^\infty \left\langle \overline{P}_{yx}(0,t)\, \overline{P}_{yx}(0,0) \right\rangle dt$$

Equation (3.1) now follows from substituting (1.14). Note that the order in which the limits are taken is important, [Evans (1981)] as

$$\eta = \lim_{\omega\to 0} \lim_{k\to 0} \tilde{N}(k,\omega) \neq \lim_{k\to 0} \lim_{\omega\to 0} \tilde{N}(k,\omega) = 0$$

The shear viscosity $\tilde{\tilde{\eta}}(k,\omega)$ is a continuous but non analytic function of k and ω while the flux autocorrelation $\tilde{N}$ is discontinuous at ω=k=0.

4. THERMOSTATTED EQUATIONS OF MOTION

The Green-Kubo relations discussed in the previous section are expressions for linear transport coefficients. They have proved of enormous value in experimental and theoretical applications. If we wish to extend this to non-linear steady states, then we are faced with two problems; one being the inclusion of higher order terms in the perturbation due to the external field, and the other being thermostatting. In a dissipative process, such as shear flow, the temperature would rise in the system if the heat generated is not removed by a heat sink. This effect is second order in the external field

($dH/dt = -JF_e = LF_e^2$), and can be ignored in the linear theory, but needs to be included in non-linear theories.

In this section, we make a couple of models of the process that removes the heat generated. It turns out that provided the dissipation is not too large, the Transient Time Correlation Functions we shall derive are independent of what model thermostat is used [Evans & Morriss (1984), Evans & Holian (1985)]. The effect of the thermostat is included into Newton's equation of motion by adding a friction like term.

Gaussian Thermostat

$$\dot{\mathbf{q}}_i = \frac{\mathbf{p}_i}{m} \tag{4.1a}$$

$$\dot{\mathbf{p}}_i = \mathbf{F}_i - \alpha \mathbf{p}_i \tag{4.1b}$$

$$\alpha = \sum_i \mathbf{F}_i \cdot \mathbf{p}_i / \sum_i p_i^2 \tag{4.1c}$$

The equilibrium distribution in this case is given by

$$f_K(\mathbf{\Gamma}) = \frac{\delta(K - K_o)\, e^{-\beta\Phi(\mathbf{q})}}{\int \delta(K - K_o)\, e^{-\beta\Phi(\mathbf{q})}\, d\mathbf{\Gamma}} \tag{4.2}$$

where

$$\mathbf{F}_i = -\nabla_i \Phi, \quad K = \sum_i \frac{\mathbf{p}_i^2}{2m}, \quad K_o = \frac{3}{2} N\beta^{-1} = \frac{3}{2} Nk_B T$$

The kinetic energy in this system is a constant of motion, justifying the terminology of $f_K(\Gamma)$ to be the isokinetic distribution.

Nosé-Hoover Thermostat

$$\dot{\mathbf{q}}_i = \frac{\mathbf{p}_i}{m} \tag{4.3a}$$

$$\dot{\mathbf{p}}_i = \mathbf{F}_i - \alpha \mathbf{p}_i \tag{4.3b}$$

$$\dot{\alpha} = \left(\frac{K}{K_o} - 1 \right) / \tau^2 \tag{4.3c}$$

Here τ is an arbitrary parameter that is proportional to time. It controls the timescale for the fluctuations in K about the steady state mean value K_0. In this case, the equilibrium distribution is canonical

$$f_c(\boldsymbol{\Gamma}) = \frac{e^{-\beta(\mathbf{H}_o + \frac{1}{2}Q\alpha^2)}}{\int e^{-\beta(\mathbf{H}_o + \frac{1}{2}Q\alpha^2)} \, d\boldsymbol{\Gamma} \, d\alpha} \tag{4.4}$$

where

$$Q = 3Nk_BT\tau^2 \,, \; H_o = \sum_i \frac{\mathbf{p}_i^2}{2m} + \Phi(\mathbf{q})$$

Both these thermostats are time reversible, and zero momentum current is a fixed point, i.e.

$$\sum_i \mathbf{p}_i = 0 \Rightarrow \frac{d}{dt}\left(\sum_i \mathbf{p}_i\right) = 0$$

We can check that the isokinetic ensemble, f_K, is a solution of Liouville's equation by explicitly evaluating the equation.

$$\frac{\partial f_K}{\partial t} = -\frac{\partial}{\partial \boldsymbol{\Gamma}} \cdot (\dot{\boldsymbol{\Gamma}} f_K) \equiv -i \mathcal{L} f$$

$$\text{LHS: } \frac{\partial f_K}{\partial t} = 0 \text{ no explicit t - dependence}$$

$$\text{RHS: } -i\mathcal{L} f_K = -f_K \frac{\partial}{\partial \boldsymbol{\Gamma}} \cdot \dot{\boldsymbol{\Gamma}} - \dot{\boldsymbol{\Gamma}} \cdot \frac{\partial f_K}{\partial \boldsymbol{\Gamma}}$$

$$= 3N\alpha f_K - \dot{\mathbf{q}} \cdot \frac{\partial f_K}{\partial \mathbf{q}} - \dot{\mathbf{p}} \cdot \frac{\partial f_K}{\partial \mathbf{p}} - \alpha$$

$$= 3N\alpha f_K - \dot{\mathbf{q}} \cdot \frac{\partial \Phi}{\partial \mathbf{q}} \frac{\partial f_K}{\partial \Phi} - \dot{\mathbf{p}} \cdot \frac{\partial K}{\partial \mathbf{p}} \frac{\partial f_K}{\partial K}$$

$$= 3N\alpha f_K + \beta\dot{\Phi} f_K - \dot{K} \frac{\partial f_K}{\partial K}$$

$$= 3N \frac{\sum_i \mathbf{F}_i \cdot \mathbf{p}_i/m}{\sum_i p_i^2/m} f_K - \beta \sum_i \frac{\mathbf{F}_i \cdot \mathbf{p}_i}{m} f_K = 0$$

$$\text{as} \quad \sum_i \frac{p_i^2}{2m} = \frac{3}{2} N \beta^{-1} = K = K_o$$

In the second line of the above α is ignored because it is a intensive quantity. Also $\dot{K} = 0$ because the kinetic energy is a constant of motion.

5. Formal solution of Liouville's equation

5.1 Liouville's equation.

Liouville's equation arises from consideration of the motion of particles making up the fluid. We work in the phase space of all particles, or Γ-space, and describe the system by means of a distribution function $f(\mathbf{q}_1,...\mathbf{q}_N, \mathbf{p}_1,...\mathbf{p}_N, t)$. By considering the number of phase points entering and leaving an infinitesimal volume of phase space (Tolman 1962), we get a generalized form of Liouville's equation.

$$\frac{\partial f}{\partial t} = -\sum_{j=1}^{N} \left(f \frac{\partial}{\partial \mathbf{q}_j} \cdot \dot{\mathbf{q}}_j + f \frac{\partial}{\partial \mathbf{p}_j} \cdot \dot{\mathbf{p}}_j + \frac{\partial f}{\partial \mathbf{q}_j} \cdot \dot{\mathbf{q}}_j + \frac{\partial f}{\partial \mathbf{p}_j} \cdot \dot{\mathbf{p}}_j \right) \tag{5.1}$$

For convenience, we shall denote the position in Γ-space by the 6N-vector $\mathbf{\Gamma} = (\mathbf{q}_1,...\mathbf{q}_N, \mathbf{p}_1,...\mathbf{p}_N)$. Liouville's equation becomes in this notation

$$\frac{\partial f(\mathbf{\Gamma},t)}{\partial t} = -\frac{\partial}{\partial \mathbf{\Gamma}} \cdot \left(\dot{\mathbf{\Gamma}} f(\mathbf{\Gamma},t) \right)$$

$$\equiv -i\mathcal{L} f(\mathbf{\Gamma},t) \tag{5.2}$$

The operator $\mathcal{L}$ is called the f-Liouvillean. Equation (5.2) can be solved formally by integrating with respect to t.

$$f(\mathbf{\Gamma},t) = f(\mathbf{\Gamma},0) - i\mathcal{L} \int_0^t f(\mathbf{\Gamma},t')dt'$$

This integral equation can be solved by iteratively substituting for $f(\mathbf{\Gamma},t')$ on the right hand side, i.e

$$f(\Gamma,t) = f(\Gamma,0) - i\mathcal{L}t f(\Gamma,0) + (-i\mathcal{L})^2 \int_0^t \int_0^{t'} f(\Gamma,t'')\, dt''\, dt'$$

$$= f(\Gamma,0) - i\mathcal{L}t f(\Gamma,0) + \frac{(-i\mathcal{L}t)^2}{2!} f(\Gamma,0) + \ldots$$

$$= \exp(-i\mathcal{L}t)\, f(\Gamma,0) \tag{5.3}$$

We can also consider the time dependence of phase variables as we follow a phase point through its trajectory.

$$\frac{dB}{dt} = \dot{\Gamma} \cdot \frac{\partial B}{\partial \Gamma} \equiv iL\, B(\Gamma)$$

The operator L is called the p-Liouvillean. This equation can be solved formally in the previous fashion.

$$B(\Gamma(t)) = \exp(iLt)\, B(\Gamma_0) \quad \text{where } \Gamma_0 \equiv \Gamma(0) \tag{5.4}$$

The p-Liouvillean and the f-Liouvillean are hermitian adjoints of each other

$$\int B(-i\mathcal{L})\, f\, d\Gamma = \int B\left(-\frac{d}{d\Gamma} \cdot (\dot{\Gamma}\, f)\right) d\Gamma$$

$$= \int \dot{\Gamma} . \frac{\partial B}{\partial \Gamma}\, f\, d\Gamma = \int iLB\, f\, d\Gamma \tag{5.5}$$

If the system were described by a Hamiltonian, then Liouville's theorem, $(\partial/\partial\Gamma)\cdot\dot{\Gamma} = 0$, would hold, and eq (5.2) takes the form of the usual Liouville equation. The existence of a Hamiltonian is a sufficient, but not necessary condition for this condition to hold. In this case, the p- and f-Liouvilleans are identical and self adjoint. Since we wish to include the effect of thermostatting (4.1 or 4.3), which is non-conservative, we must work with distinct p- and f-Liouvilleans, and the general equation (5.2).

We can form phase averages by firstly propagating the distribution function with the f-propagator, then forming the phase average over a phase variable B.

$$\langle B(t) \rangle = \int B(\Gamma)\, f(\Gamma,t)\, d\Gamma \tag{5.6}$$

$$= \int B(\Gamma) \exp(-i\mathcal{L}t)\, f(\Gamma,0)\, d\Gamma$$

This is the Schrödinger picture. Alternatively, one may form the phase average by following the phase variable along the trajectories.

$$\langle B(t)\rangle = \int f_K(\Gamma_0)\, B(\Gamma(t))\, d\Gamma \qquad (5.7)$$

$$= \int f_K(\Gamma_0) \exp(iLt)\, B(\Gamma_0)\, d\Gamma_0$$

This is called the Heisenberg picture. The equivalence of the two pictures is guaranteed by the adjointness property (5.5).

5.2 Morriss' lemma

In developing the non-linear thermostatted response theory, we need a relation between the f-propagator and the p-propagator. This is accomplished by means of Morriss' lemma, which reads

$$e^{i\mathcal{L}t} = e^{\int_0^t \Lambda(s)ds}\, e^{iLt} \qquad (5.8)$$

where

$$\Lambda(s) = \frac{\partial}{\partial\Gamma} \cdot \dot{\Gamma}\,(s) = -\frac{1}{f}\frac{\partial f}{\partial t}$$

is the phase space compression factor. To prove Morriss' lemma,we first need a result called Dyson's equation, which states

$$e^{iLt} = e^{iL_0 t} + \int_0^t e^{iL(t-s)}\, i\Delta L e^{iL_0 s} \qquad (5.9)$$

where $\Delta L = L - L_0$.

Proof of Dyson's equation:

This is an identity at t = 0. Now differentiate the left hand side with respect to t.

$$\frac{\partial}{\partial t}(\mathrm{LHS}) = iLe^{iLt} = iL(\mathrm{LHS})$$

The derivative of the right hand side is

$$\frac{\partial}{\partial t}(\mathrm{RHS}) = iL_o e^{iL_o t} + iL \int_0^t e^{iL(t-s)}\, i\Delta L e^{iL_o s}\, ds + i\Delta L e^{iL_o t}$$

$$= iLe^{iL_o t} + iL \int_0^t e^{iL_o (t-s)}\, i\Delta L e^{iL_o s}\, ds$$

The two sides of Dyson's equation satisfy the same differential equation with the same initial conditions, and hence are equal. □

Proof of Morriss' lemma:

From the product rule of derivatives, $i\mathcal{L} = \Lambda(0) + iL$, and so we can use Dyson's equation

$$e^{i\mathcal{L}t} = e^{iLt} + \int_0^t e^{i\mathcal{L}(t-s)}\, \Lambda(0)\, e^{iLs} ds$$

$$= e^{iLt} + \int_0^t e^{i\mathcal{L}s}\, \Lambda(0)\, e^{iL(t-s)}\, ds$$

We can now substitute for $e^{i\mathcal{L}t}$ on the right hand side using Dyson's equation

$$e^{i\mathcal{L}t} = e^{iLt} + \int_0^t e^{iLs}\, \Lambda(0)\, e^{iL(t-s)}\, ds \, +$$

$$+ \int_0^t \int_0^{s_1} e^{i\mathcal{L}s_2}\, \Lambda(0)\, e^{iL(s_1-s_2)}\, \Lambda(0)\, e^{iL(t-s_1)}\, ds_2\, ds_1$$

We can iterate this process to find

$$e^{i\mathcal{L}t} = e^{iLt} + \int_0^t e^{iLs}\, \Lambda(0)\, e^{iL(t-s)}\, ds \, +$$

$$+ \int_0^t \int_0^{s_1} e^{iLs_2}\, \Lambda(0)\, e^{iL(s_1-s_2)}\, \Lambda(0)\, e^{iL(t-s_1)}\, ds_2\, ds_1 \, + \, ... \qquad (5.10)$$

Now Λ is just a phase variable, and everything to the right of Λ is just an operator that acts on phase variables. The effect of e^{iLs} acting on phase variables is to give the value of the phase variable at the phase point that has evolved for time s. So (5.10) becomes

$$
\begin{aligned}
e^{i\mathcal{L}t} &= e^{iLt} + \int_0^t \Lambda(s)\, e^{iLt} ds + \\
&+ \int_0^t\int_0^{s_1} \Lambda(s_2)\, e^{iLs_1}\, \Lambda(0)\, e^{iL(t-s_1)}\, ds_2\, ds_1 + \ldots \\
&= e^{iLt} + \int_0^t \Lambda(s)\, e^{iLt} ds + \int_0^t\int_0^{s_1} \Lambda(s_2)\, \Lambda(s_1)\, e^{iLt}\, ds_2\, ds_1 + \ldots \\
&= e^{iLt} + \int_0^t \Lambda(s)\, e^{iLt} ds + \frac{1}{2!}\left[\int_0^t \Lambda(s) ds\right]^2 e^{iLt} + \ldots \\
&= \exp\left(\int_0^t \Lambda(s)\, ds\right) e^{iLt}
\end{aligned}
$$

□

6. NONLINEAR RESPONSE THEORY

6.1 Kawasaki distribution

In this section, we will give a derivation of an expression for phase averages as they evolve after an external field is switched on. At the time t=0, the distribution function is given by the equilibrium isokinetic distribution

$$
f(\mathbf{\Gamma},0) = f_K(\mathbf{\Gamma}) = \frac{\delta(K - K_o)\, e^{-\beta\Phi}}{\int \delta(K - K_o)\, e^{-\beta\Phi}\, d\mathbf{\Gamma}}
$$

We consider the impact of a steady external field, F_e, on this distribution. A discussion for the case where the external field has an arbitrary time dependence is much more complex [Evans & Morriss (1988)], and will not be given here. The external field F_e is switched on at t= 0, and changes the dynamics to

$$
\dot{\mathbf{q}}_i = \frac{\mathbf{p}_i}{m} + \mathbf{C}_i\, F_e
$$

$$
\dot{\mathbf{p}}_i = \mathbf{F}_i + \mathbf{D}_i\, F_e - \alpha \mathbf{p}_i \tag{6.1}
$$

To maintain the system at constant peculiar kinetic energy, we require that

$$\sum_i \mathbf{p}_i \cdot \dot{\mathbf{p}}_i = 0 = \sum_i \left(\mathbf{F}_i \cdot \mathbf{p}_i + \mathbf{D}_i \cdot \mathbf{p}_i \, F_e - \alpha \, p_i\right)$$

i.e. that the coefficient of friction be

$$\alpha = \frac{\sum_i \mathbf{F}_i \cdot \mathbf{p}_i + F_e \sum \mathbf{D}_i \cdot \mathbf{p}_i}{\sum_i p_i^2} \tag{6.2}$$

Note that in writing these equations we are assuming that the momenta are measured with respect to the local streaming velocity of the fluid, hence the term 'peculiar momenta'. At low Reynolds number this presents no major difficulties but in the turbulent flow regime new methods have to be applied [Evans & Morriss (1986)].

We assume that the external force F_e is properly conservative, so that in the absence of thermostatting (α=0), Liouville's theorem holds. This assumption is known as adiabatic incompressibility of phase space, or AIΓ for short. We can of course pursue the theory without invoking this assumption but it has proven to be unnecessary to do so. With AIΓ, we can compute the phase space compressibilty from (6.1)

$$\Lambda = \frac{\partial}{\partial \mathbf{\Gamma}} \cdot \dot{\mathbf{\Gamma}} = -\,3N\,\alpha(\mathbf{\Gamma}) + o(N^o)$$

The second term is intensive, and so may be ignored in the thermodynamic limit.

The time evolution of $f(\mathbf{\Gamma}_0,t)$ from its initial state $f_K(\mathbf{\Gamma}_0)$ can now be computed using Morriss' lemma

$$\begin{aligned} f(\mathbf{\Gamma}_o,t) &= e^{-i\mathcal{L}t}\, f_K(\mathbf{\Gamma}_o) \\ &= \exp\left(-\int_0^t \Lambda(s)ds\right) e^{-Lt}\, \delta(K-K_o)\, e^{-\beta\Phi(\mathbf{\Gamma}_o)} \\ &= e^{\int_o^t 3N\alpha(\mathbf{\Gamma}(-s))\,ds}\, \delta(K-K_o)\, e^{-\beta\Phi(\mathbf{\Gamma}(-t))} \end{aligned}$$

The potential Φ can be expressed in terms of its time derivative.

$$\Phi(\mathbf{\Gamma}(-t)) = \Phi(\mathbf{\Gamma}_o) - \int_0^t \dot{\Phi}(\mathbf{\Gamma}(-s))\,ds$$

Upon substituting this, we get

$$f(\Gamma_o,t) = \exp\left[\int_o^t 3N\alpha\big(\Gamma(-s)\big) + \beta\dot{\Phi}\big(\Gamma(-s)\big)\, ds\right] f_K(\Gamma_o)\ .$$

The time derivative of Φ is related in a simple way to the adiabatic heating of the fluid.

$$\begin{aligned}\dot{\Phi} &= \dot{H} \qquad \text{since } \dot{K} = 0\\ &= \dot{H}^{ad} \quad - \alpha \sum_i p_i^2/m\\ &= \dot{H}^{ad} \quad - 3N\alpha\beta^{-1}\end{aligned}$$

where dH^{ad}/dt is the heat flux dissipated in the system due to the external field F_e. If we define the dissipative flux $J(\Gamma)$ as

$$J \equiv - \dot{H}^{ad}/F_e \tag{6.3}$$

then

$$3N\alpha + \beta\,\dot{\Phi} = - JF_e$$

and so

$$f(\Gamma,t) = f_K(\Gamma)\exp\left[-\beta\int_o^t J\big(\Gamma(-s)\big)\, F_e\, ds\right] \tag{6.4}$$

This is the Kawasaki Distribution Function, [Yamada & Kawasaki (1967), Morriss & Evans (1985)] which describes the phase space distribution that has evolved under the influence of a dissipative force F_e, and Gaussian thermostatting has been used to guarantee the existence of a steady state. According to the Shrödinger picture, (5.7), we can evaluate Kawasaki phase averages for a typical phase variable B.

$$\langle B(t)\rangle = \frac{\left\langle B(\Gamma_0)\, e^{-\beta\int_o^t J(\Gamma(-s))F_e\, ds}\right\rangle_K}{\left\langle e^{-\beta\int_o^t J(\Gamma(-s))\, F_e\, ds}\right\rangle_K} \tag{6.5}$$

where $<\ ..\ >_K$ means average over phase space distribution f_K. The time evolution implicit in (6.5) is generated by the full, field-dependent, thermostatted equations of motion (6.1,2).

6.2 Transient time correlation function - derived thermodynamic quantities

The phase average can be expressed in terms of a transient time correlation function by differentiating with respect to time.

$$\frac{d}{dt}\langle B(t)\rangle = \frac{-\beta F_e\left\langle B(\Gamma_0)\, J(\Gamma(-t))\, e^{-\beta\int_0^t J(\Gamma(-s))F_e ds}\right\rangle_K}{\left\langle e^{-\beta\int_0^t J(\Gamma(-s))F_e ds}\right\rangle_K} +$$

$$+ \frac{\beta F_e\left\langle B(\Gamma_0)\, e^{-\beta\int_0^t J(\Gamma(-s))F_e ds}\right\rangle_K \left\langle J(\Gamma(-t))\, e^{-\beta\int_0^t J(\Gamma(-s))F_e ds}\right\rangle_K}{\left\langle e^{-\beta\int_0^t J(\Gamma(-s))F_e ds}\right\rangle_K^2}$$

$$= -\beta F_e \langle B(t)J(0)\rangle_K + \beta F_e \langle B(t)\rangle_K \langle J(0)\rangle_K$$

Now the average dissipative flux at t = 0 is zero, and so we get the Transient Time Correlation function

$$\langle B(t)\rangle_K = \langle B(0)\rangle_K - \beta F_e \int_0^t \langle B(s)\, J(0)\rangle_K \, ds \quad . \tag{6.6}$$

This relation is exact, regardless of the magnitude of the external field, F_e.

It is well known in equilibrium statistical mechanics that if one formally differentiates expressions for equilibrium averages of the so called first order thermodynamic quantities (e.g. p, E, V), that one derives very useful equilibrium fluctuation expressions for the second order, or derived, thermodynamic quantities (e.g. C_V, C_p, χ_T). We will now derive analogous expressions for non-equilibrium steady states using transient time correlation functions. For example, we could differentiate the Transient Time Correlation Function (6.6) with respect to the temperature of the initial (t=0), equilibrium ensemble.

$$\frac{\partial}{\partial T}\langle B(t)\rangle = k_B\,\beta^2\,\langle \Delta B(0)\,\Delta H(0)\rangle - k_B\,\beta\big(\langle B(t)\rangle - \langle B(0)\rangle\big) -$$

$\Uparrow$ equilibrium term; $\Uparrow$ steady state shift

$$- k_B\beta^3\,F_e\int_0^t \big\langle \Delta\big(B(s)\,J(0)\big)\,\Delta H(0)\big\rangle$$

$\Uparrow$ transient correlation integral

where $\Delta A = A - \langle A\rangle$ and <.. > denotes a canonical average.

A key point in this derivation is the fact that the differential nature of the Gaussian feedback thermostatting mechanism ensures that the derivative $\partial/\partial T$ commutes with the thermostatted field dependent propagator implicit in these equations. This is not true if the Nosé-Hoover thermostat is used.

If we calculate the same derivative of the Kawaski expression (6.5), and if we take B to be the internal energy H, then we would obtain an expression for the specific heat at constant volume.

$$C_v = \frac{\partial}{\partial T}\langle H(t)\rangle = k_B\beta^2\Big\langle \Delta\big(H(t)\big)^2\Big\rangle + k_B\beta^2\,F_e\int_0^t \langle \Delta H(t)\,\Delta J(s)\rangle\, ds$$

$\Uparrow$ steady state energy fluctations; $\Uparrow$ steady state time correlation function

This expression is a nontrivial generalization of the well-known equilibrium formula.

7. NON-EQUILIBRIUM MOLECULAR DYNAMICS SIMULATION

7.1 Example 1. Planar Couette Flow

We will now show how the Transient Time Correlation Function theory may be applied in computer simulations. We will first consider planar Couette flow where the steady state streaming velocity **u**, is assumed to be a linear function of position, **r**. In particular we assume $u_x = \gamma y$. Shear flow like all Navier-Stokes transport processes is driven by boundary conditions (eg. moving walls). This is inconvenient for computer simulations because of the huge surface effects that would be induced in the small systems which we are capable of simulating on computers ($100 < N \leq 100{,}000$).

Computer simulation of equilibrium systems have always employed periodic boundary conditions to minimize size dependence effects.

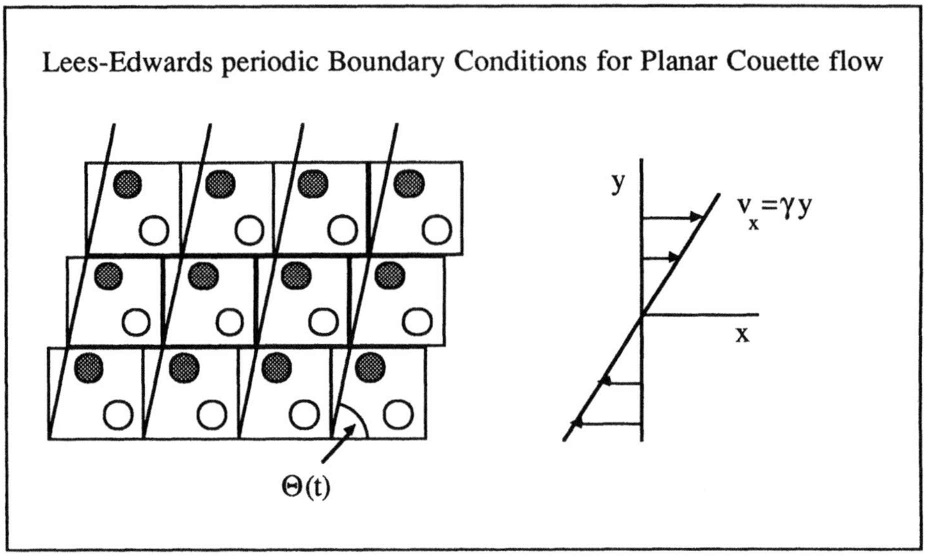

Figure 2. Lees-Edwards boundary conditions.

The figure above, illustrates a way of adapting periodic boundary conditions to planar Couette flow. The so-called Lees-Edwards periodic boundary conditions employ time-varying nonorthogonal lattice vectors. The perpendicular height of the cells remains fixed so that the shearing deformation occurs isochorically. Once $\Theta=90^0$ it may be reset to 0^0 without changing the mutual disposition of particles. If a particle exits one of the bottom cells through a top face, it is replaced by a periodic image which enters at the bottom face. This image will be positioned according to the current angle of the slewing lattice vector. The x-component of its velocity will be the old velocity minus the strain rate multiplied by the perpendicular height of the cell. Its peculiar velocity (ie. its velocity relative to the planar velocity profile) is unchanged.

This is all that is really needed to run a molecular dynamics simulation of planar Couette flow at low Reynolds number. The flow of the periodic particles above and below the primitive cell will try to induce a linear velocity profile in the system. One simply solves Newton's equations of motion for the system of N interacting particles, waits for the initial transients to decay and then accumulates statistical averages for the shear stress, the energy etc.

Of course such a system will heat up. It will never achieve a steady state. To remedy this we may employ a Gaussian or Nosé-Hoover thermostat. We must be careful, however, to ensure that it is the peculiar velocities which are thermostatted and **not** the laboratory velocities. If the Reynolds number is sufficiently small so that a linear velocity profile is stable, the streaming velocity will of course be $\hat{\mathbf{x}}\gamma y$.

The so-called SLLOD equations provide a more convenient and more elegant set of variables with which to work. The first advantage of the SLLOD equations of motion is that they only involve the coordinates and the **peculiar** momenta of the particles. Since the thermostat, the internal energy and the pressure tensor all involve peculiar, rather than laboratory velocities, this is an advantage.

A far more important advantage is that the SLLOD equations of motion transform the shearing boundary conditions responsible for driving shear flow into the form of a mechanical perturbation for which it is easy to apply the response theory described in previous sections.

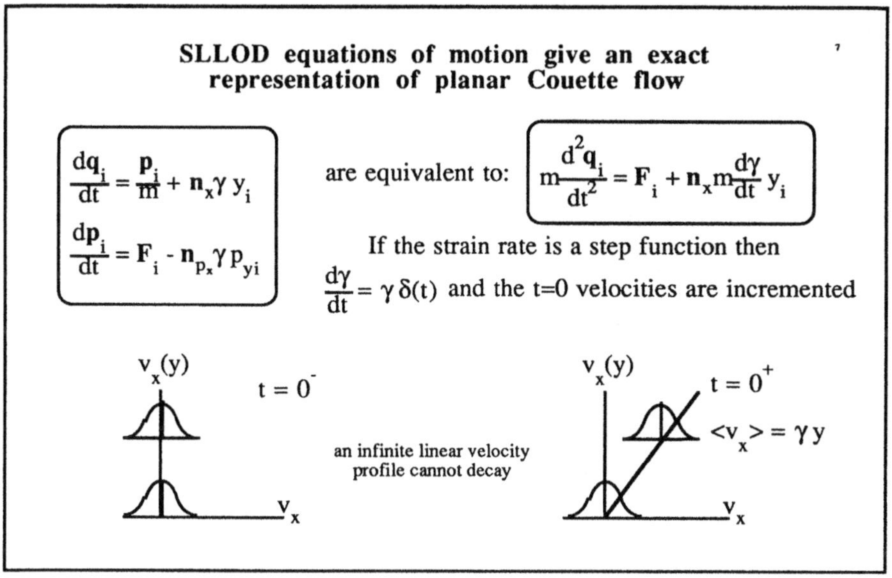

Figure 3. The SLLOD transformation.

The SLLOD equations of motion are shown in the figure 3, above. Consider subjecting a canonical ensemble of systems to the SLLOD equations at t=0. As is shown in the figure, when transformed to second order or 'Newtonian' form SLLOD dynamics have the effect of incrementing the x-velocities of every particle by an amount proportional to the product of the strain rate and the y-coordinate of that particle. This clearly induces a zero wavevector momentum gradient in the system. We know from the conservation laws that such a gradient can never decay - the system will continue shearing forever. For all times greater than zero the SLLOD equations are identical to Newton's equations and so the system follows the real time evolution for an infinite system subject initially to a planar velocity profile. (Note: The linear velocity may develop instabilities with time but the zero wavevector component of the profile cannot change. The **adiabatic** SLLOD equations of motion are exact arbitrarily far from equilibrium.)

If we look at the form of the SLLOD equations in terms of **q** and **p**, we see that the internal energy can be written as,

$$H_0 = \sum \frac{\mathbf{p}_i^2}{2m} + \Phi(\mathbf{q}_1,\dots \mathbf{q}_N)$$

where again the SLLOD momenta **p** are peculiar momenta. From the form of the SLLOD equations of motion it looks as though an external field γ, perturbs the system away from equilibrium. The equations have the form of a mechanical perturbation even though the disturbance is boundary driven. In the absence of a thermostat it is clear that the adiabatic time derivative of the internal energy is

$$\dot{H}_0^{ad} = -\gamma V P_{xy}$$

It is also clear that the condition of the adiabatic incompressibility of phase space is satisfied by the SLLOD equations of motion, $\Delta^{ad} \equiv 0$. This is in spite of the fact that the adiabatic SLLOD equations of motion are not directly deriveable from a Hamiltonian. The thermostatted form of the SLLOD equations of motion for planar Couette flow are,

$$\frac{d\mathbf{q}_i}{dt} = \frac{\mathbf{p}_i}{m} + \hat{\mathbf{x}}\gamma y_i$$

$$\frac{d\mathbf{p}_i}{dt} = \mathbf{F}_i - \hat{\mathbf{x}}\gamma p_{yi} - \alpha \mathbf{p}_i$$

where the thermostatting multiplier α, is given by the expression required to keep the peculiar kinetic energy $\Sigma\mathbf{p}_i^2/2m$ a constant, namely,

$$\alpha = \frac{\sum \mathbf{F}_i\cdot\mathbf{p}_i - \gamma p_{xi}p_{yi}}{\sum p_i^2}$$

We can now apply the Transient Time Correlation Function formalism to the thermostatted SLLOD equations of motion. If we assume that at t=0 a canonical ensemble, $f \sim \exp{-\beta H_0}$, is 'perturbed' by the 'external field' γ, the ensemble averaged value of an arbitrary phase variable, B, at time t is,

$$\langle B(t) \rangle = \langle B(0) \rangle - \beta\gamma V \int_0^t ds \langle B(s) P_{xy}(0) \rangle$$

The time dependence implicit on both sides of this equation is generated by the thermostatted, strain rate dependent equations of motion above.

In the linear regime ($\lim \gamma \to 0$) it is clear from symmetry considerations that scalar quantities are unchanged by shear flow and that the equation above reduces to the familar Green-Kubo expression for the shear viscosity when B is taken to be P_{xy}. In the linear regime the time dependence in the correlation function is generated by the field free thermostatted equations of motion. It is therefore an equilibrium time correlation function.

Figure 4 below shows that outside the linear regime the pressure , a scalar quantity, is changed by the presence of shear, an effect which is known to rheologists as shear dilatancy. Furthermore the figure shows that the Transient Time Correlation Function correctly predicts the pressure shift.

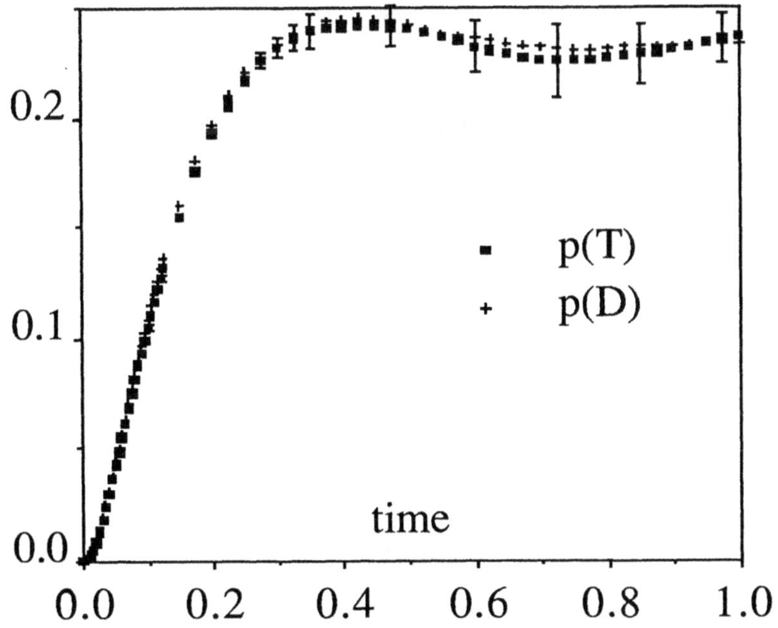

Figure 4. A comparison of NEMD computer simulation and the Transient Time Correlation Function expression for shear dilatancy in a 2-dimensional soft disc fluid.

7.2 Example 2. Isobaric Equations of motion

For the first thirty years after its invention molecular dynamics could only be used in the inconvenient microcanonical ensemble (N,V,H_0). The Nose-Hoover thermostat enables molecular dynamics to simulate the canonical ensemble ($f \sim \exp -\beta H$), while the Gaussian thermostat enables the simulation of the isokinetic ensemble ($f \sim \delta(K-K_0)\exp -\beta\Phi$, where $3N\beta^{-1}=K_0$). However, the feed-back mechanisms employed in both the Nose-Hoover and Gaussian thermostats can be used to control thermodynamic variables other than the temperature. This has lead to a revolution in the ways in which molecular dynamics may be used in statistical mechanics.

To simulate a system at constant pressure rather than at constant volume we use the SLLOD equations of motion for an isotropic dilation rather than for a shearing deformation. If this is done, it is easily seen that in the absence of a thermostat, if the dilation rate, $d\varepsilon/dt$, is treated as a Gaussian multiplier and chosen to make the hydrostatic pressure a constant of the motion, the enthalpy I, will also be a constant of the motion. This is because $dH_0/dt = - pdV/dt = -3pVd\varepsilon/dt$ and since $dI/dt = dH_0/dt + pdV/dt$, the enthalpy will be a constant of the motion. One can easily add a thermostat to simulate equilibrium systems at constant temperature and pressure.

We will now show how one applies these ideas to the simulation of a nonequilibrium system under isothermal-isobaric conditions. The general form we choose for the equations of motion is,

$$\frac{d\mathbf{q}_i}{dt} = \frac{\mathbf{p}_i}{m} + \mathbf{C}_i F_e(t) + \frac{d\varepsilon}{dt}\mathbf{q}_i$$

$$\frac{d\mathbf{p}_i}{dt} = \mathbf{F}_i + \mathbf{D}_i F_e(t) - \frac{d\varepsilon}{dt}\mathbf{p}_i - \alpha\mathbf{p}_i$$

The dilation rate and the thermostatting multiplier are chosen so that the peculiar kinetic energy $\Sigma p_i^2/2m$ and the hydrostatic pressure $(1/3V)\Sigma(p_i^2/m + \mathbf{q}_i \cdot \mathbf{F}_i)$ are constants of the motion. It is easy to show that in the absence of an external field, the isothermal-isobaric distribution, f_{Tp}, is an equilibrium solution to the Liouville equation for the system.

$$f_{Tp} = \frac{e^{-\beta I(\Gamma)}}{\int_0^\infty dV \int d\Gamma\, e^{-\beta I(\Gamma)}}$$

If we apply the time dependent external field $F_e(t)$ to an isothermal-isobaric ensemble of systems at time zero then it is easy to show that the Transient Time Correlation Function for the response of an arbitrary variable B at time t is,

$$< B(t) > = < B(0) > - \beta F_e \int_0^t ds < B(s)J(0) >$$

where if the Adiabatic Incompressibility of phase space condition is satisfied, the dissipative flux is related to the adiabatic time derivative of the enthalpy,

$$\left.\frac{dI}{dt}\right)^{ad} = - J(\boldsymbol{\Gamma})F_e$$

Using the Kawasaki representation of the nonlinear response we can derive fluctuation expressions for isobaric derivatives of first order thermodynamic quantities like the isobaric specific heat C_p.

$$C_p(t) = \left.\frac{\partial< I(t) >}{\partial T}\right)_p = < \Delta(I(t))^2 > + \frac{F_e}{k_B T^2}\int_0^t ds < \Delta I(t)\, \Delta J(s) >$$

As we have seen this derivation relies upon the use of the Gaussian feedback mechanism to maintain the temperature. Because the Gaussian feedback mechanism is differential rather than integral, as it is for the Nose-Hoover case, the temperature derivative, $\partial/\partial T$, commutes with the field dependent, Gaussian isothermal propagator.

It is possible to use the Gaussian feedback mechanism to maintain a fixed thermodynamic flux, J, rather than a fixed thermodynamic force, F_e. This introduces the possibility of simulating nonequilibrium systems in the Norton rather than the Thevenin ensemble where the dissipative flux J, rather than the external field F_e, is the independent state defining variable. Using the differential nature of the flux fixing mechanism it is then possible to compute derivatives of the thermodynamic forces required to fix the flux, with respect to the value of the flux. In this way one can derive fluctuation expressions for the inverse Burnett coefficients.

$$<F_e> = J\left.\frac{\partial<F_e>}{\partial J}\right)_{J=0} + \frac{1}{2!}J^2\left.\frac{\partial^2<F_e>}{\partial J^2}\right)_{J=0} + \;\ldots\ldots .$$

These coefficients describe a Taylor series expansion of the force in terms of the flux evaluated about equilibrium. The inverse Burnett coefficients, $\partial^n<F_e>/\partial J^n)_{J=0}$, turn out to be related to simple two-time equilibrium time correlation functions evaluated in the constant flux, Norton ensemble [Evans & Lynden-Bell (1988)].

A more detailed account of nonlinear response theory can be found in a forth coming book {Evans and Morriss (1989)].

ACKNOWLEDGEMENT

We would like to thank the Australian National University Supercomputer Facility for a generous grant of computer time.

REFERENCES

Evans D.J. (1981), *Phys Rev A* , **23**, 2622
Evans D.J. and Holian B.L. (1984), *J. Chem Phys* , **83**, 4069
Evans D.J. and Lynden-Bell R.M. (1988), *Phys. Rev.* , **A**, (to appear).
Evans D.J. and Morriss G.P. (1984), *Chem Phys* , **87**, 451
Evans D.J. and Morriss G.P. (1986), *Phys. Rev. Letts.*, **56**, 2172
Evans D.J. and Morriss G.P. (1988), *Mol Phys* , **64**, 521
Evans D.J. and Morriss G.P. (1989), "Statististical Mechanics of Nonequilibrium Liquids", to appear, (Academic Press).
de Groot S.R. and Mazur P. (1962), "Non-Equilibrium Thermodynamics", (North Holland: Amsterdam)
Morriss G.P. and Evans D.J. (1985), *Mol Phys* , **59**, 629
Temple G.F. (1961), "Cartesian Tensors", (Methuen: London)
Tolman R.C. (1962), "The Principles of Statistical Mechanics", (Dover: NY)
Yamada T. and Kawasaki K. (1967), *Prog Theo Phys* , **38**, 1030

THE PATH-INTEGRAL SIMULATION OF QUANTUM SYSTEMS

M. J. GILLAN

PHYSICS DEPARTMENT, UNIVERSITY OF KEELE,
KEELE, STAFFORDSHIRE ST5 5BG, U.K.

and

THEORETICAL PHYSICS DIVISION, HARWELL LABORATORY,
DIDCOT, OXFORDSHIRE OX11 0RA, U.K.

ABSTRACT

We explain how Feynman's path-integral formulation of quantum statistical mechanics gives a natural way of simulating quantum systems in thermal equilibrium. The theory is first outlined for a simple system, consisting of a single particle in one dimension acted on by an external potential. Starting from the standard basic expressions for the partition function and the thermal averages of observables, it is shown how a simple sequence of mathematical operations allows these expressions to be brought into Feynman's path-integral form. In this form, the quantities of interest are expressed as an integral over cyclic paths of the particle, which can be evaluated by classical simulation methods. This approach, generalized to many particles in three dimensions, gives a simulation technique for quantum many-body systems. We discuss how to calculate some important observables such as the energy and the radial distribution function. In the application of path-integral techniques to systems like liquid helium, the inclusion of quantum exchange is crucial, and we indicate how this can be achieved. We then illustrate the use of the technique by describing simulations that have been performed on (i) an electron dissolved in a molten salt; (ii) hydrogen in metals; and (iii) liquid helium-four.

1. INTRODUCTION

The simulation of condensed matter is almost always based on the complete neglect of quantum effects. The atoms or molecules are taken to follow classical trajectories, whose time evolution is governed by Newton's equation of motion. Simulation methods based on classical mechanics have been outstandingly successful, and have contributed enormously to our understanding of a wide range of systems, as is shown by the other lectures in these proceedings. The neglect of quantum effects is very often fully justified. The condition for their neglect is, roughly speaking, $\hbar\omega \ll k_B T$, where ω is a typical frequency of motion, and T is the temperature, which means that the energy of quantum fluctuations is much smaller than that of thermal fluctuations; this condition is usually satisfied, unless we are dealing with light atoms (hydrogen, helium...) or low temperatures. When we enter the quantum regime, $\hbar\omega \gtrsim k_B T$, characteristic quantum effects such as Bose condensation and the diffusion of atoms by quantum tunnelling make their appearance, and classical simulation becomes useless. In these

C.R.A. Catlow et al. (eds.), Computer Modelling of Fluids Polymers and Solids, 155–188.

lectures, I describe a technique which has become widely used to perform simulations with full (and in principle exact) inclusion of quantum effects. The technique is based on Feynman's path-integral formulation of quantum mechanics, and is known as path-integral simulation.

What I will try to do here is to present a simple explanation of how path-integral simulation works, and to give some illustrations of what can be achieved with it. I will not try to do justice to the historical development of the technique, nor to review fully the wide range of important problems to which it has been applied. To those whose work is not discussed, or is discussed only superficially, I offer my apologies. It should be stressed, though, that the idea of path-integral simulation has a long history – almost as long, in fact, as that of classical molecular dynamics and Monte Carlo methods. Feynman's path-integral formulation was first presented 40 years ago [1], and its use in simulation was already discussed in the 1960's [2]. Its exploitation as a simulation method has only attracted widespread attention in the last ten years, though. In that period, many papers have appeared on the basic principles of the method, and the problem of making efficient and reliable calculations. The reader who wishes to study the method in detail will find it useful to consult the work of Barker [3], Chandler and Wolynes [4], Ceperley and Kalos [5], Schweizer *et al.* [6], Herman *et al.* [7], Kalos [8], Parrinello and Rahman [9], and Berne and Thirumalai [10]. The last paper here gives an extensive review of applications up to 1986. It should also be emphasized that the importance of Feynman's path-integral formulation goes far beyond its use in simulation. It is one of the major formulations of quantum mechanics [11], which has proved useful not only in condensed-matter theory, but also in, e.g. elementary-particle physics.

These lectures fall into two parts. In the first part, I explain the ideas underlying path-integral simulation. For simplicity, I discuss first a single particle in one dimension, and show how the partition function can be re-expressed in terms of the partition function of a purely classical system of P beads coupled by harmonic springs, where P is a large integer. This isomorphism between the original quantum system and the corresponding classical system allows the use of classical simulation to calculate thermal averages in the quantum system. I then consider the generalization to many particles, and show how exchange effects due to indistinguishability can be incorporated. I also discuss here some important questions, such as the choice of the integer P, and methods for calculating thermal averages. In the second part, I illustrate the practical application of the technique, by describing simulations of (i) an electron in a molten salt [9]; (ii) hydrogen in metals [12]; and (iii) liquid helium [13].

2. PRINCIPLES OF THE PATH-INTEGRAL METHOD

2.1 Classical statistical mechanics

It will be useful to begin by recalling the formulae of classical statistical mechanics. Suppose we have a system of N particles, whose Hamiltonian is

$$H = \sum_{i=1}^{N} p_i^2/2m + U(\mathbf{r}_1 \ldots \mathbf{r}_N) , \tag{1}$$

where $\mathbf{p}_i$ and $\mathbf{r}_i$ are the momenta and positions, m is the particle mass and U is the total potential energy – this might be a sum of pair interactions:

$$U = \tfrac{1}{2} \sum_{i \neq j} v(|\mathbf{r}_i - \mathbf{r}_j|) , \tag{2}$$

where $v(r)$ is the interparticle potential.

A key quantity is the partition function Z, which is the integral of the Boltzmann factor over phase space:

$$Z = \frac{1}{N!h^{3N}} \int d\mathbf{p}_1..d\mathbf{p}_N d\mathbf{r}_1..d\mathbf{r}_N \; e^{-\beta H} \quad , \tag{3}$$

where $\beta = 1/k_B T$. From this, we obtain the Helmholtz free energy F:

$$F = -k_B T \ln Z \quad , \tag{4}$$

whence, by taking appropriate derivatives, we can obtain all thermodynamic quantities. The momentum integrals in equation (3) can be performed analytically, so that

$$Z = \frac{1}{N!\Lambda^{3N}} \int d\mathbf{r}_1...d\mathbf{r}_N \; e^{-\beta U} \quad , \tag{5}$$

where $\Lambda = (\beta\hbar^2/2\pi m)^{\frac{1}{2}}$ is the thermal wavelength (in thermal equilibrium, the particles have some typical momentum, and hence, by de Broglie's relation, a typical wavelength: this is Λ). This means that all thermodynamic quantities can be obtained by appropriate averages over *configuration* space. The probability density in phase space is

$$P(\mathbf{p}_1...,\mathbf{r}_1...) = (N!h^{3N}Z)^{-1} \; e^{-\beta H(\mathbf{p}_1...,\mathbf{r}_1...)} \quad . \tag{6}$$

This allows us to calculate the thermal averages of all observables. If we are interested only in observables that depend on positions, then it is enough to work with the probability density in configuration space:

$$P(\mathbf{r}_1...\mathbf{r}_N) = (N!\Lambda^{3N}Z)^{-1} e^{-\beta U(\mathbf{r}_1...\mathbf{r}_N)} \quad . \tag{7}$$

These statements form the basis of the classical Monte Carlo technique, which is a procedure for sampling configuration space with statistical weight $P(\mathbf{r}_1...\mathbf{r}_N)$, and which allows the calculation of thermodynamic quantities and static (time-independent) averages.

2.2 Quantum statistical mechanics

Now consider a quantum system having the same Hamiltonian – the momenta and positions now being operators, of course. The partition function is given by

$$Z = \mathrm{Tr}\, e^{-\beta H} = \sum_\lambda \langle\lambda|e^{-\beta H}|\lambda\rangle \quad , \tag{8}$$

where the trace can be taken over any complete orthonormal set of states $|\lambda\rangle$. If the $|\lambda\rangle$ are chosen to be eigenstates of H, then we get the familiar expression

$$Z = \sum_\lambda \exp(-\beta E_\lambda) \quad , \tag{9}$$

where E_λ are the eigenvalues of the Hamiltonian. From Z, we obtain the free energy (equation (4)), exactly as before. The thermal average of any observable A is given by

$$\langle A\rangle = Z^{-1}\, \mathrm{Tr}\, e^{-\beta H} A \quad , \tag{10}$$

which can be written as

$$\langle A\rangle = Z^{-1} \sum_\lambda e^{-\beta E_\lambda} \langle\lambda|A|\lambda\rangle \quad , \tag{11}$$

i.e. the sum of the expectation values of A in each state $|\lambda\rangle$, weighted by the probability $Z^{-1}\exp(-\beta E_\lambda)$ of finding the system in that state.

Although these expressions are easy to write down, they are impossible to compute as they stand, except for very simple systems. Part of the problem here is simply one of *representation.* Each state $|\lambda\rangle$ is a function $\phi_\lambda(\mathbf{r}_1 \ldots \mathbf{r}_N)$ defined on the configuration space of the system, i.e. a space of $3N$ dimensions. If we suppose that the wavefunction of typical single-particle system in one dimension (e.g. a single anharmonic oscillator) needs 10 basis functions for an adequate representation, then the corresponding number of basis functions for the N−body system will be of order 10^{3N}. Evidently, the number of wavefunction coefficients will be astronomical, except for very small systems.

Given this apparent intractability, it is remarkable that any progress can be made at all, and still more remarkable that it is possible to make accurate calculations of thermodynamic functions and thermal averages. This is made possible by Feynman's formulation [1,11], which we now start to explain.

2.3 A single particle in one dimension

In order to simplify the discussion as much as possible, let us consider first the case of a single particle in one dimension. The Hamiltonian is

$$H = -\frac{\hbar^2}{2m}\frac{d^2}{dx^2} + V(x) \, , \tag{12}$$

where $V(x)$ is a potential acting on the particle. We call the eigenvalues and eigenfunctions E_λ and $\phi_\lambda(x)$.

Now it will be useful to work with a quantity known as the density matrix. This is a function of two positions and of the inverse temperature β, defined as

$$\rho(x,x';\beta) = \sum_\lambda \phi_\lambda(x)\, e^{-\beta E_\lambda}\, \phi_\lambda{}^*(x') \; . \tag{13}$$

This is an important quantity, because from it all thermodynamic quantities and thermal expectation values can be calculated. For example, the partition function is given by

$$Z = \sum_\lambda e^{-\beta E_\lambda} = \int dx\; \rho(x,x;\beta) \; , \tag{14}$$

where the second equality follows from the definition, equation (13), and the fact that the ϕ_λ are normalized to unity. The probability density $P(x)$ for finding the particle at x is also related to the density matrix, since

$$P(x) = \sum_\lambda e^{-\beta E_\lambda}\, |\phi_\lambda(x)|^2 \Big/ \sum_\lambda e^{-\beta E_\lambda} = Z^{-1}\, \rho(x,x;\beta) \; . \tag{15}$$

Another significance of $\rho(x,x';\beta)$ will be shown later.

Now the density matrix has an important decomposition property. To explain what this means, consider the quantity

$$\int dx''\; \rho(x,x'';\tfrac{1}{2}\beta)\rho(x'',x';\tfrac{1}{2}\beta) \; . \tag{16}$$

Inserting here the definition, equation (13), we see that:

$$\int dx''\ \rho(x,x'';\tfrac{1}{2}\beta)\rho(x'',x';\tfrac{1}{2}\beta) =$$
$$\int dx''\ [\sum_\lambda \phi_\lambda(x)e^{-\frac{1}{2}\beta E_\lambda}\phi_\lambda{}^*(x'')]\ [\sum_\mu \phi_\mu(x'')e^{-\frac{1}{2}\beta E_\mu}\phi_\mu{}^*(x')]\ . \tag{17}$$

We can immediately perform the integration over x'', making use of the orthonormality of the ϕ_λ to obtain

$$\begin{aligned}\int dx''\rho(x,x'';\tfrac{1}{2}\beta)\rho(x'',x';\tfrac{1}{2}\beta) &= \sum_\lambda \phi_\lambda(x)e^{-\beta E_\lambda}\phi_\lambda{}^*(x') \\ &= \rho(x,x';\beta)\ ,\end{aligned} \tag{18}$$

which means that the density matrix for inverse temperature β can be decomposed as an integral over a product of density matrices for inverse temperature $\tfrac{1}{2}\beta$.

This decomposition can straightforwardly be generalized to a product of P factors at inverse temperature β/P, where P is any integer:

$$\begin{aligned}\int dx_1 \ldots dx_{P-1}\ \rho(x_0,x_1;\beta/P)\rho(x_1,x_2;\beta/P)\ldots\rho(x_{P-1},x_P;\beta/P) \\ = \rho(x_0,x_P;\beta)\ .\end{aligned} \tag{19}$$

Then, from equation (14), the partition function can be expressed as

$$\begin{aligned}Z = \int dx_1 \ldots dx_P\ \rho(x_1,x_2;\beta/P)\rho(x_2,x_3;\beta/P)\ldots \\ \ldots\rho(x_{P-1},x_P;\beta/P)\rho(x_P,x_1;\beta/P)\ .\end{aligned} \tag{20}$$

The cyclic structure of this will be noticed. Starting from position x_1, the first factor takes us to x_2; the second takes us from x_2 to x_3; and so on, until the Pth factor returns us from x_P to our starting point x_1.

2.4 Temperature and time

What is the meaning of this decomposition, and how can it help us? The answer to the first question has to do with the interpretation of $\beta\hbar$ as a kind of time, and this will indicate the answer to the second question.

To see why $\beta\hbar$ functions like a time, let us take the derivative of $\rho(x,x';\beta)$. From equation (13), we have:

$$\begin{aligned}\frac{\partial}{\partial\beta}\rho(x,x';\beta) &= -\sum_\lambda E_\lambda\phi_\lambda(x)\exp(-\beta E_\lambda)\phi_\lambda{}^*(x') \\ &= -\left(-\frac{\hbar^2}{2m}\frac{\partial^2}{\partial x^2} + V(x)\right)\rho(x,x';\beta)\ ,\end{aligned} \tag{21}$$

where we have used the fact that $E_\lambda\phi_\lambda$ is the same as the Hamiltonian acting on ϕ_λ. But this is just the time-dependent Schrödinger equation for evolution in imaginary time.

To understand this in more detail, recall that the normal time-dependent Schrödinger equation for evolution of a wavefunction $\psi(x,t)$ is:

$$i\hbar\frac{\partial}{\partial t}\psi(x,t) = \left(-\frac{\hbar^2}{2m}\frac{\partial^2}{\partial x^2} + V(x)\right)\psi(x,t)\ . \tag{22}$$

It is common to express the time evolution in terms of the so-called propagator $K(x,x';t)$, defined so that the wavefunction at time t is given in terms of the wavefunction at time zero as

$$\psi(x,t) = \int dx' \; K(x,x';t)\psi(x',0) \; . \tag{23}$$

Since $\psi(x,t)$ evolves according to equation (22), $K(x,x';t)$ must satisfy

$$i\hbar\frac{\partial}{\partial t}K(x,x';t) = (-\frac{\hbar^2}{2m}\frac{\partial^2}{\partial x^2} + V(x))\; K(x,x';t) \; , \tag{24}$$

together with the boundary condition

$$K(x,x';t{=}0) = \delta(x - x') \; . \tag{25}$$

Then what is being said in equation (21) is that the density matrix is the propagator for evolution in imaginary time. In fact, if we put $t = -i\beta\hbar$, we have:

$$\rho(x,x';\beta) = K(x,x';-i\beta\hbar) \; . \tag{26}$$

The fact that $\rho(x,x';\beta)$ satisfies the boundary condition (25) follows from the definition (13) together with the completeness condition

$$\sum_\lambda \phi_\lambda(x)\phi_\lambda{}^*(x') = \delta(x - x') \; . \tag{27}$$

Now we can see the significance of the decomposition of ρ expressed in equations (18) and (19). We can always decompose the evolution of a wavefunction over a period $t = t_1 + t_2$ into evolution over time t_1 followed by evolution over time t_2. Explicitly:

$$\begin{aligned}\psi(x,t_1+t_2) &= \int dx'' \; K(x,x'';t_2)\psi(x'',t_1) \\ &= \int dx'dx'' \; K(x,x'';t_2)K(x'',x';t_1)\psi(x',0) \; ,\end{aligned} \tag{28}$$

which is equivalent to writing:

$$K(x,x';t_1+t_2) = \int dx'' \; K(x,x'';t_2)K(x'',x';t_1) \; . \tag{29}$$

The decompositions discussed in § 2.3 express this for evolution in imaginary time. For large P, we can regard equation (19) as breaking down the evolution from 0 to $-i\beta\hbar$ into a sequence of short time steps.

2.5 The short-time approximation

Over a short time interval, the particle will behave as though it is almost free: the potential $V(x)$ does not have time to make itself felt. This suggests how to approximate for large P in our expression (20) for the partition function Z.

The density matrix for free particles can be written down exactly. In fact, its form follows immediately from equation (21). If $V(x) = 0$, this is

$$\frac{\partial}{\partial\beta}\rho(x,x';\beta) = \frac{\hbar^2}{2m}\frac{\partial^2}{\partial x^2}\rho(x,x';\beta) \; . \tag{30}$$

This has the form of a diffusion equation, where β plays the part of the time and $D = \hbar^2/2m$ is the diffusion coefficient; $\rho(x,x';\beta)$ acts as though it were describing the probability density for finding diffusing particles at position x given that they were all concentrated at $x = x'$ at time $\beta = 0$. But we know that the solution of the diffusion equation spreads out as a gaussian as time increases. The solution is:

$$\rho(x,x';\beta) = \left(\frac{m}{2\pi\beta\hbar^2}\right)^{1/2} \exp\left[-\frac{m}{2\beta\hbar^2}(x-x')^2\right] . \tag{31}$$

In the presence of a potential V which is a constant, the solution can also be written exactly:

$$\rho(x,x';\beta) = \left(\frac{m}{2\pi\beta\hbar^2}\right)^{1/2} \exp\left[-\frac{m}{2\beta\hbar^2}(x-x')^2 - \beta V\right] . \tag{32}$$

Now we note that for small β, the function falls off very rapidly as x and x' are separated, so that even when V depends on x it can be treated as almost constant as the particle propagates from x to x'. This indicates that for large P, each of the factors in the expression (20) for Z can be approximated by

$$\rho(x,x';\beta/P) \simeq \left(\frac{mP}{2\pi\beta\hbar^2}\right)^{1/2} \exp\left[-\frac{mP}{2\beta\hbar^2}(x-x')^2 - \frac{\beta}{2P}(V(x)+V(x'))\right] , \tag{33}$$

where we have used the mean value of V at x and x'.

2.6 The isomorphism

On doing this, we find the following approximate expression for Z:

$$\begin{aligned} Z \simeq Z_P &= \left(\frac{mP}{2\pi\beta\hbar^2}\right)^{P/2} \int dx_1 \ldots dx_P \cdot \\ &\cdot \exp\left\{-\beta \sum_{s=1}^{P} \left[\frac{mP}{2\beta^2\hbar^2}(x_{s+1}-x_s)^2 + P^{-1}V(x_s)\right]\right\} . \end{aligned} \tag{34}$$

It was shown by Feynman [1,11] that this becomes exact as $P \rightarrow \infty$:

$$Z = \lim_{P\rightarrow\infty} Z_P , \tag{35}$$

in which limit it can be written as a path integral:

$$Z = \int \mathcal{D}x \exp\left\{-\frac{1}{\hbar}\int_0^{\beta\hbar} d\tau \left[\tfrac{1}{2}m\dot{x}(\tau)^2 + V(x(\tau))\right]\right\} . \tag{36}$$

For present purposes, the interesting formula is equation (34). This has the form of a purely classical partition function (see § 2.1) for a cyclic chain of P beads coupled by harmonic springs (spring constant $\kappa = mP/\beta^2\hbar^2$), each bead being acted on by the true potential $V(x)$ divided by P. Note that the spring term derives from the kinetic-energy operator in the Hamiltonian. This astonishing isomorphism between a single quantum particle and a cyclic chain of P classical particles becomes exact in the limit $P \rightarrow \infty$. The implication is that the techniques that have been developed for treating the statistical mechanics of classical systems can be applied, through the isomorphism, to quantum systems. In particular, Monte Carlo and molecular-dynamics methods of simulation, which have been used with so much success in the study of classical solids and liquids can also be used for quantum systems, although, as we shall stress, the molecular-dynamics method cannot be used directly to give dynamical information. How this works in practice will become clear below.

Up to now, we have discussed only the partition function. The implications of the isomorphism for the calculation of the thermal averages of observables will be shown in § 2.11, but it is convenient to note here how the probability distribution $P(x)$ for the quantum particle can be obtained. We have already noted (equation (15)) that $P(x)$ is given by $Z^{-1}\rho(x,x;\beta)$. The decomposition of $\rho(x,x';\beta)$, together with the short-time approximation for each factor immediately gives us the result:

$$P(x_1) \simeq \frac{\int dx_2 ... dx_P \exp\left\{-\beta \sum_s \left[\tfrac{1}{2}\kappa(x_{s+1}-x_s)^2 + P^{-1}V(x_s)\right]\right\}}{\int dx_1 ... dx_P \exp\left\{-\beta \sum_s \left[\tfrac{1}{2}\kappa(x_{s+1}-x_s)^2 + P^{-1}V(x_s)\right]\right\}} , \tag{37}$$

which again becomes exact for $P \to \infty$. Thus the probability distribution in the quantum system is identical to that of one particular bead in the classical isomorph. Equivalently, $P(x)$ is P^{-1} times the probability density for finding *any* bead at position x.

2.7 High and low temperatures

In thermal equilibrium, the springs in the cyclic chain have some mean extension, and the chain as a whole therefore spreads out. This spreading has a simple intuitive correspondence to zero-point motion, which it is worth pausing to consider.

Let us characterize the size of the chain by the root-mean-square (r.m.s.) radius of gyration Δ, which we define as

$$\Delta^2 = P^{-1} \langle \sum_{s=1}^{P} \Delta x_s^2 \rangle , \tag{38}$$

where

$$\Delta x_s = x_s - \bar{x} , \tag{39}$$

with $\bar{x}$ the instantaneous centre of mass of the chain:

$$\bar{x} = P^{-1} \sum_{s=1}^{P} x_s . \tag{40}$$

For free particles, it is a simple matter to obtain a formula for Δ (see Appendix A.1). In the limit $P \to \infty$, we obtain:

$$\Delta^2 = \beta\hbar^2/12m = \pi\Lambda^2/6 , \tag{41}$$

so that Δ is just the thermal wavelength Λ, apart a numerical factor.

One sees from this that as we go to the classical limit (high temperature, large mass, or small $\hbar$), the chain shrinks and looks just like a point (classical) particle, as we might expect. Conversely, as we go to the quantum limit (low temperature, small mass, large $\hbar$), the chain spreads out and, in the case of free particles, becomes infinitely extended.

If the particle is confined by a potential, then the low-temperature spreading is limited, and Δ goes to a constant for $T \to 0$. This can be examined analytically for the harmonic oscillator

(Appendix A.1), for which in the $P \to \infty$ limit:

$$\Delta^2 = (\beta m\omega_0^2)^{-1} [\tfrac{1}{2}\beta\hbar\omega_0 \coth(\tfrac{1}{2}\beta\hbar\omega_0) - 1] \,, \tag{42}$$

where ω_0 is the oscillator frequency. As $T \to 0$, we obtain:

$$\Delta^2 \to \hbar/2m\omega_0 \,, \tag{43}$$

which will be recognized as the mean-square amplitude of vibration in the ground state. In fact, it is worth noting (Appendix A.2) that in the harmonic oscillator, the probability distribution of the chain centroid is identical to that of the classical oscillator at the same temperature (this is not true for other potentials). In this case, then, Δ^2 is precisely the quantum contribution to the mean square displacement.

More generally, it is clear that the penetration of the quantum particle into regions of high potential is enhanced by the fact that each bead in the chain feels only $1/P$ of the full potential [4].

One of the attractive features of the path-integral approach is that it makes it particularly easy to go to the classical limit. Consider, for example, the probability distribution. Because of the contraction of the chain as $T \to \infty$, we get contributions to the integrals in equation (37) only when all x_s are essentially the same, so that we get the classical distribution:

$$P(x_1) \to \exp[-\beta V(x_1)] \Big/ \int dx \, \exp[-\beta V(x)] \,. \tag{44}$$

2.8 The momentum distribution

Before leaving the case of the single particle, it will be useful to consider the momentum distribution. In classical theory, this is given by the familiar Maxwellian distribution, but in quantum theory this is no longer so.

In fact, positions and momenta have an entirely equal status – they are conjugate variables – and the momentum distribution can be discussed in the same way as the position distribution. It is natural to define the density matrix in momentum space as

$$\hat{\rho}(p,p';\beta) = \sum_\lambda \hat{\phi}_\lambda(p) e^{-\beta E_\lambda} \hat{\phi}_\lambda^*(p') \,, \tag{45}$$

where $\hat{\phi}_\lambda(p)$ are the eigenfunctions of the Hamiltonian expressed in momentum space; they are related to the $\phi_\lambda(x)$ by Fourier transformation:

$$\hat{\phi}_\lambda(p) = (2\pi\hbar)^{-\frac{1}{2}} \int dx \, e^{ipx/\hbar} \phi_\lambda(x) \,. \tag{46}$$

The momentum distribution is then given by

$$\hat{P}(p) = Z^{-1} \hat{\rho}(p,p;\beta) \,, \tag{47}$$

which has exactly the same form as $P(x)$ (equation (15)). On inserting equation (46) into equation (45), we obtain:

$$\begin{aligned} \hat{P}(p) &= Z^{-1} \sum_\lambda \frac{1}{2\pi\hbar} \int dx_1 dx_2 \, e^{ip(x_1-x_2)/\hbar} \phi_\lambda(x_1) e^{-\beta E_\lambda} \phi_\lambda^*(x_2) \\ &= (2\pi\hbar Z)^{-1} \int dx_1 dx_2 \, e^{ip(x_1-x_2)/\hbar} \rho(x_1,x_2;\beta) \,, \end{aligned} \tag{48}$$

so that the momentum distribution is essentially the Fourier transform of $\rho(x,x';\beta)$ with respect to the difference of the positions. It is an instructive exercise to check this by using it to rederive the known result for $P(p)$ for free particles (see Appendix A.3).

2.9 Generalization to many particles

Our real interest lies, of course, not in a single particle in one dimension, but in systems of many particles in three dimensions. The required generalization is straightforward. To see this, it is enough to consider two particles in three dimensions. Let the eigenvalues and eigenfunctions of the Hamiltonian be called E_λ and $\phi_\lambda(\mathbf{x},\mathbf{y})$, where $\mathbf{x}$ and $\mathbf{y}$ denote the positions of the two particles. Then the definition of the density matrix is the natural generalization of equation (13):

$$\rho(\mathbf{x},\mathbf{y};\mathbf{x}',\mathbf{y}';\beta) = \sum_\lambda \phi_\lambda(\mathbf{x},\mathbf{y}) e^{-\beta E_\lambda} \phi_\lambda{}^*(\mathbf{x}',\mathbf{y}') \ . \tag{49}$$

The partition function is given by

$$Z = \sum_\lambda e^{-\beta E_\lambda} = \int d\mathbf{x}d\mathbf{y}\ \rho(\mathbf{x},\mathbf{y};\mathbf{x},\mathbf{y};\beta) \ , \tag{50}$$

and the probability density for finding the first particle at $\mathbf{x}$ and the second at $\mathbf{y}$ is

$$P(\mathbf{x},\mathbf{y}) = Z^{-1}\rho(\mathbf{x},\mathbf{y};\mathbf{x},\mathbf{y};\beta) \ . \tag{51}$$

The density matrix can be decomposed as in equation (19), so that Z is given by

$$\begin{aligned} Z = \int d\mathbf{x}_1 d\mathbf{y}_1 \ldots d\mathbf{x}_P d\mathbf{y}_P\ &\rho(\mathbf{x}_1,\mathbf{y}_1;\mathbf{x}_2,\mathbf{y}_2;\beta/P)\ldots \\ &\ldots\rho(\mathbf{x}_P,\mathbf{y}_P;\mathbf{x}_1,\mathbf{y}_1;\beta/P) \ . \end{aligned} \tag{52}$$

In the presence of a potential function $U(\mathbf{x},\mathbf{y})$, the short-time approximation is

$$\begin{aligned} \rho(\mathbf{x},\mathbf{y};\mathbf{x}',\mathbf{y}';\beta/P) \simeq \left(\frac{mP}{2\pi\beta\hbar^2}\right)^3 \exp\{-&\frac{mP}{2\beta\hbar^2}[(\mathbf{x}-\mathbf{x}')^2 \\ &+ (\mathbf{y}-\mathbf{y}')^2] + \frac{\beta}{2P}[U(\mathbf{x},\mathbf{y})+U(\mathbf{x}',\mathbf{y}')]\} \ , \end{aligned} \tag{53}$$

where, as usual, we have taken the mean of U at the initial and final points.

On putting this short-time form into equation (52), we obtain:

$$\begin{aligned} Z \simeq &\left(\frac{mP}{2\pi\beta\hbar^2}\right)^{3P} \int d\mathbf{x}_1 d\mathbf{y}_1 \ldots d\mathbf{x}_P d\mathbf{y}_P \cdot \\ &\exp\left(-\beta\sum_{s=1}^{P}\left[\tfrac{1}{2}\kappa(\mathbf{x}_{s+1}-\mathbf{x}_s)^2 + \tfrac{1}{2}\kappa(\mathbf{y}_{s+1}-\mathbf{y}_s)^2 + P^{-1}U(\mathbf{x}_s,\mathbf{y}_s)\right]\right) \ . \end{aligned} \tag{54}$$

The generalization thus has the form we should expect: in the classical isomorph, we have a separate cyclic chain for each of the original quantum particles. The spring constant κ is equal to $mP/\beta^2\hbar^2$, as before. There is one important point to notice. The potential acting on each bead is, as before, $1/P$ of the full potential, but each bead interacts only with that bead in the other chain having the same label s. This is what we should expect: since s labels an imaginary time in the range $(0, -i\beta\hbar)$, the condition we have mentioned expresses the fact that the interaction is instantaneous.

The generalization to many particles is immediate: we have one cyclic chain for each quantum particle, each chain having the same number P of beads coupled by harmonic springs; each bead interacts only with beads on other chains having the same s, and does so via $1/P$ of the true potential. The partition function is given by:

$$Z = \left(\frac{mP}{2\pi\beta\hbar^2}\right)^{3NP/2} \int \prod_{i=1}^{N} \prod_{s=1}^{P} d\mathbf{r}_{is} \cdot$$
$$\cdot \exp\left\{-\beta \sum_{s=1}^{P} \left[\tfrac{1}{2}\kappa \sum_{i=1}^{N} (\mathbf{r}_{is+1} - \mathbf{r}_{is})^2 + P^{-1} U(\mathbf{r}_{1s} \ldots \mathbf{r}_{Ns})\right]\right\} , \tag{55}$$

where $\mathbf{r}_{is}$ is the position of bead s on the cyclic chain representing particle i.

2.10 Exchange

The scheme we have just outlined is correct only for distinguishable particles. For indistinguishable particles, we have to incorporate the appropriate symmetry of the wavefunctions [4]. This is essential for systems like liquid helium, where exchange plays a crucial role [13].

To explain this simply, consider again the system of two particles, which we assume to be bosons. Because the Hamiltonian is invariant under interchange of the particles, its eigenfunctions must be either symmetric or antisymmetric under interchange. To distinguish between the two, let us denote the symmetric eigenfunctions by $\phi_\lambda^s(\mathbf{x},\mathbf{y})$ and the antisymmetric ones by $\phi_\lambda^a(\mathbf{x},\mathbf{y})$. The density matrix for distinguishable particles, which we now call $\rho_D(\mathbf{x},\mathbf{y};\mathbf{x}',\mathbf{y}';\beta)$ was discussed in the previous section; its defining equation (49) can now be separated into two sets of terms:

$$\rho_D(\mathbf{x},\mathbf{y};\mathbf{x}',\mathbf{y}';\beta) = \sum_\lambda \phi_\lambda^s(\mathbf{x},\mathbf{y}) e^{-\beta E_\lambda^s} \phi_\lambda^{s*}(\mathbf{x}',\mathbf{y}')$$
$$+ \sum_\lambda \phi_\lambda^a(\mathbf{x},\mathbf{y}) e^{-\beta E_\lambda^a} \phi_\lambda^{a*}(\mathbf{x}',\mathbf{y}') \cdot \tag{56}$$

The only difference when we are dealing with bosons is that we now sum only over the symmetric states:

$$\rho_B(\mathbf{x},\mathbf{y};\mathbf{x}',\mathbf{y}';\beta) = \sum_\lambda \phi_\lambda^s(\mathbf{x},\mathbf{y}) e^{-\beta E_\lambda^s} \phi_\lambda^{s*}(\mathbf{x}',\mathbf{y}') . \tag{57}$$

But from this it follows that ρ_B can be expressed in terms of ρ_D:

$$\rho_B(\mathbf{x},\mathbf{y};\mathbf{x}',\mathbf{y}';\beta) = \tfrac{1}{2} [\rho_D(\mathbf{x},\mathbf{y};\mathbf{x}',\mathbf{y}';\beta) + \rho_D(\mathbf{x},\mathbf{y};\mathbf{y}',\mathbf{x}';\beta)] . \tag{58}$$

If the particles were fermions, we would obtain:

$$\rho_F(\mathbf{x},\mathbf{y};\mathbf{x}',\mathbf{y}';\beta) = \tfrac{1}{2} [\rho_D(\mathbf{x},\mathbf{y};\mathbf{x}',\mathbf{y}';\beta) - \rho_D(\mathbf{x},\mathbf{y};\mathbf{y}',\mathbf{x}';\beta)] . \tag{59}$$

The corresponding formulae for the partition functions follow immediately. For bosons, we have

$$Z_B = \int d\mathbf{x}d\mathbf{y}\, \rho_B(\mathbf{x},\mathbf{y};\mathbf{x},\mathbf{y};\beta)$$
$$= \tfrac{1}{2}\left[\int d\mathbf{x}d\mathbf{y}\, \rho_D(\mathbf{x},\mathbf{y};\mathbf{x},\mathbf{y};\beta) + \int d\mathbf{x}d\mathbf{y}\, \rho_D(\mathbf{x},\mathbf{y};\mathbf{y},\mathbf{x};\beta)\right] \cdot \tag{60}$$

When we express ρ_D in the usual way in terms of discrete chains, we therefore have Z_B as the sum of two terms, one of which involves the integral over configurations of two separate chains each of P beads, and the other an integral over configurations of one large chain of $2P$ beads, the whole thing being multiplied by the factor $\frac{1}{2}$. Schematically:

(61)

where the white circles indicate the positions integrated over in equation (60) and the arrows show the sense of increasing bead index ('time') s.

If we go through the same argument for three particles, we find a sum of three terms, the first involving three separate chains each of P beads, the second a chain of P beads and another of $2P$ beads, and the third a single chain of $3P$ beads, with associated factors:

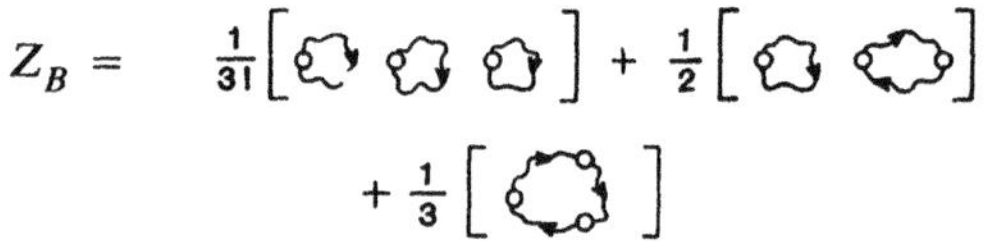

(62)

Later, it will be useful to have a name for cyclic chains of different lengths. We shall refer to the simple chain of P beads as a 1-cycle, a double-length chain of $2P$ beads as a 2-cycle, and in general an n–fold chain of nP beads as an n–cycle.

2.11 Thermal averages

So far, we have worked mainly with the partition function. But in practice, we are concerned to calculate the thermal averages of observables. The general principle is that any thermal average in the original quantum system can be re-expressed as the thermal average of some corresponding quantity in the classical isomorph. We now want to show how this works for some quantities of interest.

2.11.1 The radial distribution function.

We have already seen how to calculate the single-particle density $n_1(\mathbf{r})$, i.e. the probability density for finding a particle at position $\mathbf{r}$. According to equation (37), this is simply P^{-1} times the probability density for finding any bead in the classical isomorph at this position. In a homogeneous system such as a liquid, this quantity is simply a constant, whose value we usually know in advance. But we shall see later that $n_1(\mathbf{r})$ is of great interest when the density is non-uniform, as, for example, in metal-hydrogen systems.

Just as we derived a path-integral expression for $n_1(\mathbf{r})$, we can obtain a similar expression for the probability density $n_2(\mathbf{r},\mathbf{r}')$ of simultaneously finding a particle at $\mathbf{r}$ and another particle at $\mathbf{r}'$ [4]. To see what is involved, consider again the two-particle system. In the definition of $n_2(\mathbf{r},\mathbf{r}')$, we do not care which particle is at $\mathbf{r}$ and which is at $\mathbf{r}'$; it is therefore *twice* the probability of finding particle 1 at $\mathbf{r}$ and particle 2 at $\mathbf{r}'$, so that we have:

$$\begin{aligned} n_2(\mathbf{r},\mathbf{r}') &= 2Z^{-1} \sum_\lambda |\phi_\lambda(\mathbf{r},\mathbf{r}')|^2 \exp(-\beta E_\lambda) \\ &= 2Z^{-1} \rho(\mathbf{r},\mathbf{r}';\mathbf{r},\mathbf{r}';\beta) \cdot \end{aligned} \tag{63}$$

On going through the usual decomposition of the density matrix, we find that this is just twice the probability density, in the classical isomorph, for finding bead 1 of chain 1 at $\mathbf{r}$ and bead 1 of chain 2 at $\mathbf{r}'$. The factor of 2 can be removed again by saying that this is the probability density for finding bead 1 of either chain at $\mathbf{r}$ and bead 1 of the other chain at $\mathbf{r}'$. Since all beads are equivalent, this is P^{-1} times the probability density for finding any bead (index s) of either chain at $\mathbf{r}$ and the bead having the same s on the other chain at $\mathbf{r}'$. The generalization to many particles is immediate: the density $n_2(\mathbf{r},\mathbf{r}')$ is now equal to P^{-1} times the probability density for finding *any* bead (index s) on *any* chain at $\mathbf{r}$ and a bead of the same s on some other chain at $\mathbf{r}'$.

In a homogeneous system, it is more convenient to consider the radial distribution function $g(r)$, defined by

$$n_2(\mathbf{r},\mathbf{r}') = n_1^2 g(|\mathbf{r}-\mathbf{r}'|) \ , \tag{64}$$

with n_1, as before, the bulk (single-particle) density. From what we have just said, $g(r)$ in the quantum system is just the r.d.f. in the classical isomorph between beads having the same s on different cyclic chains. (The significance of beads having the same s, it will be recalled, is that it is only these beads that interact with each other.)

2.11.2 Average potential energy. The formula for the thermal average of the interparticle potential energy $\langle U \rangle$ in the original quantum system has exactly the same form as in a classical system:

$$\langle U \rangle = \tfrac{1}{2} \langle \sum_{i \neq j} v(|\mathbf{r}_i - \mathbf{r}_j|) \rangle = \tfrac{1}{2} N n_1 \int d\mathbf{r}\ g(r) v(r) \ . \tag{65}$$

But then, since $g(r)$ is equal to the r.d.f. between beads in the classical isomorph, it follows that $\langle U \rangle$ is exactly the thermal average of the interaction potential energy in the classical isomorph:

$$\langle U \rangle = \tfrac{1}{2} P^{-1} \langle \sum_{s=1}^{P} \sum_{i \neq j} v(|\mathbf{r}_{is} - \mathbf{r}_{js}|) \rangle \ , \tag{66}$$

where $\mathbf{r}_{is}$ is the position of bead s on chain i.

2.11.3 Average total and kinetic energies. An expression for the average total energy (i.e. the thermal average of the Hamiltonian) is conveniently obtained from the partition function, by using the Helmholtz equation [3,7,9]:

$$\langle H \rangle = \partial(\beta F)/\partial\beta = -Z^{-1} \partial Z/\partial\beta \ , \tag{67}$$

the derivative being taken at constant volume. We can calculate this derivative directly from equation (55) for Z:

$$\begin{aligned} \langle H \rangle = 3NP/2\beta - \langle \sum_{i=1}^{N} \sum_{s=1}^{P} \tfrac{1}{2}\kappa(\mathbf{r}_{is+1} - \mathbf{r}_{is})^2 \rangle \\ + P^{-1} \langle \sum_{s=1}^{P} U(\mathbf{r}_{1s} \ldots \mathbf{r}_{Ns}) \rangle \ . \end{aligned} \tag{68}$$

The last term here will be recognized as the thermal average of the potential energy discussed a moment ago. This means that the thermal average of the kinetic energy $\langle T \rangle$ is given by

$$\langle T \rangle = 3NP/2\beta - \langle \sum_{i=1}^{N} \sum_{s=1}^{P} \tfrac{1}{2}\kappa(\mathbf{r}_{is+1} - \mathbf{r}_{is})^2 \rangle \ . \tag{69}$$

The correctness of this expression is readily confirmed by using it to re-derive the mean kinetic energy of free particles, or of the harmonic oscillator (Appendix A.4).

Although formula (69) is correct, it suffers from a serious drawback as a practical method of calculating $\langle T \rangle$. Recall that the isomorphism becomes exact only in the limit $P \to \infty$. We are therefore required to calculate $\langle T \rangle$ as a difference of two large quantities, the second of which must be estimated by statistical sampling. The consequence of this is that the statistical variance will increase with P – in fact consideration of the harmonic oscillator indicates that the variance will be proportional to P [7]. This problem does not arise for the other averages we have discussed, and there has therefore been considerable interest in finding other routes to the calculation of $\langle T \rangle$. It has been shown by Herman *et al.* [7] that an exactly equivalent – but statistically preferable – expression for $\langle T \rangle$ may be obtained by invoking the virial theorem. The result is:

$$\langle T \rangle = 3N/2\beta - \tfrac{1}{2} \langle \sum_{i=1}^{N} \sum_{s=1}^{P} \delta \mathbf{r}_{is} \cdot \mathbf{F}_{is} \rangle \ , \tag{70}$$

where $\delta \mathbf{r}_{is}$ may be chosen to be

$$\delta \mathbf{r}_{is} = \mathbf{r}_{is} - \mathbf{r}_{i1} \ , \tag{71}$$

and $\mathbf{F}_{is}$ is the force exerted on bead s of chain i by beads on other chains and by the external potential (if such a potential is present). This expression has the pleasant feature that the first term gives the classical result, so that the second represents the quantum correction. Consideration of the harmonic oscillator indicates that with this formula the variance will be only weakly dependent on P [7]. In spite of this, it appears that equation (70) may not always be preferable [14,15], and the question would probably repay further study.

2.12 Dynamical quantities

Although the methods we have just discussed allow the calculation of static averages by path-integral simulation, dynamical quantities are unfortunately very much harder. By a dynamical quantity we mean anything involving a time-correlation function. The diffusion coefficient is one such quantity. The essential difficulty is easy to explain. Suppose we want the autocorrelation function $\langle A(t)A(0) \rangle$ of some dynamical variable A. The formal expression for this is:

$$\langle A(t)A(0) \rangle = Z^{-1} \operatorname{Tr} \left[e^{iHt/\hbar} A e^{-iHt/\hbar} A e^{-\beta H} \right] . \tag{72}$$

The manipulations that were applied to the Boltzmann factor $\exp(-\beta H)$ must now be applied also to the time evolution operators $\exp(\pm iHt/\hbar)$, which will be evaluated by the usual decomposition procedure:

$$K(x_0, x_P; t) = \int dx_1 \ldots dx_{P-1} \; K(x_0, x_1; t/P) \ldots K(x_{P-1}, x_P; t/P) \ . \tag{73}$$

For the matrix elements, we use the short-time approximation, as before. This is:

$$K(x, x'; t/P) = \left(\frac{mP}{2\pi i t \hbar} \right)^{1/2} \exp\left[-\frac{mP}{2it\hbar}(x-x')^2 - \frac{it}{2\hbar P}(V(x)+V(x')) \right] . \tag{74}$$

However, because the argument of the exponentials is imaginary, we are now faced with calculating a multidimensional phase integral. The large phase cancellations make this extremely difficult. Although some progress has recently been reported for simple models [16-19], the calculation of dynamical quantities for real systems remains an unsolved problem.

2.13 Simulation methods

There is a substantial literature on simulation methods for calculating thermal averages in the classical isomorph, and hence in the original quantum system [10]. What we are concerned with is statistical averages over the *configuration* space of the chain system, and any simulation method that samples this configuration space with the correct statistical weight is in principle satisfactory.

The most direct method is Metropolis Monte Carlo, which generates a canonical ensemble of configurations. However, there is no reason why the configuration space should not be sampled by molecular dynamics simulation. It is true that if we use the usual constant-energy method, we shall not obtain a canonical distribution, but this will make almost no difference for most thermal averages of interest, to order N^{-1}. However, there may sometimes be a problem with this method. For large P, the high-frequency modes of each chain will be only weakly coupled to other degrees of freedom, and equilibration will therefore be extremely slow [20]. This problem can be avoided by using instead Langevin dynamics: the normal molecular dynamics equations are modified by adding a friction term and a random force for each bead. Further details of this method are given by Gillan [12].

There is a general feature of the classical isomorph which tends to make sampling of the configuration space inefficient. This can be appreciated by considering a free chain, for which the energy function can be decomposed into independent normal modes. For large P, the spring constants (or frequencies) of these modes span a wide range. This means that if the random displacement (in Monte Carlo) or time step (in molecular dynamics) is chosen to deal effectively with the higher modes, the exploration of the configuration space of the lower modes will be painfully slow. One method that has been proposed to overcome this problem is to use normal mode coordinates as the variables [21,22]. This allows one to use different random displacements or masses for different modes. An alternative approach is the so-called 'staging' method [23]. However, it should be stressed that, for many systems, experience shows that the naive use of Monte Carlo sampling or Langevin dynamics appears to work perfectly well.

2.14 How many beads?

The equivalence of the classical isomorph to the original quantum system is only exact in the limit $P \to \infty$. In a practical simulation, we must, of course, choose some finite number of beads on each chain. How large should this be?

One answer to this question is that P must be large enough so that increasing it further makes no significant difference to the calculated averages (the meaning of 'significant' depends, of course, on the required accuracy). This means that in practice the simulations must be repeated for larger values of P, until further variations of the averages are acceptably small.

Nevertheless, it is possible to tell in advance what value of P is likely to be needed. The most important quantity here is the r.m.s. length of the springs, which we call λ. There are two separate requirements: (i) λ must be smaller than the relevant length scale of the external or interparticle potential; (ii) it must also be smaller than the r.m.s. radius of gyration Δ of the chain. In estimating λ, we note that the spring constant κ is proportional to P, so that for large P, λ will be determined almost entirely by κ, and will be little affected by the potential U. Then λ can be estimated as the r.m.s. displacement of a free oscillator of spring constant κ, so that:

$$\lambda^2 \simeq k_B T/\kappa \; . \tag{75}$$

Requirement (i) can therefore be written:

$$\kappa = mP/\beta^2\hbar^2 \gg k_B T/d^2 \tag{76}$$

or:

$$P \gg \beta\hbar^2/md^2 \ , \tag{77}$$

where d is the length scale of the potential.

In order to see the implications of requirement (ii), consider the case of a single quantum particle in a harmonic oscillator potential

$$V(x) = \tfrac{1}{2}Kx^2 \ . \tag{78}$$

We know the r.m.s. radius of gyration of the chain for this case from equation (42), so that the condition

$$\lambda^2 \lessdot \Delta^2 \tag{79}$$

takes the form:

$$k_B T/\kappa \lessdot (\beta m\omega_0^2)^{-1} \left[\tfrac{1}{2}\beta\hbar\omega_0 \coth(\tfrac{1}{2}\beta\hbar\omega_0 - 1\right] \ , \tag{80}$$

or, if we insert the formula for κ:

$$\beta\hbar^2/mP \lessdot (\beta m\omega_0^2)^{-1} \left[\tfrac{1}{2}\beta\hbar\omega_0 \coth(\tfrac{1}{2}\beta\hbar\omega_0) - 1\right] \ , \tag{81}$$

where $\omega_0 = (K/m)^{\frac{1}{2}}$ is the oscillator frequency. At high temperatures, the right hand side goes to the free-particle value $\beta\hbar^2/12m$, so the requirement is $P \gtrdot 12$. At low temperatures, the right hand side goes to $\hbar/2m\omega_0$, so that the requirement is:

$$P \gtrdot \beta\hbar\omega_0 \ . \tag{82}$$

This has a simple physical meaning. The right hand side is the ratio of the vibrational quantum to the thermal energy $k_B T$. As we expect, the required P increases as this ratio increases, and in particular goes to infinity as temperature goes to zero. The requirement can also be rewritten as

$$\beta\hbar/P \lessdot 1/\omega_0 \ . \tag{83}$$

The meaning of this can be seen by recalling the interpretation of $\beta\hbar$ as a 'time'. In this sense, $\beta\hbar/P$ plays the role of a time step in the simulation. Just as in classical molecular-dynamics simulation, then, the time step is required to be much less than the typical vibration period.

3. APPLICATIONS

In the short space available, I cannot do justice to the many interesting problems that have been studied using path-integral simulation. But in order to give an impression of the scope of the method, I list here some of the systems that have been investigated:

(1) Electron in a molten salt (Parrinello and Rahman [9])

(2) Electron in dense helium gas (Bartholomew *et al.* [24])

(3) Electron in water (Sprik *et al.* [25], Wallqvist *et al.* [26])

(4) Hydrogen and muonium in water (De Raedt *et al.* [27])

(5) Electron in water clusters (Wallqvist *et al.* [28], Thirumalai *et al.* [29])

(6) Electron in ammonia (Sprik *et al.* [30])

(7) Hydrogen in metals (Gillan [12])

(8) Liquid and solid neon and argon (Thirumalai *et al.* [31], Singer and Smith [32])

(9) Bulk water (Kuharski and Rossky [33], Wallqvist and Berne [34])

(10) Liquid helium (Ceperley and Pollock [13])

(11) The exchange frequency in solid ^{3}He (Ceperley and Jacucci [35]).

The systems fall broadly into two groups: firstly (1-7), those in which a single quantum particle is immersed in a system of atoms or molecules which are treated classically; and secondly (8-11), those in which the whole system is treated quantum mechanically.

In order to illustrate what has already been achieved by path-integral simulation, I describe in the following a small selection from the work listed above. The choice I have made is more or less arbitrary.

3.1 The electron in a molten salt

There has been interest for many years in molten mixtures of alkali metals with alkali halides, one of the reasons being that they display a metal-insulator transition as a function of concentration [36,37]. When the concentration of metal is low, it is generally accepted that the metal atom ionizes to produce a metal ion and an electron. This means that the system can be regarded as consisting of electrons dissolved in a molten salt. Discussion of the states occupied by the electrons has been based mainly on a comparison with the analogous situation in a crystal. It is well known that the incorporation of excess metal in an alkali halide crystal occurs by the formation of vacancies on the anion sublattice, and that the electrons produced by ionization of the metal atoms occupy these vacancies. The localized state of an electron in an anion vacancy is known as an F-centre [38]. Such centres are readily detected by optical absorption experiments, in which the electron is excited from its ground state. The optical absorption peak associated with the lowest excitation persists in the liquid, though somewhat broadened and shifted in frequency [39]. The indication is that the electron in the molten salt occupies a hole rather analogous to an anion vacancy.

A simulation study of an electron in molten KCl has been performed by Parrinello and Rahman [9] using the path-integral technique. Because of the high temperature involved and the large mass of the ions, quantum effects for the molten salt itself can be assumed negligible. Only the electron itself is represented by a cyclic chain, the ions being represented as normal classical point particles.

A word is in order here about the idea of a mixed system of quantum and classical particles, since it arises in many of the applications mentioned at the beginning of this section. It goes without saying that any fundamental discussion of a quantum system must be based on the proposition that the entire system is described by quantum mechanics. Logically, then, when we apply the path-integral formalism, every particle must be represented in the classical isomorph by the appropriate cyclic chain. But we have already seen that the r.m.s. radius of gyration Δ of each chain in thermal equilibrium depends on the mass of the particle. An estimate of Δ for the ions in KCl at 1000 K based on equation (41) shows that $\Delta \simeq 0.015$ Å, which is far smaller than all other length scales. Consequently, one makes only a negligible error if all the beads on each cyclic chain for the ions are regarded as being concentrated at a single point. But this is just equivalent to representing the ions by classical point particles. This kind of argument gives a perfectly precise sense to the notion of a mixed quantum-classical system, at least when we are considering static averages in thermal equilibrium.

Parrinello and Rahman modelled the interaction $V_{\alpha\beta}(r)$ between ions of types α and β by a Born-Mayer potential:

$$V_{\alpha\beta}(r) = z_\alpha z_\beta e^2/r + A_{\alpha\beta} \exp(-r/\rho_{\alpha\beta}) , \tag{84}$$

where z_α is the charge on ions of type α in units of the elementary charge e, and the parameters $A_{\alpha\beta}$ and $\rho_{\alpha\beta}$ have the numerical values given by Fumi and Tosi [40]. The potential between the electron and the anions is taken to be purely Coulombic:

$$V_{e-}(r) = e^2/r , \tag{85}$$

the reasoning being that this Coulombic repulsion will keep the electron away from the anion core. The electron-cation interaction evidently cannot be treated in this way, and this is modelled by an empirical pseudopotential [41]:

$$\begin{aligned} V_{e+}(r) &= -e^2/R , \quad r < R \\ &= -e^2/r , \quad r > R , \end{aligned} \tag{86}$$

where R characterizes the cation core radius. For most of the calculations, R was taken equal to 1.96 Å. The total potential energy of the classical isomorph is thus:

$$\begin{aligned} U_{\text{iso}} = &\sum_{s=1}^{P} \frac{mP}{2\beta^2\hbar^2}(\mathbf{r}_{s+1}-\mathbf{r}_s)^2 + P^{-1} \sum_{\alpha} \sum_{i=1}^{N_\alpha} \sum_{s=1}^{P} V_{e\alpha}(|\mathbf{r}_s-\mathbf{R}_{\alpha i}|) \\ &+ \tfrac{1}{2} \sum_{\alpha\beta} \sum_{i=1}^{N_\alpha} \sum_{j=1}^{N_\beta}{}' \, V_{\alpha\beta}(|\mathbf{R}_{\alpha i}-\mathbf{R}_{\beta j}|) , \end{aligned} \tag{87}$$

where m is the electron mass, P is the number of beads on the electron chain, $\mathbf{r}_s$ is the position of bead s on the chain, $\mathbf{R}_{\alpha i}$ is the position of ion i of type α and N_α is the number of ions of this type. The simulations were performed with the usual periodic boundary conditions, the Coulomb interactions being handled by the Ewald technique [42]. The simulated system consisted of a single electron together with 150 cations and 149 anions, in order to ensure electrical neutrality. The calculations employed constant-pressure molecular dynamics and were carried out at 1000 K.

One of the important questions studied in the simulations was the tendency of the electron to become localized. The procedure adopted to examine this was as follows. The electron chain was first artificially stretched so that it was dispersed over the entire system. With the chain constrained to this configuration, the ions were allowed to follow the normal equations of motion long enough to ensure equilibration. The constraint was then released, and the entire system was allowed to evolve freely. Very rapidly, the chain contracted until it formed a compact bundle having a radius of about 2 Å. This fully confirms the view that the electron occupies a highly localized state in molten salts.

Parrinello and Rahman also used their simulations to calculate the kinetic and potential energies of the electron, and the sum of the two, which they refer to as the binding energy. The calculations were done for a wide range of bead numbers P, and are instructive from the point of view of the methodology of path-integral simulation. We pointed out in § 2.11.3 that the kinetic energy $\langle T \rangle$ can be calculated from two separate but equivalent formulae, which are expected to have very different statistical properties. The kinetic energy values from the two formulae, together with the potential energy and the binding energy are reproduced in table 1. The results for $\langle T \rangle$ indicate that, at least for this system, the virial formula (equation (70)) is definitely superior. The statistical variance is independent of P for this formula, but increases markedly with P for the other formula, as expected. The variance on the potential energy seems to be independent of P. It is also clear that the results have converged satisfactorily for P greater than about 20. It is interesting to note that this is what would be expected from the criteria given in § 2.14. The criterion (77) based on the ratio of the r.m.s. spring length to the length scale d of the potential, if we set d equal to the core radius R in the electron-cation interaction, gives $P \gtrsim 20$. The criterion based on the ratio of excitation energy to thermal energy can be evaluated by setting $\hbar\omega_0$ equal to the energy of the main absorption peak, which is about 1.3 eV [39]. This gives $P \gtrsim 15$, which is almost the same as the length-scale criterion, as it happens.

Systems consisting of electrons dissolved in various solvents are of considerable interest and importance, and the list given at the beginning of this section provides several more examples of path-integral simulation in this area.

P	K.E. eqn. (69)	K.E. eqn. (70)	P.E.	B.E.
12	0.091 (0.008)	0.077 (0.024)	−0.549 (0.068)	−0.462
25	0.100 (0.022)	0.103 (0.026)	−0.417 (0.054)	−0.304
50	0.109 (0.034)	0.108 (0.024)	−0.401 (0.050)	−0.283
100	0.120 (0.060)	0.111 (0.029)	−0.396 (0.062)	−0.275
201	0.194 (0.093)	0.086 (0.026)	−0.410 (0.064)	−0.314

TABLE 1. Path-integral results for the kinetic energy (K.E.), potential energy (P.E.) and binding energy (B.E.) of an electron dissolved in molten KCl from ref. [9]. Energies are given in Ry units and the statistical variance is shown in brackets.

3.2 Hydrogen in metals

It has been well known for over a hundred years that hydrogen dissolves readily in transition metals [43] (for reviews, see refs. [44-47]). Metal-hydrogen systems are of great technological importance, both because of applications such as hydrogen storage and the passivation of donors and acceptors in semiconductors, and also because of the problems that can be caused by the presence of hydrogen, such as embrittlement and swelling. The light mass of hydrogen means that its typical vibrational frequencies are high – generally much higher than those of the metal lattice – and this implies that quantum effects will be very important, except at high temperatures. Because of this, it has proved very difficult to develop reliable quantitative theories. The arrival of path-integral simulation opens up completely new avenues for theoretical study of these systems. I will summarize here some of the results that have emerged from recent exploratory simulations.

Before discussing the simulations, it will be useful to mention briefly some relevant features of metal-hydrogen systems. In solution, hydrogen occupies well-defined interstitial sites (figure 1), which can be determined by neutron-diffraction and channelling experiments [48,49]. The frequencies of vibration of the hydrogen about these sites can be measured by inelastic neutron scattering [50]. Representative values of the measured frequencies are given in table 2. It can be seen from this that the vibrational quantum divided by k_B is typically in the range 1000 – 2000 K, so that quantum effects should certainly be dominant at room temperature and below. Diffusion of the hydrogen occurs by transitions between the interstitial sites. The temperature-dependent diffusion coefficient has been measured in many metals by a variety of techniques [51]. As an illustration, results are shown for the isotopes of hydrogen in niobium in figure 2. A number of points are noteworthy. Firstly, D rises to very high values (of order 10^{-4} $cm^2\ sec^{-1}$) at high temperatures. Such values are even somewhat greater than typical diffusion coefficients for atoms in liquids. Secondly, D follows the Arrhenius law $D \sim \exp(-\beta \Delta E)$ over

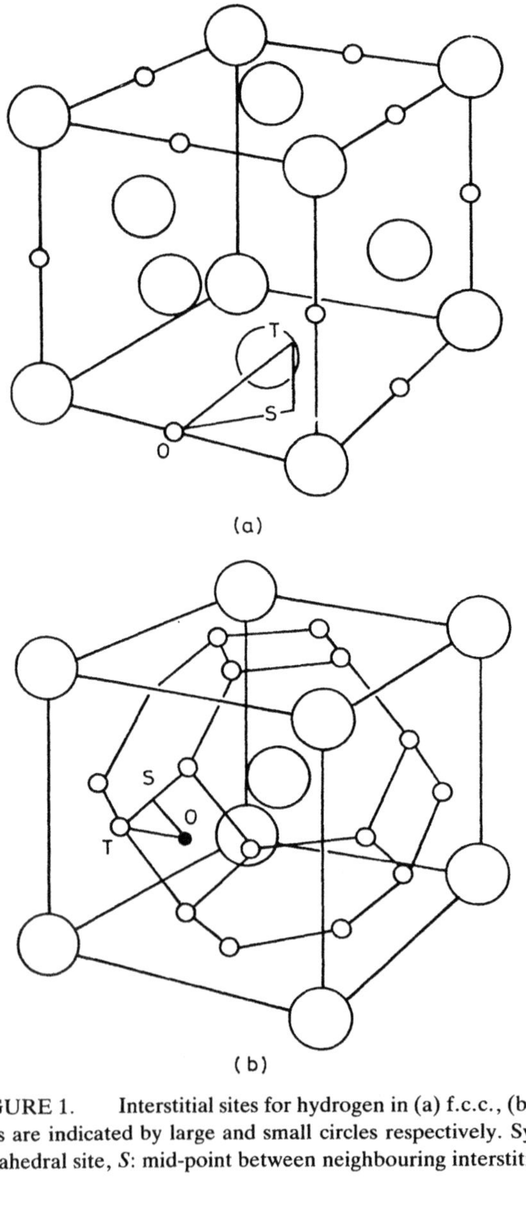

FIGURE 1. Interstitial sites for hydrogen in (a) f.c.c., (b) b.c.c. metals. Metal atoms and interstitial sites are indicated by large and small circles respectively. Symmetry points are: O: octahedral site, T: tetrahedral site, S: mid-point between neighbouring interstitial sites.

considerable temperature ranges. This indicates that transitions between sites are thermally activated. Examples of measured activation energies ΔE are given in table 2. Thirdly, one notes a substantial change of ΔE in Nb at about room temperature, an effect also observed in some other b.c.c. metals. This indicates a qualitative change in the character of the transition process at this temperature. Fourthly, the diffusion coefficients of the different isotopes differ but little above the transition, but differ greatly below it. This is a strong indication of the importance of quantum effects at low temperatures: in classical mechanics, we should expect D to depend on the isotopic mass roughly as $m^{-\frac{1}{2}}$, and much larger variations would be very hard to account for.

Metal	Structure	$\hbar\omega_0$ (eV)	ΔE (eV)
Pd	f.c.c.	0.066	0.23
Ni	f.c.c.	–	0.41
Nb	b.c.c.	0.11, 0.18	0.068, 0.106
Ta	b.c.c.	0.12, 0.17	0.040, 0.140
V	b.c.c.	0.12, 0.17	0.045

TABLE 2. Vibration frequencies and activation energies for hydrogen in various metals. In the b.c.c. metals, there are two independent frequencies, because of the non-cubic symmetry of the interstitial site. Where two activation energies are given, they refer to the low- and high-temperature regions respectively. Values are taken from refs. [50] and [51].

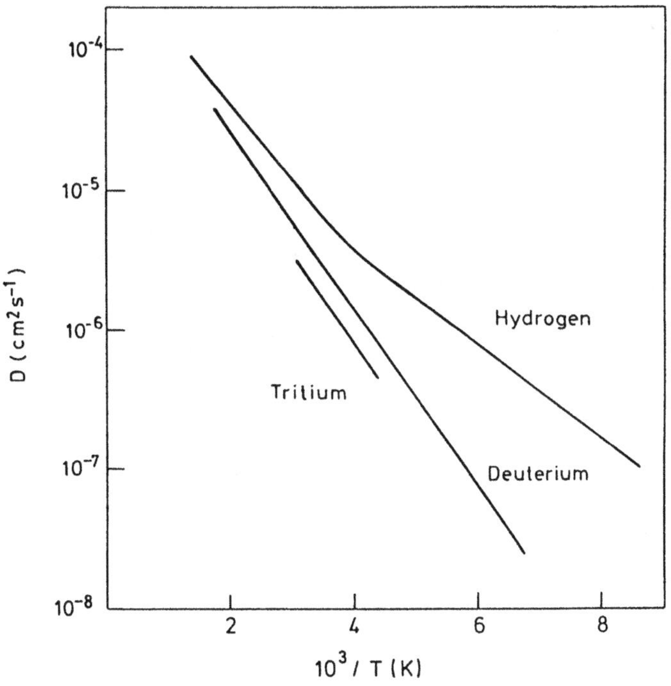

FIGURE 2. Experimental diffusion coefficient for isotopes of hydrogen in Nb, from ref. [51].

In fact, the standard theory of hydrogen diffusion, due to Flynn and Stoneham [52], assumes that intersite transitions occur entirely by tunnelling through energy barriers. The assumption is that at low temperatures, the vibrational motion is entirely frozen out, so that the hydrogen is in the vibrational ground state on each site. Transitions are taken to occur by tunnelling between these ground states. This kind of description has proved very difficult to generalize to high temperatures, and particularly as one approaches the classical limit, because of the rapidly increasing number of excited quantum states that come into play.

As a final preliminary remark, we note that a crucial factor in understanding hydrogen in metals is known to be the effect of lattice relaxation. The interaction between hydrogen and a metal atom is predominantly repulsive, and this causes a strong relaxation of the lattice surrounding an occupied interstitial site. The relaxation energy, which is known to be of order 0.3 eV, is one of the largest energies in the problem. Lattice relaxation plays an essential role in the diffusion process. This many-body aspect of the problem is one of the main things that makes it difficult, and also makes it particularly suitable for study by simulation methods.

The path-integral simulations that we now describe were performed on a model for hydrogen in niobium [12]. When hydrogen dissolves in a metal, it loses its electron, which becomes one of the itinerant electrons in the metal. When we speak of hydrogen, then, we are really referring to a proton, whose charge is screened by the surrounding sea of electrons. The electrons are, of course, not treated explicitly: the simulations are based on a form for the total energy as a function of the positions $\mathbf{R}_i$ of the metal atoms and the position of the screened proton. The energy of the metal subsystem $U(\{\mathbf{R}_i\})$ is represented in the form developed by Finnis and Sinclair [53] and other authors [54], namely:

$$U_M(\{\mathbf{R}_i\}) = -A \sum_i f(\rho_i) + \tfrac{1}{2} \sum_{i \neq j} V(|\mathbf{R}_i - \mathbf{R}_j|) \ , \tag{88}$$

where V is a short-range pair potential and the 'densities' ρ_i are themselves a sum of pair terms:

$$\rho_i = \sum_{j(\neq i)} \phi(|\mathbf{R}_i - \mathbf{R}_j|) \ . \tag{89}$$

For an explanation of this energy function, including the form of the various functions $\phi(r)$ and $V(r)$, the reader is referred to the paper of Finnis and Sinclair, which also presents a parametrization for a number of transition metals, including niobium. These parameters were slightly modified for the simulation work, as described by Gillan [12].

The metal-hydrogen interaction is represented by a simple exponential $A_H \exp(-r/\sigma)$, and the two parameters here were fixed by fitting to the two measured hydrogen vibration frequencies (there are two independent frequencies because of the tetragonal symmetry at the occupied interstitial site in niobium).

The path-integral simulations have been used to study the probability distribution $n_1(\mathbf{r})$ for hydrogen as a function of position $\mathbf{r}$ in the unit cell. As explained in § 2.6, this can be obtained from the probability distribution for the beads of the cyclic chain. Calculations were made for the isotopes H and D and for the positive muon μ^+ in niobium at 300 K, the simulations being performed on a system of 54 Nb atoms and 5 quantum particles in the case of H and D, and a single quantum particle in the case of μ^+. Bead numbers $P = 20$ were found to be sufficient for H and D and $P = 80$ for μ^+. Results for $n_1(\mathbf{r})$ are shown in figure 3, where for comparison we also give simulation results in the classical limit, which corresponds to $P = 1$. It should be recalled that in the classical limit, static distributions like $n_1(\mathbf{r})$ are rigorously independent of the masses of the particles, so that the differences between the curves are entirely due to

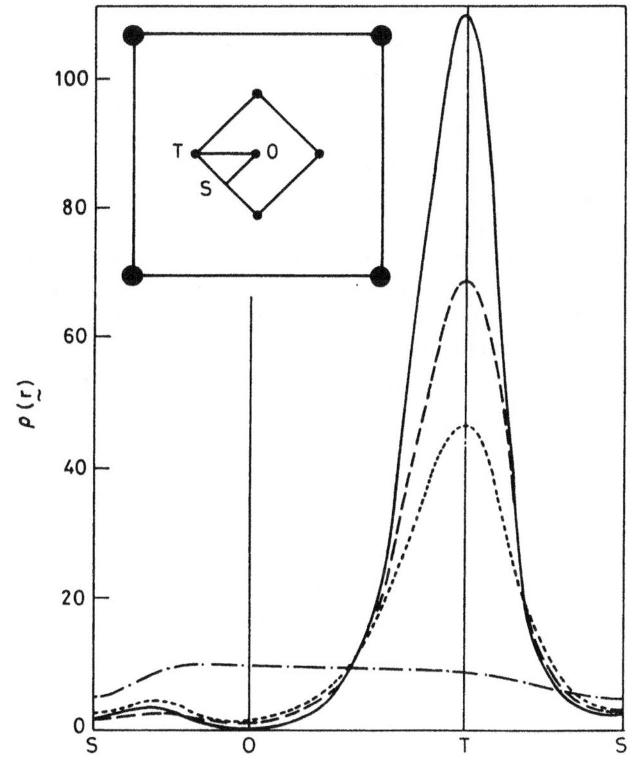

FIGURE 3. Simulated density distribution (normalized to the bulk density) for D (dashed line), H (dotted line) and μ^+ (chain line) in Nb at 300 K; solid line shows the classical prediction. Inset: points *T*, *O* and *S* on a face of the b.c.c. cube.

quantum effects. These effects are substantial for D and very important for H. For μ^+, the effects are so huge that the distribution is spread almost uniformly over a large region of the unit cell.

We expect that results of this kind will be important because $n_1(\mathbf{r})$ can be probed by neutron diffraction experiments [48], though measurements of sufficient accuracy have not yet been performed. The distribution $n_1(\mathbf{r})$ depends in a rather direct way on the metal-hydrogen interaction. A comparison of quantum simulation results with diffraction measurements thus offers a new way of investigating the metal-hydrogen interaction.

We turn now to the diffusion coefficient *D*. As we have seen in § 2.12, no satisfactory method has yet been discovered for calculating dynamical quantities by path-integral simulation, so that rigorous results for *D* cannot yet be obtained.

In spite of this, path-integral simulation can be used to obtain an estimate for the temperature variation of *D*. The reason why this is possible can be seen by considering the transition rate Γ between sites according to classical mechanics. Conventional transition-state theory shows that Γ follows approximately the Arrhenius law $\Gamma = \Gamma_0 \exp(-\beta\Delta E)$, and that ΔE is the difference of energy for the fully-relaxed system at the saddle point separating stable sites and at the stable site itself. This is equivalent to saying that Γ is proportional to the ratio between the probability for finding the system at the saddle point and the probability for finding it at the stable site, this ratio being proportional to the Boltzmann factor $\exp(-\beta \Delta E)$. The implication is that the transition rate (a dynamical quantity) is closely related to the energy difference, or the probability ratio (time-independent quantities).

What is needed, then, is a quantum generalization of classical transition-state theory. We have argued elsewhere [12] that the appropriate generalization involves the ratio Q of probabilities for finding the *centroid* of the quantum chain at the saddle point and at the stable site. For details of the argument, the reader should consult Gillan [12], where a technique for calculating this ratio in path-integral simulation is also described. Such a calculation yields not D or Γ themselves, but only something proportional to these quantities, the proportionality factor being weakly dependent on temperature.

In order to explore these ideas, we performed a series of simulations on H and D in niobium over a wide range of temperatures, in which the probability ratio Q was calculated. The results, shown in figure 4, are extremely interesting. For both isotopes, there are two temperature regimes characterized by different activation energies, just as in the experiments (figure 2). The crossover temperature for H is at about 200 K, which is quite close to the experimental value. The activation energies, too, are in semiquantitative agreement with experiment. In the high temperature regime studied, ΔE for H and D is 0.084 and 0.094 eV respectively, compared with the experimental values 0.106 and 0.127 [51]. At low temperatures, the calculated activation energy is 0.041 eV, compared with the experimental value of 0.068 eV [51].

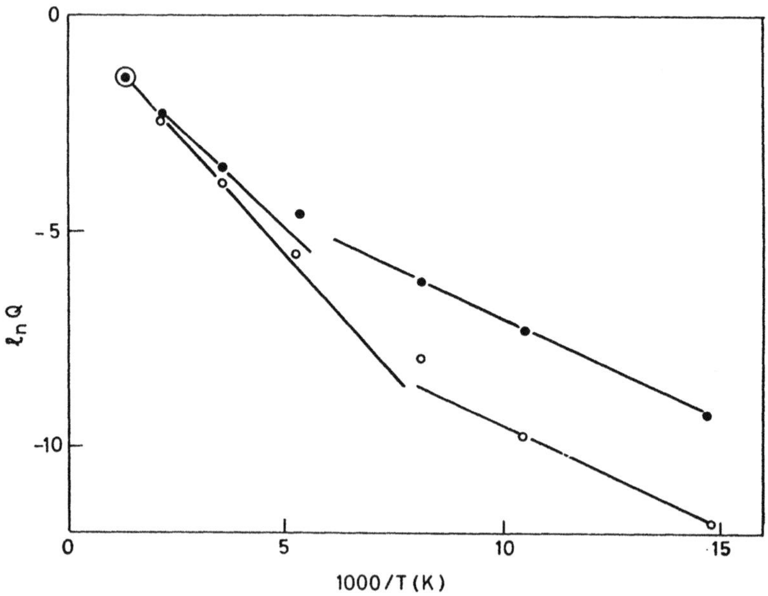

FIGURE 4. Ratio Q of probabilities for finding hydrogen at the saddle point and at the stable site, from simulations of H (filled circles) and D (open circles) in Nb. Slopes of the straight lines give the activation energies for diffusion at low and high temperatures.

Why does this transition occur? From the path-integral viewpoint, the simulations show that it is associated with a qualitative change in the typical conformation of the quantum chain when its centroid is at the saddle point. At high temperatures, where the r.m.s. radius of gyration Δ is small, the whole chain is concentrated at the saddle point. But as the temperature is lowered, Δ begins to increase rapidly and the chain extends down the two sides of the potential barrier, until at very low temperatures the chain spreads across the entire region connecting the two stable sites. We interpret this spreading as a manifestation of intersite tunnelling. The meaning is thus that the observed crossover is associated with a change from quantum tunnelling through the barrier to classical hopping over the barrier.

It should be stressed that this quantum simulation work on hydrogen in metals is still only in the early stages, but the results obtained so far are encouraging, and it seems that the path-integral method has much to offer in this area.

3.3 Liquid ^{4}He

Of all quantum systems, liquid helium-four must certainly rank as one of the most fascinating, because of its spectacular superfluid properties. The path-integral simulations of Ceperley and Pollock [13,35] on this system therefore merit special attention. The superfluidity is closely linked with the condensation of a macroscopic fraction of particles into the zero-momentum state, and arises because of the Bose statistics of the ^{4}He atoms; this is a case where exchange must be included in the path-integral formulation. We have seen how to do this in § 2.10, but we need now to indicate how Bose condensation manifests itself in this approach [4].

Recall first that for distinguishable quantum particles, where each particle corresponds to a chain of P beads, the partition function is given as an integral over all configurations of these chains. For indistinguishable particles, we have to consider also all ways the chains can be re-linked into a mixture of 1-cycles, 2-cycles, etc., where an n–cycle is a composite chain made by stringing together segments, each of which contains P beads. There is a close analogy here with a system in which chemical association can occur. To take a familiar example, consider a gaseous system of iodine and hydrogen molecules, which can associate to form hydrogen iodide molecules:

$$H_2 + I_2 \rightleftharpoons 2HI \ . \tag{90}$$

If we want to write down the partition function Z for such a system, we can express it as a sum of contributions, which are classified according to the numbers of H_2 and I_2 molecules and the number of molecules HI. From this way of writing Z, we can derive the law of mass action, which relates the concentrations of the various species:

$$[HI]^2/[H_2][I_2] = K \ , \tag{91}$$

where the equilibrium constant K is determined by the free energy of the reaction. It was demonstrated by Chandler and Wolynes [4] that this analogy is quite precise: there is a law of mass action which gives the number per unit volume ρ_n of n–cycles in the system of cyclic chains for a Bose system.

For non-interacting bosons, the equilibrium constant governing the ρ_n can be calculated analytically. It is convenient to define

$$K_n = \rho_{n+1}/\rho_n\rho_1 \ , \tag{92}$$

which determines the ratio of densities of $(n+1)$–cycles and n–cycles. It is readily shown that K_n can be expressed as

$$K_n = z_{n+1}/z_n z_1 \ , \tag{93}$$

where z_n is the *internal* partition function of an n–cycle, i.e. the integral of its Boltzmann factor over the coordinates of its nP beads, but excluding centre-of-mass translation, together with the appropriate symmetry factor (§ 2.10). These z_n are thus given by

$$z_n = \frac{1}{n}\rho(\mathbf{r},\mathbf{r};n\beta) \ . \tag{94}$$

From equation (53), we have

$$\rho(\mathbf{r},\mathbf{r};\beta) = (m/2\pi\beta\hbar^2)^{\frac{3}{2}} = \Lambda^{-3} \ , \tag{95}$$

from which we obtain

$$\rho_{n+1}/\rho_n = \left(\frac{n}{n+1}\right)^{\frac{5}{2}} \rho_1 \Lambda^3 \ . \tag{96}$$

It follows from this that as $\rho_1\Lambda^3 \rightarrow 1$, the ratio goes to unity for large n–cycles, which implies that we get cycles of macroscopic length. This is the path-integral manifestation of Bose condensation. To see that this agrees with the usual criterion for Bose condensation in the non-interacting system, note that equation (96) is equivalent to

$$\rho_n = n^{-\frac{5}{2}}(\rho_1\Lambda^3)^{n-1}\rho_1 \ , \tag{97}$$

so that for $\rho_1\Lambda^3 = 1$, the number density of n–cycles is $n^{-\frac{5}{2}}\rho_1$. Hence the total number of particles per unit volume, which we call ρ_0, is given by

$$\rho_0 = \sum_{n=1}^{\infty} n\rho_n = \rho_1 \sum_{n=1}^{\infty} n^{-\frac{3}{2}} = 2.612\rho_1 \ . \tag{98}$$

Bose condensation therefore occurs when the thermal wavelength is given by

$$\rho_0\Lambda^3 = 2.612 \ , \tag{99}$$

which is the usual result (see e.g. [55]).

Exactly why this corresponds to Bose condensation can be seen more clearly by considering the momentum distribution of the particles. We noted in § 2.8 that for a single-particle system the momentum distribution is the Fourier transform of the density matrix with respect to the difference of the two positions. Now for an N–particle system the full density matrix involves N pairs of positions. The single-particle momentum distribution is the Fourier transform of the quantity $n(r)$ defined by

$$n(r) = \frac{\int d\mathbf{r}_1 \ldots d\mathbf{r}_N \ \rho(\mathbf{r}_1,\mathbf{r}_2\ldots\mathbf{r}_N;\mathbf{r}_1+\mathbf{r},\mathbf{r}_2\ldots\mathbf{r}_N;\beta)}{\int d\mathbf{r}_1\ldots d\mathbf{r}_N \ \rho(\mathbf{r}_1\ldots\mathbf{r}_N;\mathbf{r}_1\ldots\mathbf{r}_N;\beta)} \ . \tag{100}$$

This corresponds to one open chain beginning at $\mathbf{r}_1$ and ending at $\mathbf{r}_1 + \mathbf{r}$, the remaining chains being closed. As $\rho_1\Lambda^3$ approaches unity, the mean number of segments in the open chain diverges, and $n(r)$ decays at large r more and more slowly. It can be shown that for $\rho_1\Lambda^3 > 1$,

$n(r)$ goes to a constant n_0 as $r \rightarrow \infty$. When we Fourier transform, this constant yields a δ-function at zero momentum, which signifies a macroscopic number of particles in the zero-momentum state, and it turns out that n_0 is precisely the 'condensate fraction', i.e. the fraction of particles in the zero-momentum state.

We turn now to the simulations themselves. The interaction between helium atoms was modelled by a pair potential due to Aziz *et al.* [56]; this is a carefully constructed potential, which accurately reproduces experimental thermodynamic and transport data for helium gas, and also agrees closely with *ab initio* calculations at small and large separations. The

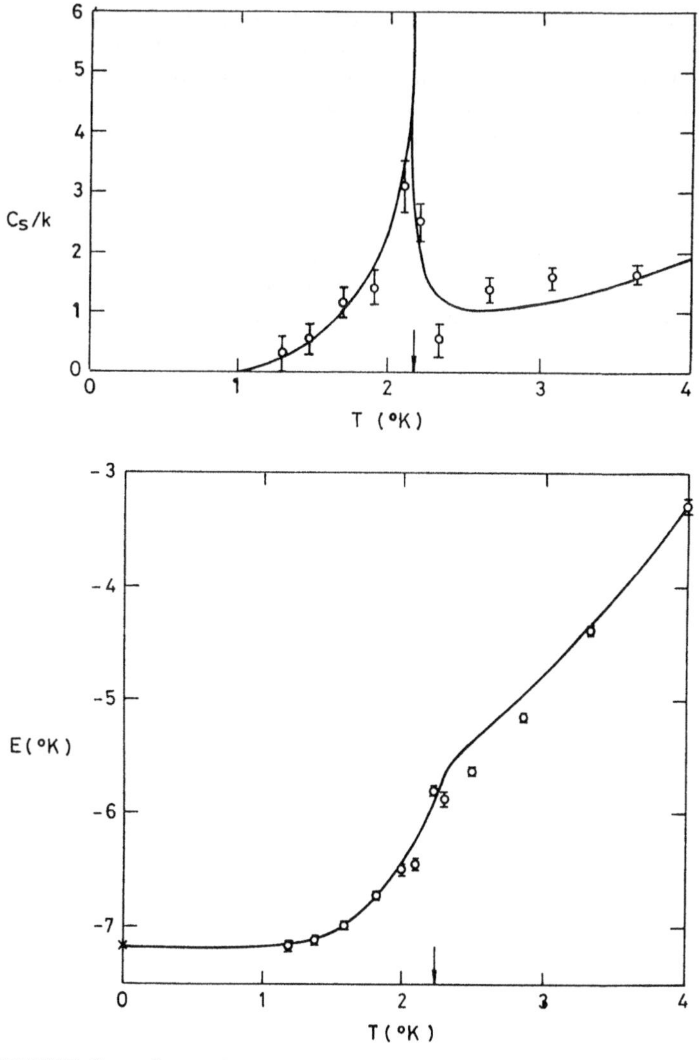

FIGURE 5. Internal energy E and specific heat C_s of liquid helium near T_λ (arrow), from simulations of Ceperley and Pollock [13]. Solid lines and circles show experimental and simulation results respectively.

calculations were perfomed on a system of 64 atoms for a range of temperatures from 1 K to 4 K, which spans the experimental λ–temperature of 2.17 K. Special sampling techniques were developed for constructing the thermal-equilibrium ensemble of chain linkages; these are described in ref. [13].

The computed internal energy and specific heat compared with experimental results are shown in figure 5. The reproduction of the observed thermodynamic singularity at the λ–point is quite remarkable, although there are slight deviations near the transition itself, which the authors attribute to finite-size effects.

The simulations were also used to calculate the off-diagonal density matrix $n(r)$ and hence the condensate fraction n_0, which is shown as a function of temperature in figure 6. It is noticeable that n_0 appears to deviate from zero appreciably above T_λ, but this is attributable to the difficulty of determining the true asymptotic value of $n(r)$ in a relatively small system. The low-temperature value of n_0 is about 9 %, which is within a percent or so of the value deduced from neutron-scattering experiments [57].

4. CONCLUSION

Much has already been achieved with path-integral simulation. The three examples discussed in the previous section illustrate how it has opened up previously intractable problems to quantitative study. But much also remains to be achieved, and I conclude this

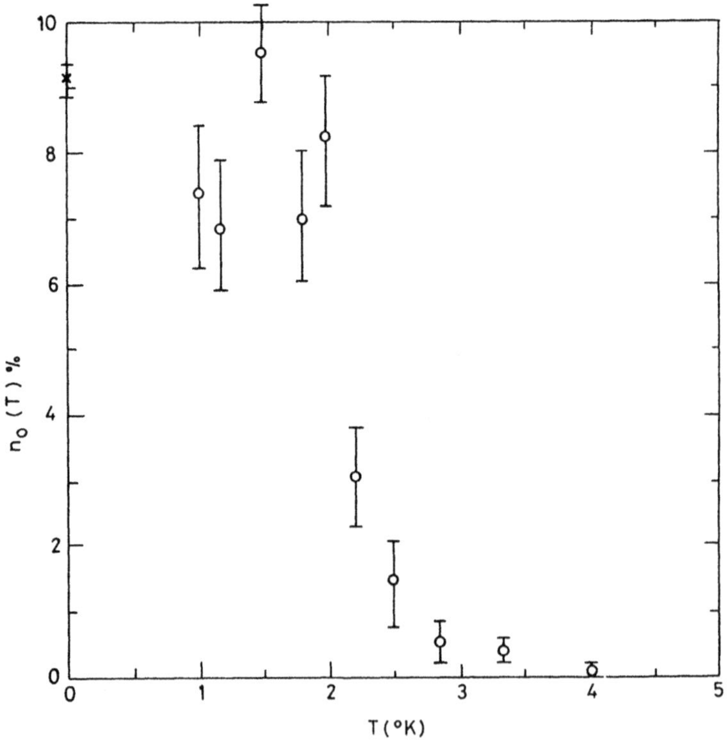

FIGURE 6. Simulation results of Ceperley and Pollock [13] for condensate fraction n_0 as a function of temperature in liquid helium.

survey by pointing to two of the major problems facing quantum simulation. The first has already been mentioned in § 2.12: that of calculating dynamical quantities. The other is that of treating fermion systems. Both are very difficult, and in both cases the difficulty stems from massive cancellation.

There has been recent encouraging progress with the calculation of time correlation functions, based on ideas of Miller, Filinov, Doll and others, and there is hope that it may soon be possible to calculate dynamical quantities at least for simple idealized (but non-trivial) systems. The importance of being able to calculate such quantities can be illustrated from the work on hydrogen in metals. The hydrogen diffusion coefficient is one of the most important quantities, but the path-integral technique is at present capable of giving only an approximate estimate for it. Quantum effects are also very important for the dynamics of protons in many other problems, such as the diffusion of hydrogen ions in water, or the rates of chemical reactions involving proton transfer. The same is even more true of problems involving electrons. An example here is the much studied $Fe^{2+} - Fe^{3+}$ reaction in aqueous solution. The discovery of an effective technique for calculating dynamical quantities would open up important new possibilities.

The difficulty with fermion systems arises from the effect of exchange. As we have seen in § 2.10, contributions from different linkages of the cyclic paths enter with opposite signs, so that the different contributions cancel each other to a large extent. This means that thermal averages are completely overwhelmed by statistical noise. The importance of progress in this area cannot be overestimated. To see this, one has only to think, for example, of the implications for high-temperature superconductivity of an effective technique for simulating electrons in ceramics.

ACKNOWLEDGMENT

The preparation of this paper has been supported in part by the Underlying Research Programme of the United Kingdom Atomic Energy Authority.

REFERENCES

[1] R. P. Feynman, *Rev. Mod. Phys.,* **20**, 367 (1948)

[2] L. D. Fosdick, *J. Math. Phys.,* **3**, 1251 (1962); L. D. Fosdick and H. F. Jordan, *Phys. Rev.,* **143**, 58 (1966); *ibid.,* **171**, 128 (1968)

[3] J. Barker, *J. Chem. Phys.,* **70**, 2914 (1979)

[4] D. Chandler and P. G. Wolynes, *J. Chem. Phys.,* **74**, 7 (1981)

[5] D. M. Ceperley and M. H. Kalos, in *Monte Carlo Methods in Statistical Physics,* ed. K. Binder, Berlin: Springer-Verlag, p. 145 (1981)

[5] K. S. Schweizer, R. M. Stratt, D. Chandler and P. G. Wolynes, *J. Chem. Phys.,* **75**, 1347 (1981)

[7] M. F. Herman, E. J. Bruskin and B. J. Berne, *J. Chem. Phys.,* **76**, 5150 (1982)

[8] M. H. Kalos (ed.), *Monte Carlo Methods in Quantum Problems,* Dordrecht: Reidel (1984)

[9] M. Parrinello and A. Rahman, *J. Chem. Phys.*, **80**, 860 (1984)

[10] B. J. Berne and D. Thirumalai, *Ann. Rev. Phys. Chem.*, **37**, 401 (1986)

[11] R. P. Feynman and A. R. Hibbs, *Quantum Mechanics and Path Integrals,* New York: McGraw Hill (1965); R. P. Feynman, *Statistical Mechanics,* New York: Benjamin (1972)

[12] M. J. Gillan, *Phys. Rev. Lett.*, **58**, 563 (1987); M. J. Gillan, *Phil. Mag. A,* **58**, 257 (1988)

[13] D. M. Ceperley and E. L. Pollock, *Phys. Rev. Lett.*, **56**, 351 (1986)

[14] G. Jacucci, in *Monte Carlo Methods in Quantum Problems,* ed. M. H. Kalos, Dordrecht: Reidel (1984)

[15] K. Singer and W. Smith, *Chem. Phys. Lett.*, **140**, 406 (1987)

[16] E. C. Behrman and P. G. Wolynes, *J. Chem. Phys.*, **83**, 5863 (1985)

[17] N. Makri and W. H. Miller, *Chem. Phys. Lett.*, **139**, 10 (1987)

[18] J. D. Doll, R. D. Coalson and D. L. Freeman, *J. Chem. Phys.*, **87**, 1641 (1987)

[19] J. D. Doll, D. L. Freeman and M. J. Gillan, *Chem. Phys. Lett.*, **143**, 277 (1988)

[20] R. W. Hall and B. J. Berne, *J. Chem. Phys.*, **81**, 3641 (1984)

[21] J. D. Doll, *J. Chem. Phys.*, **81**, 3536 (1984)

[22] J. D. Doll, R. D. Coalson and D. L. Freeman, *Phys. Rev. Lett.*, **55**, 1 (1985)

[23] M. Sprik, M. L. Klein and D. Chandler, *Phys. Rev. B,* **31**, 4234 (1985)

[24] J. Bartholomew, R. Hall and B. J. Berne, *Phys. Rev. B,* **32**, 548 (1985)

[25] M. Sprik, M. L. Klein and D. Chandler, *Phys. Rev. B,* **32**, 545 (1985)

[26] A. Wallqvist, D. Thirumalai and B. J. Berne, *J. Chem. Phys.*, **86**, 6404 (1987)

[27] B. de Raedt, M. Sprik and M. L. Klein, *J. Chem. Phys.*, **80**, 5719 (1984)

[28] A. Wallqvist, D. Thirumalai and B. J. Berne, *J. Chem. Phys.*, **85**, 1583 (1986)

[29] M. Sprik, R. W. Impey and M. L. Klein, *J. Chem. Phys.*, **83**, 5802 (1985)

[30] M. Sprik, R. W. Impey and M. L. Klein, *Phys. Rev. Lett.*, **56**, 2326 (1986)

[31] D. Thirumalai, R. W. Hall and B. J. Berne, *J. Chem. Phys.*, **81**, 2523 (1984)

[32] K. Singer and W. Smith, *Molec. Phys.*, in press

[33] R. A. Kuharski and P. J. Rossky, *Chem. Phys. Lett.*, **103**, 357 (1985); *J. Chem. Phys.*, **82**, 5164 (1985)

[34] D. Wallqvist and B. J. Berne, *Chem. Phys. Lett.*, **117**, 214 (1985)

[35] D. M. Ceperley and G. Jacucci, *Phys. Rev. Lett.*, **58**, 1648 (1987)

[36] M. A. Bredig, in *Molten Salt Chemistry,* ed. M. Blander, New York: Interscience (1964), p. 367

[37] W. W. Warren, in *The Metallic and Non-Metallic States of Matter,* eds. P. P. Edwards and C. N. R. Rao, London: Taylor and Francis (1985)

[38] W. B. Fowler, *Physics of Colour Centres,* London: Academic (1968)

[39] W. Freyland, K. Garbede, H. Heyer and E. Pfeiffer, *J. Phys. Chem.,* **88**, 3745 (1984)

[40] F. G. Fumi and M. P. Tosi, *J. Phys. Chem. Solids,* **25**, 31 (1964); *ibid.,* **25**, 45 (1964)

[41] V. Heine, *Solid State Phys.,* **24**, 1 (1970)

[42] M. J. Sangster and M. Dixon, *Adv. Phys.,* **25**, 247 (1976)

[43] T. Graham, *Phil. Trans. R. Soc. Lond.,* **156**, 399 (1866)

[44] G. Alefeld and J. Völkl (eds.), *Hydrogen in Metals,* Berlin: Springer-Verlag (1978)

[45] T. Ezawa, K. Yamakawa and F. E. Fujita, *Hydrogen in Metals,* Sendai: Japan Institute of Metals (1980)

[46] P. Jena and C. B. Satterthwaite (eds.), *Electronic Structure and Properties of Hydrogen in Metals,* New York: Plenum (1983)

[47] Y. Fukai and H. Sugimoto, *Adv. Phys.,* **34**, 263 (1985)

[48] G. Nelin, *Phys. Stat. Sol. (b),* **45**, 527 (1971)

[49] H. D. Carstanjen, *Phys. Stat. Sol. (a),* **59**, 11 (1980)

[50] T. Springer, in *Hydrogen in Metals,* Vol. 1, eds. G. Alefeld and J. Völkl, Berlin: Springer-Verlag (1978), p. 75

[51] J. Völkl and G. Alefeld, in *Hydrogen in Metals,* Vol. 1, eds. G. Alefeld and J. Völkl, Berlin: Springer-Verlag (1978), p. 321

[52] C. P. Flynn and A. M. Stoneham, *Phys. Rev. B,* **1**, 3966 (1970)

[53] M. W. Finnis and J. E. Sinclair, *Phil. Mag. A,* **50**, 45 (1984)

[54] M. S. Daw and M. I. Baskes, *Phys. Rev. B,* **29**, 6443 (1984)

[55] C. Kittel, *Elementary Statistical Physics,* New York: Wiley (1958), chapter 21

[56] R. A. Aziz, V. P. S. Nain, J. S. Carley, W. L. Taylor and G. T. McConville, *J. Chem. Phys.,* **70**, 4330 (1979)

[57] H. R. Glyde and E. C. Svensson, in *Methods of Experimental Physics,* Vol. 23B: Neutron Scattering, New York: Academic (1987), p. 303; T. R. Sosnick, W. M. Snow, P. E. Sokol and R. N. Silver, Los Alamos Report LA-UR-88-505 (1988)

APPENDIX

At various points in the text, we have alluded to side calculations which help to fill out the details of the argument. These calculations are collected together here, with outlines of how they can be carried through.

A.1 Radius of gyration of the chain

We show that the mean square radius of gyration Δ^2 for the quantum chain representing a free quantum particle in one dimension is, in the limit $P \to \infty$, given by equation (41), and that for an oscillator of frequency ω_0 it is given by equation (42). The two cases can be treated together by considering first the oscillator and then taking the limit $\omega_0 \to 0$ to get the free particle.

The chain in the oscillator potential $V(x) = \frac{1}{2}Kx^2$ is a purely harmonic system, and can be decomposed into normal modes. The normal mode coordinates can be taken as the ξ_n defined by

$$x_s = \sum_{n=0}^{P-1} \exp(-2\pi ins/P)\xi_n \ . \tag{A1}$$

The coordinate ξ_0 is the position of the chain centroid, and is real, but the others are complex, and the condition that x_s be real requires that

$$\xi_n{}^* = \xi_{P-n} \ , \quad 1 \leqslant n \leqslant P-1 \ . \tag{A2}$$

The square S of the radius of gyration for a given chain configuration is

$$S = P^{-1} \sum_{s=1}^{P} \Delta x_s^2 = \sum_{n=1}^{P-1} |\xi_n|^2 \ , \tag{A3}$$

the mode $n = 0$ being absent. The potential energy of the classical isomorph is

$$\begin{aligned} U_{\text{iso}} &= \tfrac{1}{2}KP^{-1} \sum_{s=1}^{P} x_s^2 + \frac{mP}{2\beta^2\hbar^2} \sum_{s=1}^{P} (x_{s+1}-x_s)^2 \\ &= \sum_{n=0}^{P-1} \left[\tfrac{1}{2}K + \frac{mP^2}{\beta^2\hbar^2}(1-\cos(2\pi n/P))\right]|\xi_n|^2 \ . \end{aligned} \tag{A4}$$

Now Δ^2 is the average of S:

$$\Delta^2 = \int dx_1 \ldots dx_P \ S \, e^{-\beta U} \Big/ \int dx_1 \ldots dx_P \ e^{-\beta U} \ . \tag{A5}$$

The integrals can be performed by changing variables from the x_s to the ξ_n. Here, we use the fact that the ξ_n are not all independent. If P is odd, everything can be written in terms of the P variables ξ_0, $\xi_1{}'$, $\xi_1{}''$, $\ldots\xi_{\frac{1}{2}(P-1)}{}'$ and $\xi_{\frac{1}{2}(P-1)}{}''$, where $\xi_n{}'$ and $\xi_n{}''$ denote real and imaginary parts. For even P, the details are slightly different. In either case, after doing the integrals we get:

$$\Delta^2 = (2\beta)^{-1} \sum_{n=1}^{P-1} \left[\tfrac{1}{2}K + \frac{mP^2}{\beta^2\hbar^2}(1-\cos(2\pi n/P))\right]^{-1} \ . \tag{A6}$$

Now take the limit $P \to \infty$. The only important contributions come from $\cos(2\pi n/P) \simeq 1$, and we can replace $(1 - \cos(2\pi n/P))$ by $2\pi^2 n^2/P^2$. Therefore:

$$\Delta^2 = (\beta\hbar^2/2\pi^2 m) \sum_{n=1}^{\infty} [n^2 + \beta^2\hbar^2 K/4\pi^2 m]^{-1} \ . \tag{A7}$$

To get the final result, equation (42), we have to use the formula

$$\sum_{n=1}^{\infty} [n^2 + \alpha^2]^{-1} = \frac{1}{2\alpha^2}[\pi\alpha \coth \pi\alpha - 1] \ . \tag{A8}$$

A.2 Distribution of chain centroid

We show that for a quantum particle in an oscillator potential, the probability distribution for the chain centroid is the same as that for a single classical particle in the potential.

The probability density for finding the chain centroid at x is

$$P_c(x) = \int dx_1 \ldots dx_P \; \delta\Big(x - P^{-1} \sum_{s=1}^{P} x_s\Big) \, e^{-\beta U} \Big/ \int dx_1 \ldots dx_P \; e^{-\beta U} \; . \tag{A9}$$

The result follows immediately by re-writing this in terms of the normal coordinates ξ_n and noting that the chain centroid position is ξ_0. The integrals involving ξ_n for $n \neq 0$ cancel top and bottom, and we get left with

$$P_c(x) = \int d\xi_0 \; \delta(x-\xi_0) \exp(-\tfrac{1}{2}\beta K \xi_0^2) \Big/ \int d\xi_0 \exp(-\tfrac{1}{2}\beta K \xi_0^2) \; , \tag{A10}$$

which is the same as for a classical oscillator. The result does not, unfortunately, extend to systems other than the oscillator.

A.3 Momentum distribution for a free particle

We show how equation (48) for the momentum distribution yields the usual Maxwellian distribution for a free particle. In this case:

$$\rho(x_1, x_2; \beta) = \left(\frac{m}{2\pi\beta\hbar^2}\right)^{\frac{1}{2}} \exp\left[-\frac{m}{2\beta\hbar^2}(x_1 - x_2)^2\right] , \tag{A11}$$

so that the partition function is

$$Z = \int dx \; \rho(x,x;\beta) = L(m/2\pi\beta\hbar^2)^{\frac{1}{2}} \; , \tag{A12}$$

where L is the length of the system. The Fourier transform appearing in equation (48) is

$$\begin{aligned} &\int dx_1 dx_2 \left(\frac{m}{2\pi\beta\hbar^2}\right)^{\frac{1}{2}} \exp\left[ip(x_1 - x_2)/\hbar - m(x_1 - x_2)^2/2\beta\hbar^2\right] \\ &\qquad = L \exp(-\beta p^2/2m) \cdot \end{aligned} \tag{A13}$$

Hence:

$$\hat{P}(p) = \frac{1}{2\pi\hbar}\left(\frac{2\pi\beta\hbar^2}{m}\right)^{\frac{1}{2}} \exp(-\beta p^2/2m) = \left(\frac{\beta}{2\pi m}\right)^{\frac{1}{2}} \exp(-\beta p^2/2m) \; , \tag{A14}$$

which is the usual Maxwellian distribution. This result holds only for free particles, of course.

A.4 Kinetic energy of an oscillator

We verify that equation (69) gives the correct result for the kinetic energy $\langle T \rangle$ of an oscillator in thermal equilibrium. As before, we call the spring constant of the oscillator K and its frequency ω_0. Equation (69) relates $\langle T \rangle$ to the potential energy of the springs in the cyclic chain. In the case of an oscillator, this is closely related to the mean square radius of gyration. To see this, note first that since the classical isomorph is purely harmonic, its mean potential energy is $\frac{1}{2}Pk_BT$:

$$\tfrac{1}{2}Pk_BT = \frac{K}{2P}\sum_{s=1}^{P}\langle x_s^2\rangle + \tfrac{1}{2}\kappa\sum_{s=1}^{P}\langle(x_{s+1}-x_s)^2\rangle\ . \tag{A15}$$

Now write

$$\sum_{s=1}^{P} x^2 = P\bar{x}^2 + \sum_{s=1}^{P}\Delta x_s^2\ , \tag{A16}$$

where $\bar{x}$ is the chain centroid and $\Delta x_s = x_s - \bar{x}$. Hence:

$$\tfrac{1}{2}Pk_BT = \tfrac{1}{2}K\langle\bar{x}^2\rangle + \tfrac{1}{2}K\Delta^2 + \tfrac{1}{2}\kappa\sum_{s=1}^{P}\langle(x_{s+1}-x_s)^2\rangle\ , \tag{A17}$$

where Δ is the r.m.s. radius of gyration. We showed in A.2 that the distribution of $\bar{x}$ is the same as that of a classical particle, so that $\tfrac{1}{2}K\langle\bar{x}^2\rangle = \tfrac{1}{2}k_BT$. It follows that according to equation (69):

$$\langle T\rangle = \tfrac{1}{2}Pk_BT - \tfrac{1}{2}\kappa\sum_{s=1}^{P}\langle(x_{s+1}-x_s)^2\rangle = \tfrac{1}{2}k_BT + \tfrac{1}{2}K\Delta^2\ . \tag{A18}$$

Taking the expression for Δ^2 given in equation (42), we obtain:

$$\langle T\rangle = \tfrac{1}{4}\hbar\omega_0\coth\tfrac{1}{2}\beta\hbar\omega_0\ , \tag{A19}$$

which is the standard result for the mean kinetic energy of an oscillator.

THE METHOD OF CONSTRAINTS : APPLICATION TO A SIMPLE N-ALKANE MODEL

J.P. Ryckaert
Pool de Physique,CP223,
Université Libre de Bruxelles,
Boulevard du Triomphe, B1050,
Brussels (Belgium)

ABSTRACT. We consider a specific semi rigid model of n–alkane where methyl and methylene atomic groups, represented as single spherical beads, are connected through rigid bonds. The model considers explicitly the CCC bendings and the torsional vibrations around the CC bonds. Such alkane model, representative of most models used in chain molecules molecular dynamics simulations, helps us to illustrate the method of constraints presented otherwise in its general form. In this method, the dynamics of the chain is expressed in terms of the cartesian coordinates of the beads. The total force acting on a bead is expressed as the sum of the force deriving from the potential energy and the forces of constraint which act along the rigid bonds. These last forces involve a set of Lagrange multipliers: the different methods to incorporate them into the numerical integration algorithms are reviewed and explicitly illustrated for the trimer case (propane with two rigid bonds and a flexible angle). For long chain molecules, the need of an iterative technique which treats constraints individually in succession (SHAKE method) will particularly be emphasised. The paper ends with some considerations on the coupling of a system of molecules with constraints to a heat bath.

1. INTRODUCTION

The computer simulation of macromolecules is still in its infant stage: if one considers melts and solids of chain molecules, it is rather impressive to realise how fast the situation has evolved since the first Molecular Dynamics (MD) simulation of n–butane, the shortest flexible chain of the n–alkane series [1]. Super-computers perform today MD simulations of a few hundreds of chain molecules composed of a few tens of segments each, over a period of time of typically fifty to hundred picoseconds. In a work in progress, we are able to consider a system of 180 semi-rigid molecules of tricosane (C_{23} H_{48}) with a full atomic model (12780 atoms): it requires a bit more than one VP200 (Siemens-Fujitsu) hour to run 3 picoseconds of real time so that an experiment requires a few tens of CPU hours. Using a simpler chain model, K. Kremer et al. [2] used 1500 Cray-XMP hours to

C.R.A. Catlow et al. (eds.), Computer Modelling of Fluids Polymers and Solids, 189–201.

study the beads diffusional behaviour in polymer melts over many millions of time steps and for many different chain lengths (typically 20 chains of 100 monomers) in order to analyse their long time diffusive behaviour and compare it to the predictions of the reptation model [3]. For comparison, 10 years ago, the standard experiments on liquid n–alkanes (using CH_2 and CH_3 bead models) treated about 100 butane molecules or 27 decane molecules over 10-20 pico-seconds [4, 5].

This evolution of MD simulation capabilities in the field suggests that the polymer range is the next step for the coming years. Both the theoretical framework of polymer dynamics [3] and the huge industrial interest to understand the microscopic aspects of rheology (polymers in flows), rubber elasticity, plasticity,... will certainly enhance this trend.

Before considering some applications of computer simulations to polymeric materials, this first paper is intended to introduce the relevant MD techniques in this area on the basis of a rather simple n–alkane model. Paraffins are indeed used as the chain molecule reference system on which new potentials and techniques have been continuously tested. The reasons are easy to find: i) from the point of view of simulations, the atomic structure is simple and long-range (electrostatic) forces can be neglected while, ii) comparison with experimental data is favored by the huge amount of data accumulated on hydrocarbons.

The simulation of chain molecules at atomistic level is usually based on semi-rigid models treated by classical mechanics. As the fastest motions in which one is normally interested are the local conformational changes (gauche-trans isomerisation around CC bonds), all motions occuring on time scales shorter than the torsional vibrations around CC bonds (frequencies around 200 cm^{-1}) are usually eliminated by geometrical constraints. The obvious advantage here is to keep the time step as large as possible given the mentioned objectives but the "freezing" of these degrees of freedom is further justified by the fact that all the vibrational modes at higher frequencies are basically in their ground state at 300K. In section 2, we summarise two very close versions of a rather simple model used in early liquid experiments in order to illustrate in the following the MD techniques: a more detailed discussion on paraffins potentials is delayed to the next papers on applications. In section 3, the cartesian equations of motion of the beads are derived while the various methods to integrate them numerically are summarised in the next section. Section 5 is concerned with thermalisation methods for molecules with constraints, methods which are relevant either to accelerate the thermal equilibration between inter and intramolecular degrees of freedom in polymer systems prior to a standard MD experiment, or to control temperature in polymeric systems under dissipative non-equilibrium situations.

2. A SIMPLE SEMI-RIGID MODEL FOR N–ALKANES

In the two closely related models (A and B) [5,6] we introduce in this first paper, methylene or methyl groups are represented as single point particles (beads) located at the carbon nuclei positions. The chain's geometry is illustrated in figure1, which also serves to define the symbols used in the discussion of intra and intermolecular potentials. All specific values of the model parameters are gathered in table 1.

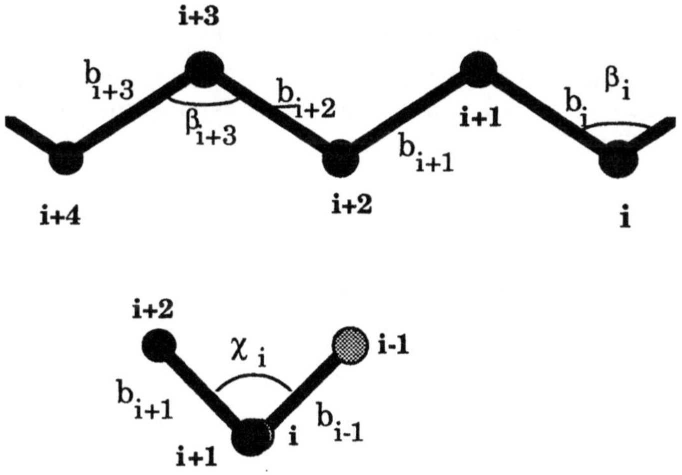

Figure 1 : illustration of the geometry of the chain

For an arbitrary C_nH_{2n+2} chain, $\mathbf{r}_i$ and $\dot{\mathbf{r}}_i$ (i=1,n) denote the cartesian coordinates and velocities of the beads (the molecular index is not indicated as there is never ambiguity). Let m' and m represent the masses of methyl and methylene groups respectively. Let b_i denote the bond joining beads i and i+1 and β_i the bending angle between bonds b_{i-1} and b_i. The torsional angle around b_i, χ_i , is defined as the dihedral angle between planes defined respectively by (b_{i-1} , b_i) and (b_i , b_{i+1})).

The CC strechings (about 700 cm^{-1}) are frozen at their average distance d_{cc} by (n-1) rigid bond constraints

$$(\mathbf{r}_{i+1} - \mathbf{r}_i)^2 - d_{cc}^2 = 0 \qquad i=1,n-1 \qquad (1a)$$

Bending vibrations of β_i (about 300 cm^{-1}) around the mean β^* are frozen by constraints in model A but considered explicitly in model B, this point being the only distinction between the two models discussed in this first paper. In model A, (n-2) rigid bond constraints between second neighbours i and i+2 are added to the CC constraints (1a) to assure the rigidity of the CCC triangles, i.e.

$$(\mathbf{r}_{i+2} - \mathbf{r}_i)^2 - (2\, d_{cc} \sin(\beta^*/2))^2 = 0 \qquad i=1,n-2. \qquad (1b)$$

In model B [6], one considers an harmonic potential in terms of angle β_i

$$Vb\,(\beta_i) = \frac{1}{2} k_b\, (\beta_i - \beta^*)^2 \qquad i=2,n-1 \qquad (2)$$

The torsional motions around CC bonds (about 150 cm^{-1}) follow a potential adjusted on n-butane data with 3 minima corresponding to trans and gauche conformations (see figure 2).

$$Vr\,(\chi_i) = \sum_{l=0}^{5} a_l \cos^l \chi_i \qquad i=2,n-2 \qquad (3)$$

The gauche minima of Vr occur at $\chi_i = \pm\, 2\pi/3$ at Vr=2.93 kJ mol^{-1} and the trans-gauche barriers at $\chi_i = \pi/3$ culminate at 12.34 kJ mol^{-1}

Symbol	Definition	Unit	Value(s)
m	methylene mass	K $mole^{-1}$	0.014
m '	methyl mass	K $mole^{-1}$	0.015
d_{cc}	C C bond length	m	$1.53\,10^{-10}$
β^*	av. CCC bending angle	radian	1.91063
k_b	bending force constant	kJ mol^{-1}	520
a_l (l=0,5)	Torsional potential (eq.3)	kJ mol^{-1}	9.279, 12.156, -13.120, -3.060, 26.241, -31.495.
ε	bead-bead LJ energy depth	kJ mol^{-1}	0.5986
σ	bead-bead LJ core parameter	m	$3.92\,10^{-10}$

Table 1: Alkane potential parameters for models A and B.

The intermolecular potential consists of Lennard-Jones (LJ) interactions between all bead sites. Such interactions apply also as covolume effects within the same chain when beads are separated by more than three bonds.

The dynamics in cartesian coordinates requires the cartesian forces acting on all beads. As we see in the next section, the total force divides itself into two terms, i.e. **F**, the force deriving from the potential and **G**, the force of constraint due to the geometrical constraints implying that bead. **F** is itself splitted into LJ, bending and torsional contributions.

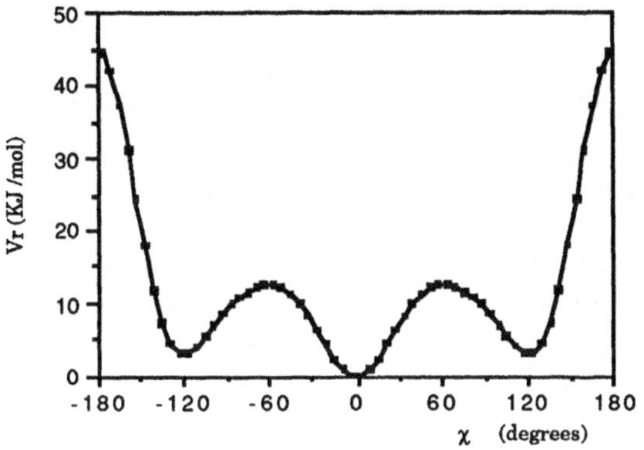

Figure 2 : torsional potential around CC bond.

To get these two last contributions of **F** explicitly, the corresponding potential terms must be derived respect to atomic coordinates. We leave the derivation of torsional forces for models A and B as exercices, the first one being simpler to handle as β_i is constant (see [7] for B). Here, we consider as an illustration the bending forces appearing in model B. Clearly, the potential for β_i (equation (3)) involves only beads j=i-1,i and i+1. We have

$$\mathbf{F}_j = -\frac{\partial Vb}{\partial \mathbf{r}_j} = -\frac{\partial Vb}{\partial (\cos \beta_i)} \frac{\partial (\cos \beta_i)}{\partial \mathbf{r}_j}$$

with
$$\cos \beta_i = \frac{(\mathbf{r}_{i+1} - \mathbf{r}_i) \cdot (\mathbf{r}_{i-1} - \mathbf{r}_i)}{d_{cc}^{\;2}} \tag{4}$$

Thus, one has
$$\mathbf{F}_{i\pm 1} = \frac{A(\beta_i)}{d_{cc}^{\;2}} (\mathbf{r}_{i\mp 1} - \mathbf{r}_i) \;,\; \mathbf{F}_i = -\mathbf{F}_{i+1} - \mathbf{F}_{i-1}$$

where
$$A(\beta_i) = k_b (\beta_i - \beta^*) \sin^{-1} \beta_i \tag{5}$$

3 THE CARTESIAN EQUATIONS OF MOTION: THE PROPANE CASE.

We consider a set of n point particles of mass m_i and coordinates $\mathbf{r}_i$ (i=1,n) interacting according to a potential energy V(**r**) and subject to a set of l geometric (holonomic) constraints, written formally as

$$\sigma_k(\mathbf{r}) = 0 \quad k=1,l. \tag{6}$$

The Lagrange equations of the first type [8] for the individual particles are expressed in classical mechanics as

$$m_i \ddot{\mathbf{r}}_i = - \partial V / \partial \mathbf{r}_i - \sum_{k=1}^{l} \lambda_k \, \partial \sigma_k / \partial \mathbf{r}_i \qquad (7)$$

where the set of $\{\lambda_k\}$ (k=1,l) are Lagrange multipliers to be determined by requiring that the solution of equations (7) satisfies the constraint relations (6). To obtain the Lagrange multipliers explicitly, equations (7) are substituted in the second time derivative of equations (6).

Instead of discussing the general case, it is always easier to follow the mathematical developments and numerical treatments on an example: we will take the propane case, C_3H_8 on the basis of the model B (with explicit bending potential) of section 2.

The two constraints per molecule labelled 1 and 2 in the following, are the bond constraints (1a) on $\mathbf{r}_{12} = \mathbf{r}_1 - \mathbf{r}_2$ and $\mathbf{r}_{23} = \mathbf{r}_2 - \mathbf{r}_3$ respectively. The forces $\mathbf{F}_i$ on the three beads contain contributions from LJ interactions with beads belonging to different molecules (standard expression not made explicit here) and intramolecular bending forces given by equation (5) with i=2.

Using the notations of section 2, the equations of motion (7) read

$$\begin{aligned} m' \, \ddot{\mathbf{r}}_1 &= \mathbf{F}_1 - 2\lambda_1 \mathbf{r}_{12} \\ m \, \ddot{\mathbf{r}}_2 &= \mathbf{F}_2 + 2\lambda_1 \mathbf{r}_{12} + 2\lambda_2 \mathbf{r}_{32} \\ m' \, \ddot{\mathbf{r}}_3 &= \mathbf{F}_3 - 2\lambda_2 \mathbf{r}_{32} \end{aligned} \qquad (8)$$

The second time derivatives of the constraints are simply

$$\begin{aligned} \ddot{\mathbf{r}}_{12} \cdot \mathbf{r}_{12} + \dot{\mathbf{r}}_{12}{}^2 &= 0 \\ \ddot{\mathbf{r}}_{32} \cdot \mathbf{r}_{32} + \dot{\mathbf{r}}_{32}{}^2 &= 0 \end{aligned} \qquad (9)$$

Sustitution of equations (8) into (9) yield after straightforward algebra

$$\begin{pmatrix} (\frac{1}{m'} + \frac{1}{m}) & \frac{\cos(\beta_2)}{m} \\ \frac{\cos(\beta_2)}{m} & (\frac{1}{m'} + \frac{1}{m}) \end{pmatrix} \begin{pmatrix} \lambda_1 \\ \lambda_2 \end{pmatrix} = \begin{pmatrix} 0.5\, d_{cc}^{-2} \, (\dot{\mathbf{r}}_{12}{}^2 + (\frac{\mathbf{F}_1}{m'} - \frac{\mathbf{F}_2}{m}) \cdot \mathbf{r}_{12}) \\ 0.5\, d_{cc}^{-2} \, (\dot{\mathbf{r}}_{32}{}^2 + (\frac{\mathbf{F}_3}{m'} - \frac{\mathbf{F}_2}{m}) \cdot \mathbf{r}_{32}) \end{pmatrix} \qquad (10)$$

In equations (10), as it is the case for any semi-rigid molecule, the l*l matrix multiplying the λ vector (containing all Lagrange multipliers) is dependent upon the intramolecular configuration (specified here by β_2, the flexible bending angle). The r.h.s. contains all kinematic and external forces effects tending to disrupt the constraints and which must be balanced by appropriate constraint forces along the bonds. If required, the solution of equations (10) $\lambda(\mathbf{r},\dot{\mathbf{r}})$ can trivially be obtained by matrix inversion.

4 NUMERICAL INTEGRATION OF DYNAMICAL EQUATIONS

4.1 *Method using the constraint forces expression in terms of coordinates and velocities*

Equation (10) leads to an expression of the Lagrange multipliers in terms of positions and velocities. When incorporated back into the equations of motion (8), the n explicit accelerations $\ddot{\mathbf{r}}_i$ $(\mathbf{r},\dot{\mathbf{r}})$ consist of a set of 3n coupled second order differential equations. Constraints are now implicitly satisfied.

This approach was originally the first one to be attempted [9,10] but was then almost immediatly abandonned in favor of the approach treating Lagrange multipliers as ad-hoc parameters as explained in the next paragraph. The straightforward numerical integration of the explicit equations of motion, because of inherent algorithm and round-off errors, produces a trajectory where constraints are not rigorously but only approximately satisfied: although rather slowly, the discrepancy appears to diverge systematically. Very recently however, R. Edberg et al. [11] proposed an algorithm based on this approach. Tested on n-alkanes represented according model A, this numerical integration algorithm does however not appear as efficient as the ones mentioned below. Note in particular that it was necessary to correct cartesian positions and velocities of a given molecule every 2000 steps approximately [11].

4.2 *Method treating Lagrange multipliers in constraint forces as parameters.*

For a given molecule with l constraints, all constraint forces are expressed originally (equations (7) and (8)) in terms of l Lagrange multipliers. The idea here is to consider them as free parameters which are fixed by requiring that all l constraints are effectively satisfied by the numerical solution (resulting from the adopted integration algorithm) instead of requiring this property for the exact solution : by implication, the constraint forces themselves are only approximate but it can be shown [10] that the order of accuracy of the integration algorithm is not altered. In order to consider a specific case, we use here the Verlet algorithm [12] (which is most often adopted in the practise) and apply it to the propane equations (8). We then write, with a time step h,

$$
\begin{aligned}
\mathbf{r}_1(t+h) &= \mathbf{r}_1'(t+h) - 2\frac{\lambda_1\, h^2}{m'}\mathbf{r}_{12}(t) \\
\mathbf{r}_2(t+h) &= \mathbf{r}_2'(t+h) + 2\frac{\lambda_1\, h^2}{m}\mathbf{r}_{12}(t) + 2\frac{\lambda_2\, h^2}{m}\mathbf{r}_{32}(t) \qquad (11) \\
\mathbf{r}_3(t+h) &= \mathbf{r}_3'(t+h) - 2\frac{\lambda_2\, h^2}{m'}\mathbf{r}_{32}(t)
\end{aligned}
$$

where the $\mathbf{r}_i'(t+h)$ i=1,3 are in fact the predicted positions in absence of constraint forces, i.e.

$$
\mathbf{r}_1'(t+h) = -\,\mathbf{r}_1(t-h) + 2\,\mathbf{r}_1(t) + \frac{h^2}{m'}\mathbf{F}_1(t) \qquad (12)
$$

for atom 1 and similar expressions hold for the the other beads.

What is now required is to determine the λ such that the $\mathbf{r}_i(t+h)$ do satisfy the two constraints (equation (1a)) exactly. This can be numerically done in two different but ultimately equivalent iterative techniques, using the <u>matrix method</u> or the <u>SHAKE method</u> more specially developed for large molecules.

4.2.1 The matrix method

From equation (11), it comes that the bond vectors at the new time step are respectively

$$
\begin{aligned}
\mathbf{r}_{12}(t+h) &= \mathbf{r}_{12}'(t+h) - 2\,h^2\left(\frac{1}{m}+\frac{1}{m'}\right)\lambda_1\,\mathbf{r}_{12}(t) - 2\,h^2\frac{1}{m}\lambda_2\,\mathbf{r}_{32}(t) \\
\mathbf{r}_{32}(t+h) &= \mathbf{r}_{32}'(t+h) - 2\,h^2\frac{1}{m}\lambda_1\,\mathbf{r}_{12}(t) - 2\,h^2\left(\frac{1}{m}+\frac{1}{m'}\right)\lambda_2\,\mathbf{r}_{32}(t) \qquad (13)
\end{aligned}
$$

Substitution of equations (13) into the two bond constraint relations (equation (1a)) yields a system of algebraïc equations of the second order in λ which we write as a vectorial equation with linear terms grouped in the l.h.s.,i.e.

$$
\mathbf{M}(\mathbf{r}(t),\mathbf{r}'(t+h))\,\lambda = \sigma'(t+h) + \mathbf{O}\,(h^2\lambda)^2 \qquad (14)
$$

where the elements of $\sigma'(t+h)$ are the constraint expressions (equations (1a)) evaluated with predicted coordinates $\mathbf{r}'(t+h)$ (equation (12)), the elements of $\mathbf{O}\,(h^2\lambda)^2$ are the square of the sum of linear terms in λ in equations (13). The matrix $\mathbf{M}$ reads

$$
\mathbf{M}\,(\mathbf{r}(t),\mathbf{r}'(t+h)) = -\,4\,h^2 \begin{pmatrix} \left(\frac{1}{m}+\frac{1}{m'}\right)\mathbf{r}'_{12}\cdot\mathbf{r}_{12} & \frac{1}{m}\mathbf{r}'_{12}\cdot\mathbf{r}_{32} \\ \frac{1}{m}\mathbf{r}'_{32}\cdot\mathbf{r}_{12} & \left(\frac{1}{m}+\frac{1}{m'}\right)\mathbf{r}'_{32}\cdot\mathbf{r}_{32} \end{pmatrix} \qquad (15)
$$

Equation (14) can now be solved by iteration according to

$$\lambda^{(n+1)} = \mathbf{M}^{-1}\,\sigma'(t+h) + \mathbf{M}^{-1}\,\mathbf{O}\,(h^2\lambda^{(n)})^2 \quad (16)$$

in which $\lambda^{(n)}$ represents the estimate of the lagrange parameters after n iterations, starting with $\lambda^{(0)}=0$. Note that this procedure requires a single matrix inversion and then, it is usually found that three iterations are needed to achieve a sufficient accuracy on the λ.

4.2.2 The SHAKE method

This alternative iterative technique consits in solving the basic equation (14) in an iterative way, line by line, i.e. considering all constraints in succession and cycling many times over the whole set. For molecules with many tens or hundreds of constraints, the SHAKE method is indicated as it avoids the tedious construction of the matrix (15) and the time consuming resolution of equation (16). Moreover, note that this method is even sometimes adopted for molecules with a small number of constraints given its simple coding.

i) *The propane case*

To explain how it works in the practise, it is convenient to introduce the notation $\mathbf{r}_i^s(t+h)$ (or more simply $\mathbf{r}_i^s$) to denote the cartesian atomic coordinates based on the running estimate λ^s obtained at some stage of the iterative process, say after completing s single constraint operations. In our propane example, using equations (11), we have thus

$$\mathbf{r}_1^s\,(t+h) = \mathbf{r}_1'(t+h) - 2\frac{\lambda_1^s\,h^2}{m}\,\mathbf{r}_{12}\,(t)$$

$$\mathbf{r}_2^s\,(t+h) = \mathbf{r}_2'(t+h) + 2\frac{\lambda_1^s\,h^2}{m}\,\mathbf{r}_{12}\,(t) + 2\frac{\lambda_2^s\,h^2}{m}\,\mathbf{r}_{32}\,(t) \qquad (17)$$

$$\mathbf{r}_3^s\,(t+h) = \mathbf{r}_3'(t+h) - 2\frac{\lambda_2^s\,h^2}{m}\,\mathbf{r}_{32}(t)$$

Suppose now the $\mathbf{r}_i^s$ are the current positions after N full cycles over the two constraints and after the treatment for the bond constraint 1 (on $\mathbf{r}_{12}$) at cycle N+1 so that s amounts 2N+1. Seeking for an updated λ_2^{s+1}, while keeping the estimate of the other lagrange multipliers unchanged (i.e. $\lambda_1^{s+1} = \lambda_1^s$), we define the correction δ_2^s as:

$$\lambda_2^{s+1} = \lambda_2^s + \delta_2^s \qquad (18)$$

in terms of which, combining equations (18) and (17), we have

$$\mathbf{r}_1^{s+1}(t+h) = \mathbf{r}_1^s(t+h)$$
$$\mathbf{r}_2^{s+1}(t+h) = \mathbf{r}_2^s(t+h) + 2\frac{\delta_2^s h^2}{m}\mathbf{r}_{32}(t) \tag{19}$$
$$\mathbf{r}_3^{s+1}(t+h) = \mathbf{r}_3^s(t+h) - 2\frac{\delta_2^s h^2}{m'}\mathbf{r}_{32}(t)$$

Applying the second bond constraint (on $\mathbf{r}_{32}$) to the new positions (equation (19)), we get a single variable equation in δ_2^s which yields

$$\delta_2^s = \frac{d_{cc}{}^2 - \mathbf{r}_{32}^{s\,2}}{4\,h^2\left(\frac{1}{m} + \frac{1}{m'}\right)\mathbf{r}_{32}^s \cdot \mathbf{r}_{32}} \tag{20}$$

if second order terms are neglected. Substitution of this last solution in equations (19) provides the updated positions. Note that only atomic coordinates implied in the constraint under consideration (here k=2) are modified at this iterative step. The procedure continue with the first constraint using the corresponding expressions to equations (19) and (20) for the 12 bond and so on. The iterative procedure is initiated with the predicted positions $\mathbf{r}_i'(t+h)$ based on an unconstrained time step (12) and is stopped when all constraints are satisfied within some preset tolerance value.

ii)*The general case*

Going back to equations (6) and (7), it is useful to mention at this point the algebra needed by the SHAKE iterative process for the treatment of a molecular model having an arbitrary number of holonomic constraints of any functional form.

Cycling over the whole set of constraints many times, each of them, $\sigma_k(\mathbf{r})=0$ say, is treated in the same way to update the atomic coordinates. On the basis of those obtained after s such elementary operations, the new set is obtained according to

$$\mathbf{r}_i^{s+1} = \mathbf{r}_i^s - h^2\frac{\delta_k^s}{m_i}\frac{\delta\,\sigma_k(\mathbf{r})}{\delta\,\mathbf{r}_i} \tag{21}$$

from which it is obvious that only cartesian coordinates involved in $\sigma_k(\mathbf{r})$ are affected within this iterative step. The value δ_k^s is fixed by solving, up to first order, the equation obtained by substituting the new coordinates expression (21) into the kth constraint. It results after trivial algebra that

$$\delta_k^s = h^{-2}\,\sigma_k(\mathbf{r}^s)\,/\big(\sum_i m_i^{-1}\big(\frac{\delta\,\sigma_k(\mathbf{r}^s)}{\delta\,\mathbf{r}_i^s}\big).\big(\frac{\delta\,\sigma_k(\mathbf{r})}{\delta\,\mathbf{r}_i}\big)\big) \qquad (22)$$

It is useful to realise at this point that equations (19) and (20) in the propane example are just particular cases of equations (21) and (22) respectively, namely the expressions valid for bond constraints. This helps realizing that the last two formulae are the basic expressions to apply, within the SHAKE iterative technique, to the MD of molecular models where the constraint scheme is more complex than a network of rigid bonds. Illustrative examples of such models will be given later in applications.

5 TEMPERATURE CONTROL FOR MOLECULES WITH CONSTRAINTS

Recently developped 'constant temperature' ('constant T') MD methods are nowadays often used to control the temperature in non-equilibrium computer simulations (to counteract heat dissipation in presence of gradients, to simulate quenching experiments,..). However these methods are also extremely useful simply to prepare an 'equilibrated' sample to initialize a standard (constant energy) MD run at equilibrium. In a majority of cases indeed, the thermodynamic state to be simulated is specified by its temperature (together with its density or pressure): performing a preliminary run at a controled temperature is the easiest way to drive the system towards the required state. According to our experience in preparing thermalised solid samples of n–alkanes prior to standard MD experiments [13,7], we found that the only method able to achieve the expected repartition of the kinetic energy over all degrees of freedom within a few hundreds steps was the stochastic collisions method of H.C. Andersen [14,15].

In this constant T method, the heat bath coupling consists in changing from time to time the velocity of a particle from its running value (during a normal run) to a value sampled at random in the canonical ensemble distribution corresponding to the temperature of interest. In the case of systems with a broad spectrum of vibrational modes (like in solid alkanes), it appears that thermal equilibrium is established in an optimised way by this method in comparison with all other constant T methods where velocities are affected uniformly on the basis of a total kinetic energy control. In these sub-class of methods indeed, the physical coupling between the various modes assures alone the repartition the total kinetic energy among all degrees of freedom. As we expect Andersen's thermal coupling to be quite generally useful for macromolecules, we give here only a few specific comments on its non trivial application to molecules with constraints.

In the cartesian coordinates approach, geometrical constraints impose indeed restrictions on the cartesian velocities. The explicit relations, which are simply the time derivative of the geometrical constraints themselves (equation (6)), require that velocities are orthogonal to the constraint surfaces, i.e.

$$\sum_{i=1}^{n} \left(\frac{\delta \sigma_k(\mathbf{r})}{\delta \mathbf{r}_i} \cdot \dot{\mathbf{r}}_i \right) = 0 \qquad k=1,l \qquad (23)$$

For the bond constraint linking beads 1 and 2, equation (23) would simply gives $\mathbf{r}_{12}.\dot{\mathbf{r}}_{12} = 0$, which means that the relative velocity of the two beads has no component along the bond vector.

The procedure to follow is sufficiently transparent to be exposed without the detailed justification that can be found elsewhere [16,17]. The idea is to sample atomic velocities $v_{i\alpha}$ (any atom i, any cartesian component α) as if there was no constraint, i.e. in a gaussian distribution of zero mean and standard deviation $(k_B T/m_i)^{1/2}$. In a second step, the velocity components orthogonal to the constraints must be removed. This can be done in the following way suited to be combined with an existing MD program.

a) Perform a virtual dynamical step of time h^* starting from the actual coordinates at the time t of the stochastic collision, according to a linear trajectory

$$r'_{i\alpha}(t+h^*) = r_{i\alpha}(t) + h^* \, v'_{i\alpha} \qquad (24)$$

yielding a new atomic positions {r'} based on the sampled velocities {v'}.

b) Apply one form of the constraint method of section 4.2 to the {r'} (equations (24)) so that the new coordinates $\{r''(t+h^*)\}$ satisfy the constraints.

c) The required velocities $\{v''_{i\alpha}\}$ are simply obtained as

$$v''_{i\alpha} = \frac{(r''_{i\alpha}(t+h^*) - r_{i\alpha}(t))}{h^*} \qquad (25)$$

provided h^* is sufficiently small, typically one tenth of the real time step h [16,17].

To restart the dynamics with these new velocities, one cannot apply the constraint algorithm of section 4.2 based on the traditional Verlet algorithm. One has to use another integration scheme: a simple Taylor expansion of **r**(t+h) around time t truncated after the force term is usually sufficient. Practical details about these techniques are given in [12,16,17] .

References

[1] Ryckaert J.P. , Bellemans A., (1975) Chem. Phys. Lett. , **30**,123.
[2] Kremer K, Grest G.S., Carmesin I., (1988) Phys. Rev. Let. **61**, 566
[3] Doï M., Edwards S.F., (1986) 'The theory of Polymers Dynamics', Clarendon Press-Oxford)
[4] Weber T.A.,(1978),J. Chem. Phys., **69**, 2347.
[5] Ryckaert J.P. , Bellemans A., (1978) Faraday Discuss. Chem. Soc., **66**, 95.
[6] Van der Ploeg, P., Berendsen H.J.C.,J. Chem Phys.(1978), **76**, 3271.
[7] Ryckaert J.P., Mc Donald I.R., Klein M.L. submitted to Molec. Phys.
[8] Bradbury T.C.,(1968), Theoretical Mechanics (Wiley-Interscience), chap. XI.
[9] Orban J., Ryckaert J.P.,(1974) 'Methods in Molecular Dynamics'. Cecam Report of activities.
[10] Ryckaert J.P., Ciccotti G., Berendsen H.J.C., (1977),J. Comput. Phys. **23**, 327.
[11] Edberg R., Evans D.J., Morriss G.P., J. Chem. Phys.(1986), **84**, 6933.
[12] Allen M.P., Tildesley D.J. (1987), Computer Simulation of Liquids (Clarendon Press-Oxford)
[13] Ryckaert J.P., Klein M.L., Mc Donald I.R., Phys. Rev. Lett.(1987) **58**, 698.
[14] Andersen H.C.,(1980) J Chem Phys **72**, 2384.
[15] Andrea T.A., Swope W.C., Andersen H.C., J. Chem. Phys.(1983) **79**, 4576.
[16] Ciccotti G., Ryckaert J.P., (1986), Comput. Phys. Rep. , **4**, 345.
[17] Ryckaert J.P., Ciccotti G.(1986), Molec. Phys. **58**,1125.
[18] Ryckaert J.P., (1985), Molec. Phys. **55**, 549.

MOLECULAR DYNAMICS OF CHAIN MOLECULES

JULIAN H.R. CLARKE
Chemistry Department
U.M.I.S.T
Sackville Street
Manchester
M60 1QD
U. K.

ABSTRACT. Molecular dynamics studies of liquids and solids composed of short chain molecules are reviewed. The factors that need to be considered in choosing a set of interaction potentials for such systems are discussed; the Hamiltonian for such systems involves two three and four-body intramolecular terms which define the flexibility of the molecule and although one or more of these can, if necessary, be eliminated by using constraints ultimately the choice of interaction potential depends on the system being studied and the properties that are of interest. In the case of fluids much can be still learned from studies using the "bead" model in which methylene groups, for instance, are treated as single interaction sites. Topics discussed here include the chemical transformation rates between gauche and trans conformers, the evidence available regarding solvent effects on molecular conformations for small molecules and also the ways in which molecular flexibility can directly influence transport properties. The need to ensure the attainment of equilibrium in such systems where there are slowly relaxing degrees of freedom will be emphasised.

I. Introduction

In this article we review the application of the MD techniques discussed in the previous article to simulation studies of systems, primarily liquids, containing simple short chain molecules. We shall consider only alkanes since very little simulation work has been performed on chemically more complex molecules.

These systems are important in several respects. Firstly they provide a vehicle for introducing the effects of important degrees of freedom associated with molecular flexibility; there is an immediate challenge for computer simulation to reproduce experimentally observed behaviour and hopefully to provide insight and understanding into physical phenomena. Secondly these small molecules are the building blocks of long chain polymers and macromolecules to be considered in later lectures. If the idea of transferable potentials is to be used for complex molecules then we must first understand how different terms in the model potentials affect the properties of small molecules.

C.R.A. Catlow et al. (eds.), Computer Modelling of Fluids Polymers and Solids, 203–217.

We shall first survey the choices that have been made for the intermolecular and intramolecular potentials in such systems. We will then discuss the characterisation of the slow conformational fluctuations in these systems emphasising the need for long simulation times in the preparation of samples and proposing a criterion for configurational equilibrium. In section 4 we survey the current evidence for "solvent effects" on the distribution of conformers in liquid alkanes and briefly discuss new evidence that any such effect is extremely small in butane, the simplest of flexible hydrocarbons. Lastly in section 5 we discuss the effect of flexibility of dynamic properties of liquids.

II. Choice of Potential

In general one can identify two prevailing philosophies in choosing the interaction potential and both of these can be recognised in the the various approaches that have been made to simulating chain molecules. The first approach aims to use the simplest potential which incorporates the essential features of interest. The emphasis here is gaining an understanding of the basic physics underlying some macroscopic phenomenon or on the comparison with theoretical predictions. An example is the use of the "bead" model to represent inter- and intra-chain van der Waals interactions. In this case CH_2 or CH_3 groups are considered as single interaction centres, the beads on different molecules interacting via a Lennard-Jones potential with suitably chosen parameters. This procedure has been quite satisfactory as far as liquid state simulations are concerned. The second approach concentrates attention on reproducing some properties of the real system as accurately as possible in which case the chosen form for the interactions can be crucial. For simulations of solid alkanes, for instance, the bead model is unsatisfactory and it has been found necessary to use a full atomic desription with carefully chosen parameters in order to reproduce correct structures and phase transformations [1,2].

Some of the different schemes that have been used over the past ten years are compared in Table I. A few of the differences are relatively trivial but other terms can have quite subtle effects on the observed properties as will be explained below. In general the connectivity and flexibility of a chain molecule can be defined in terms of 2- 3- and 4-body intramolecular potentials relating to bond stretching, valence angle bending and dihedral angle rotation respectively. As explained in the previous article in practice it is usually not necessary to consider explicitly the bond stretching motions since these are of very high frequency and are largely decoupled from the other slower molecular degrees of freedom. It has also been fairly common practise to eliminate angle bending motions by using constraints but this is a much less justifiable procedure for the following reasons. Firstly the introduction of any constraints alters the dihedral angle population densities [3] and for angle bending constraints it is a quite significant effect (we shall return to this in a later section). Secondly it is known that in longer chain molecules collective angle bending motions make an important contribution to the overall chain flexibility and to the mechanical properties [4,5] so that it is desirable that they be included explicitly.

Simple harmonic potentials are probably satisfactory for the valence angle bending motions since the amplitudes are usually small. This potential can be parametrised either directly in θ [6]

$$\Phi(\theta) = \frac{1}{2}k_\theta(\theta-\theta_0) \qquad (1)$$

TABLE I: SOME POTENTIAL MODELS USED IN MOLECULAR DYNAMICS STUDIES OF ALKANE CHAINS

reference	A	B	C	D	E
system	butane	decane bilayer	butane n-octane	poly- ethylene	tricosane crystal
bonds	rigid	rigid	harmonic	rigid	rigid
bond length/nm	0.153	0.153?	0.1539	0.153	c
bond angle	rigid	harmonic in θ	harmonic in $\cos\theta$	harmonic in $\cos\theta$	harmonic in $\cos\theta$
k_θ / kJ mol^{-1}	-	520 rad^{-2}	130	520	520 rad^{-1}
θ_0/degrees	109.47	?	112.15	112.81d	112
torsional potential	$f(\cos\alpha)^e$	$f(\cos\alpha)^e$	$f(\mathbf{r}_{12},\mathbf{r}_{34},\mathbf{r}_{14})$	$f'(\cos\alpha)^d$	$f(\cos\alpha)^e$
intermolecular potential	bead LJ(12-6)	bead LJ(12-6)	bead LJ(9-6)	bead [illegible](12-6)	atomic exp-6
$\frac{\varepsilon}{k_B}$ /K	72	51.7^a, 77.3^b	50.4	72	c
σ/nm	0.3923	0.374^a 0.274^b	0.3494	0.428	c

A. Ryckaert J-P. and Bellemans A., 1978 Faraday Disc. Chem. Soc., **66**, 95
B. van der Ploeg P. and Berendsen H.J.C. , 1982 J. Chem. Phys., **76**, 3271
C. Weber T.A., 1978, J. Chem. Phys., **69**, 2347
D. Brown D. and Clarke J.H.R., 1988, Molecular Simulation (in press)
E. Ryckaert J-P, Klein M.L. and I.R. McDonald, 1987, Phys. Rev. Lett., **58**, 698

a - CH_3 sites, b - CH_2 sites, c - Williams potential parameters
d - parametrisation due to Steele D., 1985 J. Chem. Soc. Farad. Trans. II, **81**, 1077
e - parametrisation from ref. A

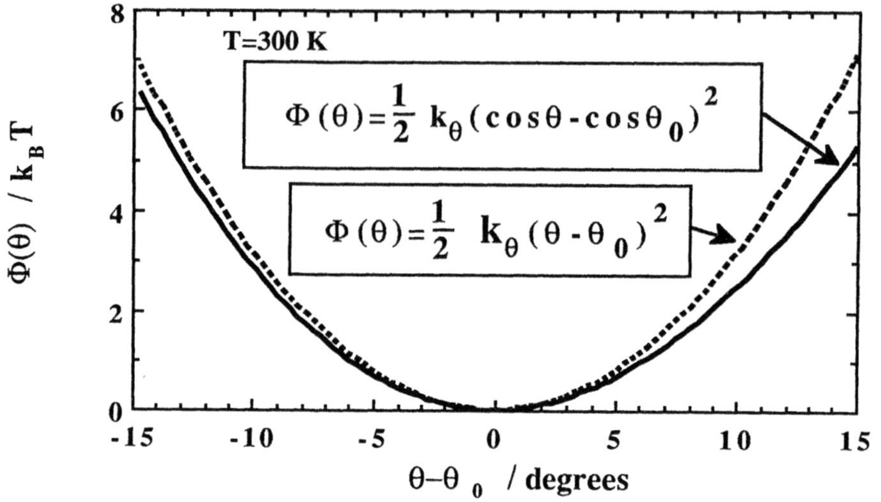

Figure 1. The forms of harmonic potentials used for the valence angle bending motions in alkane chains.

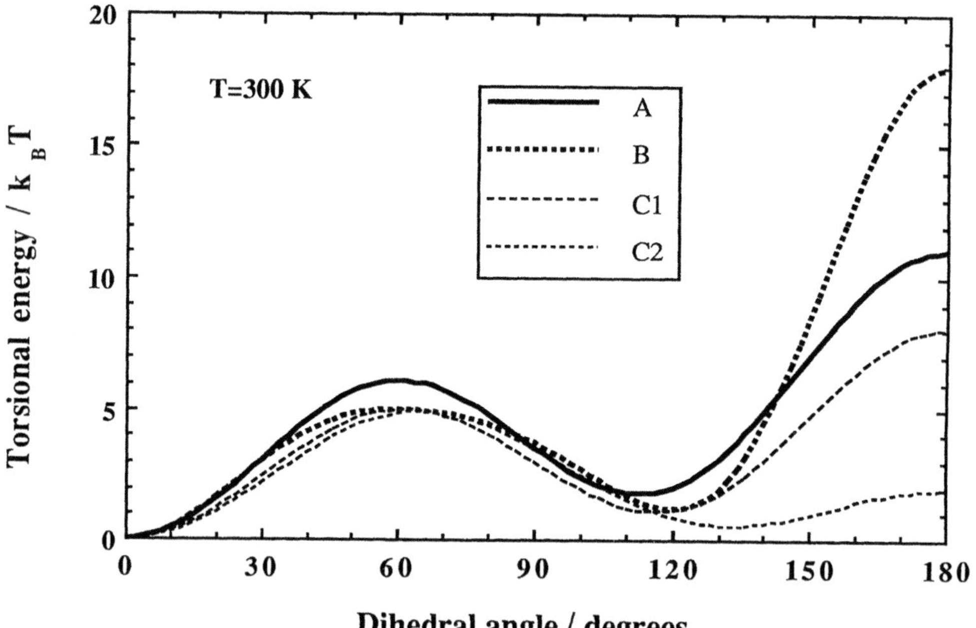

Figure 2. Torsional potential energy functions at 300 K for alkane chains. Curve A is from ref.9 and B from ref.8. Curves C1 and C2 are for the potential proposed in ref. 7 with the valence angle set to the equilibrium value and to 120° respectively. Note the large reduction in the gauche-gauche barrier [5].

or else in terms of cos θ [7].

$$\Phi(\theta) = \frac{1}{2}k_\theta(\cos\theta - \cos\theta_0) \tag{2}$$

As shown in Fig. 1 there is little difference in the form of these potentials for small displacements so that it is quite satisfactory to use the computationally much more convenient cosine form. The choice of k_θ has in many studies been somewhat arbitrary. Weber used a value of 130 kJ mol^{-1} taken from a theoretical analysis of spectral data on alkanes. The value is in fact far too low since equation (2) was mistakenly assumed to represent the complete angle bending potential (a direct interaction between next nearest neighbours should also have been included for this particular parametrisation). A value of 520 kJ mol^{-1} has recently been chosen for k_θ in equation (2) for long chains on the basis of fits to the Young's modulus of polyethylene [5]. Independently the same value was chosen in a simulation of alkane chains in a lipid bilayer [6] although reasons for this choice were not disclosed.

A comparison of some of the available dihedral potentials is shown in Fig. 2. Many years ago Ryckaert and Bellemans [8] produced a parametrised dihedral potential fitted to available spectroscopic and thermodynamic data and this has been widely adopted in computer simulations. More recently an improved parametrisation has been proposed on the basis of *ab initio* calculations and more accurate experimental data [9].
Weber [7] proposed a more general approach to modelling the flexibility of an alkane by using a potential which explicitly incorporates coupling between angle bending and torsional motions. Although some coupling is perhaps to be expected on physical grounds it has been suggested [5] that this model strongly overestimates the effect in giving a very large reduction in the magnitude of the gauche-gauche torsional barrier as θ increases, as shown in Fig.2. In the absence of experimental data on the nature of the coupling it seems safer at present to use potentials which treat the angle bending and dihedral angle motions independently.

III. Preparation of samples

One basic requirement of any simulation if it is to be at all representative of bulk behaviour is that the size of the sample and length of the simulation must be such as to contain all the important fluctuations in the system. The length required is therefore determined by the longest characteristic relaxation time in the system. For liquid alkanes at moderate densities this is likely to be the time for gauche-trans conformational fluctuations to come to equilibrium. We outline below a simple method by which this time can be measured and discuss the extent to which equilibrium has been achieved in various simulations of liquid butane which are summarised in Table II.

Three methods have been used to characterise the chemical reaction. Firstly one can compute the normalised correlation function [10] for the fraction of, for instance, the trans conformers in the liquid as Weber did [7] for butane

$$C_T(t) = \frac{<[X_T(t)-<X_T>][X_T(0)-<X_T>]>}{<[X_T(0)-<X_T>]^2>} \tag{3}$$

For a simple first order process this is expected to be an exponential decay [10]

$$C_T(t) = \exp(-kt) \tag{4}$$

where k is the overall rate constant.

Another approach to measuring the relaxation time is to start the simulation with all molecules in, say, the trans state and to monitor the change in the distribution of dihedral angles amongst the molecules as a function of time. If the molecule is symmetric about a particular dihedral angle then at equilibrium the proportions of gauche+ (G^+) and gauche- (G^-) conformers should be on average identical. Edberg *et al* used non-equilibrium molecular dynamics in this way to study the decay directly after preparing a sample of butane with all the molecules in the trans state [11]. A third method is to use the "hazard" analysis as described by Helfand [12] and utilised by Weber [7] and Marechal [13].

It has been shown recently however that there is a very straightforward and direct method of following the conformational kinetics by observing the decay of populations within the *equilibrium* liquid [14]. The method is physically transparent and provides an easy link with the mechanism of the reaction. At various times τ_i the number of molecules in a particular conformational state, say the trans configuration (T), is identified as constituting a subset i of the equilibrium liquid. One then measures the subsequent decay of the trans population within this subset towards the equilibrium distribution. The decay is averaged over many subsystems i as defined by a sequence of start times τ_i to define a relaxation function $R_{TT}(t)$,

$$R_{TT}(t) = \frac{< N_T^i\ (\tau_i+t) >_i}{< N_T >} \tag{5}$$

where $R_{TT}(t)$ is unity at $t = 0$ and $R_{TT}(t) = <X_T>$ at $t = \infty$. The method relies on the fact that at equilibrium the behaviour of a representative subsystem is on average identical to that of the whole system. The relationships between the relaxation function and the correlation functions for conformational fluctuations are discussed in detail elsewhere [14] but it is readily shown that $R_{TT}(t)$ measures the *self* correlations for decay of the trans population. In fact the total correlation function expression equation (3) overcomplicates and even confuses the issue since it includes both self and cross correlation terms in the averaging. For simple liquids such as butane cross correlations do not make a significant contribution to the mechanisms of conformational fluctuations and rate equations are expressed in terms of the rate of change of *single molecule* properties. Cross correlations between different molecules may exist in more complex systems and would be included as extra terms in the rate equation.

Some results for liquid butane are shown in Fig. 3 for the decay of both trans and gauche conformers. Previous studies of the conformational kinetics have assumed a two state mechanism for the trans-gauche (T-G) equilibrium in butane defined as follows.

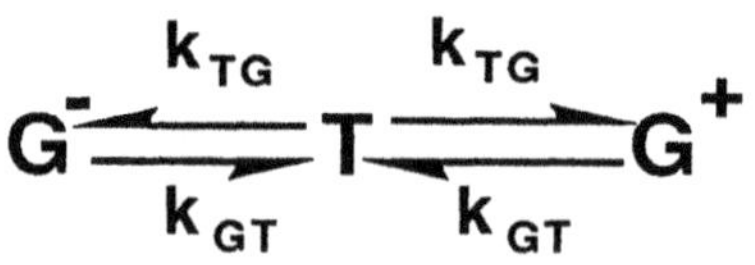

This is in fact an oversimplification and there is a good deal more to learn about the mechanism than has hitherto been disclosed. From the most recent analysis [14] it turns out to be impossible to fit the data accurately unless a *direct* mechanism for G+ ↔ G- interchange is incorporated into the scheme. This can be written in the following form

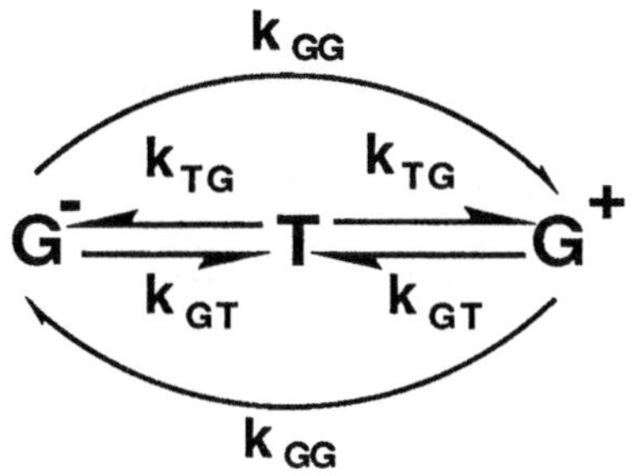

It must be emphasised, however, that these G+ ↔ G- transitions do *not* occur over the barrier between the two gauche states but as a result of a molecular conformation passing from one gauche state straight through trans to the other gauche state as illustrated in Fig. 4. First order kinetic equations can be easily written for this scheme [14].

$$\frac{dX_T(t)}{dt} = -2k_{TG}X_T(t) + k_{GT}(X_{G-}(t) + X_{G+}(t)) \tag{6}$$

$$\frac{dX_{G-}(t)}{dt} = k_{TG}X_T(t) - k_{GT}X_{G-}(t) - k_{GG}X_{G-}(t) + k_{GG}X_{G+}(t) \tag{7}$$

$$\frac{dX_{G+}(t)}{dt} = k_{TG}X_T(t) - k_{GT}X_{G+}(t) - k_{GG}X_{G+}(t) + k_{GG}X_{G-}(t) \tag{8}$$

Considering the decay of X_T for a subset i

$$X_T(t) = \frac{k_{GT}}{k}(1-\exp(-kt)) + X_T(0)\exp(-kt) \tag{9}$$

where

$$k=2k_{TG}+k_{GT} \tag{10}$$

There is a consequent rise in G^- (or G^+) and

$$X_{G-}(t) = \frac{1}{2}\left(\frac{k_{GT}}{k} - X_T(0)\right)\left(\exp(-kt) - \exp(-\lambda t)\right) + X_{G-}(0)\exp(-\lambda t) + \frac{k_{TG}}{k}\left(1 - \exp(-\lambda t)\right) \tag{11}$$

where

$$\lambda = 2k_{GG} + k_{GT} \tag{12}$$

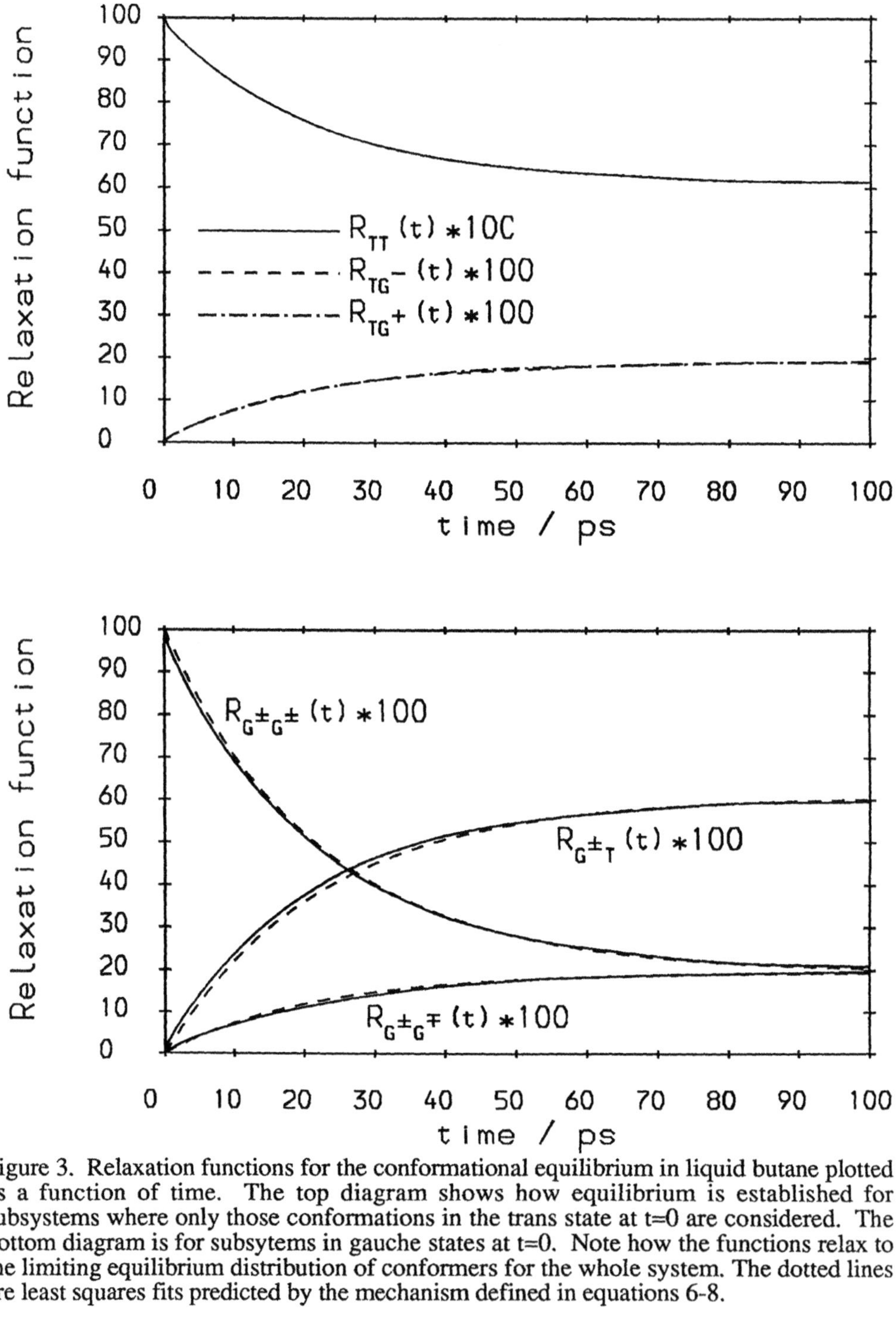

Figure 3. Relaxation functions for the conformational equilibrium in liquid butane plotted as a function of time. The top diagram shows how equilibrium is established for subsystems where only those conformations in the trans state at t=0 are considered. The bottom diagram is for subsytems in gauche states at t=0. Note how the functions relax to the limiting equilibrium distribution of conformers for the whole system. The dotted lines are least squares fits predicted by the mechanism defined in equations 6-8.

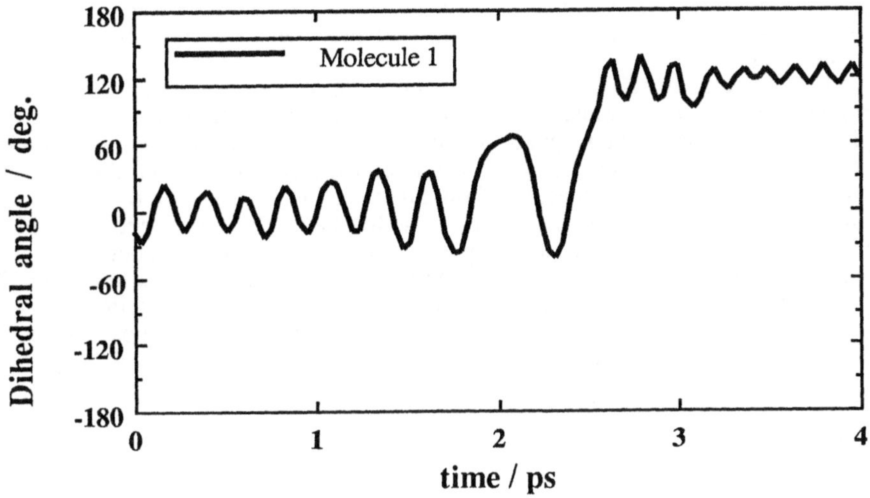

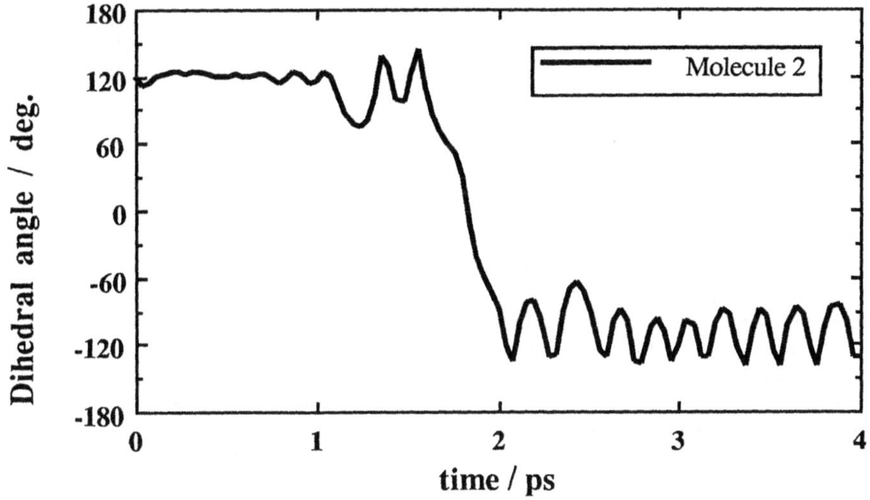

Figure 4. Variation of the dihedral angle α with time for two selected molecules in liquid butane. For gauche states $\alpha = 120^o$ whilst for trans states $\alpha = 0$. This illustrates typical T→G and G→G direct transitions. Note also the large amplitude of torsional oscillations in the region of these transitions.

and the best fit values of the rate constants are shown in Table II. The predicted rates of change of concentration are seen in Fig. 3 to provide an extremely good fit to the simulation data.

The above analysis shows that the time scale of conformational fluctuations is of the order of 100 ps and this must set a lower limit to the preparation time for an equilibrium sample of liquid butane. It also points to a useful general criterion for equilibrium in any system of flexible molecules; the condition is that the conformational distribution of an arbitrarily chosen subset of molecules should decay to an average value which is identical to that evaluated for the whole system. On this basis the most recent liquid butane simulations [14] should be fully sampling the equilibrium fluctuations but at least as far as the conformational properties are concerned some of the results of earlier work [7,8,11] must be viewed with some circumspection. The direct method is quite general and is easily adapted to monitor conformational fluctuations in more complex chain molecules. In the case of dense polymers close to the glass transition on a time scale of 100ps there is a substantial variation of $<X_T^i>$ about $<X_T>$ and in this case much longer times are necessary to satisfy the conformational equilibrium condition [15].

IV. "Solvent" effects on conformational equilibria

Gauche and trans conformers of even the simplest of flexible molecules butane have a significantly different shape. Torsional motions can cause even more dramatic changes in more complex molecules such as peptide chains or proteins [16]. In the case of butane the gauche form has a more compact structure the trans form and some time ago the suggestion was made that molecular interactions in the dense liquid should favour the gauche configuration and that this should cause the ratio of trans to gauche isomers to be different in the equilibrium liquid than in the dilute gas [10].

Early simulations of liquid butane seemed to support this prediction as is seen from Table II. It will be explained below however that the most recent analysis taken with much longer simulations show that to within fairly small limits of error solvent effects on conformations in the case of liquid butane are extremely small. Considerable confusion arose in previous studies as a result of incorrect calculation of the expected value of $<X_T>$ for the dilute gas itself which often lead to the erroneous claim of the existence of solvent effects [11]. In molecules where constraints are substituted for intramolecular forces the simple Boltzmann averaged result has to be modified [3]. In the case of bond constraints this is a relatively small effect but it is quite a substantial correction in the case of valence angle constraints, which were used for all the simulations reported in Table II. Each configuration contributing to the Boltzmann average has to be weighted using the determinant of the metric tensor of the particular conformation as indicated in the Table. Calculation of this determinant is straightforward for butane but is certainly a non-trivial task for any longer chain molecule [3]. As mentioned previously this is a substantial reason for choosing explicit angle bending forces rather than angle constraints to describe chain molecules.

In Fig. 5 we show the equilibrium distribution of dihedral angles in liquid butane from the latest simulation [14]. It is seen that the liquid and ideal gas results are almost superimposable over the entire distribution. The small discrepancies evident at the peak maxima are consistent with a ~1% excess of gauche states in the liquid for one of the gauche peaks is purely a residual statistical error in the simulation data. There are few well documented analyses for more complex single component hydrocarbon liquids. Although solvent effects are to be expected and indeed have been demonstrated using molecular

TABLE II: RESULTS FROM SIMULATIONS OF LIQUID BUTANE USING THE POTENTIAL MODEL DESCRIBED IN REF. A

property	A	B	C	D	E
ensemble	N,V,E	N,V,T_m	N,V,E	N,V,T_m	N,V,E
N	64	64	64	64	108
ρ_m^*	0.365	0.365	0.365	0.365	0.365
t/ps	14	333	249	203	1800
$<T>$/K	292	291	285	292	291±1
$<p_{corr}>$/bar	34	59	-15	115	17±5
$<\Phi_\alpha>$/J mol^{-1}	3.08	2.76	2.69	2.73	2.67±0.02
$<\Phi_{inter}^{corr}>$/kJ mol^{-1}	-19.5	-19.5	-19.6	-18.1	-19.6±0.01
%trans	54	61	59	61	61±1
$\%trans_{ideal}^{corr}$ (a)	62	62 (67)[b]	63	62	62
transition rate/ns^{-1}	49	45	-	45	38
k_{TG}/ns^{-1}	-	-	-	9.5	8.8±0.4
k_{GT}/ns^{-1}	-	-	-	29	27.2±1.1
k_{GG}/ns^{-1}	-	-	-	-	9.2±0.6

references overleaf

TABLE II (CONT)

A. Ryckaert J-P. and Bellemans A., 1978 Faraday Disc. Chem. Soc., **66**, 95
B. Edberg R., Evans D.J. and Morriss G.P., 1986, J. Chem. Phys., **84**, 6933
C. P.A. Wielopolski and Smith E.R. , J. Chem. Phys., 84, 6940 (1986)
D. Edberg R., Evans D.J. and Morriss G.P., 1987, J. Chem. Phys., **87**, 5703
E. Brown D. and Clarke J.H.R., submitted for publication in J. Chem. Phys.
a - from E using the expression

$$x_T = \int_{-\pi/3}^{\pi/3} d\alpha\ [g^{1/2}(\alpha)\exp(-\beta\phi(\alpha))] \Big/ \int_{-\pi}^{\pi} d\alpha\ [g^{1/2}(\alpha)\exp(-\beta\phi(\alpha))]$$

g, the determinant of the metric tensor for the butane model was calculated using the expansion formula given by van Gunsteren W.F. 1980, Mol. Phys., **40**,1015.

b - value quoted in B

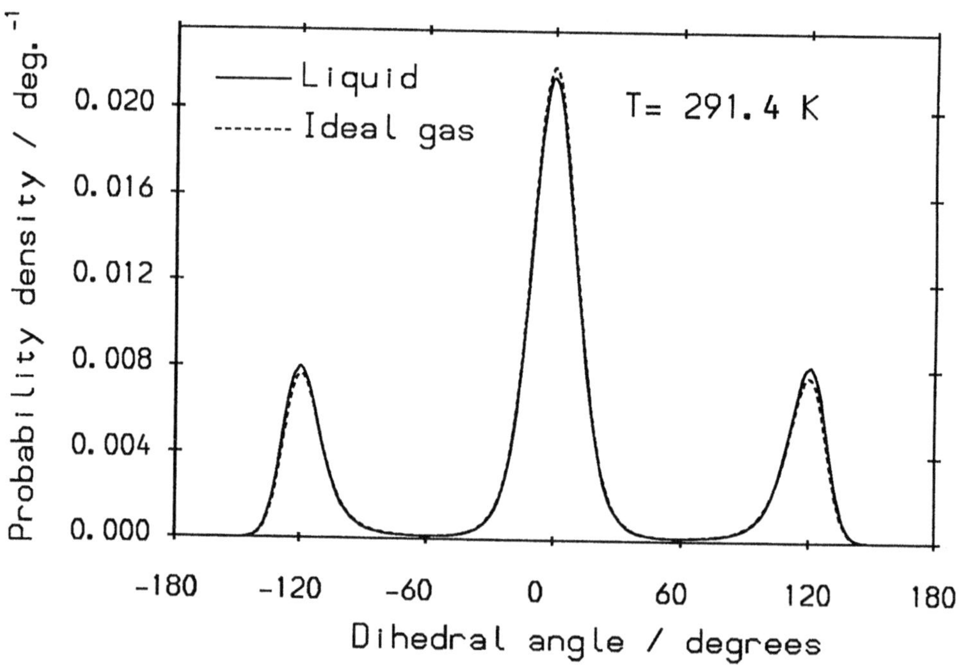

Figure 5. The dihedral angle distribution in liquid butane from an 800 ps simulation at 291 K [14] . The dotted line shows that predicted for the ideal gas at the same temperature.

dynamics for *solutions* of chain molecules [17], it has long been suggested that the pure liquid is a special case where the surrounding molecules act like a theta-solvent in which there is no net effect on conformational distibutions as a result of intermolecular interactions [18].

V. Molecular Flexibility and Transport Properties

There are two ways in which conformational effects might couple to transport properties in liquids composed of chain molecules. Firstly since the rate of conformational change is relatively slow it is possible that different conformers will have for instance different diffusion coefficients. The liquid is then to all intents and purposes considered as a simple mixture of conformers. Attempts have been made to examine this effect for short time molecular diffusion in liquid butane [13] but the results show no statistically significant difference in the behaviour of the two conformers. The second effect is a more subtle one in which the higher frequency intramolecular motions resulting from conformational degrees of freedom might *dynamically* couple with the intermolecular motions giving rise to the transport property. The importance of such mechanisms has been examined in the case of liquid hexane for the case of diffusion [19] and shear viscosity [20,21].

In this work close comparisons were made between the behaviour of hexane (using a model similar to A in Table I) and a second liquid called "flexane" which was composed of molecules identical in all respects with the hexane model except that the potentials restraining the torsional motions were set to zero. This kind of comparison would be

TABLE III. THERMODYNAMIC, STRUCTURAL AND TRANSPORT DATA FOR MODEL HEXANE AND FLEXANE FROM A COMBINATION OF CONSTANT TEMPERATURE AND CONSTANT ENERGY SIMULATIONS. E IS THE TOTAL *INTERMOLECULAR* POTENTIAL ENERGY.

Property	Hexane		Flexane	
T / K	200	300	200	300
p / bar	560±70	2370±20	290±50	2120±40
$-E$ / J mol^{-1}	48800±60	46200±90	48410±80	45960±60
G^{∞} / GPa	4.3	-	4.1	-
$\langle r_{16}^2\rangle^{1/2}$ / nm	0.557±.001	0.556±.002	0.520±.002	0.517±.001
D / 10^{-9} m^2 s^{-1}	0.35±.03	0.97±.08	0.6±.1	1.12±.08
$\eta(0)$ / mPa s	2.9±.01	1.5±.01	1.5±.01	1.5 [b]

a. data from Clarke J.H.R., and Brown D., 1986, Mol. Phys., **58**, 815; 1987, J. Chem. Phys. **86**,1542.

b. estimated value

extremely difficult to achieve with real experimental liquids and is an example of the usefulness of computer simulation in demonstrating the effect that a certain molecular feature can have on bulk properties. A selection of the properties of these two liquids is shown in Table III.

Removal of the torsional barriers in hexane caused a decrease in the viscosity and an increase in the diffusion coefficient by a factor of about two at 200 K. Although there are small differences in the static properties of the two liquids it is extremely difficult to explain the observed changes as due to structural effects. Flexane does adopt a somewhat more compact intramolecular structure but the rms end-to-end distance $<r_{16}^2>^{1/2}$ is only about 7% less than that of hexane and the effect on the bulk thermodynamic properties such as the pressure is quite small. Also the value of $<r_{16}^2>^{1/2}$ in hexane is very similar at the two temperatures.

The results are very suggestive of strong coupling between internal and molecular centre of mass motions. Comparison of the velocity autocorrelation functions at 200 K shows that the difference in diffusion coefficients can be accounted for in terms of the short time behaviour. The more rigid hexane molecules show a pronounced "cage effect" as shown in Fig. 6 whereas in the case of flexane the facile torsional motions soften themolecular collisions. The second important result was that the transport coefficients change much more rapidly for hexane as the temperature is raised so that at 300 K both the diffusion coefficient and visocosity have very similar values for the two molecules. This

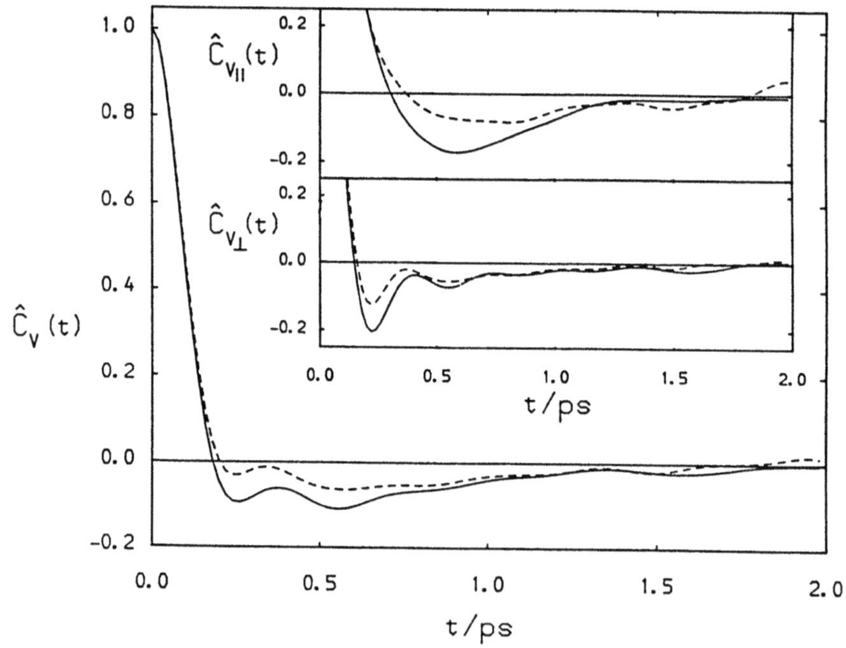

Figure 6. Normalised centre-of-mass velocity autocorrelation functions for hexane (solid lines) and flexane (broken lines). Inset is shown a resolution of the correlation function into components parallel and perpendicular to the molecule end-to-end vectors [19].

shows that the torsional motions in liquid hexane are already sufficiently excited at 300 K for it to behave as a "flexible" molecule. It was concluded that flexibility in this context is best described in terms of an increased amplitude of torsional vibrations. It certainly has little to do with actual conformational transitions since these occur at a rate which is two orders of magnitude slower. Detailed analysis showed that an rms angular displacement of about 30^0 was sufficient to provide efficient coupling. Transitions between distinct trans and gauche conformational states are not however involved in the coupling since they occur at too slow a rate.

Acknowledgement

The author is grateful to Dr. D. Brown for many useful discussions.

References

1. Ryckaert J-P and Klein M.L. , 1986, J. Chem. Phys., **85**, 1613
2. Ryckaert J-P, Klein M.L. and I.R. McDonald, 1987, Phys. Rev. Lett., **58**, 698
3. van Gunsteren W.F., 1980, Mol. Phys., **40**,1015
4. Brown D. and Clarke J.H.R., 1986 J. Chem. Phys., **84**,2858
5. Brown D. and Clarke J.H.R., 1988, Molecular Simulation (in press)
6. van der Ploeg P. and Berendsen H.J.C. , 1982 J. Chem. Phys., **76**, 3271
7. Weber T.A., 1978, J. Chem. Phys., **69**, 2347
8. Ryckaert J-P. and Bellemans A., 1978 Faraday Disc. Chem. Soc., **66**, 95
9. Steele D., 1985 J. Chem. Soc. Farad. Trans. II, **81**, 1077
10. Chandler D., 1978, J. Chem. Phys., **68**, 2959
11. Edberg R., Evans D.J. and Morriss G.P., 1986, J. Chem. Phys., **84**, 6933
12. Helfand E., 1978, J. Chem. Phys., **69**, 1010
13. Marechal G., Ph.D. Thesis (Universite Libre de Bruxelles, 1988)
14. Brown D. and Clarke J.H.R., submitted for publication in J. Chem. Phys.
15. Brown D. and Clarke J.H.R., submitted for publication in Macromolecules
16. McCammon J. and Harvey S.C., "Dynamics of Proteins and Nucleic Acids", Cambridge University Press, 1987
17. Smit B.P., Van der Put A., Peters C.J., de Swaan Arons J. and Michels J.P.J., 1988, Chem. Phys. Lett., **144**, 555
18. Flory P.A., "Principles of Polymer Chemistry", Cornell University Press, Ithaca, New York, 1953
19. Clarke J.H.R., and Brown D., 1986, Mol. Phys., **58**, 815
20. Brown D. and Clarke J.H.R.,, 1983, Chem. Phys. Lett., **98**, 579
21. Clarke J.H.R., and Brown D., 1987, J. Chem. Phys. **86**,1542

COMPUTER MODELLING OF OXIDE SURFACES AND INTERFACES

P. J. Lawrence and S. C. Parker, School of Chemistry, University of Bath, U.K.

1. Introduction

The aim of this article is to describe an approach for modelling the structures and energetics of surfaces and interfaces of inorganic solids using computer simulation techniques. The wide range of applications of these methods on surfaces and grain boundaries has recently been reviewed by Mackrodt [1]. Much of this work has been confined to the cubic rocksalt oxides MgO, CaO and NiO [2,3], and only recently has been extended to more complicated materials such as UO_2[4] and α-Al_2O_3[5]. We will illustrate the application of these techniques by considering both perfect surfaces and point defects at surfaces in the corundum-structured oxides. The principal surface defects we shall consider are impurities as they are of fundamental importance in determining the properties of the materials i.e. impurities will affect sintering, densification and the catalytic properties of the materials [6,7,8]. A comprehensive knowledge of the effects of impurity level concentrations on behaviour is therefore of critical importance in the design and implementation of technologically important materials and processes. The advent and improvement of surface sensitive optical techniques such as AES, SEM and LEED has allowed more extensive experimental investigations of the role of impurities to take place. These techniques can now be coupled with theoretical calculations using static simulation methods to give a better picture of the effects of both iso- and aliovalent impurities on ceramic surfaces.

Before discussing the approach for modelling impurities we shall describe the method for obtaining the structures and energies of the most stable surfaces.

2. Surface Methodology

The low energy and hence the most common surfaces of a crystal are generally those of low Miller index. These planes are the closest packed with large interplanar spacings. However in ionic crystals other constraints also apply. If the Madelung sums are not to diverge with increasing crystal size then the crystal must not only be electrically neutral but also have no net dipole moment perpendicular to

C.R.A. Catlow et al. (eds.), Computer Modelling of Fluids Polymers and Solids, 219–248.

the surface [9]. Bertaut [10] demonstrated that when there is a dipole moment perpendicular to the surface, the surface energy diverges and is infinite. Such surfaces are therefore unstable, and cannot occur naturally without the adsorption of foreign atoms or surface roughening [11], which will have important consequences in the interpretation of, for instance Low Energy Electron Diffraction (LEED) data.

The strategy is to use a two region approach for calculating the energies of extended planar defects such as free surfaces and grain boundaries. The crystal is considered as a stack of planes and assumed to be periodic in two dimensions, with the stack being divided into two regions: a region I adjacent to the interface where the ions are allowed to relax to their equilibrium configurations and a region II in which the ions are held fixed relative to each other but the region as a whole may move (Figure 1).

This allows for dilation or contraction of the crystal. Energies calculated by this method are therefore energies at constant area. The surface energy U^s per unit area can thus be defined as;

$$U^s = \sum_{\lambda=1}^{m} \left[U(\lambda) - U(m) \right] / A \qquad (1)$$

where the energy is summed over all ions in plane λ, m is a bulk plane and U(m) is the bulk energy.

2.1 The electrostatic contribution

A second important feature of modelling ionic crystals is that the long range Coulombic interactions need to be evaluated, and because they act over a long range suitable algorithms need to be used. The treatment of the Coulombic interactions in a 2-dimensional lattice was first considered by Parry [12,13] who considered a limiting case of the Ewald summation used for 3-dimensional bulk problems (see Jackson, this volume). Heyes *et al* [14] give a fuller treatment of the long range Coulombic potential and forces, including the effect of a resultant dipole moment perpendicular to the surface which was initially neglected by Parry.

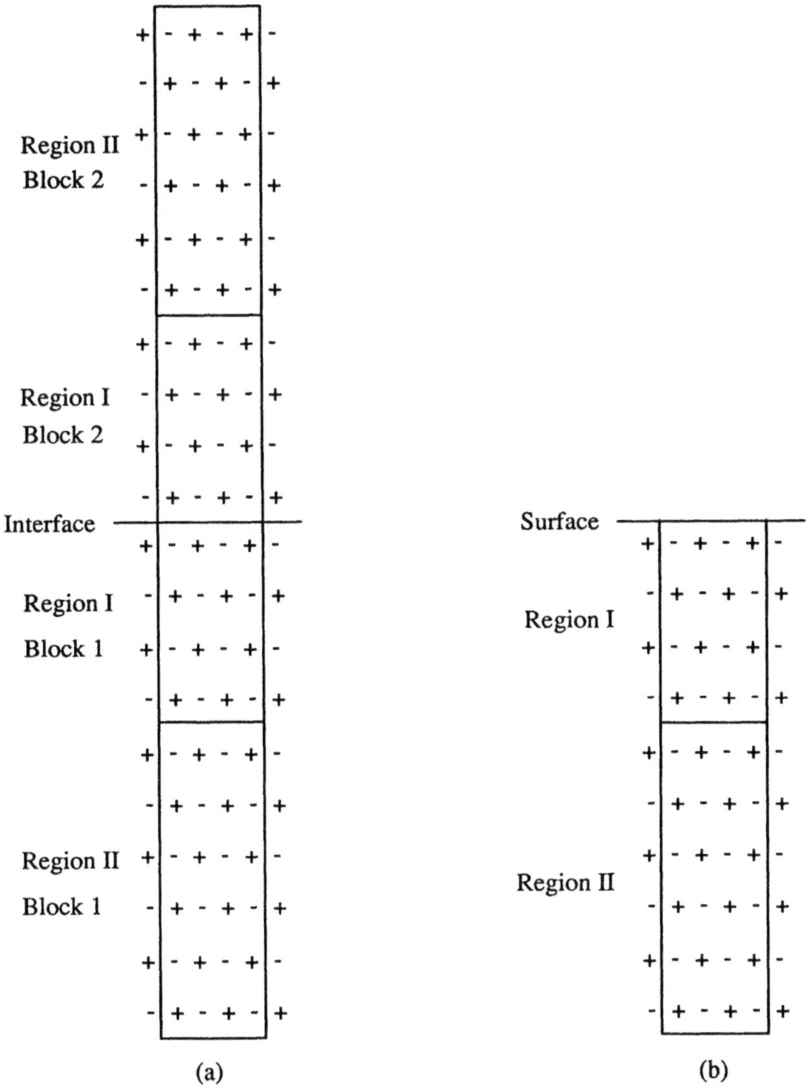

Figure 1. Schematic representation of the crystal regions for a) a block calculation, b) a surface calculation

The Coulomb contribution to the energy (Madelung) of the block is given by;

$$U^{MAD} = \frac{1}{2}\sum_{i}^{N_1} q_i \sum_{j}^{N_2} q_j \left(|\underline{r}_{ij} + \underline{l}|\right)^{-1} \tag{2}$$

N_1= the number of ions in region I

N_2= the number of ions in regions I and II

$q_i$and q_j are the charges of ions with positions r_i and r_j

$\underline{r}_{ij} = \underline{r}_i - \underline{r}_j$

$\underline{l}$ = two dimensional lattice vector

By taking advantage of the two-dimensional periodicity of the real space we can transform the summation of the first term into reciprocal space and the second into real space. The reciprocal space contribution is summed over all reciprocal lattice vectors $\underline{k}$, and consists of two terms, a term equivalent to $\underline{k}\neq 0$ and a $\underline{k}=0$ term. The latter term can be neglected in the Ewald method due to the constraint that the charge distribution in the crystal must be neutral. For a surface the $\underline{k}=0$ term will always have a finite value, although in certain cases e.g. an unrelaxed ⟨100⟩ MgO surface, it will be independent of the perpendicular distance u_{ij}. The magnitude of this term will be directly affected by the degree of relaxation i.e. the change in u_{ij}, at the interface, but it <u>must</u> be implicitly included in the relaxation. The reciprocal lattice contribution to the electrostatic energy is given by:-

$$\phi_{ij}^{G} = \frac{\pi}{A}\left[-2\, u_{ij}\, \mathrm{erf}(Gu_{ij}) - \frac{2\exp(-G^2 u_{ij})^2}{\pi^{1/2} G}\right]$$

$$+ \frac{\pi}{A}\sum_{\underline{k}\neq 0} \frac{\exp(i\underline{k}.\underline{p}_{ij})}{k}(B+C) \tag{3}$$

where A is the area of the unit cell, G is an arbitrary parameter, chose to speed convergence, and

$$B = \exp(ku_{ij})\ \mathrm{erfc}\left(\frac{k}{2G} + Gu_{ij}\right) \qquad (4)$$

$$C = \exp(-ku_{ij})\ \mathrm{erfc}\left(\frac{k}{2G} - Gu_{ij}\right) \qquad (5)$$

The term $\underline{r}_{ij}$ is resolved into two components, $\underline{u}_{ij}$ perpendicular to the interface and $\underline{p}_{ij}$ in the plane:

$$\underline{r}_{ij} = \underline{u}_{ij} + \underline{p}_{ij} \qquad (6)$$

erf(x) and erfc(x) are the standard error and complementary error functions. The real space contribution to the electrostatic energy is:

$$\phi_{ij}^{R} = \sum_{\underline{l}} R_{ij}^{-1}\,\mathrm{erfc}(GR_{ij}) \qquad (7)$$

where $\underline{R}_{ij} = \underline{r}_{ij} + \underline{l}$.

The self interaction term is calculated and then removed by carrying out the real lattice summation without the singular term $\underline{l}=0$, and by subtracting from the reciprocal lattice contribution the term $2G/\pi^{1/2}$.

The expressions used, give the electrostatic potential at any plane in terms of a potential due to a planar sublattice of charge q at a perpendicular distance z. For distances z greater than a few interplanar spacings the contribution from a planar sublattice reduces to a particularly simple form:

$$V(z) = 2\pi\left(\frac{q.z}{A}\right) \qquad (8)$$

Thus the potential will be infinite at infinite distances from a plane, although it should be noted that when the crystal is constructed as a neutral block the infinities will cancel e.g. Figure 2a. This is because the crystal will be symmetric about a central point in the block. If there is a dipole moment perpendicular to the surface, figure 2b, the self cancellation will not take place and the potential will have a finite value at large distances from the surface.

This result can have important consequences in determining the relative stability of different crystal faces. For example, the stacking sequence of the ⟨111⟩ rocksalt surface as shown in Figure 2b consists of alternate charged planes producing a dipole moment perpendicular to the surface. This leads to a charged surface which can be stabilised if charge is transferred from the lower to the upper face [15], as illustrated in Figure 2c. The electric field at the centre planes will then be zero, due to the self cancellation of the charges above and below the centre plane. This gives a finite surface energy for any block size. The polar surface of ionic crystals therefore cannot be stable without substantial reconstruction or adsorption of additional charge. This confirms the conclusion of Levine and Mark [16,17] that the charged surfaces of an ionic crystal of general formula MX will be less stable than the uncharged surfaces. We can use this result to identify the most stable surfaces without recourse to calculate every possible permutation explictly.

This can also be used to determine which ions terminate at the surface in the case of the ⟨111⟩ fluorite surface, e.g. UO_2. There are three possible stacking sequences for this surface, with either a cation surface or one of two anion surfaces, two of which are shown in Figure 3. However, by inspection only one surface will have no net dipole perpendicular to the surface and hence it will be the most stable.

2.2 Surface energies

One of the difficulties in this field is finding reliable experimental data with which to compare. This is particularly true for complex oxides. However in the case of the rocksalt structured materials Tasker [4,5] showed that the agreement is good especially for the more ionic materials. This approach has recently been extended to more complex oxides e.g. the hexagonally close packed structure of α-Cr_2O_3 which has a large number of low index, and potentially low energy surfaces. The corundum structured transition metal oxides have no simple cleavage planes such as the ⟨100⟩ rocksalt surface, though they have been reported to cleave along the ⟨1012⟩ plane [18] which may indicate that this surface is the most stable.

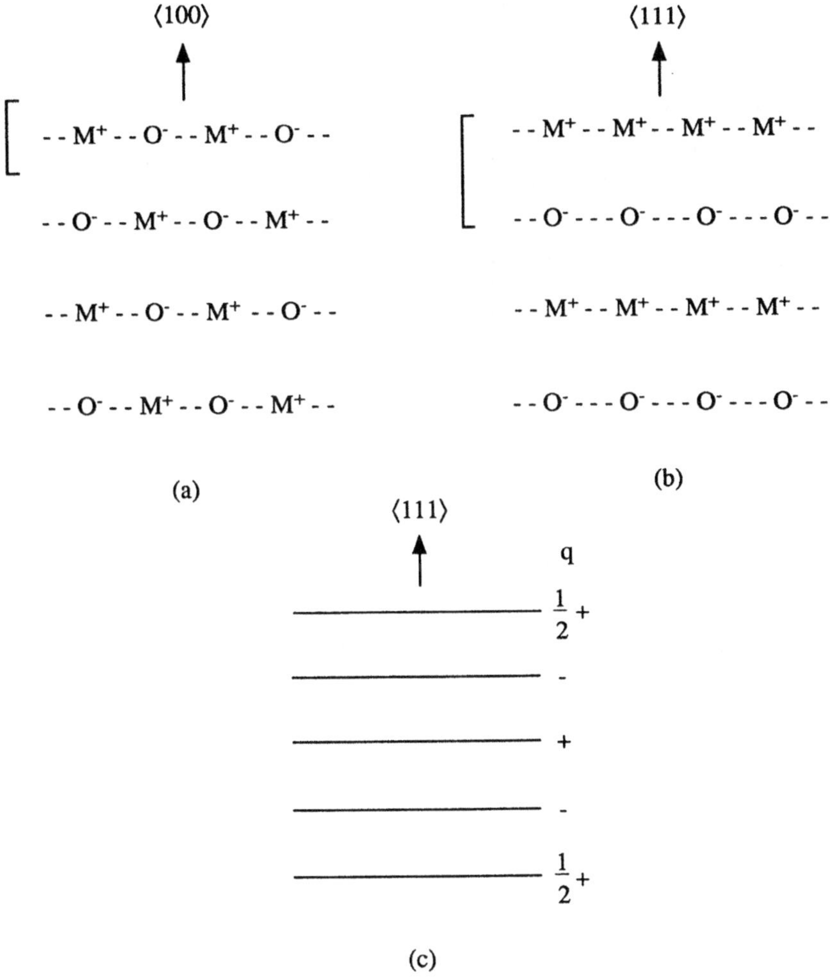

Figure 2: Distribution of Charges on Planes for Stacking Sequences Parallel to Surface. (a) ⟨100⟩ Rocksalt Surface; (b) ⟨111⟩ Rocksalt Surface; (c) Stabilised ⟨111⟩ Rocksalt Surface

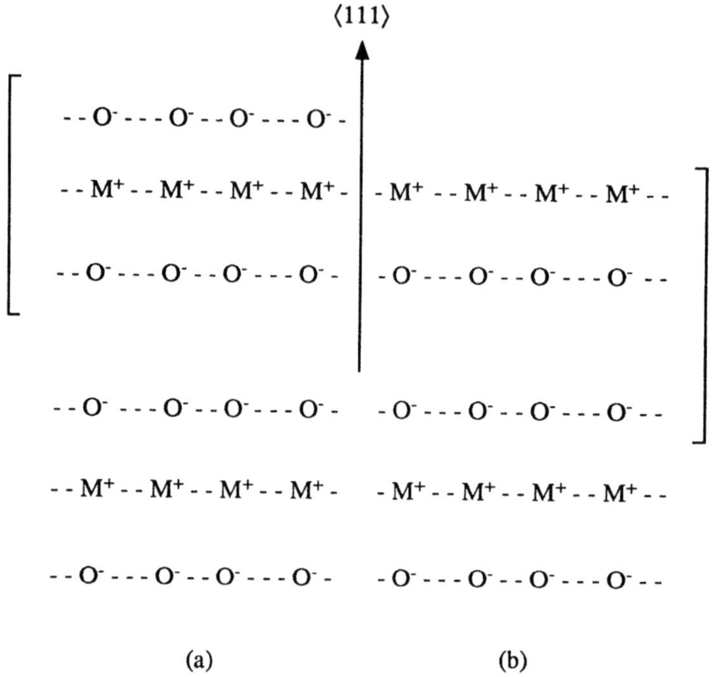

Figure 3: Schematic Representation of the Fluorite ⟨111⟩ Surface, (a) with terminating anion plane; (b) with terminating cation plane

The surface energies for the five lowest index planes of Cr_2O_3 are given in table 1. Both the unrelaxed and relaxed surface energies are given for both empirical and non-empirical potential (see the chapter of Jackson and chapters (1) and (2)). As can be seen from Table 2 the relaxed energies are appreciably lower than the unrelaxed energies for both potentials. The values for the unrelaxed energies for each plane are approximately equal for both potentials but there is a significant change upon relaxation from the simple bulk termination. The empirical potential in general gives surface energies 0.4 Jm^{-2} less than the non-empirical potential except for the $\langle 10\bar{1}0\rangle$ prism plane for which the difference between the potentials is only 0.17 Jm^{-2}. The magnitude of the energy changes on relaxation can be seen by considering the non-empirical potential. In the case of the $\langle 10\bar{1}0\rangle$ plane there is a reduction in energy of 2.72 Jm^{-2} or 54.5% whereas for the $\langle 10\bar{1}2\rangle$ plane the reduction in energy is 0.76 Jm^{-2} or 27.4%. The change in surface energy is seen to vary with each plane for both potentials, although the relative ordering of stability before relaxation is the same for both potentials:

$$\langle 10\bar{1}2\rangle < \langle 11\bar{2}0\rangle < \langle 10\bar{1}1\rangle < \langle 0001\rangle < \langle 10\bar{1}0\rangle$$

The ordering of the relative stability of the surfaces changes with relaxation, giving different orders for the two potentials:

Non-empirical

$$\langle 10\bar{1}2\rangle < \langle 0001\rangle \leq \langle 10\bar{1}0\rangle \leq \langle 11\bar{2}0\rangle < \langle 10\bar{1}1\rangle$$

Empirical

$$\langle 0001\rangle \leq \langle 10\bar{1}2\rangle < \langle 11\bar{2}0\rangle < \langle 10\bar{1}1\rangle \leq \langle 10\bar{1}0\rangle$$

For the non-empirical potential the $\langle 10\bar{1}2\rangle$ plane is predicted to be the most stable with the difference between the $\langle 0001\rangle$ and $\langle 10\bar{1}0\rangle$ planes being inherently negligible within the limitations of these calculations. This is in contrast to the empirical potential which calculates the $\langle 0001\rangle$ and $\langle 10\bar{1}2\rangle$ planes to be at least 0.2 Jm^{-2} more stable than any other plane. This result also illustrates the advantage of using independently derived potentials as this gives a guide to the sensitivity of the results to changes in the potentials and allows us to set the limits of our uncertainty, particularly when there is little experimental work with which to compare. As noted above, experimental determination of the surface energies are often difficult. An alternative approach for comparing the relative stability of oxide surfaces is to compare the equilibrium morphology.

Surface	Unrelaxed Energy (Jm^{-2})	Relaxed Energy (Jm^{-2})
0001	4.59	1.61
$10\bar{1}0$	5.02	2.10
$11\bar{2}0$	3.42	1.90
$10\bar{1}1$	4.34	2.05
$10\bar{1}2$	2.77	1.70

Table 1: Calculated Perfect Surface Energies (Empirical Potential) for Cr_2O_3

Surface	Unrelaxed Energy (Jm^{-2})	Relaxed Energy (Jm^{-2})
0001	4.77	2.21
$10\bar{1}2$	2.77	2.01
$10\bar{1}0$	4.99	2.27
$11\bar{2}0$	3.53	2.35
$10\bar{1}1$	4.41	2.72

Table 2: Calculated Perfect Surface Energies (Non-empirical Potential) for Cr_2O_3

2.3 Crystal Morphology

The equilibrium crystal morphology of a material can be determined by applying Wulff's Theorem [19], although it was Gibbs [20] who first proposed that the equilibrium form of a crystal should posses minimal total surface energy for a given volume i.e.

$$\gamma_{cryst} = \sum_i \gamma_i A_i = \text{minimum at constant volume} \qquad (9)$$

where γ_i and A_i are the specific surface energy and specific area if the ith crstallographic face. Wulff proposed that the shape would be defined such that h_i, the face normal vector from a point within the crystal would be proportional to γ_i or

$$h_i = \lambda\gamma_i \qquad (10)$$

where λ is a constant depending on the absolute size of the crystal. This relationship only holds for small crystals, and can typically be compared with experimentally grown crystals which are less than one micron in diameter. This has been implemented by Davies [21] to calculate the equilibrium crystal morphologies. Figures 4a and 4b show the calculated crystal morphology of Fe_2O_3 before and after surface relaxation compared with that determined experimentally. The calculated morphology assuming that the surfaces are simply bulk terminations is totally unlike the experiment while the calculated morphology which includes relaxation of the surface is quite similar to experiment. The major difference is that the basal plane is much more noticeable on the experimentally grown crystal than on the calculated material.

3. Impurity segregation to surfaces

One of the major problems facing the modeller is intrepreting the experimental results; one approach is by comparison with experimental isotherms. The commonly used Langmuir/Arrhenius isotherm assumes the heat of segregation is independent of temperature. We will show that this is merely a special solution of a non-Arrhenius isotherm which allows for the variation of the heat of segregation with surface coverage/concentration of the impurity ion. We will then demonstrate that the

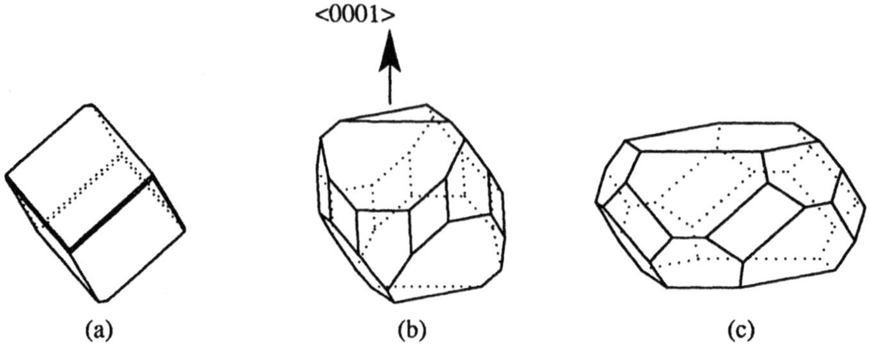

Figure 4: The morphology of Fe_2O_3 (a) calculated assuming that the surfaces were simple bulk terminations, (b) calculated after relaxation and (c) determined experimentally.

segregation of the isovalent impurities Y^{3+}, La^{3+}, and Al^{3+} to low index planes of Cr_2O_3 depends strongly on which surface is considered. From the derived heats of segregation it is possible to evaluate the effect of temperature and bulk impurity concentration on the equilibrium surface coverages. The problem is that the isotherms are often experimentally difficult to measure and hence we suggest an alternative strategy, namely comparison with the equilibrium surface morphologies of crystals.

3.1 Isovalent segregation to the surfaces of chroimia

In this section we show the heats of segregation of Al^{3+}, Y^{3+}, and La^{3+} to the $\langle 0001 \rangle$, and $\langle 11\bar{2}0 \rangle$ in Cr_2O_3 and how the equilibrium surface coverage varies with impurity concentration and the equilibrium crystal morphology.

Figure 5 shows the calculated heats of segregation for Al^{3+}, Y^{3+}, and La^{3+} at the $\langle 0001 \rangle$ plane of Cr_2O_3. These energies are obtaining by performing the surface energy minimisation as described for the perfect surface except that some of the ions are replaced with impurities. The segregation energy is the energy difference between placing the impurity at the surface compared to the bulk. For all the impurities the heats of segregation are approximately constant for all coverages i.e. the heat of segregation is independent of coverage. There is a small increase in the heat of

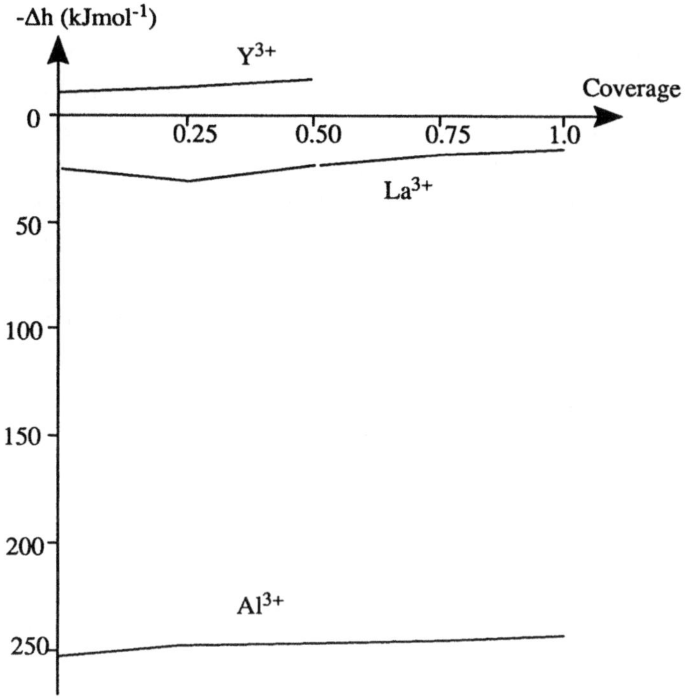

Figure 5: Calculated Heats of Segregation as a function of Coverage

at the Basal Plane

segregation from -27 $kJmol^{-1}$ to -12 $kJmol^{-1}$ for La^{3+}, but within the accuracy limits of both experiment and calculation we can say that this is constant. The two interesting features are the positive segregation energies for Y^{3+}, and the surprisingly large negative heats of segregation of Al^{3+}. Following the simple argument that the main driving force to segregation is the reduction in elastic strain, we might expect Y^{3+} and La^{3+} both substantially larger cations than Cr^{3+} to have a strong driving force for segregation, with there being no driving force to segregation for Al^{3+}. This would be in accordance with the segregation of Be^{2+}, Ca^{2+}, Sr^{2+}, and Ba^{2+} to the ⟨001⟩ surface of MgO and K^+ at the ⟨111⟩ surface of Li_2O [1]. Considering first the segregation of Y^{3+}. As mentioned in the previous section there is appreciable relaxation in the surface region of the basal plane of Cr_2O_3. This will serve to reduce the volume of the surface region so that impurities bulkier than Cr^{3+} will be more restricted. In contrast La^{3+} has a negative heat of segregation. Thus the heat of segregation involves a delicate balance between differences in steric strain caused by the size of the impurity and surface relaxation, which involves electrostatic and short range interactions. Mackrodt has concluded that only trivalent ions larger than Y^{3+} will segregate to the basal plane of alumina [22], which would give some support to our observations for Y^{3+} and La^{3+} but not those for Al^{3+}. Trivalent aluminium ions are calculated to have a large heat of segregation of ≈245 $kJmol^{-1}$. This would seem strange as one would expect aluminium ions to dissolve into the bulk as they should not induce any elastic strain into the host lattice. If the heat of segregation is calculated before relaxation of the crystal it is ≈220 $kJmol^{-1}$, indicating that before relaxation, aluminium ions do prefer to be in the bulk of the crystal. Al^{3+} ions have a radius approximately 0.1Å less than that of Cr^{3+}, so Al^{3+} ions segregating to the surface will cause less steric hindrance. We therefore conclude that the negative heat of segregation is solely due to the dominance of surface relaxation.

Figure 6 shows the calculated values of $\Delta h(x)$ to the $\langle 11\bar{2}0\rangle$ prism plane, as a function of coverage. Here the calculations suggest that at low coverages all three impurities will segregate from the bulk to the surface, which is more open when relaxed than the ⟨0001⟩ plane. As coverage increases especially for La^{3+}, steric factors take over at the surface. This is qualitatively similar to the calculated behaviour of La^{3+} at the $\langle 11\bar{2}0\rangle$ plane of alumina [22]. Y^{3+} appears to have a minimum in the heat of segregation at approximately half a monolayer coverage, suggesting the possible formation of a coherent ordered second phase, although further calculations are required to determine more accurately the exact coverage at which the minimum occurs. Again Al^{3+} ions segregate to the surface, which we again ascribe to surface relaxation overcoming steric hindrance.

There is currently little experimental data on the segregation of impurities to many complex oxides such as chromia. An alternative approach to simulating the isotherm is to compare the morphologies with known concentrations of impurities. The effect of impurities on segregation energies can be illustrated by considering Al^{3+} impurities in chromia. As the Al^{3+} coverage increases the basal plane is seen to become more and more dominant until at 100% or full monolayer coverage the crystal is predicted to be plate like as shown in Figure 7.

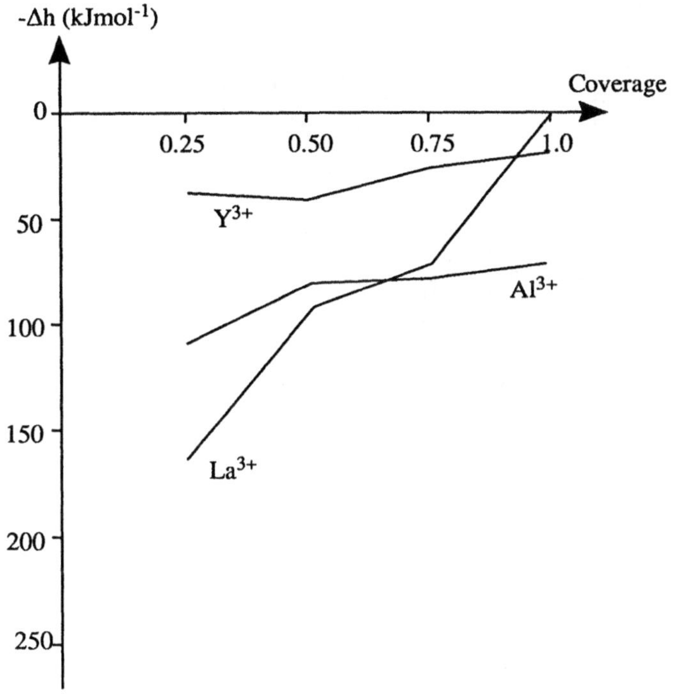

Figure 6: Calculated Heats of Segregation to the ⟨11$\bar{2}$0⟩ plane as a function of coverage

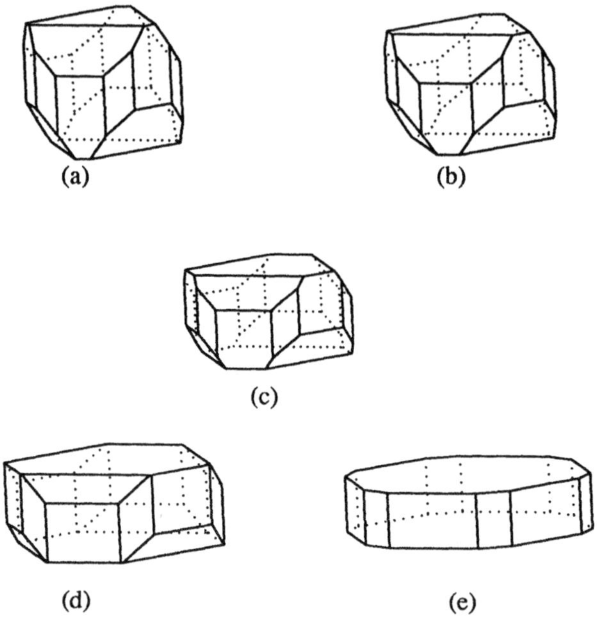

Figure 7: Equilibrium Crystal Morphology of Cr_2O_3 with increasing Al^{3+} surface coverage; (a) 0%; (b) 25%; (c) 50%; (d) 75%; (e) 100%

3.2 Segregation Isotherms

The experimental techniques used for surface analysis, measure surface concentrations of impurity that are assumed to be equilibrated with the bulk as a function of temperature. Heats or enthalpies of segregation are extracted from plots of log(surface concentration) against 1/T. These are often linear and assumed to be indicative of Arrhenius or Langmuir type behaviour:

$$x_s \propto x_b \exp\left({}^{-\Delta H}\!\big/_{k_B T} \right) \qquad (11)$$

where x_s is the surface atomic ratio i.e. the ratio of impurity occupied sites to unoccupied sites, x_b the bulk atomic ratio, ΔH the coverage independent heat of segregation, k_B Boltzmann's constant, and T the temperature. The assumption of coverage independent ΔH values is a crude approximation except at very low coverages, if simple notions of steric hindrance are applied. This has recently been extended by Mackrodt and Tasker [23] in which ΔH is allowed to vary with surface coverage. This assumes monolayer coverage, allowing a direct relationship between surface coverage and surface atomic ratio x_s.

The approach is to assume the system consists of two types of site, one surface and one bulk. The total free energy of the system G is written as:

$$G = \sum_i n_i^b g_i^b + \sum_i n_i^s g_i^s - kT\ln\Omega \qquad (12)$$

where n_i^b and n_i^s are the number of bulk and surface sites of ion/atom type i, with free energies g_i^b and g_i^s. The third term represents the configurational entropy. If the number of surface sites are assumed to be small a modified expression for x_s, which allows for the variation of Δg with surface coverage can be obtained:

$$x_s \propto x_b \exp\left[-\left(\Delta h + x_s (x_s + 1) \frac{d\Delta h}{dx_s} \right) \Big/ k_B T \right] \qquad (13)$$

The first step in modelling the isotherm is to calculate to the segregation energy as a function of coverage.

3.3 Aliovalent Impurity Segregation

Frenkel [24] first pointed out that pure ionic crystals in which Schottky disorder predominates should in thermal equilibrium possess a charged surface, and a region of space charge of opposite sign to the surface, as the free energies of formation of anion and cation vacancies differ. The accumulation of one type of point defect at a surface e.g. cation vacancies will result in a negatively charged surface and a region of positive charge beneath the surface. This charge distribution will further retard the formation of cation vacancies while enhancing the formation of anion vacancies. This was further confirmed by the work of Lehovec [25] on NaCl. Eshelby *et al* [26] noted that edge dislocations, which act as sources and sinks for vacancies, would also give rise to such charging effects. The analogue of the surface charge is a charge on the dislocation line itself with the space charge distributed with cylindrical symmetry around the dislocation line.

When a crystal contains a small concentration of impurities which possess an absolute value of charge different from that of the host ions in the lattice, there also arises a space charge region near the surface. Considering a rocksalt structured oxide containing divalent cation impurities, at temperatures where the vacancy concentration is essentially determined by the divalent cation concentration, the surface develops a negative charge and the space charge is positive. The resulting potential distribution serves to enhance the cation vacancy concentration while reducing that anion vacancy concentration hence maintaining electroneutrality in the bulk of the crystal.

A detailed treatment of the space charge is given by Yan *et al* [27], and Kleiwer and Koehler [28]. Their approach makes four major assumptions which are: (i) there are only two types of defect site available, one at the surface and one in the bulk, (ii) defect concentrations are low and hence non-interacting, (iii) impurities are located only on the cation sub-lattice and (iv) the surface is perfect with no surface irregularities. The free-energy is then minimised where the total free energy of a crystal width L is therefore given by:

$$F = \int_0^L dx \left[n_+(x)F^+ + n_-(x)F^- + n_B(x)(F^+ + F^- - B) + \frac{1}{2}\rho(x)\Phi(x) \right] - TS_c \qquad (14)$$

where $n_+(x)$ is the density of cation vacancies at x, $n_-(x)$ the density of anion vacancies at x, $n_B(x)$ the density of bound pairs at x, F the analogous free energies and S_c the configurational entropy. The charge density $\rho(x)$ is given by:

$$\rho(x) = e\{n_-(x) - n_+(x)\} \qquad (15)$$

and $\Phi(x)$ the electrostatic potential is given by Poisson's equation:

$$\nabla^2\Phi(x) = \frac{\partial^2\Phi(x)}{\partial x^2} = \frac{-4\pi\rho(x)}{\epsilon} \qquad (16)$$

where ϵ is the static dielectric constant. The free energy is then minimised with respect to the defect densities, $n_+(x)$, $n_-(x)$, and $n_B(x)$ to obtain the defect concentration at temperature T. On minimising the free energy and assuming that n_+/N, n_-/N and n_B/N are much less than one, the defect densities are found and are of the form:

$$n_+(x) = N\exp\left\{ -(F^+ - e\Phi(x))/kT \right\}$$

$$n_-(x) = N\exp\left\{ -(F^- - e\Phi(x))/kT \right\} \qquad (17)$$

$$n_B(x) = Nz_n\exp\left\{ -(F^+ + F^- - B)/kT \right\}$$

where N is the number of lattice sites per unit volume, z_n the number of nearest unlike neighbours of an ion, B the binding energy and k is Boltzmann's constant. Solving this problem for more complicated defect schemes is often difficult and requires many approximations. However, recent calculations of aliovalent defect segregation has shown that often we need only consider the segregation of neutral species, for example Mg^{2+}, Ca^{2+} and Fe^{2+} impurities in Cr_2O_3 [29] and Al_2O_3 [30] as $(M_M'-V_O^{\cdot\cdot}-M_M')^x$ i.e two cation substitutions with an oxygen vacancy. The justification for this lies in the increased binding energy of the neutral cluster at the surface due to the interaction of the opposing image charges. For example the binding energy of the magnesium based cluster is calculated to be 389 $kJmol^{-1}$ at the basal plane. The advantage is that by considering neutral defects we can neglect space charge effects. However, when we wish to extend the simulations to non-stoichiometric systems we will need to solve for the space charge.

The availability of experimental data for the segregation of Mg to Al_2O_3 [31-33] meant that a qualitative comparison could be attempted. Baik *et al* [33] have studied the segregation of Mg to the ⟨0001⟩ plane of doped sapphire using AES and LEED. They concluded that in a "closed system" magnesium segregates to the surface above 1473K with an effective Langmuir heat of segregation of ≈-183 $kJmol^{-1}$ in the temperature range 1573 to 1873K. LEED evidence for the formation of a two dimensional overlayer was also reported which may be due to the formation of an ordered spinel-like phase. Recent work by Kenway et al [30] on the segregation of Mg to the prism plane of alumina calculated non-Langmiur type behaviour for the segregation versus coverage, as shown in Figure 8. The pronounced minimum is at a coverage which is consistent with the enhanced stability of a spinel-like phase.

The variation of surface Mg coverage with temperature is then plotted in figure 9. The results are compared with experiment [34] by choosing a value for the pre-exponential factor that gives the experimental value at an observed coverage and then using this value of A over the given temperature range. This approach is necessary because the vibrational entropy terms in the free energy are not calculated. From figure 9, the qualitative agreement between the experimental and calculated coverages is good. As the temperature increases from 1573K to 1773K the logarithm of the surface defect coverage decreases in a non-linear fashion.

4. Grain boundary modelling

In this section we shall show that the techniques and models discussed above

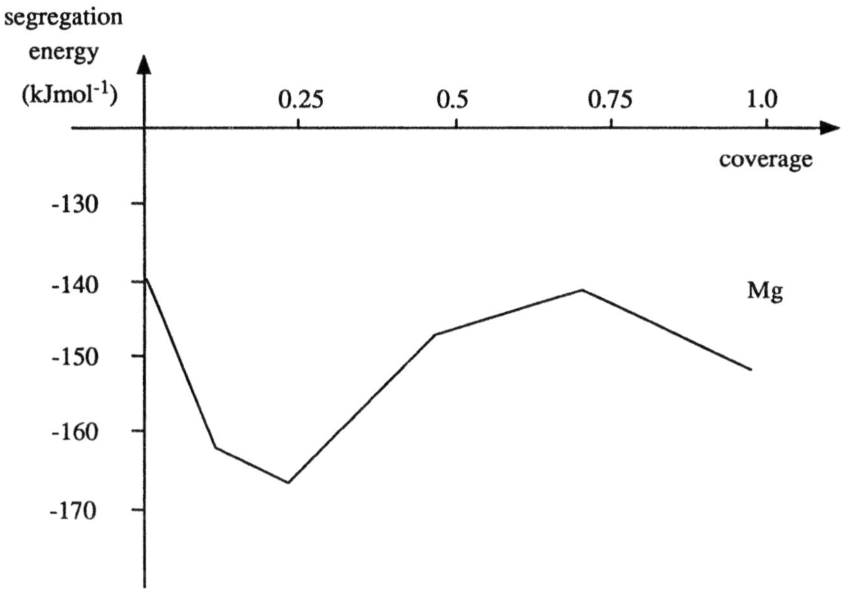

Figure 8: Plot of variation of segregation energy with coverage for Mg'_{Al}-V_O^x-Mg'_{Al} at the prism surface of α-Al_2O_3

can also be employed for modelling grain boundaries. Many materials are polycrystalline and the boundaries between the grains dominate many of the technologically important properties. For example, the electrical properties of semiconductors [35] and ceramics [36] are strongly influenced by internal interfaces. Material strength and toughness are functions of grain size and processes such as creep, sintering and corrosion are generally dominated by the presence of grain boundaries.

A grain boundary is the interface between two crystalline grains with different orientations. Nine parameters (three describe the misorientation, two the boundary inclination, and three the relative translation; the ninth is the boundary plane location which may pass between non-equivalent planes) are required for a complete macroscopic description of a particular interface. Such a model does not give any information about the detailed atomic configuration of the boundary, nor the relationship between misorientation angle and properties such as interfacial energies. Consequently three models have been proposed to provide additional information

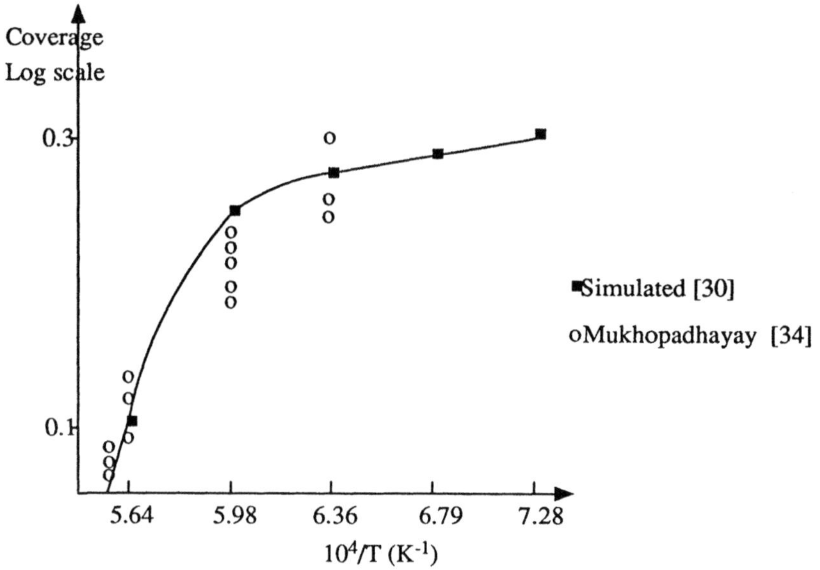

Figure 9: Calculated Temperature Dependence of the Equilibrium Mg surface coverage at the prism surface of Al_2O_3

about the boundary structure and properties.

The first of these is the dislocation model which describes low-angle boundaries as regular networks of lattice dislocations [37,38]. This model is successful in describing the variation of a large number of interfacial properties with misorientation angle, although it's success is limited to the low angle regime where the atom cores of the boundary ions are resolvable. The second model that we will consider is the coincident site lattice (CSL) model, which results from two interpenetrating crystal lattices that have relative orientations such that a finite number of lattice sites coincide in both crystals. The inverse of the ratio of coincident sites to lattice sites is denoted Σ. Coincidence boundaries are always periodic in the boundary plane, but the unit cell may be large. Certain low Σ boundaries have particularly low interfacial energies and diffusivities and these are often referred to as 'special' boundaries. The *O* lattice introduced by Bollmann [39] is a generalisation of the CSL in which the interior cell points which coincide in the two lattices are designated *O* points. This terminology arises from the fact that a rotation of one lattice about any *O* point will bring the lattices into full coincidence. When the transformation between the two lattices is a pure rotation, such as a mirror twin plane, the *O* lattice is an array of lines running parallel to the rotation axis. By introducing local rotations about each of the lines the mismatch between the lattices can be localised on the cell walls between the *O* lines. The third model that we will consider is the structural unit model which provides a means of extrapolation between structures calculated by simulation techniques. Sutton and Vitek [40] carried out an extensive examination of calculated configurations of tilt boundaries in both BCC and FCC metals. They managed to identify a number of common densely packed structural units, from which it was possible to go continuously between favoured boundaries. However the problem arises that the identification of the basic structural unit is not unique. For a given series of boundary structures there is a hierarchy of choices of favoured or delimiting boundaries. Consequently this model also has only limited predictive powers.

The theoretical investigation of grain boundaries in ionic crystals has mainly been limited to NiO and MgO. The development of the code MIDAS, allowed Duffy and Tasker [41] to investigate a number of low Σ boundaries in NiO, which have high two dimensional periodicities. They calculated the structure and energy of a series of tilt grain boundaries. Earlier simulations [42] of twist boundaries in rocksalt structured oxides indicated that boundaries with the twist axis orientated along the $\langle 001 \rangle$ direction are only barely stable with respect to dissociation into two free surfaces. This is somewhat surprising as stable $\langle 001 \rangle$ twist boundaries have been observed experimentally in MgO [43]. The calculated instability can however be removed if the density of ions in the boundary plane is reduced. Therefore to find the true minimum energy configuration the calculation should be initiated with different densities of ions in the boundary plane. Such calculations have been carried out for the $\langle 001 \rangle$ twist boundaries in NiO [44], with the proposed stable structures being analysed in terms of the models outlined above.

Twist boundaries with $\langle 011 \rangle$ rotation axes were found to be stable with the full density of ions in the boundary plane. However even in this case the interfacial energy can be lowered by reducing the density of ions in the boundary plane [45]. This emphasises the importance of examining the variation in interfacial energy with the number of ions in the boundary plane during all simulation studies.

Duffy and Tasker have also calculated the energies of point defects at grain boundaries [46] and the activation energies for ionic motion along grain boundaries in NiO [47]. For the coincident tilt boundaries in NiO they concluded that all defects and impurities segregate to grain boundaries and were also able to estimate the boundary defect concentrations.

More recently Mackrodt [1] has calculated the energies of symmetric twin boundaries in MgO, CaO, SrO, and α-Al_2O_3, plus the segregation energies of Ca^{2+} to the [310]/⟨001⟩ tilt boundary in MgO and Y^{3+} to the ⟨0001⟩ mirror twin in alumina. Mackrodt reported appreciable dilation of the crystal at the alumina twin boundaries, and boundary energies of the relaxed structures substantially less than those of the unrelaxed species.

4.1 Grain Boundary Structures and Energetics

We have calculated the energies of the mirror twin boundaries at the five low index planes of Cr_2O_3, which are given in table 3. From this we can see that there are appreciable differences between the unrelaxed and relaxed twin energies particularly for the $\langle 10\bar{1}1\rangle$ and $\langle 11\bar{2}0\rangle$ planes. The surprisingly low energy of 0.59 Jm^{-2} for the $\langle 10\bar{1}0\rangle$ twin suggests that the boundary is relaxing to a configuration that is very close to the bulk. This is due the change in cation coordination which allows the two blocks to fit together, not unlike a jigsaw, producing a configuration that is qualitatively similar to that found in the bulk. Figure 10 shows the relaxed and unrelaxed structures for the ⟨0001⟩ and $\langle 10\bar{1}2\rangle$ mirror twins. The points of particular interest to note here are the changes in interplanar spacing upon relaxation; indeed for all the boundaries considered there are appreciable changes in interplanar spacing.

4.2 Impurity Segregation

The work of Yurek *et al* [48] showed that yttrium is present at grain boundaries in chromia scales. Consequently we decided to investigate segregation of yttrium to the $\langle 10\bar{1}2\rangle$ and $\langle 10\bar{1}0\rangle$ mirror twins in Cr_2O_3. The heat of segregation as a function of coverage for the $\langle 10\bar{1}2\rangle$ boundary is given in Figure 11 together with the Δh values derived for the free surface. The two boundaries reveal contrasting

Twin Boundary	Unrelaxed Energy (Jm^{-2})	Relaxed Energy (Jm^{-2})
$10\bar{1}2$	8.32	2.72
0001	8.05	1.24
$10\bar{1}1$	50.54	1.56
$11\bar{2}0$	15.90	3.88
$10\bar{1}0$	3.46	0.59

Table 3: Mirror Twin Boundary Energies (Jm^{-2}) Before and After Relaxation for Cr_2O_3

behaviour which is both interesting and may be useful in interpreting some of the observed experimental behaviour. At the $\langle 10\bar{1}2\rangle$ boundary the values of Δh are calculated to be twice those of the free surface at low coverages, i.e. less than half a monolayer coverage, but approximately the same at coverages in the order of a full monolayer. This suggests the formation of an ordered second phase of approximate composition '$YCrO_3$' at half monolayer coverage. This is behaviour in direct contrast to that at the $\langle 10\bar{1}0\rangle$ boundary.

We have previously calculated a small driving force to segregation at the $\langle 10\bar{1}0\rangle$ surface for yttrium with the possibility of the formation of a stable phase. At the twin boundary we calculated that there is no driving force to segregation. We suggest that this could be linked to the low boundary energy and boundary structure. As previously mentioned the boundary structure is very close to that of a pure bulk configuration, which as our segregation model compares the free energy of the impurity at the interface and in the bulk, might explain the observations made here.

The relevance of these results to experimental data can be seen from the work of Przybylski *et al* [49], who report both coherent and incoherent perovskite phases,

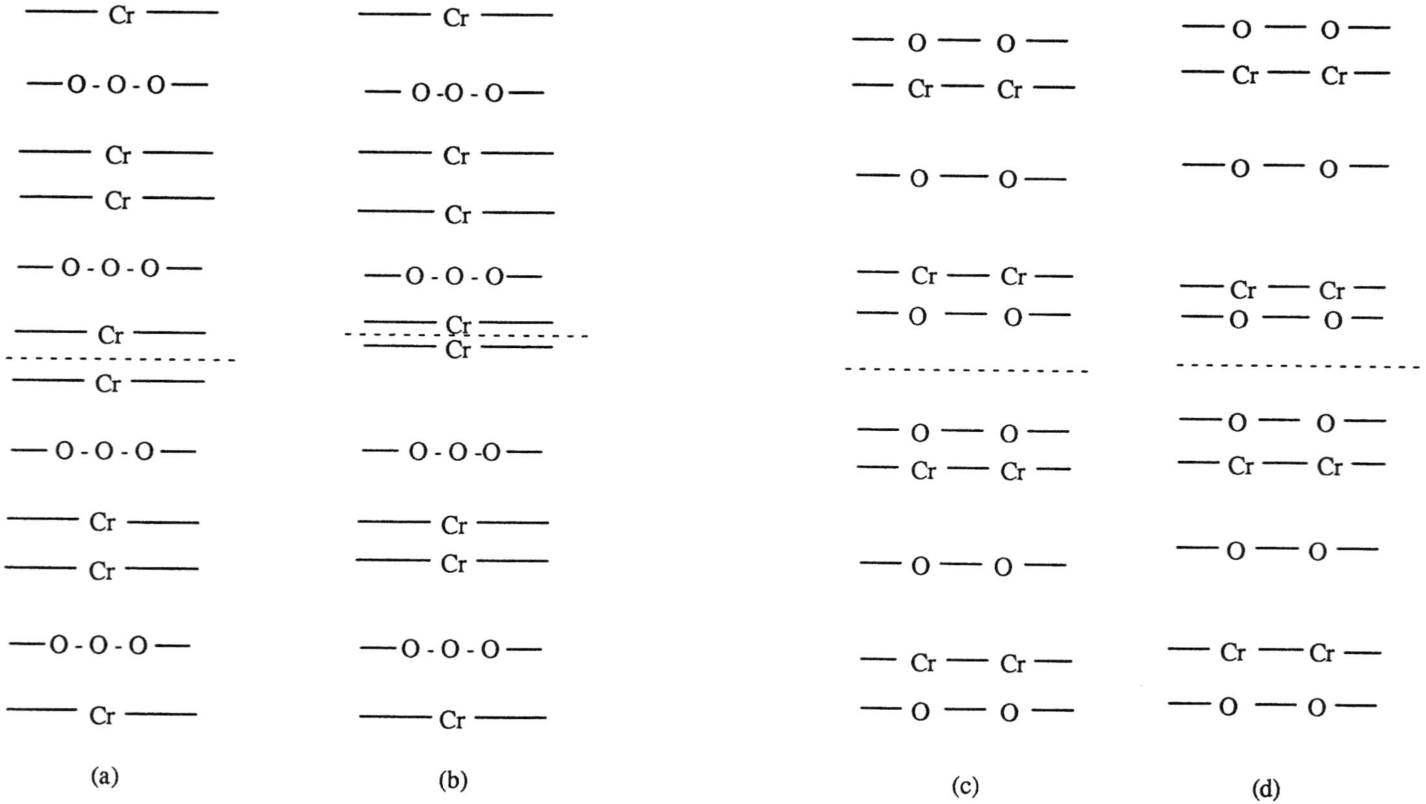

Figure 10: Schematic Representation of Twin Boundary Structures Before and After Relaxation; (a) ⟨0001⟩ Before; (b) ⟨0001⟩ After; (c) $\langle 10\bar{1}2 \rangle$ Before; (d) $\langle 10\bar{1}2 \rangle$ After

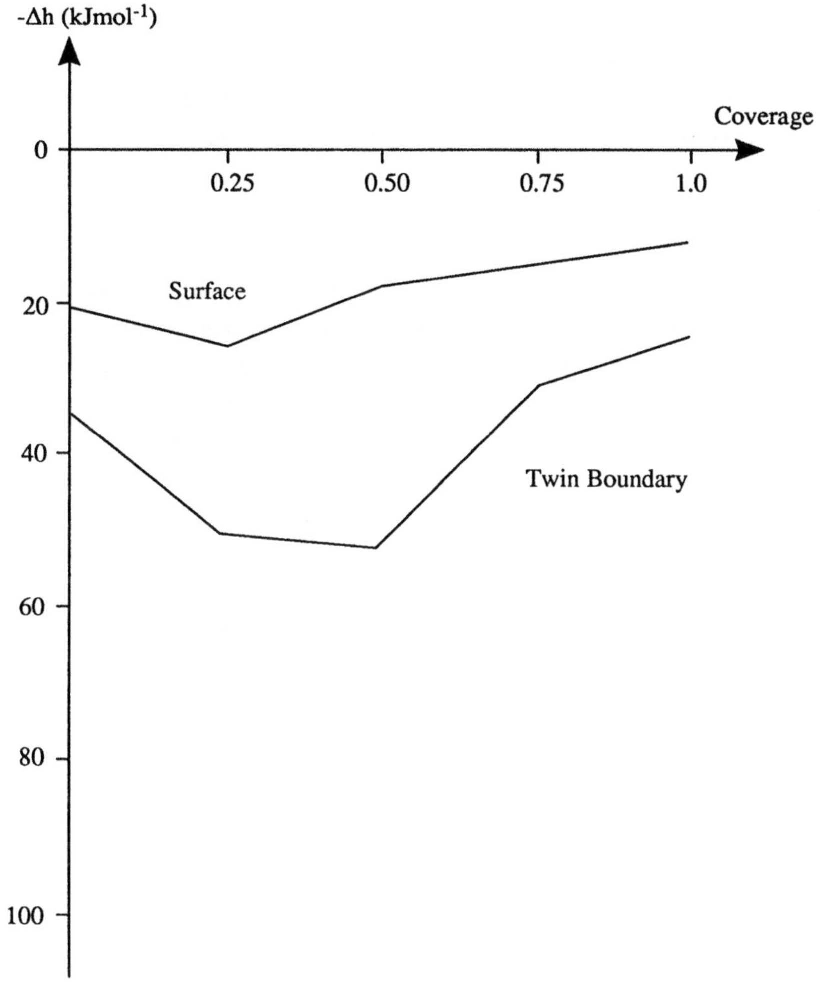

Figure 11: Calculated Heats of segregation of Y^{3+} as a function of Coverage (⟨$10\bar{1}2$⟩); (a)free surface; (b) mirror twin boundary

of stoichiometric composition $YCrO_3$ in their chromia scales. These phases take up only a small volume of the scale which suggests that the segregation of yttrium to grain boundaries is not uniform throughout the scale. This would be in agreement with our calculations which show non-uniform segregation to boundaries.

5. Conclusion

This article has been concerned with the application of computer simulation techniques to surface properties. Impurities in ceramic oxides will strongly influence the material properties and the surface behaviour. Until recently it has been generally assumed that the impurity heat of segregation is independent of surface coverage, and that impurities smaller than the host lattice cation will dissolve into the lattice while those larger than the host lattice cation will segregate. As grain boundaries control many surface phenomena, one of the key areas of development will be simulation of general grain boundaries. Future calculations including entropic effects and explicit consideration of space charge can only enhance the power of these tools in atomistically interpreting current experimental data, and directing future research in profitable directions.

References

1. W.C. Mackrodt, to be published in "Ceramic Microstructures '86"
2. P.W. Tasker and D.M. Duffy, Surf. Sci., 137 91 (1984)
3. P.W. Tasker, E.A. Colbourn and W.C. Mackrodt, J. Am. Ceram. Soc., 68 74 (1985)
4. P.W. Tasker, Surf. Sci., 87 313 (1979)
5. P.W. Tasker, Adv. Ceram. vol.10, p176, Ed. W.D.Kingery, (The American Ceramic Society, Columbus, Ohio, 1984)
6. D.W. Budworth, "An introduction to Ceramic Science", (Pergammon Press, Oxford, 1970)
7. R.L. Coble, J. Appl. Phys., 32 793 (1961)
8. R.C. Johnson and R.L. Coble, J. Am. Ceram. Soc., 61 110 (1978)
9. P.W. Tasker, J. Phys. C, 12 4977 (1979)
10. F. Bertaut, Compt. Rendu., 246 3447 (1958)
11. G.L. Benson and K.S. Yon, "The Solid Gas Interface", p203, Ed. E.A. Flood, (Arnold, London, 1967)
12. D.E. Parry, Surf. Sci., 49 433 (1975)

13. D.E. Parry, Surf. Sci,, 54 195 (1976)
14. D.M. Heyes, M. Barber, J.H.R. Clarke, J. Chem. Soc. Faraday Trans.II, 10 1485 (1977)
15. J.T. Kummer and Y.Y. Yao, Can. J. Chem., 45 421 (1967)
16. R.W. Nosker, P. Mark and J.D. Levine, Surf.Sci., 19 291 (1970)
17. J.D. Levine and P. Mark, Phys.Rev., 144 751 (1966)
18. V.E. Henrich, Prog. Surf. Sci., 14 175 (1983)
19. G. Wulff, Z. Kristallogr. Kristallgeom., 34 949 (1901)
20. J.W. Gibbs, "Collected Works", (Longman, New York, 1928)
21. M.J. Davies, P.J.Lawrence and S.C.Parker, to be published
22. W.C. Mackrodt, Adv. Ceram. vol.23, p293, Eds. C.R.A. Catlow and W.C. Mackrodt, (The American Ceramic Society, Westerville, Ohio, 1987)
23. W.C. Mackrodt and P.W. Tasker, to be published in J. Am. Ceram. Soc.
24. J. Frenkel, "Kinetic Theory of Liquids", p37, (OUP, New York, 1946)
25. K. Lehovec, J. Chem. Phys., 21 1123 (1953)
26. J.D. Eshelby, C.W.A. Newey, P.L. Pratt and A.B. Lidiard, Phil. Mag., 3 75 (1958)
27. M.F. Yan, R.M. Cannon and H.K. Bowen, J. Appl. Phys., 54 2 (1983)
28. K.L. Kliewer and J.A. Koehler, Phys. Rev. A, 140 1226 (1965)
29. A. Atkinson, private communication
30. P.R. Kenway, S.C. Parker and W.C. Mackrodt, Molecular Simulation, in press (1989)
31. R.I. Taylor, J.P. Coad and R.T. Brook, J. Am. Ceram. Soc., 57 539 (1974)
32. S. Baik and C.L. White, J. Am. Ceram. Soc., 70 682 (1987)
33. S. Baik, D.E. Fowler, J.M. Blakely and R. Raj, J. Am. Ceram. Soc., 68 281 (1985)
34. S.M. Mukhopadhyay, A.P. Jardine, J.M. Blakely and S. Baik, J. Am. Ceram. Soc. 71 358 (1988)
35. C.R.M. Grovener, J. Phys. C, 21 4097 (1985)
36. L.M. Levinson (Ed.), Adv. Ceram. vol.1, (The American Ceramic Society, Columbus, Ohio, 1981)
37. J.M. Burgers, Proc. K. Ned. Akad. Wet. Ser. B, 42 293 (1939)
38. W.L. Bragg, Proc. Phys. Soc., 52 54 (1940)
39. W. Bollmann, "Crystal Defects and Interfaces", (Springer-Verlag, New York, 1970)
40. A. Sutton and V. Vitek, Phil. Trans. Roy. Soc. A, 309 1 (1983)
41. D.M. Duffy and P.W. Tasker, Phil. Mag. A, 50 155 (1984)

42. D. Wolf and R. Benedeck, Adv. Ceram. vol.1, p107, Ed. L.M. Levinson, (The American Ceramic Society, Columbus, Ohio, 1981)

43. C.P. Sun and R.W. Balluffi, Phil. Mag. A, 46 49 (1982)

44. D.M. Duffy and P.W. Tasker, Phil. Mag. A, 48 155 (1983)

45. D.M. Duffy, J. Phys. C, 19 4393 (1986)

46. D.M. Duffy and P.W. Tasker, Phil. Mag. A, 50 143 (1984)

47. D.M. Duffy and P.W. Tasker, Phil. Mag. A, 54 759 (1986)

48. G.J. Yurek, K. Przybylski and A.J. Garratt-Reed, J. Electrochem. Soc., 134 2643 (1987)

49. K. Przybylski, A.J. Garratt-Reed and G.J. Yurek, J. Am. Ceram. Soc., 69 C264 (1986)

HARDWARE ISSUES IN MOLECULAR DYNAMICS ALGORITHM DESIGN

D.C. Rapaport,
Department of Physics,
Bar-Ilan University,
Ramat-Gan, Israel

Abstract

The diversity of architectures on which modern supercomputers are based presents a serious obstacle for the designer of software that is to operate efficiently in a variety of environments. As case studies in how to live with processor hardware, two very different kinds of contemporary computer design are introduced, one being the familiar vector computer, the other an experimental loosely-coupled network of processors. The basic concepts of each kind of computer that are relevant to algorithm development are discussed and the manner in which the machines have been harnessed to the task of large-scale molecular dynamics simulation are described. Only by taking into account the specific features of each machine was it possible to carry out what are the most extensive simulations of this kind to date, involving as many as 200,000 particles. These techniques apply to models based on differentiable potentials; an approach very different from direct numerical integration of the equations of motion is required for models that employ step potentials. A method derived from concepts of event-driven simulation and list processing that proves to be highly effective for problems of this type is presented. New results obtained by applying these techniques to the study of fluid dynamical problems at the microscopic level are briefly mentioned.

1 Introduction

"Simulation" is a word that is frequently encountered in theoretical science, far more than in the past, a fact which reflects the growing dependence on the computer for problem solving. Once it used to be necessary to carry out extensive analytical simplification of a problem in order to reduce the numerical complications to a minimum; the appearance of the computer made the idea of *ab initio* study feasible – in which the computations address the original problem and not some simplified version of

C.R.A. Catlow et al. (eds.), Computer Modelling of Fluids Polymers and Solids, 249–267.

it. Of course even the original problem tends to be an idealization of reality, in which only what are regarded to be the salient features of the problem are included. Statistical methods often supply the basis for a simulational study; the broad range of techniques that go by the name of Monte Carlo methods are widely used as a means for sampling the characteristic states of many kinds of system. In the case of dynamical systems there is a choice: they can be approached via the Monte Carlo route or, alternatively, can be modeled directly by solving the equations of motion of the elementary constituents. The latter approach is known as Molecular Dynamics (MD). While Monte Carlo can only study the equilibrium properties of dynamical systems, no such limitation applies to MD.

The underlying principle of MD is extremely simple: Prescribe the geometric and other structural features of the constituents of the system together with the interactions between them, and then solve the Newtonian equations of motion. This numerical approach to what amounts to an analytically intractable, classical N-body problem has proved remarkably successful in a wide variety of contexts, including astrophysics, biophysics, chemical and material physics, fluids, plasmas and polymers. With each successful calculation the MD method gains a new foothold, and the growing role of MD is reflected in the numbers of publications reporting results obtained using the method. A volume covering a broad selection of applications of MD appeared recently [1].

As the problems become larger and more complex the computational requirements expand accordingly. The eventual consequence is that the MD practitioner encounters the supercomputer, but, unlike many supercomputer users who find that their computations are of a kind that the vector-oriented architecture of modern processors handles effectively, quickly comes to appreciate that MD algorithms are less than ideal candidates for vectorization.

Vector processing is only one step along the road to large-scale computing; the other is distributed processing. Given that any specific device technology has an upper limit to its performance, the guaranteed way to higher throughput will always be parallelism, assuming the algorithm in question is one that can be suitably adapted. The fact that computers consisting of coupled vector processing units are now available commercially, and the realization that high-performance microprocessors can now be linked together into large networks, are indications of the kind of computing milieu in which large-scale MD simulation will soon be carried out.

There are a wide range of different kinds of problems to which MD can be applied, and the nature of the computational requirements varies according to application. One key criterion for classifying problems is whether the interactions are of short or long range. In the case of long-range interactions the forces are often decomposed into two components – a short-range part which is treated in detail and a long-range part which is simplified by regarding it as a slowly varying background effect. Short-range interactions present a special kind of problem in that they are local in nature and a significant part of the computational task is determining which particles in the system actually interact; in cases where the interaction range is small compared with the overall size of the system this can indeed be the heaviest part of the whole calculation. The absence of any regularity in the identities of particles which lie within interaction range, in a fluid whose structure is in a continual state of flux, creates a problem that is not obviously adapted for efficient vector processor implementation.

In these lectures, we describe schemes by which the short-range MD problem can adapted for both vector and distributed processors. We also describe a method for handling the ultimate short-range interaction problem, namely hard spheres – or any other model involving purely step potentials. Unlike the basic form of the MD algorithm, which is simple to formulate and trivial to program, the price to be paid in mapping the algorithms onto processors of more complex organization, such as those to be described here, are algorithms that are also more complex. While the scientist often tends to give the impression that computational aspects of a problem are best not discussed in public – witness the dearth of technical descriptions in papers whose results derive from non-standard calculations – the increasing complexity of the hardware, and the fact that the state of the tools available for software development lags substantially behind the hardware, means that algorithms for dealing effectively with large problems will have to be developed with the actual processor architecture in mind. This far from idyllic situation will persist until software tools that are capable of automating the mapping process can be developed; while vectorization of comparatively straightforward problems can now be carried out automatically, no generally useful solution exists for vectorizing more complex problems, or for reworking anything beyond trivial examples into a form suitable for distributed processing.

In the sections which follow we outline some of the architectural features of modern supercomputers that are important in the design of efficient algorithms. We then summarize the basic MD approach as a prelude to describing what must be done in order to reformulate the algorithms to perform effectively on these machines. The totally different approach needed for step potentials is also described; it is an example of a problem that has yet to benefit from anything other than the conventional scalar computer, but alternative developments in processor design might be of help. Finally, we discuss the performance of the algorithms when actually put to work, and briefly mention applications of these techniques in which qualitatively new results have been obtained. In order to keep the length of this piece reasonable the mathematical details have not been included, but they are to be found in print elsewhere [2]

2 Supercomputer architecture

In the early days of computing, programming was an activity that required an intimate understanding of the workings of the machine; programming languages as we now know them did not exist, and programs were expressed directly in terms of machine instructions. With the advent of higher level languages, of which an early example was Fortran – still (in improved form) the most familiar language in scientific and engineering environments – the link between programs and hardware was effectively severed; algorithms no longer had to take into account the idiosyncracies of individual machines, and programs were to a considerable extent portable between computers of different kinds. Hardware developments that occurred over a significant period following the appearance of programming languages were largely invisible to the user; substantial improvements in processing speed were obtained not only by using faster device technology but also by introducing such concepts as instruction lookahead, high-speed registers, cache storage, multiple function units, and so on. None of these technical advances affected the user who simply enjoyed the performance gains, and even those individuals who resorted to assembly language programming in order to go beyond the limited optimization capabilities of available compilers were limited in the hardware features they could actually access; the principal advantage of assembly language was that it allowed a more judicious use of the high-speed registers, thereby reducing the need to access the slower main memory.

Vector and distributed processing have once again forced the algorithm/software designer to consider hardware – if more than minimal gains are to be extracted from the new computer architectures [3]. While many proposals for new kinds of processor architecture exist, and in some instances prototype or even commercial machines based on these ideas built [4], the mainstream of supercomputer architecture is based on the use of vector processing, and of late, on distributing computational tasks across networks of communicating processors. Other less "conventional" approaches that will not be addressed here include massive bit-serial parallelism, dataflow and systolic arrays.

Vector processing resembles a factory assembly line. The arithmetic (or other) functional units are divided into several segments, each responsible for a distinct part of the computation. Data are input to the first segment and emerge eventually from the last; while a scalar (i.e., non-vector) processor would perform a single computation at a time, the vector pipeline is capable of handling several calculations at once because each pipeline segment contains partial results for a different calculation. Thus, even though the time required to complete a single arithmetic operation may be no faster than for a scalar computer, the throughput of a vector processor is in theory increased by a factor equal to the number of segments in the pipeline. The theory is imperfect however, since it assumes that the vector pipeline is always full; the time required to complete a vector instruction depends on the number of operands to be processed but it also includes a fixed overhead needed to begin the instruction and to initially fill the pipeline with data. If the number of operands is small it will be the overhead which dominates, and the somewhat absurd situation, in which the use of vector processing can actually slow down a calculation, is not impossible. Depending on the actual machine, the length requirements to achieve a performance that approaches the theoretical limit may be quite severe; often vector lengths of 50–100 are needed just to approach half speed.

Another consequence of the design of a typical vector processor is that data to be processed in vector fashion must be stored contiguously, or at least with constant intervals (or *stride*) between consecutive elements to be processed. Not all algorithms lead to such a convenient form of data organization, MD included, and most vector processors provide reasonably effective methods for rearranging data to facilitate the construction of vectors. The *gather* operation is a vector instruction for assembling a vector out of disorganized data items that are referenced by means of a vector of indices; there is a complementary *scatter* operation for achieving the opposite effect. While the gather/scatter operations alleviate some of the difficulties resulting from the constraints on the way vectors can be stored, they tend to be more time-consuming operations than regular vector arithmetic and thus should not be used to excess. Another pair of operations that also provide more flexibility in vector handling are *compress* and *expand*. The former constructs a vector out of a subset of elements extracted from another vector by means of a bit mask – also a vector but of a special kind – that can itself be constructed by (for example) a vectorized series of comparison operations; the latter operation reverses the process. Once again, there can be a penalty associated with the use of these operations, since the total time may well depend on the longer of the two vectors.

While the basic principles of vector processing are comparatively straightforward (it is the application of these principles which may present problems) there are a number of substantially different approaches taken in actual processor design. It is very difficult to make general statements as to how these design features can impact performance and how an algorithm can be constructed to take advantage of them, or at least to avoid violating any implied operating assumptions inherent in the hardware.

One frequently encountered potential trap is memory interleaving. Since hardware economics dictates that memory speed be several times slower than the processor it serves, memory is divided into a number of independent banks the phases of whose cycles are staggered. Consecutive memory words are stored in different banks with the result that full-speed vector processing is not impeded by the slower memory. The problem arises when the interval between successive data items processed vectorially is such that they lie in the same memory bank; when this happens a serious performance degradation is likely to result.

Other processor design features can sometimes be utilized in order to improve performance. Some machines are register oriented: data for vector processing are initially transferred in batches to sets of high-speed vector registers, processing occurs using the register contents, and the results that are initially returned to the registers are then transferred to memory. The more computation that can be carried out using only the data in the registers the better the performance. The use of registers can set a comparatively small upper limit to the allowed vector length; longer vectors must be subdivided to fit into the registers, with a possible performance penalty. The alternative is direct memory to memory processing which eliminates the need for vector registers, as well as (within reason) the length restrictions. The functional units that carry out the processing can be present in varying quantities, and may be either general purpose or dedicated to particular operations only. Where multiple functional units exist there may be provision for *chaining*, a procedure by which the results of one unit are fed directly to another to improve throughput wherever a series of operations are associated with each vector input. Another feature is the *control vector* (similar to the bit mask) which regulates storage of results; this fills the role of the logical *if* in permitting conditional operations in a manner compatible with vector processing – the control vector determines which particular outputs of a vector operation are generated and which are suppressed.

There has been considerable progress in compiler design to provide automatic vectorization of programs. A reasonably competent compiler can recognize loops that are candidates for vectorization and carry out the needed reorganization. Some compilers are clever enough to replace conditional operations by the control vector alternative. Subdivision of vectors (a process known as *stripmining*) to satisfy length limitations should be no problem. But as with all mechanized procedures there are limits to what can be done; if the compiler fails to recognize the fact that code can be rearranged to facilitate vectorization, then the result will be a significant (an order of magnitude is not unknown) drop in performance. Sometimes a simple manual resequencing of the steps in a calculation will help the compiler, but on occasion a more drastic reformulation of the entire algorithm is needed in order to effectively utilize the capabilities of the vector processor.

Distributed processing places before the algorithm designer an even greater variety of possibilities. The familiar taxonomy of distributed processing divides the field into two classes: SIMD (single instruction with multiple data) and MIMD (multiple instructions and data) [3]. In the former, all processors execute the same sequence of instructions but on their own versions of the data; a limited degree of independence exists in that each processor can decide whether to store a particular result, depending on local conditions. Computer systems in the latter class are sets of almost independent processors, each executing its own version of the program (which may or may not be common to some or all of the processors) and working with its own private data. Both schemes imply a communication capability, but this is not specified in the taxonomy.

The MIMD approach offers the greater flexibility and is more suited for the kinds of calculations associated with Molecular Dynamics; although for other kinds of simulation, in particular systems embedded in lattices, SIMD machines may be more cost- effective. Advances in VLSI have now led to single chip microprocessors which outperform the mainframe computers of not too many years ago; the assembly of significant numbers of such processors can lead to an MIMD machine of considerable power.

Communication between the individual processing elements is an important ingredient in the design of a multiprocessor system. Even though it may be possible to partition a computation so that different pieces execute concurrently, the amount and frequency of data transfer between processors is a key factor in establishing the overall efficiency. The typical communication transaction extends over a time proportional to the amount of data to be transported, but in addition involves a fixed overhead to initiate and coordinate the transfer. A computation in which large amounts of data must be moved between processors in the course of the work, or in which large numbers of transfers of even small quantities of data occur, may well turn out to be rate-limited by communication. Both the nature of the algorithm and the capacity of the communication network are contributors to the performance, and each case must be considered individually; the optimal situation is of course one in which the amount of data transferred is small in comparison to the amount of computation that is carried out, together with a network having high transfer rates and low fixed overheads.

Ideally, the user of a multiprocessor system should view the act of communication as just another service provided by the operating system or, better still, an extension to the language (on the assumption that the communication cannot be made totally invisible). While such a situation may in due course become the norm, the user is often not shielded from the need to deal with communication directly. Conceptually, processors appear to pass data between one another by direct point-to-point links, but at the physical level communication may proceed by alternative means. In the case of a large number of communicating processors it is no longer feasible to have links between all possible pairs. Links between processors can be organized in many ways; examples of possible schemes include a grid (or hypercube) layout with nearest neighbor connections, a common communication bus, or a crossbar switch. Both cost and throughput determine what should be used, but if the bandwidth is not adequate for the mean data transfer rates required, the system is doomed to failure. Fluctuations occur in transfer rate, and a network that does not have enough excess capacity is also not going to be of much use. With shared communication resources there is the potential for access contention; more than a minimal amount will again degrade performance.

Another approach to communication is based on the concept of shared memory. In some multiprocessor systems all memory is shared, and this represents yet another aspect of the taxonomy mentioned above. For communication purposes one (or more) separate memory units of comparatively small size (depending on the needs) can be used. All processors can communicate via the shared memory with some means of hardware arbitration used to ensure fair and consistent access. The shared memory is divided logically into a set of *mailboxes* allowing every processor to talk "directly" to any other. This approach permits considerable flexibility in that it can emulate point-to-point networks with any connection topology; one disadvantage is that the speed of the memory determines the combined bandwidth of the entire network. The nature of the problem to be solved is a factor that cannot be ignored: if all processors need to be have frequent (read/write) access to the bulk of the data used to describe the system then shared memory may be essential, as the only means of making the data available to the processors that need it.

The above discussion has been deliberately kept general. The particular applications to be introduced later were carried out on two kinds of system, one a single-processor vector supercomputer, the other a set of coupled high-performance processors linked via shared memory. The vector machine was a CDC Cyber 205, no longer the latest in supercomputers but a design that has reappeared in the new ETA processors. One of the less popular features of this particular design (all processors tend to have their weak points) is the need for comparatively long vectors in order to achieve reasonable performance; this in fact turned out to be the motivation for some of the algorithmic developments described later. The distributed processing was carried out using a set of FPS 264 scientific processors controlled by an IBM 3081 computer that constituted the experimental LCAP system. Further details of implementation and performance will be given later.

3 Basic molecular dynamics approach

The more familiar applications of MD are readily characterized and implemented [1]. If the constituents are taken to be structureless particles and the forces between particles assumed to be functions of mutual separation only, then the equations of motion follow immediately and are easily integrated by some standard scheme such as the "leapfrog", or by a predictor-corrector method of suitable order. Boundary conditions need to be taken into account, and unless the physics of the system demands otherwise, periodic boundaries are used; if ambiguity arises as to which periodic image of a particle should be used in computing interactions it is the nearest one that is selected. For long-range forces Ewald sums can be employed in an effort to compensate for the finite-range cutoff enforced by the boundaries. The system is allowed to evolve, and at regular intervals measurements are made of bulk thermodynamic properties, structural features, coordinate and velocity combinations for computing correlation functions and transport coefficients, and so on. For more complex systems, extra work is required, but the basic idea remains the same. In the case of rigid structures representing molecules there may be several force centers to be considered, and the Newton-Euler equations for rotational motion must also be solved, or alternatively, rotation may arise as a consequence of solving constraint equations that are introduced to preserve the structural integrity of the molecule. Extensions of the MD approach are used, for example, to permit studies of fluids at constant pressure rather than the usual constant volume. Flexible molecules such as polymers can be treated within the framework already described; the intramolecular interactions are simply chosen to be sufficiently strong to prevent the molecule from decomposing.

As system size grows the possibility of introducing computational shortcuts merits consideration. In the case of interactions whose range is significantly less than the size of the box containing the system, a lot of effort is wasted in looking at particle pairs that turn out to separated by more than the cutoff distance. For an N-particle system there are a total of $N(N-1)/2$ interacting pairs, but for a cutoff range r_c significantly less than the box size the number of interacting pairs is actually proportional to N, with a coefficient proportional to r_c^D, where D (= 2 or 3) is the spatial dimension of the problem. The bulk of the unwanted computation is eliminated by subdividing the box into cells of edge not less than r_c and only computing the interactions for particles occupying the same or neighboring cells. Both parts of this calculation (subdivision and interactions) require amounts of computation that are proportional to N only. An effective means of organizing the data describing which particles belong to which cells is based on the *linked list.* With each cell is associated a pointer to one of the particles in the cell; this particle has a pointer indicating the next one in the same cell, and so on; when there are no further particles in that cell the list terminates with a null pointer.

Not all particles lying within adjacent cells (or even in a single cell) need be within interaction range, and a further refinement entails the construction of *neighbor lists* in which each particle is supplied with a list of those of its neighbors that actually lie within interaction range. If the size of the shell used to determine which particles go into the neighbor list is enlarged a little beyond r_c then the list need not be updated at each time step, thus leading to additional savings.

A further gain in speed can be achieved by tabulating the interactions and performing a table lookup based on the separation of the interacting particles, possibly improved by interpolation between the tabulated entries. Use of cells also reduces the work needed to handle periodic boundaries, since the knowledge of which cells are being processed provides information as to whether the coordinates need be adjusted for periodicity.

On scalar processors these techniques prove highly effective. The situation becomes less rosy when a vector processor is to be used for the computations since the cell or neighbor-list reorganization of the particle data does not result in vectors of length adequate for effective processor performance. There is little that even the best of vectorizing compilers can achieve if the algorithm is fundamentally unsuitable for vectorization. An alternative approach is clearly required, preferably one which builds on gains already achieved by the data reorganization described above.

4 Vectorized MD

The MD algorithm as originally formulated (without cell lists) is easily and effectively vectorized. The fact that all pairs of particles are considered makes it possible to construct long loops over particle pairs, and it is precisely these long loops that provide the fodder needed by the vector processor. The problem is of course that there is no way that the several-fold speedup resulting from vectorization can compensate for a calculation that now grows at an N^2 rate rather than linearly with N; the starting point for vectorization must follow the reduction provided, for example, by the cell-list reorganization.

The technique developed for this purpose is actually quite simple. It involves introducing the concept of *layers*: since the cells can contain varying numbers of occupants, layers are used to provide a systematic organization of the cell contents. This is done in the following manner. The first particles in each occupied cell (the order in which particles are placed in the cells has no significance) are all assigned to the first layer, which is simply an array in which one element is associated with each cell. The second particles in each of those cells that are multiply occupied are assigned to the second layer, and so on. The first layer will contain a significant fraction of the particles (normally more than half, although the distribution depends on both fluid density and cell size), the second layer substantially less, with the occupancy distribution tapering off very rapidly for a dense fluid and cells that are only slightly larger than r_c. Given the substantial fraction of particles occupying at least the first and second layers it is apparent that the bulk of the interaction computations can be dealt with efficiently using vectors of more than adequate length, since it is now possible to process pairs of heavily populated layers rather than, as was the case previously, pairs of cells that are minimally occupied.

The actual layer assignment process is readily adapted for vectorization: rather than scanning the particles one by one and assigning them to whichever layer is appropriate, a procedure which does not lend itself to vectorization, the process is inverted and it is the layers that are populated one at a time. As the filling of each layer is completed the set of remaining particles that have yet to be assigned is compressed, so each pass deals with fewer particles; the total number of passes (i.e., layers) equals the maximum cell occupancy, so that for the dense fluid mentioned above (which forms the subject of the simulations) only a small number of passes (typically four) are needed.

The reformulation of the interaction calculation requires that the loop that examines all pairs of adjacent cells be replaced by one that considers all layer pairs. The interaction calculation based on cell lists in fact consists of several nested loops. The loop over cell pairs is actually a pair of loops, one over the first cell of each pair, the other over the offsets to the second cell; within this pair is a second pair that selects the individual particles from each of the cells. The revised version consists of a pair of loops that produce all possible layer pairings; nested within is the loop that generates all possible offsets between neighbor cells (including the null case of a cell with itself); the innermost of the nested loops is over all cells, and actually does the interaction calculation for all pairs of particles to be found in the two relatively offset layers. This innermost loop proves an ideal candidate for vectorization.

The special-purpose instructions introduced earlier that aid in rearranging data in a form suited for vector processing play an important part in these computations. Prior to carrying out the interaction evaluations, the particle coordinate data must be rearranged in a form that will permit vector processing based on the layer contents. The vectorized *gather* operation is used for this task; this rearrangement need only be performed once for each time step even though particles will typically be considered several times in the course of the computation as partners in different interacting pairs. At the conclusion of the calculation the resulting particle accelerations must be reorganized to undo the effect of the layer structuring; here the *scatter* operation is used. In the course of the interaction calculation proper, only pairs of cells that are occupied from the point of view of both of the layers being considered are involved in the processing. Since the appearance of gaps due to empty cells (in the respective layers) would reduce the benefits of vectorization, if not prevent it entirely, the crucial innermost interaction loop is preceded by the use of a *compress* operation to collect all pairs of particles that are actually present, and following the calculation, the results are *expand*ed back to their correct locations. It turns out that there a number of alternative ways of organizing the details of the computation; which is the most effective may well depend on what kind of instructions the hardware handles most efficiently.

These data-moving operations consume at least as much, if not more, processing time as a similar number of arithmetic operations. The implication is, that due to the stringent requirements for efficient vector hardware utilization, the processor is forced to spend a significant amount of time on "non-productive" data manipulation. There is little that can be done to rectify this state of affairs given the hardware limitations. What really counts however, is the overall performance; while the MFlop rate for the MD calculation may fall significantly below peak, if the vector processor is still able to carry out the simulations several times faster (per unit hardware cost) than any alternative machine, then the effort spent in algorithm redesign is more than justified.

There are several additional aspects of the revised algorithm still to be covered. Periodic boundary conditions are dealt with by the simple expedient of making duplicate copies of particles close to the box boundaries; these particles participate in the interaction calculations, but care is taken to ensure that they are discarded after the calculation is complete. The extra work introduced by these additional particles can be written off as a "surface" effect. Another issue is the amount of storage required. The data reorganization used to make vectorization possible consumes significant amounts of temporary storage, with the potential for even doubling the amount of data needed in the course of the calculation. For large simulations this could well lead to problems caused by a lack of adequate processor storage. The solution is to subdivide the computation by considering only those particles belonging to a given region within the box at one time; temporary storage requirements are then proportional to the number of particles in the region and not to the total system size. These regions must of course overlap by a limited amount to ensure that all interactions are properly accounted for. The subdivision not only reduces the amount of temporary storage needed but also eliminates problems due to vector length exceeding some impassable upper bound (extant on certain machines) that even the compiler may not be able to work around. Finally, it should be noted that the vectorized algorithm is based on cell lists rather than on neighbor lists. Even though the latter lead to fewer interaction calculations it is not apparent that the additional work needed to rearrange data in this form would compensate for the reduction in the number of interacting pairs; if anything, the calculation is likely to prove less efficient in vectorized form than that employing cells.

5 Distributed MD

A distributed implementation of the MD computation poses a new set of problems. Whether the individual processors participating in the effort are vector or scalar machines has little bearing on the additional tasks that must be introduced into the simulation.

There are two very different ways of dividing the computational load among the processors, and it is the nature of the simulated system that determines which is the more appropriate. The first is based on subdividing the system on the basis of particles, with each processor being responsible for a fixed subset of the total. Since in a fluid subject to continual structural change there is no long-term relationship between particles that happen momentarily to be neighbors, a particle in a given processor will typically be interacting with other particles that are to be found in each of the processors, even in the case of short-range forces. Thus, in order to complete an interaction computation, the particle coordinates must be made available to all other processors, and the corresponding contributions to the interaction retrieved. This results in an amount of communication that is directly proportional to system size. In the case of long-range interactions there would appear to be no alternative, but if just short-range interactions are involved a different approach can be followed. The alternative means for subdividing the calculation is to make the different processors responsible for separate spatial regions of the system, by requiring them to process whichever particles happen to be in those regions at each instant. Interprocessor communication is necessary in this case to allow computation of the interactions between particles lying close to, but on opposite sides of the region boundaries, and also to permit particles to move between the regions; the amount of communication is proportional to the size of the interface between regions, an amount that is much smaller than the overall system size. It is the latter scheme that is dealt with here.

The distributed processing implementation of MD follows either the cell- or layer-based approach, depending on whether the processors used are of scalar or vector type, and adds the necessary communication and synchronization framework. The computations for each time step of the simulation begin by transferring copies of the coordinates of those particles that are close to region boundaries to the processors responsible for the adjacent regions. This is done in order that the interaction computations in each processor can proceed independently and completely in parallel; the amount of data involved is relatively small since the thickness of the interface region need be no larger than the cutoff r_c. As was the case with the subdivision for vector processing described earlier, the copied particles are used only to help in computing the forces and are then deleted. Following the interaction calculations and the coordinate updating, a few particles will no longer be in their correct regions. Further transfers are required to move all data pertaining to these particles to the correct neighboring processor; again only a comparatively small number of particles are affected at each time step.

At intervals throughout the course of the simulation data must be collected, both to permit evaluation of various properties of interest and for monitoring the behavior of the system. Whatever calculations are involved in this task would be either carried by one of the processors belonging to the network or else by a processor acting in a supervisory role. It will also be necessary to periodically dump the entire state of the system to disk storage, both to permit runs to be broken into segments and to allow recovery after hardware or other failure. The communication and control mechanisms must provide support for these and similar functions; some may be standard features of whatever operating system controls the processors and the network, but if not provided they must be incorporated into the application.

In the simulations that were based on these ideas the network contained only a small number of processing nodes (typically four), so that the amount of communication was minimal and had very little impact on overall performance. As the number of particles per node decreases (corresponding to the use of more nodes), the importance of efficient communication increases, and there will be some minimum occupancy value below which the cost-effectiveness of the network begins to diminish. There are two contributory factors here, namely the increasing surface to volume ratio which determines the fraction of particles that interact across region boundaries, and the proliferation of communication paths.

A further point which can sometimes be important is the matter of load balancing. Provided all processors are required to perform a similar amount of work there should be no problem in this respect, but in the case of a spatially inhomogeneous system, the subdivision into regions of equal size could lead to some processors having to handle more particles than others. In such a situation a dynamical resizing of the regions would be required to restore equal loading.

6 Step potentials

The numerical techniques used in solving differential equations require continuity of the functions involved. Step potentials violate this condition and must therefore be treated by some alternative procedure. Potentials of this kind were in fact used in the earliest MD computations, [5] and have the advantage that there are no numerical integration errors involved since particles move in straight lines between collisions. When particles do collide the event is instantaneous and the outcome can be determined exactly. The computer of course introduces a certain amount of error into these calculations due to the finite-precision arithmetic employed, but such errors are many orders of magnitude smaller than the integration errors that arise when solving differential equations numerically. Though the algorithms are technically more complicated, the use of step potentials has the advantage not only of accuracy, but also of efficiency in certain instances when a scalar processor is used; if the detailed form of the interaction is not of great consequence then the technique is definitely an alternative worth considering. Unfortunately the methods used for step potentials do not appear amenable to vectorization, but that does not rule out the possibility of using some other form of non-conventional processor architecture to enhance the performance.

The simplest model fluid based on a step potential is a system of hard spheres or disks. Square wells – both attractive and repulsive – can be incorporated, and it is not especially difficult to construct compound objects such as polymer chains or rings, with or without restrictions as to the degree of flexibility, by an extension of the same ideas (a barrier of infinite height is used to prevent neighbor monomers in the chain escaping from each other). The term *collision* is then used to include the case of a pair of particles reaching a separation at which some change in their mutual potential energy occurs; for hard spheres this is a collision in the familiar sense, but for a polymer it could also mean that a bond has reached its maximum extension. The velocity changes of the particles participating in the collision follow simply from the equations of energy and momentum conservation.

The difficulty associated with this kind of simulation is neither the processing of the actual collisions, nor the prediction of when such collisions are due to occur – itself also a simple exercise in algebra, but dealing with the large quantities of information concerning collisions that may occur in the future. It is clearly uneconomical in the extreme to follow each collision with a calculation to predict the next collision (anywhere in the system) based on considering every pair of particles in the fluid. Most of the information gained during one such exhaustive consideration of all the pairs will still be valid a good many collisions later and thus should not be discarded. However, information relating to particles that do collide ceases to be valid; this must be erased and replaced by revised predictions.

The most efficient means for approaching such a problem involves two distinct aspects. The first is the recognition that distant particles are unlikely to collide until after they have undergone many intermediate collisions and, therefore, that collision prediction should be limited to nearby particles. But the possibility of distant particles colliding cannot be ignored entirely and so a cell subdivision is once again introduced, but now with the purpose of selecting those pairs of particles which should be examined to see whether collision is imminent. Use of cells does mean that another class of event, in addition to the collision, must be considered, namely that of a particle moving from one cell to another.

The situation, as it now stands, is that throughout the course of the simulation all potential collisions between particles in the same and adjacent cells must always be known. When a collision occurs, all future collisions listed for the particles involved are cancelled and replaced by a newly computed set. Similarly when a particle moves to a new cell there are new potential collision partners which must be taken into account. Following both kinds of events, the time of the next cell crossing(s) must also be computed. The cells can be kept sufficiently small to hold very few particles (as little as one on average); computing future collisions then involves considering only a few immediate neighbors. On the average there will be several future events scheduled per particle at any given time and thus the continual scheduling and cancelling of future events can entail a substantial amount of data handling. An inappropriate choice of method for organizing the data can therefore result in a very inefficient simulation.

This leads to the second aspect involved in simulations of this kind. The apparently optimal way of organizing the calendar of future events is based on the binary tree. In the simplest of data structures involving pointers – the linear list – each entry points to its successor, and the average number of steps to locate a particular entry is therefore half the total length of the list. The binary tree [6] is a modified form of this structure in which each entry has provision for two pointers, allowing for up to two successors. If the tree is well balanced, in the sense that all branches are of essentially the same length, then the number of steps to locate an entry drops to a value proportional to the logarithm of the number of entries, clearly an enormous gain if the tree size is 10^4 or more. In the case of MD simulation, the almost random nature of the collisions (at least from the point of view of event scheduling) ensures that on average the tree remains balanced, and the access algorithm performance remains close to theoretical.

Basing the algorithm on the tree data structure ensures that the operations of scheduling a future event, which calls for insertion of an entry at a particular place in the time-ordered tree, and determining the next scheduled event, both require processing that depends logarithmically on tree (and hence system) size. Cancellation of an event, which implies the deletion of an entry from the tree, can be carried out with an amount of work that is essentially independent of tree size; however a prerequisite for efficient deletion is that the entry concerned can be quickly located. In order to assist in locating entries, additional data structures are introduced: all events in the tree that involve a particular particle are linked together, which means that if one instance of an event associated with a particle is available (i.e., the currently occurring event) the remainder can be readily found. The actual situation is a little more complicated; two such lists are in fact needed for each particle, and for efficiency, the list structure used is circular with pointers in either direction.

As indicated above, the approach used for step potentials does not lend itself to vectorization; the reason for this should by now be obvious. In order to achieve the substantial (orders of magnitude) reduction in the amount of work required, a complex data structuring has been introduced; if this is replaced by something simpler, that is more amenable to vectorization, the additional effort required will far outweigh the gains that could be achieved by vectorization. It might well be possible to construct special purpose hardware to handle the event processing. Using what are known as content-addressable or associative arrays, the work associated with the addition, removal and location of events in the tree could be handled in near-instantaneous fashion, but such an effort has yet to be attempted.

7 Performance and results

The techniques introduced so far were all developed for use in large-scale applications of MD, especially problems in fluid dynamics where system sizes need to be large by comparison with the majority of MD simulations carried out in the past. The reason for the increased size is that the observation of phenomena that correspond to what occurs in the real world of laboratory fluid dynamics requires length scales that are orders of magnitude greater than those needed for more familiar MD studies of liquid structure, thermodynamics and transport coefficients, where a system whose linear extent is some ten atomic diameters can be perfectly adequate. Thus, while the "average" MD computation may nowadays involve of the order of 1000 particles, on occasion perhaps 5000, some of the fluid flow simulations have reached sizes of 200,000 and the end is not yet in sight.

The measure of the success of a design for an algorithm is how well it performs. In order to demonstrate the effectiveness of the layered approach several performance figures will be given; these were derived from a two-dimensional system of particles at liquid density in which the interactions were of very short range – this corresponds to the case where the potential gain from using layers is greatest. (The interaction is in fact the familiar Lennard-Jones potential, truncated at the minimum, and shifted to make it positive throughout.) On an IBM 4381/13, a scalar machine rated at slightly over 1 MFlop/s [7] (note that such measures are notoriously unreliable outside the context for which they were originally obtained – this particular number comes from the Linpack performance test), the cell-list method required 126 μs for the interaction calculations associated with a single atom. The FPS 264 processor, a high-performance attached processor with a theoretical peak performance of 38 MFlop/s but with actual performance as much as an order of magnitude slower, required approximately 50 μs for the same work; a coupled set of machines of this type were used in the distributed processor studies. Finally, the vector computer used was a two-pipe CDC Cyber 205, with computations being carried out in half (32 bit) precision; while the unreachable peak performance is 200 MFlop/s (or half this for 64 bit arithmetic – relevant because not all the data manipulation involved can benefit from the higher rate) the measured time per atom was 4.1 μs/atom. Comparison of the Cyber and IBM performance shows that the former is some 30 times faster; a similar speed factor is to be found in one of the Linpack tests, thus the performance of the MD algorithm on the vector machine is not unrespectable.

The contribution to the performance arising from the use of layers was also measured. In the absence of layers the computations ran approximately a factor of seven slower on the Cyber, a marked difference in rate which more than justifies the effort required in developing the layered version of the algorithm. Layer construction required approximately 20% of the processing time, integration of the equations of motion and other assorted minor calculations 5%, with the remainder being devoted to the interaction calculations proper. Timings on the vector processing varied by less than 10% for a series of systems ranging in size from 5,000 to 250,000 particles; the additional work required for the larger systems, in that they had to be subdivided to avoid excessive vector lengths and temporary storage, had minimal impact on performance. The distributed studies used four processors linked via a single shared memory with sufficiently high bandwidth to effectively handle the communication demands. The overheads required for interprocessor data transfer amounted to only about 3% of the total run time; on the other hand, when communication was routed through the supervisory host computer (an IBM 3081) having a much lower bandwidth and with other responsibilities as well, the run time grew by approximately 50%. This should serve as an object lesson in that communication performance can sometimes be critical.

The simulations carried out on the processor network involved systems of up to 170,000 particles and were aimed at observing the nature of fluid flow past a fixed obstacle. [8] While not *a priori* obvious that anything of interest ought to be expected in such small systems, the sizes of which are many orders of magnitude less than the smallest flow experiments that could ever be conducted, the simulations soon revealed behavior that strongly resembled (both qualitatively and quantitatively) what is seen experimentally. Phenomena such as eddy formation, wake oscillations downstream of the obstacle, and the appearance of a vortex street, were all reproduced.

In order to continue these studies on a more systematic basis using more accessible hardware, the vector version of the algorithm was developed, and a series of simulations of systems containing as many as 200,000 particles begun. This work is still underway, and even larger systems are contemplated in order to widen the scope of the study.

The performance of the event-driven simulation for modelling step potentials is measured in terms of collisions per unit processing time; a comparison of the speeds of event- and time-driven simulation turns out to be strongly dependent on the nature of the potentials and the density of the system (which in turn determines the number of collisions the particles experience in the course of their diffusive motion). In studies of the simplest fluids, such as hard disks, the use of step potentials does seem to be more efficient on a scalar machine. The measured performance on the IBM 4381/13 is close to 2×10^6 collisions/hr. The method has been used [9] for studying thermal convection in fluid layers; the familiar convective rolls seen experimentally have also been reproduced at the microscopic level, again a somewhat surprising result given the enormous disparity in scales between the macroscopic systems where these effects are all too familiar and the microscopic systems studied by MD – a mere 14,000 particles were used in these simulations.

These two sets of comparatively recent results strongly hint at the potential use of MD simulation in the study of fluid flow. One might hope that more large-scale modelling of this type will prove helpful in understanding the microscopic origins of some of these hydrodynamic phenomena (for example, the nature of the correlations between particles in the region where an indication of flow instability first appears); as it is the field of fluid dynamics tends to follow an engineering approach and, owing to its complexity, is fraught with empiricism.

8 Prognosis

The increasing availability of what are today called *supercomputers* should see a rapid spread in the range of problems to which MD is applied. More exciting is the prospect of new generations of machines with vastly improved performance; from the MD practitioner's point of view this will allow studies of much larger systems over longer periods of simulated time, and will also support the additional effort needed for greater realism.

The thorn among these roses is that it may no longer be possible to conceal the complexity of the hardware from the algorithm developer. No two modern multi-vector-processor systems will look alike, a standard programming language that can handle such hardware issues in a transparent but effective manner is not on the horizon, and thus, on the assumption that efficiency is a relevant issue, development of simulation algorithms will have to be oriented towards the hardware on which production is to be carried out. Portability of software will be the major victim; customized program development will be required for each kind of computer, not an especially rewarding occupation but essential if one is to take advantage of the latest developments in computer technology.

Acknowledgements

Portions of the work described here were carried out during visits to IBM, Kingston NY, and the Center for Simulational Physics at the University of Georgia, Athens GA.

References

[1] *Molecular Dynamics Simulation of Statistical Mechanical Systems*, Proceedings of the International School of Physics "Enrico Fermi", Course XCVII, Varenna, 1985, eds. G. Ciccotti and W.G. Hoover (North-Holland, Amsterdam, 1986).

[2] D.C. Rapaport, Computer Phys. Repts. (to appear).

[3] R.W. Hockney and C.R. Jesshope, *Parallel Computers*, 2nd edn., (Adam Hilger, Bristol, 1988).

[4] K. Hwang, IEEE Computer **18**(6), 57 (1985).

[5] B.J. Alder and T.E. Wainwright, J. Chem. Phys. **31**, 459 (1959).

[6] D.E. Knuth, *Sorting and Searching* (Addison-Wesley, Reading, 1973).

[7] J.J. Dongarra, Argonne National Lab. Math. and Comp. Sci. Tech. Memo no. 23 (unpublished, 1988).

[8] D.C. Rapaport, Phys. Rev. A **36**, 3288 (1987); D.C. Rapaport and E. Clementi, Phys. Rev. Lett. **57**, 695 (1986).

[9] D.C. Rapaport, Phys. Rev. Lett. **60**, 2480 (1988).

PARALLEL COMPUTERS AND THE SIMULATION OF SOLIDS AND LIQUIDS

David Fincham

Computer Centre and Physics Department
University of Keele, Staffordshire, ST5 5BG, U.K
and
SERC Daresbury Laboratory
Warrington, Cheshire, WA4 4AD, U.K

ABSTRACT

Parallel computers are a very cost-effective way of providing high performance computational power, though at present it is not always easy to use them effectively. Molecular dynamics simulation is intrinsically highly parallel in nature, both in the force evaluation phase of the timestep and in the motion integration phase. Methods are described which implement molecular dynamics simulation on two parallel computers: the Distributed Array Processor (DAP), and networks of Transputers. On the DAP, lattice dynamics can be studied by means of a geometrical decomposition which gives one processing element (PE) to each ion or molecule. The method can be modified for liquids by exchanging molecules between PEs when necessary. An alternative approach for molecular liquids maps pair interactions, rather than molecules, onto PEs. In the Transputer case, processors are connected in a ring and molecule data circulate in such a way that all required pair interactions may be evaluated. The data movements take place in parallel with the computation so little loss of efficiency arises from communication delays.

Introduction

These lectures introduce the concepts of parallel computing with particular reference to molecular dynamics simulation. We will define a parallel computer as one having a number of processors which can work together to solve a single application problem. My aim is to convince you that parallel computing is important; that is is interesting; and that it is useful. I need, however, to warn you that it is not always easy. Although there is a natural mapping between the parallelism in the physical world and that inside the computer, the tools to turn this mapping into working computer programs exist in only the most rudimentary form. We still have to consider every computing problem afresh for every new parallel computer which is invented. These lectures are in this spirit, and will concentrate on how to do molecular dynamics on two particular computers: the Distributed Array Processor (DAP) and a Transputer network. I have previously [1] reviewed the application of parallel computers in molecular

C.R.A. Catlow et al. (eds.), Computer Modelling of Fluids Polymers and Solids, 269–288.

simulation and would refer you to that review for more details, particularly of the use of the DAP. The methods I will describe for Transputers are newer and further details are in a preprint [2].

Parallel computers

In many intensive scientific calculations we are in some way simulating the behaviour of the physical world: that is, we are evaluating a computational model whose equations encapsulate the behaviour of the physical system, and using the computer to follow in detail the evolution of that model. Two good examples are weather forecasting and molecular dynamics. In the real world, lots of things are happening simultaneously ("in parallel" or "concurrently"). The weather in Bath is happening at the same time as the weather in Keele: this molecule is moving at the same time as that other molecule. From this point of view it seems sensible to have a computer with many processors working on the simulation. One processor can work on the weather around Bath; another on the weather around Keele. Since the weather in the two towns is independent only on a short time scale, these processors will need to periodically exchange information. Similarly in molecular dynamics we could have a processor solving the equations of motion of each molecule: they will also need to exchange information since the molecules are interacting with each other. This concept of a number of processors, linked together so they can exchange information, is what is meant by a parallel (or concurrent) computer.

Parallel computing as a concept has been around almost as long as sequential or single-processor computing. The fact that the latter has been so dominant until recently is largely a historical accident. Until the advent of semiconductor technology the speed with which processors could access memory and communicate data was a more serious limit than their raw speed of computation, and there was little incentive to couple processors together. This has left us at a disadvantage now that parallel computers are a reality. The strict sequential model of computation (one calculation after another) underlies all widely used scientific programming languages and, indeed, the very concept of a computational algorithm. Much rethinking is necessary before we can use parallel computers effectively.

Why do we need parallel computers? It is a question of cost- effectiveness. It is much cheaper to make powerful computers by replicating many simple components than by pushing the single processor design to the technological limits. I would like to contrast two ways of using this greater cost-effectiveness. We can either build machines at supercomputer cost, but many times more powerful, and solve very big computational problems that are impossible with current supercomputers. Or we can build machines with supercomputer power but at much smaller cost so we can afford to have one in our own laboratory (or even office). Then we can solve problems in a different way; interacting with the calculation as it progresses and so using the number-crunching power of the computer as an aid to our own thought processes. As computational scientists I am sure we will find uses for both approaches.

There are three main types of parallel computer in use at the present time. The first has a number of processors sharing (that is, directly addressing) a common memory. One way in which such a computer can be used is to have different individual jobs running in each processor: this is usually called multiprocessing. Only when a given application program uses more than one processor do we have true parallel computing. The latest generation of supercomputer (e.g the Cray X/MP) has a small number (up to eight) of powerful vector processors. Another interesting example is the Sequent which has up to 30 Intel 80386 processors (the same as in the latest IBM PC). The advantage of a shared memory is that there is no need to have direct processor-to-processor communications, since one processor can simply read from memory a variable written by another. Of course, it needs some careful programming in order to control and synchronise such reading and writing. The disadvantage of such machines is that it is not possible to have more than a handful of processors because the single memory cannot provide data quickly enough to support large numbers of processors. You therefore do not get the power and cost-effectiveness of massively parallel machines. I will accordingly say no more about shared memory designs.

A second type of parallel computer is the single-program distributed- memory machine. Each processor (better called a processing element, PE, in this case) has its own data memory. However, it does not run its own program. Instead it receives instructions broadcast to it from a master processor. Each PE executes the same instruction, in lockstep with the others, but on its own data. The pioneering example of this kind of machine is the Distributed Array Processor (DAP). Because the PEs do not run a program themselves they can be simple and there can be many of them. The DAP has 1024 or 4096 PEs connected as a square array (32 x 32 or 64 x 64 respectively). Each PE can thus send data directly to each of its four nearest neighbours. The master processor communicates with the PEs by means of row and column highways.

A third type of parallel computer is the distributed-memory multiple- program machine. Each processor has its own memory and runs its own program: it is a computer in its own right. (Note that if there are large numbers of processors they are likely to be running the same program. This is still different from the single-program case because the path through the program may be data dependent). There must also be a means by which data can be passed from one processor to another. On the grounds of cheapness this type of machine is usually constructed from microprocessors. An example specifically designed for building parallel computers is the Transputer. As well as being a powerful microprocessor this has circuits for floating-point arithmetic and a small amount of memory on the same chip. (Its memory can be extended by attached memory chips.) Also on the chip are circuits which drive four communication links, by means of which Transputers can be connected together. There are obviously many different networks that can be constructed. In some machines based on Transputers, such as the Meiko Computing Surface and the Niche Computing Platform the connections are made by electronic switches under software control so that the network topology can be changed, depending on the program being run.

Molecular Dynamics

Molecular dynamics is very simple in principle. On each timestep we first evaluate the force on each particle due to its interaction with other particles.

$$\mathbf{F}_i = \sum_{j \neq i} \mathbf{f}_{ij} \tag{1}$$

and usually calculate global quantities such as potential energy at the same time

$$U = \sum_{i<j} u_{ij} \tag{2}$$

Then we integrate the equations of motion of the particles to get new positions and velocities ready for the next timestep.

There is no problem in finding calculations which can be done in parallel. The pair interactions are independent of each other and can be evaluated simultaneously. Then, when the force on each particle is known, its motion can be integrated independently of the others. However, there are a number of problems which must be overcome in realising this intrinsic parallelism in a working program.

A processor calculating a particular pair interaction needs to know the coordinates of the two particles involved. This exemplifies a general problem in any parallel method: how do we make sure that the right data is in the right place at the right time? (The *data mapping* problem.)

There are many more pair interactions than there are particles. Thus the motion integration phase of the calculation cannot efficiently make use of as many processors as can the pair evaluation phase. (The *degree of parallelism* problem.)

Summations are involved in obtaining forces from pair forces and in obtaining global quantities. The summation of data from different processors is not a highly parallel procedure, as is explained in Figure 1. (The *summation* problem.)

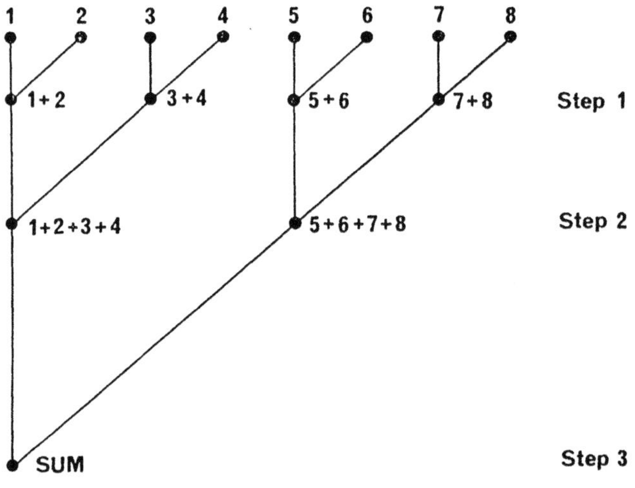

Figure 1. Parallel summation of data values taken from N different processors. The summation proceeds in a series of stages, in each of which values are added in pairs. In the later stages most of the processors are idle, so the method is not very efficient.

A fourth problem is revealed if we consider a simple and obvious approach to parallel molecular dynamics. Suppose we have one processor per particle, and this processor knows the coordinates of all the particles. Then it can simply go ahead and use equation (1) to calculate the total force on the particle (and also calculate a partial energy sum) and then integrate its motion, all completely independently of the other processors. The drawback with this approach is that every interaction is calculated twice, and so the maximum efficiency with which we can utilise processors is 0.5. Since parallel computers are very cost-effective we could perhaps live with this *unique-pairs* problem, but fortunately we will see that it is possible to avoid it.

We will now go on to consider how molecular dynamics has been implemented on the DAP and on Transputer networks, and see how these four problems have been overcome. We will also gain insight into these particular computers and how they are programmed.

Molecular dynamics on the DAP

The first method we will describe was developed for solid-state problems. To see how it works consider a simplified example of a 2- dimensional 32 x 32 square lattice of spherical molecules simulated on a 32 x 32 DAP so there is one molecule per PE (Figure 2). Each PE holds the centre-of-mass coordinate of its molecule. This is declared in Fortran Plus, the DAP programming language, as

```
real x(,),y(,)
```

Also let us suppose that each molecule interacts with a certain number of near neighbours, and that we have cyclic boundary conditions. The force on every molecule due to its interaction with the neighbour on its immediate "north" can be calculated by the following Fortran Plus statements

```
xij = x - shsc (x)
yij = y - shsc (y)
fors = force (xij,yij)
fxij = fors * xij
fyij = fors * yij
ftotx = ftotx + fxij - shnc (fxij)
ftoty = ftoty + fyij - shnc (fyij)
```

This is a very terse language which needs some explanation. Every quantity here is a 32 x 32 array, declared in the same way as x(,), and each statement applies to every PE (i.e to every molecule) in parallel. The function shsc(x) ("shift south cyclic") moves its array argument south by one place, using the direct connections between neighbouring PEs. This is a function built into the language: it automatically takes into account the cyclic boundary conditions. Thus the statements

```
xij = x - shsc(x); yij = y - shsc(y)
```

calculate, in each PE, the vector separating the particle in that PE from its northern neighbour. From this vector the pair force is calculated in the normal way. The pair force is than added on to a total force accumulator, and shifted north one place to be subtracted from the force accumulator for the other particle of the pair. Interactions with other neighbours can then be successively accumulated, out to the range of the potential. If they are not nearest neighbours, multiple shifts will have to be made to bring data from the appropriate pair into one PE for the force evaluation.

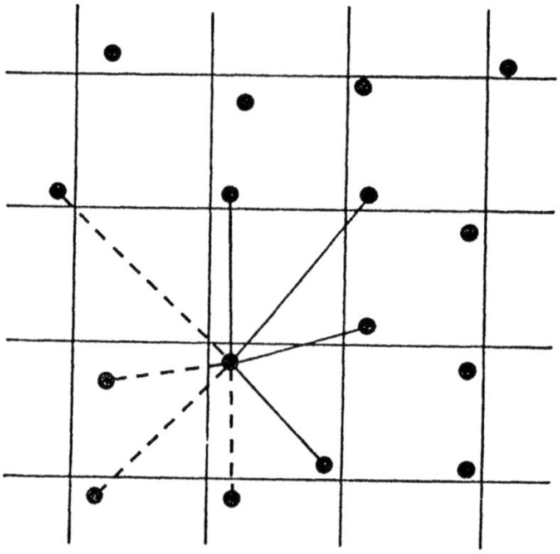

Figure 2. Molecular dynamics simulation can be used to study lattice dynamics on the DAP. The lattice structure is mapped onto the DAP with one molecule per PE. Because particles do not move far from their lattice sites each molecule (PE) interacts with a fixed set of other molecules (PEs). Interactions with a particular neighbour are evaluated for all molecules simultaneously.

This method is very efficient for the type of problem it is designed to handle, essentially limited to lattice dynamics. Molecules can move around their lattice sites, but not so far that they change neighbours, since it is assumed they interact only with other molecules in a fixed set of nearbye PEs. The method solves the unique pairs problem: each lattice direction is considered only once and the pair force added to the accumulator for one member of the pair and subtracted from the accumulator for the other member, using shifts. Because the mapping of the problem onto PEs is geometrical, the evaluation of inter-molecular interactions involves only short data movements which can make use of the direct connections between PEs. The different pair forces acting on a particular molecule are accumulated sequentially in that molecule's PE, so the main summation problem is overcome. (There is still the question of global summations: this is discussed further below). The parallelism in the method, however, is over molecules, not pair interactions. This means that we cannot efficiently have fewer molecules than PEs (of course, we can have more, by having more than one molecule per PE).

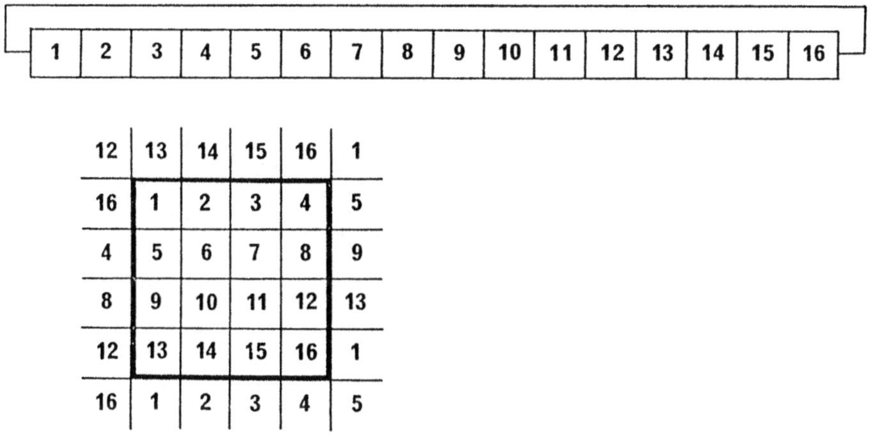

Figure 3. The hardwired boundary condition of the two-dimensional DAP can be used to implement three-dimensional skew-periodic boundaries. To illustrate this simply we show instead a two-dimensional lattice mapped onto a one-dimensional "DAP" with cyclic boundaries.

In our simple example we have considered only translational motion, but there is no problem in including orientational motion in the model, making use of quaternion parameters. This is what was done in work at Edinburgh and Cambridge on plastic crystals [3]. These molecular crystals show phases in which there is orientational disorder (the molecules rotate) but the translational order is maintained. A good example is SF_6 (for obvious reasons, this phenomena tends to appear in crystals where the molecules are reasonably spherical). Clearly the parallel method we have just described is ideal for such a study. The use of a large number of molecules (4096) was important in this work because the transition to the plastic phase is induced by long wavelength fluctuations. A three-dimensional simulation was mapped onto the two-dimensional DAP. This is possible in general [4], but we will not go into this further here, except to point out that it is still possible to use the two-dimensional hard-wired cyclic boundaries of the DAP to produce three-dimensional periodic boundaries, albeit skew periodic. This is illustrated in Figure 3. The skewness is not a problem, since anyway the periodic boundaries are an artefact intended to mimic an infinite system. Another relevant point to make here is that the DAP is also a useful graphics engine [5], and graphics has an important role to play in elucidating the complex dynamical phenomena in these systems [6].

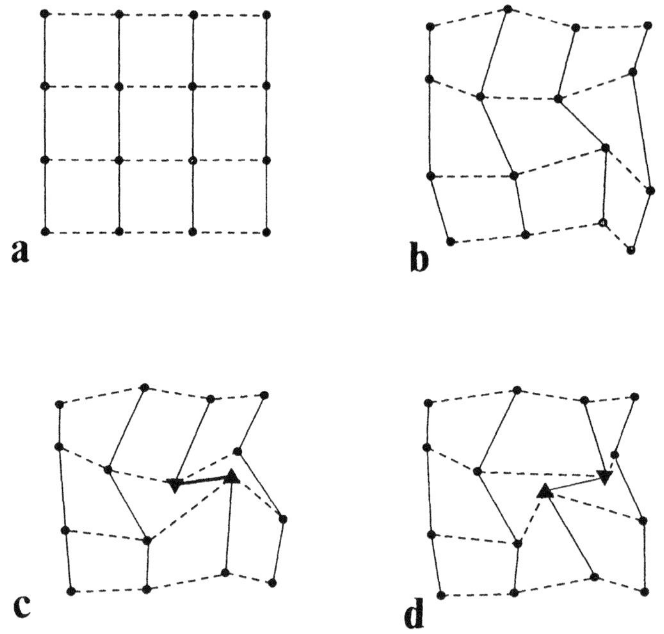

Figure 4. The Monotonic Grid method of performing liquid simulations on the DAP. The molecules start on a regular grid, (a), which becomes distorted as they move, (b). When two molecules interchange the order of one of their Cartesian coordinates (c) the molecules are swapped between their respective PEs, thus restoring the monoticity of the grid, (d).

What about the simulation of fluids? The problem here is that the molecules move around, complicating this kind of geometric decomposition. An approach can be developed based on the Monotonic Logical Grid of Boris [7]. The idea here is to still assign one molecule to each processor, and to ensure that a given processor always contains the data for a molecule in roughly the same region of space. This is achieved as follows. Imagine for a moment that the molecules start on lattice sites (Figure 4) and then move away from them as the dynamics progresses and the lattice begins to melt. In a strictly geometrical decomposition each PE would be associated with a fixed region of space, and the molecules would move from one PE to another as they cross the boundaries from one cell to another. This would not be appropriate on the DAP since it would result in variable numbers of molecules (0, 1, 2, or more) being present in each PE. Instead, molecules remain in the same PE until such time as two molecules have interchanged the order of one of their Cartesian coordinates. When such an event happens the molecules are interchanged between the two PEs. (All the data pertinant to the molecules is exchanged: if it is required to follow the history of individual molecules to evaluate time correlation functions this will include an identifying index.) This procedure ensures that the molecules remain at the nodes of a monotonic grid that is a distorted version of the original lattice. This means that each PE always contains a single molecule, and nearby molecules will be found in nearby PEs, though it may be necessary to loop over a larger set of nearby PEs to be sure of picking up all interactions within range than in the solid-state case. Although we assumed the molecules start out on a lattice this is not necessary since the initial (distorted) monotonic grid can be set up for a liquid-like configuration by standard sorting techniques.

I will now turn to another approach. This is based on the observation that, for many liquid state problems, the DAP has too many processors. In studying the structure, thermodynamics and short-time dynamics at most liquid state points quite small numbers of molecules are adequate, because the correlations to be studied are of quite short range, not much larger than the range of the potential. Rather than a geometric distribution of molecules to PEs, it is sensible to map pair interactions onto PEs.

To see how this is done, consider first a simulation of 32 molecules in a 32 x 32 DAP. The essential code for this is as follows.

```
real x(),y(),z(),fx(),fy(),fz()
real xij(,),yij(,),zij(,),rsq(,),rsqr(,),fors(,)
logical offdiagonal (,)
:
:
xij = matc (x) - matr (x)
yij = matc (y) - matr (y)
zij = matc (z) - matr (z)
call periodic (xij,yij,zij)
rsq = xij*xij + yij*yij + zij*zij
rsqr = 0.0
rsqr (offdiagonal.and.rsq.le.rcutsq) = 1./rsq
fors = force (rsqr)
fxij = fors * xij
fyij = fors * yij
fzij = fors * zij
fx = sumc (fxij)
fy = sumc (fyij)
fz = sumc (fzij)
```

x,y and z, fx,fy and fz are now vector quantities (32 elements); other quantities are 32 x 32 arrays (usually called matrices in DAP documents, though not necessarily matrices in the mathematical sense). The function matc(x) replicates a column vector across the array using the row highways: matr(x) replicates a row vector. These functions are used to produce a matrix of pair separation vectors, as illustrated in Figure 5. From the matrix of pair separations a matrix of pair forces is obtained. The columns of this are summed to give a vector of forces, using the sumc(x) function. The energy of the system can be obtained at the same time as a global sum over PEs. Despite the non-parallel aspect of such summations the DAP is remarkably efficient at performing

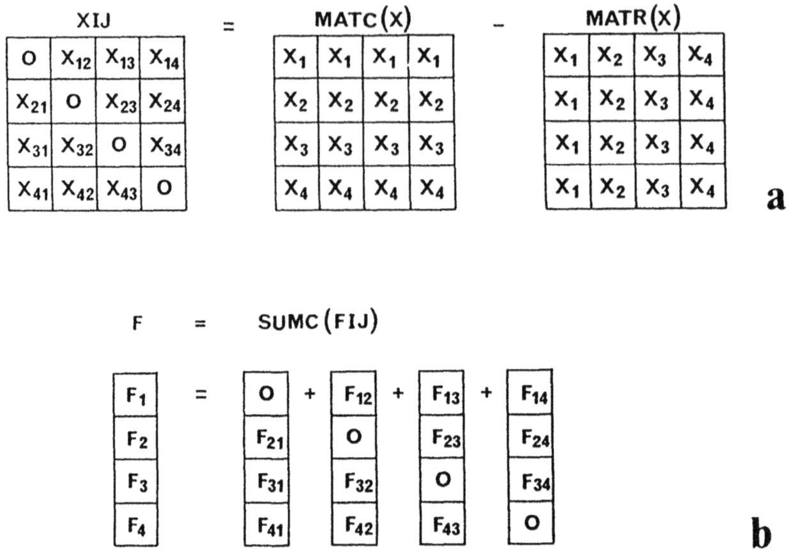

Figure 5. (a) In Fortran-Plus on the DAP, the built-in replicator functions MATC and MATR can be used to produce a matrix of pair separations from a vector of coordinates. (b) The built-in SUMC function can be used to produce a vector of forces from a matrix of pair forces.

them. To give some insight into the reason it is necessary to mention an interesting and important point about the DAP that we have skipped over so far. It has 1024 PEs, but these are single-bit processors, all higher-level arithmetic being provided in software. This software is able, in conjunction with the nearest neighbour PE connections, to make the DAP work in two basic modes when dealing with multi-bit numbers. In matrix mode it operates as 1024 processors acting in parallel on matrices of 1024 elements: in vector mode it acts as 32 processors acting in parallel on vectors of 32 elements. The summation functions built into the language manage to utilise the PEs (and also make use of the master processor) with great effectiveness.

In the code given above we have applied a spherical cut-off by means of a logical expression which, in Fortran Plus, is used instead of subscripting to select a sub-set of the array elements. We do not save any computer time by doing this: each PE performs the same calculation, but those for whom the logical expression evaluates to .false. do not store the result in memory. To achieve maximum efficiency it is therefore advisable to have as few interactions outside the range of the potential as possible, and this means using a computational box which is as nearly spherical as possible. The truncated octahedron is a popular choice. It is also worth noting that in systems of molecules with electric dipole or multipole moments it is advisable not to apply a spherical cut-off. The effect is then to convert the usual nearest- image convention into an approximation to the Ewald sum [8].

The major disadvantage of the method just described is that each interaction is calculated twice. However, in most liquid simulations one would require more than 32 molecules. One then divides the pair interaction space into 32 x 32 blocks to be evaluated in parallel. Only blocks in the upper triangular region of pair space, and along the diagonal, need to be considered, and the loss of efficiency due to double-counting in the diagonal blocks becomes quite small. This "all- pairs" mapping then becomes a very practical technique which has been used to good effect in the study of the thermodynamics of mixing of molecular fluids [9], and dielectric effects in polar fluids [10].

Molecular dynamics using Transputers

We will consider in this section traditional liquid or solid state molecular dynamics where the size of the simulated system is not much greater than the range of the potential, and it is customary to use a spherical cutoff or neighbour list. Typically such simulations involve up to 1000 molecules. The study of some phenomena requires the use of very large numbers of particles, and computational boxes with dimensions many times larger than the range of the potential. In these circumstances a geometrical decomposition of the problem becomes essential, and the parallelisation of such cell-based techniques is discussed in the lectures by Professor Rapaport [11], albeit using processors very unlike Transputers.

There are two main differences between computers constructed from Transputers and the DAP. First, Transputers are multiple-program rather than single-program machines; second, because the processors are more complex we are unlikely to be able to afford so many of them. In particular, it is unlikely that the number of processors will be greater than the number of molecules, so we will be able to parallelise over molecules, not over pair interactions.

Can we find a data mapping that distributes the work evenly over processors and yet solves the unique-pairs problem? Yes, in fact several [2]. We describe first a method known as SLD (Systolic Loop Double) and suppose for simplicity that we have P processors and P particles. (A particle can be a spherical molecule or ion, or a rigid polyatomic molecule if orientational coordinates are included in the data.) The data representing each particle is stored in its individual "home" processor. This data consists of coordinates, velocities, force accumulators, and possibly other information such as type indices. At the start of each time step, a copy of the particle coordinates is taken; there is also a duplicate set of force accumulators, both sets being initialised to zero. The duplicate coordinates and forces constitute a data packet which subsequently circulates round the ring of processors in a series of "pulses", as illustrated in Figure 6. On each pulse, each processor evaluates the pair interaction between its fixed home particle and the circulating particle. The pair force is added onto the fixed force accumulator, and subtracted from the circulating accumulator. After $(P-1)/2$ pulses (P must be odd) all the pair interactions have been considered. The circulating force

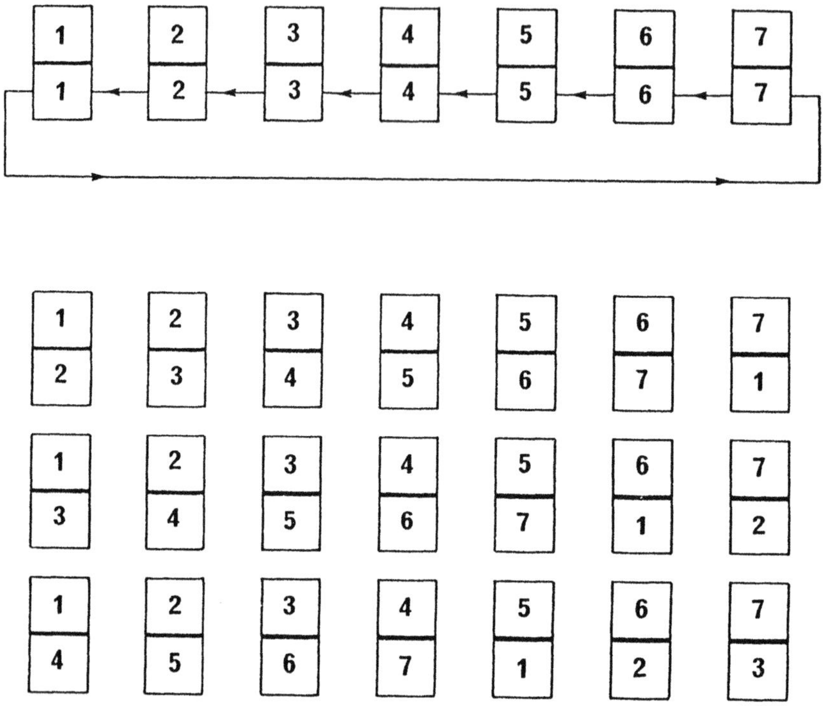

Figure 6. The SLD method of evaluating pair interactions on multiple Transputers. The processors are conected in a ring. One set of particle coordinates and force accumulators remain fixed in their "home" processor: a second set circulate round the ring in a series of systolic "pulses" in such a way that all pairs are considered.

accumulators must then be rewound by $(P-1)/2$ pulses, or wound on by $(P+1)/2$ pulses, until they reach their home processor. The fixed and circulating accumulators are then summed to give the total force on the particle. Its motion can then be integrated, ready for the next time step.

In practice, the number of particles, N, is likely to be greater than the number of processors, P. The same basic method can be used, but with a group of particles having its home in each processor, and circulating as a packet. There needs to be a preliminary stage in which the interactions between particles within a group are considered. Then, on each pulse, each processor evaluates the interactions between particles in the fixed group and those in the circulating group. This is called the SLD-G method.

For the global quantities such as potential energy, virial and kinetic energy, each processor maintains a sub-total for each pair or particle it considers. At the end of the step, these sub-totals are circulated round the ring so that each processor forms a total sum, and then independently calculates temperature, pressure etc. and carries out velocity scaling, if required, on its own particles. One processor requires communication with the outside world so that values may be printed, written to file or displayed graphically. Apart from this, every processor is equivalent to every other and they all run the same program. Indeed, if no spherical cutoff or neighbour list is used, the processors perform identical calculations and this method could easily be implemented on a single-program computer such as the DAP.

There are two aspects of this method that are worth looking at in more detail: load balancing and communication overheads. Obviously in any parallel computer we would like each processor to do the same amount of work. Otherwise some will be waiting idly for others to catch up, leading to a loss of efficiency. If we suppose the groups of particles contain equal numbers, and that there is no cutoff, then we see that each processor evaluates the same number of interactions, and integrates the motion of the same number of particles. In this case the method achieves perfect load balancing, which is the main reason for its introduction. In the case of liquid simulations the exact number of particles used is not significant and it is usually possible to ensure that the number of particles is an exact multiple of the number of processors ($n = N/P$) so that the groups are of equal size. We have already mentioned that it may be advantageous not to apply a cutoff if the molecules possess multipole moments. In other cases, we would want to apply a cutoff. Because each processor runs its own program it can simply skip over interactions which lie out of range of the potential, just as in a conventional sequential program, and thereby save on computer time. Also, because each processor deals with the same set of pairs on each time step, it is possible for each processor to maintain and use a neighbour list for this set of pairs. If a cutoff or neighbour list is used, however, there will be statistical fluctuations in the number of in-range pairs on each processor on each pulse. There will therefore be a certain loss of efficiency as processors have to wait until the one with the largest number of in-range pairs has completed its work before they can proceed with the next pulse. The worst case occurs when $n=1$ i.e $N=P$. It is almost certain that at least one processor will have an in-range interaction on each pulse, and the calculation will then proceed at the same pace as if all the interactions were in range. The loss of efficiency decreases as n increases. This effect is analysed in detail in [2], and it is shown that for $n>16$ almost 100% load-balancing efficiency is achieved.

Another possible source of inefficiency lies in the time taken to communicate data between processors. The amount of data communicated between the processors on each pulse depends on the number of particles, n, in the data packet. The time taken to perform the force evaluation on each processor depends on n^2 in the all-pairs case, because all the interactions between the n particles in the fixed group and those in the circulating group need to be considered. Even if a cutoff is applied, there is an n^2 component because all pairs need to be examined to see if they are in range. (In the neighbour-list case there is also an n^2 component due to the need to periodically reform the list.) Thus the ratio of communication time to compute time decreases as n increases. A detailed examination shows again that there is no significant loss in efficiency for n greater than about 16. This type of analysis is valid for any type of parallel computer. For those based on Transputers it is possible to reduce this communication overhead by rearranging the calculation to utilise the ability of Transputers

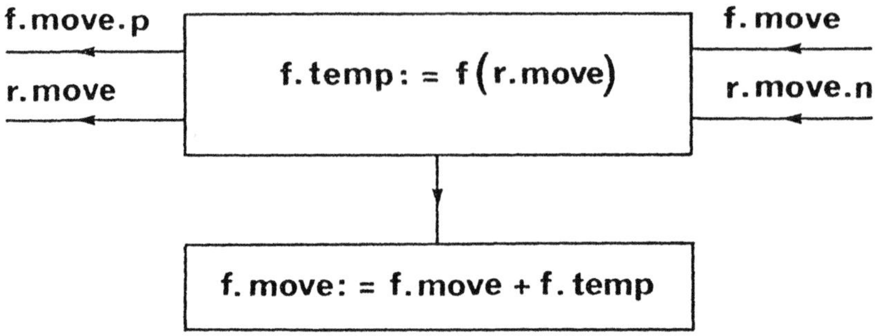

Figure 7. A modification of the SLD method allows the computation of pair forces to proceed at the same time as the communication of data. One processor is shown, for a particular pulse: all processors are equivalent. The processor simultaneously: (i) calculates the pair force, using the fixed coordinate and the current moving coordinate, r.move, into a temporary force variable f.temp; (ii) receives the moving force accumulator required for this pulse, f.move; (iii) receives the coordinate which will be required for the next pulse, r.move.n; (iv) transmits the force accumulator from the previous pulse, f.move.p; (v) transmits the current coordinate, r.move.

to communicate data while continuing to perform numerical calculations. This is illustrated in Figure 7.

The accumulation of the thermodynamic quantities is done separately in each processor. Although this is wasteful, the amount of arithmetic involved is so small as to be insignificant. Of greater importance is the communication time involved in distributing the sub-totals around the processors. This is a fixed quantity for a given number of processors, whereas we have just seen that the force calculation time depends on n^2. Again, the loss of efficiency decreases with n, the ratio of number of particles to number of processors.

The SLD-G method is actually not the best to use, though it is the easiest to explain. A better method is called SLS-G (S for Single) in which the Transputers are connected in a chain, there is only one data packet per group, and there is no need for a rewind or wind-on stage. This is illustrated in Figure 8. The method is somewhat more complicated to program because not all processors are equivalent. One processor, the head of the chain, is home to one group of particles rather than two, and takes no part in the force evaluation except in calculating intra-group interactions.

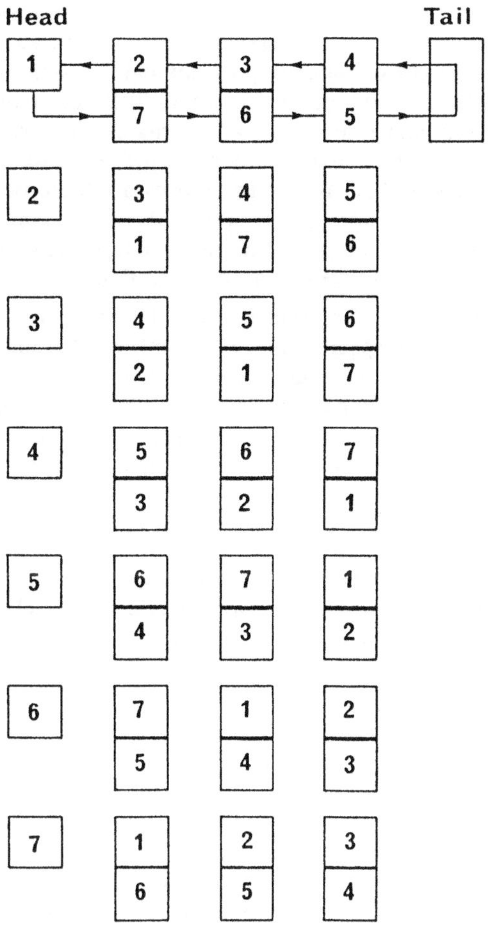

Figure 8. The SLS method of evaluating pair interactions. If P is the number of particles, or groups of particles, there are (P-1)/2 processors (plus the Head and Tail) and all interactions are considered in P systolic pulses.

Programming Transputers

As we saw from the examples above, the DAP is easy to program, because it is a single-program machine. A DAP program is very like a conventional Fortran program: a single instruction stream is obeyed sequentially. The difference is that some statements perform operations on complete vectors or complete matrices. A Fortran Plus program is therefore simpler than an ordinary Fortran program performing the same operations on arrays element-by-element by means of DO loops. Of particular utility is the use of logical expressions to select subsets of array elements: much clearer to read and less error-prone than picking up the correct indices individually. Again, many difficult operations, such as periodic boundaries, which can be tricky to program, are built into the language.

Transputers and other multiple-program computers are more flexible but consequently more difficult to program. It is necessary to provide separate and possibly different programs for each processor, and to program data-passing between them explicitly. Actually, to avoid confusing ourselves, let us call the software running inside one Transputer a process, and reserve the word program to describe the complete set of cooperating processes which, together, achieve the result we want. There is a further complication: Transputers, like most other processors these days, can time-share; that is, run several independent processes, apparently at the same time, but in reality giving them very small slices of processor time in turn. Thus there may be several processes running on a single Transputer at the same time.

The first method of programming Transputers we will look at is by means of a special language, Occam, designed specifically to support parallel multiple-process computing. Occam and the Transputer were designed together, and Transputers implement the language very efficiently. The basic concepts in Occam are the process and the channel. A process in Occam is a unit of work which the processor executes to completion, but it can be a more fine-grained concept than in conventional terminology: what in other languages would be a statement

```
a := b + c
```

is, in Occam an elementary process, assignment. Processes are built up in Occam hierarchically by means of constructors. One constructor is called SEQ, which means "make a process out of component processes by executing them sequentially". For example

```
SEQ
  a := b + c
  d := a * f
```

is a process with two components executed sequentially: components are indented by two spaces relative to their constructor.

Another important constructor is PAR, which means "execute the component processes in parallel (i.e simultaneously)". The compound processes built up by means of constructors can themselves become components of higher-level processes.

Processes running in parallel with each other do not share variables, but can communicate by means of channels. To do this, two more elementary processes are defined in the language: input

```
chan ? variable
```

and output

```
chan ! expression
```

Two parallel processes communicate by means of a shared channel; one outputs the variable and the other inputs it. Either process will wait for the other before sending or receiving as necessary, so that synchronisation is automatic. To give some flavour of Occam without going into too many details we outline two examples in Figures 9 and 10.

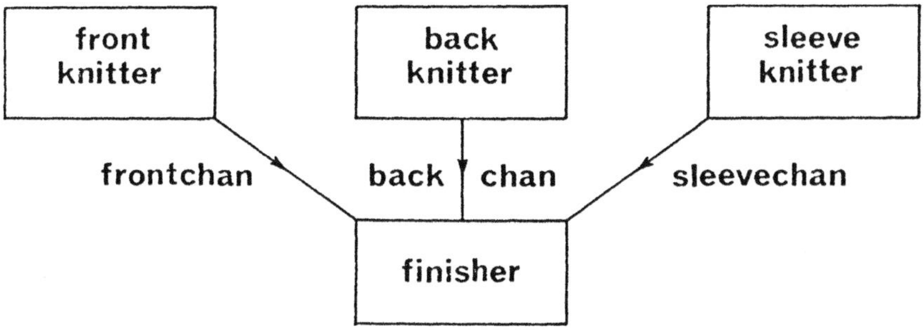

a

```
PAR
  SEQ                    -- front knitter
    front := knitfront (wool)
    frontchan ! front
  SEQ                    -- back knitter
    back := knitback (wool)
    backchan ! back
  SEQ                    -- sleeve knitter
    l.sleeve := knitsleeve (left,wool)
    sleevechan ! l.sleeve
    r.sleeve := knitsleeve (right,wool)
    sleevechan ! r.sleeve
  SEQ                    -- finisher
    PAR
      frontchan ? front
      backchan ? back
      SEQ
        sleevechan ? l.sleeve
        sleevechan ? r.sleeve
    sweater := sew (front,back,l.sleeve,r.sleeve)
```

b

Figure 9. Four workers cooperate together to knit a sweater: they are shown diagramatically in (a) as processes connected by channels, and (b) is the Occam equivalent. Note how synchronisation is automatic, and how the Finisher receives the parts of the sweater in a PAR process with three components: the PAR is completed, and the sweater sewn, only when the three component processes are complete, i.e all the parts of the sweater have been received.

A very important point as far as Occam is concerned is that the parallel processes may be on the same processor or on different processors. If they are on the same processor the channel is implemented by a location in memory. In reality the processes are timesliced, but logically they are independent and simultaneous. If parallel processes are on different processors, the channels are implemented over the physical links between processors. The great advantage of this is that Occam programs can be developed on a single Transputer, and behave in a logically identical manner when distributed over a network of Transputers. All you have to do to the Occam program is to

```
PAR process = 1 FOR N                  -- N parallel processes.
  SEQ step =1 FOR number.of.steps      -- loop over steps
    SEQ
      r.move := r                      -- copy coordinates
      f := 0                           -- zero fixed and moving
      f.move := 0                      -- force accumulators
      SEQ pulse = 1 FOR (N-1)/2        -- loop over pulses
        SEQ
          PAR
            in ? r.buffer, f.buffer    -- systolic
            out ! r.move,f.move        -- data
          r.move := r.buffer           -- movement
          f.move := f.buffer           --
          pairforce (r - r.move, f.pair)
          f := f + f.pair
          f.move := f.move - f.pair
      SEQ pulse = 1 FOR (N+1)/2        -- wind-on
        SEQ
          PAR
            in ? f.buffer
            out ! f.move
          f.move := f.buffer
      SEQ
        f := f + f.move          -- sum fixed and moving accumulators
        dynamics (r, v, f)       -- update positions and velocities
```

Figure 10. An Occam outline of the SLD method (see Figure 6 for the diagrammatic equivalent). Note how the systolic data movement is achieved, by receiving new values into a buffer while simultaneously transmitting the old. This means that there is synchronisation of the data passing right round the ring: this is the meaning of the term "systolic".

add a few statements indicating which process goes on which processor, and which channel goes on which link.

A second method of programming Transputers which I would like to discuss is the use of the Trollios (formerly called Trillium) system developed at Cornell University [12]. This is a general purpose parallel programming system that has been implemented for Transputers by several manufacturers. The idea is that the application program consists of a number of processes, written in a standard high-level language (Fortran or C) which are distributed over the processors. These processes communicate with each other by calling a library of message-passing routines. These routines interact via a kernel, a sort of mini-operating system which sits on each processor. There are three different levels of message-passing routines. In the datalink layer messages are connected between directly-connected processors over their links, just as in Occam. In the network layer messages may be addressed to any node in the network, and are automatically routed by the kernel. Obviously there is some cost in this routing, and a useful extra facility, known as virtual circuiting, allows a route to be held open for subsequent messages once it has been found. At the network level there is some danger of the system clogging up with messages being sent faster than they are received, and a further layer, the transport layer, is also provided. These routines do not send messages until they have checked that the destination process is ready to receive them. The main use of this layer is during program development. Other facilities provided by the Trollios kernel are the ability to directly read and write files from any node to a common filestore, the intermediate message-passing being transparent to the programmer; and the ability to dynamically load and kill processes while the program is running. The Trollios approach is processor-independent, and particularly good use of this feature is made by the Niche Advanced Computing Platform [13]. The kernel is implemented on the hosting Sun workstation as well as the network of Transputers, so a single multi-process program can make use of all the Sun facilities (mouse, graphics, windows, networking) as well as Transputer number-crunching power.

To conclude, I should mention briefly several other approaches to programming Transputer networks. It is possible to incorporate Fortran (or C) processes within an Occam program, using the Fortran to perform the calculation and Occam to do the communications, when the Occam part is known as a harness. Such harnesses can be quite elaborate and provide many of the same message-passing facilities as Trollios. However, there is a difference. Harnesses are still part of the application program and loaded with it. The Trollios kernel is booted onto the network processors separately and the host operating system can interrogate it independently of the application program while the latter is running. This is very useful in debugging, particularly in determining if the application has reached a deadlock condition in which it cannot proceed because all processes are waiting for other processes to send them data. Nor do harnesses provide the dynamic process management available with Trollios.

Parallel Fortran is becoming available for Transputers: this is Fortran with extensions to allow the creation of parallel threads (a sort of sub-process) within a Fortran process, as well as message-passing between processes. Finally, a true distributed operating system, Helios [14] has recently been launched.

Conclusion

I hope I have convinced you that parallel computers are an effective way of doing molecular dynamics. Parallel decompositions exist both for the DAP and Transputer networks, though no general methodology exists for producing such decompositions. The effort expended in designing and programming them is worthwhile because of the great cost effectiveness of parallel computers. For a hundred thousand pounds one can now obtain a machine with the computational power of a Cray-1.

REFERENCES

[1] D. Fincham, *Molec. Simulation* 1 (1987) 1

[2] A.R.C. Raine, D. Fincham and W. Smith, Daresbury Laboratory preprint (1988): to appear in *Comput. Phys. Commun.*

[3] M.T. Dove *et al*, *Molec. Phys.* 57 (1986) 865; M.T. Dove, *J. Phys. C: Solid State Physics* 19 (1986) 3325

[4] G.S. Pawley and G.W. Thomas, *J. Computat. Phys.* 47 (1982) 165

[5] R.E. Hubbard and D. Fincham, *J. Molec. Graphics* 3 (1985) 12

[6] M.T. Dove, D. Fincham and R.E. Hubbard, *J. Molec. Graphics* 4 (1986) 79

[7] J. Boris, *J. Computat. Phys.* 66 (1986) 1

[8] D.J. Adams and G.S. Dubey, *J. Computat. Phys.* 72 (1987) 156

[9] D. Fincham, N. Quirke and D.J. Tildesley, *J. Chem. Phys.* 84 (1986) 4535 and 87 (1987) 6117

[10] M. Neumann, O. Steinhauser and G.S. Pawley, *Molec. Phys. 52 (1984) 97*

[11] D.J. Rapaport, these proceedings

[12] D. A. Brown and G. D. Burns, report from Cornell Theory Centre, Cornell University, Ithaca NY,USA

[13] Manufacturer's information from Niche Technology Ltd., Ropemaker Court, 11-12 Lower Park Row BS1 5BN.

[14] Information from Distributed Software Ltd., 24 Brewmaster Buildings, Charlton Trading Estate, Shepton Mallet, Somerset BA4 5QE, U.K

MOLECULAR SIMULATIONS OF PROTEIN STRUCTURE, DYNAMICS AND THERMODYNAMICS

CHARLES L. BROOKS III

DEPARTMENT OF CHEMISTRY
CARNEGIE MELLON UNIVERSITY
PITTSBURGH, PA 15213

I. INTRODUCTION

Proteins display a rich variety of structural, dynamic and chemically functional character. It is the wide range of chemical functionality provided by the twenty amino acids composing proteins which is responsible for the importance of these macromolecules in biology. The chemical nature of protein function is controlled by the structural arrangement and motions of chemically labile groups. Thus, the detailed investigation of the motions and structure of a single protein molecule is of importance in understanding the reactive nature of protein chemistry.

Protein function in biological systems is linked to physical properties as well as chemical ones. The secondary and tertiary structures of globular protein, and the fluctuations about these structures, are key to the control of processes such as protein folding and substrate binding. In this capacity, proteins behave thermodynamically, and the statistical characterization of their complex and inhomogeneous many-body behavior is necessary for a complete understanding of protein function.

It was factors related to both the chemical and statistical characterization of protein motion which motivated the first application of molecular simulation methods to proteins eleven years ago (1977).[1] The methods of molecular dynamics were first applied to a study of the protein bovine pancreatic trypsin inhibitor by McCammon, Gelin and Karplus.[1] This initial work sparked the development of a new field now known as protein dynamics.

The aim of studies in protein dynamics is often focused on the collective or many-body nature of proteins. In this sense the methods of protein dynamics have adopted techniques for simulating a large number of particles for the purpose of exploring thermodynamic and dynamic properties: these are methods which evolved from the pioneering work of Alder and Wainwright[2] on the dynamics of hard spheres and that of Rahman on the simulation of soft-sphere fluids.[3] However, in the study of chemical processes such as catalysis, the analysis of individual trajectories is of interest. For example, in studies of activated processes in proteins it is the individual trajectories which are important in determining the reactive flux across a transition state.[4]

Protein dynamics has evolved from, and bridges, two distinct fields of molecular simulation: classical trajectory analysis and computational statistical

C.R.A. Catlow et al. (eds.), Computer Modelling of Fluids Polymers and Solids, 289–334.

mechanics. It draws tools from both of these areas to investigate the nature of protein structure, dynamics and reactivity. Many of these tools are well developed and established in their respective areas. It is not our intent in this chapter to describe such methods in detail (the general methods of molecular simulation are reviewed in other chapters within this volume). Instead, our purpose is to briefly review the range of dynamical methods utilized in the simulation of protein motion, the techniques used to analyze protein structure, dynamics, thermodynamics and reactivity and to present some of the results which have emerged to guide us in our understanding of protein function. We hope to leave the reader with a qualitative, and in some instances quantitative, understanding of both the techniques and the significance of their application to understanding protein function. Also, we wish to provide a base for further studies in this area. On this point, it is noteworthy that there are a number of important reviews of methods and applications in protein dynamics which precede this work, many of which are more comprehensive than the present chapter. They include recent books by Brooks, Karplus and Pettitt[5] and McCammon and Harvey[6] as well as a large number of reviews by researchers in the field of protein dynamics. A partial list of recent reviews include works by Brooks and Karplus,[7] Careri, Fasella and Gratton,[8] McCammon and Karplus,[9] Karplus and McCammon,[10] Levitt,[11] van Gunsteren and Berendsen,[12] and Edholm et al.[13] For more extensive discussions of methods and applications the interested reader is urged to explore the above referenced works.

The development of this chapter will begin with a brief overview of the potential energy functions used in protein simulations. Next we will discuss the algorithms necessary to propagate atomic motion and the methods used to analyze protein dynamics trajectories for equilibrium thermodynamic properties and dynamical information. Following this we will review a representative number of results which have emerged from protein simulations – focusing first on structure and fluctuations and then on dynamical properties. The general features of protein structure and dynamics will be emphasized as well as observations which pertain to the mechanics of how simulations are carried out. The role of solvent in biological media will be considered. Results which illustrate the important influence of solvation on protein structure and dynamics will be given for examples involving enzyme catalysis and protein fluctuations. Finally, we will present and discuss thermodynamic simulation methods for proteins, and their application to protein–substrate binding and protein stability. The chapter will close with a perspective on the growing field of protein dynamics and a discussion of some of the future problems to be addressed.

II. The Protein Potential Function and Its Implementation. To carry out a dynamical simulation of the atomic motions in a protein it is necessary to develop a relationship between the structure of the molecule (molecular conformation) and its energy. From such a relationship the classical forces acting on individual atoms can be obtained, and with these forces Newton's equations of motion may be solved to produce a time series of molecular configurations for the protein. Many properties may be evaluated from the molecular trajectory; for example, the average structure of the protein,[14] NMR relaxation times,[15] or fluorescence anisotropy decay signals.[16] In addition, the molecular trajectory can serve as an "ideal" experimental data set with which methods of experimental data processing can be examined and developed.[17,18] Thus, the utility of protein dynamics rests on the ability to develop a structure–energy relation.

The development of a structure–energy relation, generally referred to as the potential energy function, is guided by a balance between computational restrictions (or limitations) and the desire to represent the energetic penalty of molecular distortions at the greatest level of accuracy possible. Of course, ab initio quantum chemical calculations can yield a very accurate description of molecular geometry and energetics when sufficiently high levels of theory are applied.[19] However, the detailed description of molecular systems at high levels of accuracy is practically limited to small molecules, generally consisting of one to eight non-hydrogen atoms; this is due to the computational effort involved in ab initio calculations. Therefore, the repeated evaluation of energies and forces using ab initio quantum chemistry methods is presently unfeasible (we note, however, that small portions of a larger molecular framework may be treated using quantum chemical methods when such a description is necessary[20]). Instead, what is required is a simple functional form which may be rapidly evaluated and differentiated, yet is sufficiently accurate to reproduce a large number of experimental (and theoretical) observations. The empirical potential energy functions initially developed for molecular mechanics studies of small organic molecules[21] satisfy these criteria. They have been adapted for the study of proteins and other biologically important molecules.[22,23]

Empirical potentials generally express the structure–energy relation as an additive sum of "physically intuitive" interaction terms which include energy penalties for: distorting pairs of bonded atoms (bonds); splaying the valence angle defined by triples of atoms (angles); rotating around a central bond in quadruples of atoms or distorting the chirality around a chiral center (dihedral angles); and for the pairwise through-space interactions between atoms due to permanent or induced electrostatic effects, nuclear or exchange repulsion (nonbonded). The form of this potential most often used in protein simulations is shown in eqn. (1) below.

$$U = (1/2) \sum_{\text{bonds}} K_b(b-b_o)^2 + (1/2) \sum_{\text{angles}} K_\theta(\theta - \theta)^2 +$$

$$(1/2) \sum_{\substack{\text{dihedral}\\ \text{angles}}} K_\varphi[1 + \cos(n\varphi - \delta)] + \sum_{\substack{\text{nonbonded}\\ \text{pairs}}} \left[\frac{A}{r^{12}} - \frac{C}{r^6} + \frac{q_1 q_2}{\epsilon r}\right] \quad (1)$$

In this expression the total potential energy of a molecular system is related to its molecular configuration through the introduction of localized internal coordinates and nonbonded van der Waals and electrostatic terms.

The parameters which describe the "chemical interactions", i.e., collectively the bond, angle and dihedral terms, are most often derived by fitting individual terms from a target potential function such as eqn. (1) to experimental data from vibrational and NMR spectroscopy on small subunits contained in the biological macromolecules. For example, the carbonyl bond force constant (K_b) could be derived from the vibrational spectrum of a small molecule, or the dihedral potential parameters for rotation around the C –C bond in a valine sidechain would come from consideration of the gauche/trans equilibrium measured via NMR. The empirical nature of the potential thus guarantees that certain experimental properties are reproduced quite accurately; provided the chosen form of the function fits the experimental data well, and that the experiments on small fragments are transferable to larger systems. In general, both of these caveats must be kept in mind when the results from a given calculation are being interpreted. For example, to reproduce vibrational spectra of small molecules potential forms which are much more sophisticated than that given by the first three terms in eqn. (1) are used. These most often include higher order terms in the bond and angle potentials as well as coupling between the various internal coordinates.[24] This level of sophistication has not, for the most part,[25] been introduced into protein dynamics simulations due to the increased (computational) effort necessary to evaluate the energy and derivatives. However, protein dynamics has not (yet) been used extensively to examine such detailed atomic behavior. Therefore, in most cases the simpler potential form is adequate.

Parameterization and evaluation of the through-space nonbonded interactions in eqn. (1) presents the greatest difficulty in the development of empirical potentials for biological macromolecules. There are a number of reasons for this difficulty. First, the A and C parameters, which represent the repulsive "van der Waals" radius and attractive dispersion interactions, are not directly observable by experimental measurement. In fact, they represent a complicated and non-uniquely decomposable set of high level quantum chemical interactions (Pauli exclusion, nuclear repulsion, induced electronic effects, etc.) which are not readily amenable to theoretical analysis. A second problem arises in attempts to represent the electrostatic interactions due to a permanent charge distribution on particular atomic centers i.e., the q's in eqn. (1). Again, such constructs are not uniquely defined observable properties. Finally, a number of problems arise in relation to the a) inclusion of explicit solvent, b) the treatment of long-range electrostatic forces and c) the form of the dielectric screening function used for electrostatic interactions. These points will be addressed in the material which follows.

An approach to deriving the van der Waals portion of the nonbonded parameters (the A's and C's for all pairs of atoms) is to choose a basis set of atomic radii and van der Waals self-interaction strengths, based on optimization by comparison with experiment[26] or theoretical calculations,[27] and then to generate the pairs of interaction parameters using combination rules. Some of the commonly used combination rules for these parameters are:

$$A_{ij} = (A_i A_j)^{1/2} \quad \text{and} \quad C_{ij} = (C_i C_j)^{1/2} \qquad (2)$$

or

$$\sigma_{ij} = (\sigma_i + \sigma_j)/2 \quad \text{and} \quad \epsilon_{ij} = (\epsilon_i \epsilon_j)^{1/2} \qquad (3)$$

where σ_i, the Lennard-Jones radius, is related to A_i and C_i as $\sigma_i = (A_i/C_i)^{1/6}$ and ϵ_i is related to either A_i or C_i as $\epsilon_i = 1/4\ A_i/\sigma_i^{12} = 1/4\ C_i/\sigma_i^{6}$. In general, no preference has been demonstrated for the use of either set of combination rules. Both give similar values for the "mixed" potentials except when large disparities are present in the parameters being combined, e.g., in calculating van der Waals parameters for O...H nonbonded interactions eqns. (2) and (3) give slightly different interaction potentials.

To find an appropriate set of atomic charges ab initio quantum chemical calculations are often carried out on pairs of interacting molecules to generate an interaction surface.[28] If van der Waals parameters are assumed to have been chosen previously for all atoms in the pair of interacting molecules, the van der Waals interaction surface is subtracted from the ab initio surface and the residual interaction is assumed to be due to electrostatic charges permanently distributed on atomic (or other) centers. These charges are adjusted to best fit the residual surface. An alternative approach is to derive the electrostatic potential due to a single molecule of interest from experimental or ab initio data and to fit an atom centered charge distribution to best reproduce this potential.[29]

The forces of chemical bonding, the bonds, angles and dihedrals defined in eqn. (1), are of a limited range, e.g., nearest neighbor pairs, third and fourth neighbors. The van der Waals and electrostatic interactions just described are not. It, therefore, becomes necessary in the implementation of protein empirical potentials to limit the range of these interactions, thus reducing the computational expense. In protein simulations the method of choice is to include interactions between an atom at all of its neighbors within some spherical radius, r_{cut}. In this way direct interactions are only "felt" for atoms (or molecular fragment groups) with separations less than r_{cut}. Typical ranges used are 6A – 15A, with values between 7.5A – 10.0A being most common. The interaction cutoff value and the method by which the interactions are "turned-off" become an integral part of the potential function and do influence the computed properties.

To illustrate this point, and to demonstrate some of the effects which result from different treatment of long range electrostatic interactions, let us jump ahead in our discussion and consider a recent study by Brooks on the effects of long range cutoffs on protein dynamics.[30] This work systematically examines the structural differences and fluctuations associated with long range force cutoffs in the protein CO-myoglobin, and provides a nice benchmark for pseudo vacuum calculations on proteins. Brooks considered three standard forms of long range force truncation implemented in the CHARMM program.[22] These describe the modification of the simple Coulomb-law by an "electrostatic interaction cutoff" operator,[31] S(r), of the form

$$S(r) = \begin{cases} 1 & r < r_{cut} \\ 0 & r \geq r_{cut} \end{cases}, \tag{4}$$

$$S(r) = \begin{cases} 1 & r \leq r_{on} \\ (r_{off}-r)^2\,(r_{off} + 2r-3r_{on})/(r^3_{off}-r^3_{on}) & r_{on} < r < r_{off} \\ 0 & r \geq r_{off} \end{cases} \tag{5}$$

and

$$S(r) = \begin{cases} 1-2\,(r/r_{cut})^2 + (r/r_{cut})^4 & r < r_{cut} \\ 0 & r \geq r_{cut} \end{cases} \tag{6}$$

Several long simulations were carried out (150 ps each) for each operator using different values of the cutoff distance (r_{cut}) or cutoff range (r_{on}, r_{off}) and results of the dynamics were analyzed for the rms atomic deviation between the crystallographic and dynamics average structures, and the rms atomic fluctuations averaged over the dynamics. These results are summarized in Table I for several cases, most of which used the dielectric constant, ϵ in eqn. (1), equal to r (the separation of the pair of atoms).

Table I. RMS Fluctuations and Displacements for the Last 100 Picoseconds of Each Simulation (in Angstroms)[a]

Truncation Method	Cutoff Distance	RMS Fluctuations all atoms	RMS Fluctuations main chain	RMS Deviation all atoms	RMS Deviation main chain
ϵ=r, eqn. 6	4.5	2.23	1.96	5.70	5.27
ϵ=r, eqn. 6	5.5	1.34	1.10	4.59	3.72
ϵ=r, eqn. 6	6.5	0.97	0.81	2.81	2.27
ϵ=r, eqn. 6	7.5	0.74	0.59	2.87	2.34
ϵ=r, eqn. 6	8.5	0.90	0.77	2.35	1.91
ϵ=r, eqn. 6	10.5	0.70	0.59	3.05	2.44
ϵ=r, eqn. 6	13.5	0.65	0.53	2.36	1.96
ϵ=r, eqn. 6	none	0.63	0.50	2.20	1.68
ϵ=r, eqn. 5	3.5-4.5	0.58	0.46	2.89	2.54
ϵ=r, eqn. 5	4.5-5.5	0.53	0.43	2.35	1.82
ϵ=r, eqn. 5	5.5-6.5	0.63	0.49	1.97	1.63
ϵ=r, eqn. 5	7.5-8.5	0.60	0.47	2.68	2.26
ϵ=r, eqn. 5	9.5-10.5	0.62	0.48	1.88	1.39
ϵ=r, eqn. 5	12.5-13.5	0.51	0.38	2.17	1.69
ϵ=r, eqn. 5	none	0.63	0.50	2.20	1.68
ϵ=r, eqn. 5	5.0-9.0	0.77	0.63	2.20	1.75
ϵ=r, eqn. 5	6.0-8.0	0.73	0.59	2.34	1.86
ϵ=r, eqn. 5	6.5-7.5	0.58	0.44	1.84	1.43
ϵ=r, eqn. 5	6.75-7.25	0.50	0.40	1.68	1.27
ϵ=r, eqn. 5	6.875-7.125	0.42	0.29	1.56	1.16
ϵ=r, eqn. 4	4.5	1.40	1.14	5.92	4.74
ϵ=r, eqn. 4	5.5	1.44	1.21	4.49	4.04
ϵ=r, eqn. 4	6.5	1.07	0.87	2.91	2.49
ϵ=r, eqn. 4	7.5	0.81	0.66	2.23	1.78
ϵ=r, eqn. 4	8.5	0.79	0.61	2.63	2.16
ϵ=r, eqn. 4	10.5	0.77	0.64	2.26	1.70
ϵ=r, eqn. 4	13.5	0.77	0.65	2.45	1.96
ϵ=r, eqn. 4	none	0.63	0.50	2.20	1.68
ϵ=1, eqn. 6	7.5	0.71	0.59	3.23	2.57
ϵ=1, eqn. 6	none	0.65	0.53	2.79	2.24
ϵ=2, eqn. 6	7.5	0.98	0.77	3.87	3.23
ϵ=2r,eqn. 6	none	0.78	0.62	2.26	1.81
ϵ=2r,eqn. 6	7.5	0.99	0.81	2.67	2.25

a. Data from reference 30.

What emerges from this analysis by Brooks is that adverse effects occur with all cutoff schemes for small values of r_{cut} or r_{off}. Also, cutoff scheme (5) gives the best agreement with the no-cutoff results when the switching range is long, e.g., between 5-9 A or 6-8A. These calculations illustrate the sensitivity of results such as fluctuations and motion to the model potential used.

A final point worth noting in regards to this study is that a single property, such as the rms deviation from the crystal structure, cannot be used as a sole criterion in judging the quality of the potential function. For example, in Table I it is illustrated that using very small ranges for r_{on} and r_{off} can result in rms deviations which are quite small, suggesting good agreement, while producing fluctuations which are much underestimated.

We have illustrated that the protein potentials used in protein dynamics and mechanic studies are built of component parts using a combination of empirical and theoretical fitting. These potentials can be evaluated to some degree from the calculation of measurable properties. This is usually confined to examining other small units that make up a macromolecular structure; direct comparison between calculations and observations on proteins for the purpose of parameter development is in most cases not possible. It therefore becomes an important task for the researcher to evaluate his or her computed data with extreme care. The limitations of the empirical potential energy function must be kept in mind and it should be noted that the motions which result from its exploration via molecular dynamics constitute the investigation of a mathematical model, albeit a highly complex one, from which the transfer of new information must be critically judged in light of the model's ability to reproduce known and related results.

Despite these cautions, the available protein potential functions have been reasonably successful at reproducing gross experimental features of protein structure, fluctuations and dynamics, and many insights into protein motions have been gained by their qualitative investigation.[5,6] However, it can be anticipated that as more detailed questions are probed with protein dynamics the incomplete nature of the potential functions will become apparent and their further refinement will proceed.

III. Protein Dynamics: Algorithms and Methods

Given an explicit form for the potential and the corresponding parameters, the forces on individual atomic centers may be computed. The forces give rise to a classical description of the equations of motion. The numerical solution of these equations yields the desired molecular trajectory. In this section we will discuss the numerical algorithms and computational protocols used to produce such a trajectory. The methods used in the analysis of protein dynamics will be presented as well.

Before we sketch details of the algorithms used in protein dynamics, several considerations involved in choosing the algorithm which is most appropriate should be noted. If explicit account is taken of the solvent surroundings and periodic replicas of the solute–solvent system are used to mimic a bulk phase, standard molecular dynamics algorithms using periodic boundary conditions in the NVE, NVT or NPT ensembles are appropriate. The techniques of periodic boundary conditions are presented elsewhere and the reader is referred to the literature for a detailed discussion of these methods.[5,7] If pseudo vacuum calculations are to be carried out, then it is often most appropriate to couple the motion of the protein atoms to a heatbath and a Langevin dynamics algorithm is best used.[32] When localized enzyme chemistry; or thermodynamic changes associated with modification of specific amino acid sites on a protein; or chemical modification of inhibitors (drugs) bound in localized binding sites are of primary interest, the stochastic boundary molecular dynamics method is the best choice.[33] This technique combines standard molecular dynamics and Langevin dynamics algorithms to give a realistic description of local dynamic and thermodynamic properties. A brief overview of each of the methods is given below.

As discussed in Chapter (3), a very useful algorithm for the numerical integration of the large number of coupled differential equations associated with proteins was introduced for dynamical simulations of Lennard-Jones fluids by L. Verlet in 1964.[34] The Verlet algorithm may be derived from the simple central difference of the Taylor series expansion for the atomic position at time t. For the vector position of atom i, $\mathbf{r}_i$, the algorithm is

$$\mathbf{r}_i(t+\Delta t) = 2\mathbf{r}_i(t) - \mathbf{r}_i(t-\Delta t) + \Delta t^2 m_i^{-1}\mathbf{F}_i(t), \tag{7a}$$

where $\mathbf{F}_i$ is the total force on atom i at time t and Δt is the timestep. The corresponding velocity propagation equation is

$$\mathbf{v}_i(t) = [\mathbf{r}_i(t+\Delta t) - \mathbf{r}_i(t-\Delta t)]/2\Delta t. \tag{7b}$$

We note that the position propagation is of order Δt^3 and the velocity propagation is of order Δt^2, and lags a step behind the position. The Verlet algorithm has been found to provide a very stable numerical procedure for the integration of many coupled differential equations and appears to meet an appropriate balance between accuracy and timestep. The usual timestep used in solving protein equations of motion with this and the other algorithms presented below ranges from 0.5 fs to 4 fs.

The simple Verlet algorithm embodied in eqns. (7) above lends itself most readily to simulations in the NVE ensemble, since there is no simple and direct mechanism for coupling the system to a constant temperature reservoir. For this reason alternate propagation algorithms have been introduced. Among these are the leap frog[35] and Beeman algorithms,[36] which indentically reduce to the Verlet algorithm for position propagation, eqn. (7a), but differ in the velocity propagation scheme. By altering the propagation algorithm, direct coupling of the system to simple linearly (in the velocity) dissipative forces for the control of temperature is allowed. This adaptation permits simulations to be carried out in the NVT ensemble. Procedures for carrying out constant temperature simulations have been developed by Anderson,[37] Nose and Hoover,[38] Evans et al.[39] and Berendsen et al.[40] The technique most often employed in protein dynamics is that of Berendsen et al. However, a note of caution must be attached with the use of this algorithm: the final equilibrium ensemble which is approached using a Berendsen "thermostat" for temperature control is undefined. Thus, the calculation of thermodynamic properties must be called into question when connection formula appropriate for a specific statistical ensemble are used in simulations with undefined statistical properties.

To couple the system to a constant pressure reservoir it is necessary to introduce an auxiliary equation of motion for the volume of the system.[37,40] This couples volume fluctuations to the positions and momenta of the atomic centers. In its usual implementation the positions of all atomic coordinates are scaled at each step in dynamics to satisfy an external pressure constraint, thus achieving a constant pressure simulation. This technique has been successfully applied to simulations of simple and complex liquids.[41] However, its application to highly anisotropic systems, such as proteins, must be considered carefully since the isotropic scaling of distances clearly distorts bond distances and angles more than might naturally occur in volume fluctuations.

When simulations are carried out in the absence of explicit solvent it is often of interest to include the effects of solvation implicitly. This is accomplished through the addition of solvent averaged forces, i.e., by use of a potential of mean force, and by the addition of dissipative and fluctuating forces. One simple correction that was introduced as a computational expedient in early simulations of biological molecules was the modification of Coulomb's law for electrostatic forces which used $\epsilon = r$, in eqn. (1) above. It was argued that this form provided additional screening of the electrostatic forces and as such was a simplified potential of mean force correction. More sophisticated theories now exist for the calculation of solvent mediated potentials of interacting biopolymers.[42] No real changes of the numerical algorithms described above are necessary to add the equilibrium solvent corrections. However, to incorporate a balanced coupling to an external heatbath (dissipative and fluctuating forces) slight modification of the simple Verlet algorithm is necessary.

To illustrate how such an algorithm is derived, we first introduce the Langevin, as opposed to Newtonian, equation of motion for an atom,

$$\ddot{\mathbf{r}}_i(t) = m_i^{-1} \; [\mathbf{F}_i(t) + \mathbf{f}_i(t)] - \beta\dot{\mathbf{r}}_i(t), \qquad (8)$$

where $f_i(t)$ is a random force which mimics the incessant collisions between the protein atoms and solvent, $-\beta\dot{\mathbf{r}}(t)$ represents the force of dissipation which describes the flow of thermal energy into and out of the solvent bath and β is the inverse velocity relaxation time. In Langevin theory these forces are balanced by a fluctuation-dissipation theorem,[43]

$$\langle \mathbf{f}_i(t) . \mathbf{f}_i(o) \rangle = 6k_BT\beta\delta(t) \qquad (9)$$

and the random force is assumed to be distributed in a Gaussian fashion around a zero mean. In the absence of an external force, $\mathbf{F}_i(t) = 0$, the Langevin equation of motion may be solved exactly to yield

$$\mathbf{r}_i(t) = \mathbf{r}_i(o) + \beta^{-1}\, \dot{\mathbf{r}}_i(o)\, (1-e^{-\beta t}) + \int_o^t \beta^{-1}[1-e^{-\beta(\xi-t)}]\; \mathbf{f}_i(\xi)d\xi . \qquad (10)$$

By expanding this solution around $t\pm\Delta t$ in a Taylor series, adding the expansions as done for eqn. (7a) above, and substituting in the Verlet velocity expression, eqn. (7b), one finds an algorithm for propagation of Langevin dynamics which is quite similar to the Verlet algorithm

$$\mathbf{r}_i(t+\Delta t) = (1-{}^1/_2\ \beta\Delta t)^{-1}\ [\{2\mathbf{r}_i(t) - \mathbf{r}_i(t-\Delta t) + \Delta t^2\mathbf{m}^{-1}\ (\mathbf{F}_i(t) + \mathbf{f}_i(t))\} - {}^1/_2\ \beta\Delta t\ \mathbf{r}_i(t-\Delta t)], \qquad (11)$$

and which reduces to it in the limit of $\beta \to 0$.[32] This algorithm is of order $\beta\Delta t^3$ for position propagation has been shown to provide an excellent numerical counterpart to the Verlet method for use in protein dynamics.[44]

Lastly, we briefly describe the stochastic boundary molecular dynamics (SBMD) method. This theoretical approach combines aspects of both of the numerical algorithms described above, i.e., Verlet and Langevin, to provide a methodology for carrying out dynamical simulations of localized processes in solvated biopolymers.[33] It is of greatest utility in studies of the structure and dynamics associated with enzyme catalysis, and in thermodynamic calculations of enzyme–inhibitor bonding and protein stability.

In the SBMD approach a local region of biomolecule is targeted as the "active" site and the remainder of the system is partitioned into three regions with respect to it. Each region is defined using a spatial and topological criterion such that atoms near the "active" site are propagated by Newtonian dynamics while atoms in an intermediate region called the buffer region (generally 10–14 A from the "active" site) undergo Langevin dynamics and distant atoms (>14–20A) are eliminated from direct consideration. The influence of distant atoms is replaced by boundary forces which approximate a potential of mean force on atoms in the inner two regions. If the boundary force acting as atom i is denoted $\mathbf{F}_i^B$, then the equation of motion used in SBMD is defined by

$$\ddot{\mathbf{r}}(t) = \begin{cases} m^{-1}\,[\mathbf{F}_i(t) + \mathbf{F}_i^B(t)] & \text{"active" site region} \\ m^{-1}\,[\mathbf{F}_i(t) + \mathbf{F}_i^B(t) + \mathbf{f}_i(t)] - \beta\dot{\mathbf{r}}_i(t) & \text{buffer region} \end{cases} \qquad (12)$$

In eqn. (12) we see that both Newtonian and Langevin equations of motion are represented. The algorithms most suitable for SBMD are the Verlet-like, Langevin algorithm of eqn. (11) together with the standard Verlet velocity propagation, eqn. (7b).

Solvent is included in SBMD simulations by partitioning a system which initially has solvent explicitly present. Alternatively, a spherical region of solvent may be overlayed onto the protein structure and overlapping waters deleted to create a hydrated system before partitioning the protein.[45] In addition to the partitioning of protein and solvent, the SBMD method requires the input of the boundary forces and the frictional coupling parameter β. Boundary forces for the protein atoms are derived from the mean-square fluctuations of those atoms. They are assumed to be linear in the atomic displacement. The boundary forces on solvent are developed from analytical consideration of the mean solvent field.[46] The frictional parameters are formally related to the velocity correlation time. However, it has been observed that the most important aspect of choosing these parameters is that they are sufficiently large to ensure energy dissipation at the boundary.[33] For this purpose, values ranging from 60–120 ps^{-1} have been used in SBMD calculations on proteins.

With any of the molecular dynamics methods just discussed one may compute a molecular trajectory. However, it is not a simple matter to prepare a biological structure for dynamics simulation. Instead some protocol must be followed to "condition" the system in preparation for molecular dynamics. A sample of a "typical" protocol is illustrated in figure 1. The first step in this protocol requires the input of a starting structure for the biomolecule. This is most often taken from the results of an x-ray structure determination. The crystal structure usually contains some van der Waals contacts which were not eliminated in the refinement process. Also, in many cases atoms which have large mobility in the crystalline environment, e.g., charged sidechains, are unresolved by the structure determination and these must be "built" into the structure. Both of the points above suggest that initial energy refinement of the crystallographic structure is necessary. Thus, several cycles of energy minimization are carried out on the x–ray coordinates. This is most often done in the presence of harmonic restraints on the atoms, which restrain the positions to be near the initial crystallographic ones. Minimization then proceeds in several cycles with the harmonic restraint forces being reduced in each subsequent cycle until they are zero. In this way the bad contacts are relieved with a minimum movement from the initial structure. Protein structures subjected to such an energy refinement typically deviate 0.7–1.2 A, overall atomic rms, from the starting crystal coordinates. Once the initial structure is energy refined it must be equilibrated prior to the collection of the atomic trajectory for later analysis. To accomplish this, a phase of heating and equilibration dynamics is performed. In the heating portion atoms in the initial minimized (0 K) structure are given velocities chosen from a Boltzmann distribution characteristic of a low temperature system, e.g., around T_{in}=50 K. The dynamics algorithm is then propagated for several steps and the velocities are reassigned, or scaled, to an incrementally higher temperature until the system is at the final temperature. The equilibration phase then begins. During equilibration the mean kinetic energy of the system is monitored and velocities are reassigned if they drift beyond some predetermined range of the desired value. The equilibration period is carried out until averages of the potential and kinetic energies, separately, are stable. A period of dynamics to be used for analysis then proceeds. During this phase coordinates and velocities are often saved for subsequent analysis.

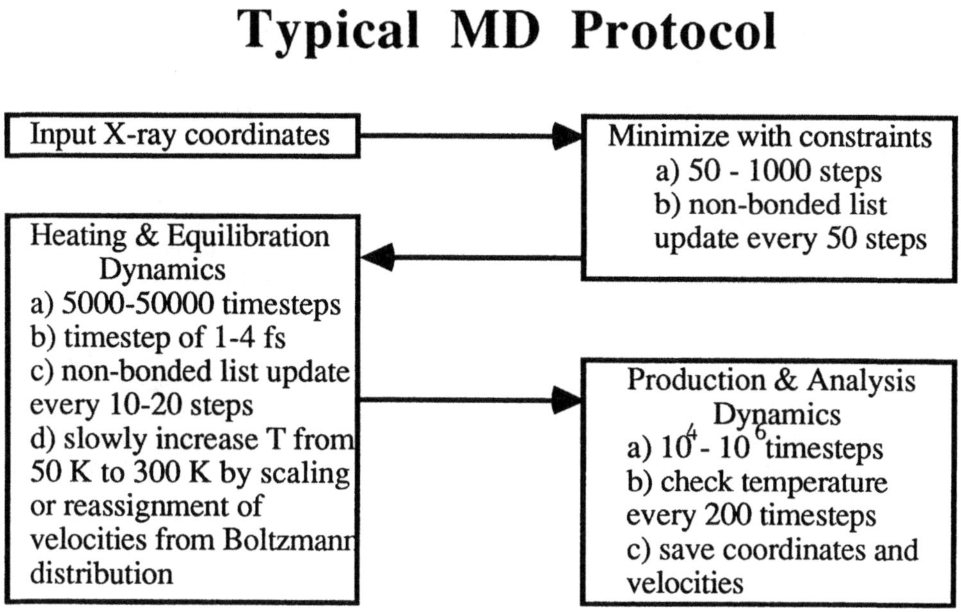

Fig. 1 A typical protocol for a molecular dynamics simulation of protein.

Finally, we note that several properties may be computed from the dynamics trajectory. These properties are most often related to an average over molecular motions. We can define two types for such an average. The first are equilibrium time-independent averages. If we let P_i represent the value of a particular time-independent property of interest at timestep i (e.g., the atomic position of an atom, or its mean-square amplitude of oscillation) and N be the duration of the protein dynamics calculation in timesteps, then the time-independent average of P is defined as

$$\overline{P} = \frac{1}{N+1} \sum_{i=o}^{N} P_i \qquad (13)$$

The other type of property most often computed from dynamics simulations is the correlation function of a dynamical variable. If we let A_i be the value of a dynamical variable at timestep i, then the autocorrelation function of A is given by

$$C_j = \overline{A_i A_{i+j}} = \frac{1}{N - j+1} \sum_{i=o}^{N-j} A_i A_{i+j} \qquad (14)$$

The correlation functions of greatest interest in probing protein dynamics include the atomic displacement and velocity autocorrelation functions, and correlation functions between vectors representing the absorption or emission dipoles for electronic, magnetic and vibrational excitations. Examples include the function $P_2\ [\mu_A(i).\mu_E(i+j)]$, which is the second Legendre polynomial of the angle between the absorption dipole direction (at time i) and the emission director (at a later time i+j). This function is related to the decay of fluorescence from an excited fluorophore.[16] A similar correlation function relating magnetic polarization between two neighboring nuclear spin dipoles is also of some interest. This function yields information regarding the degree and timescale of nuclear Overhauser enhancements (NOEs).[17]

Through the calculation of time-independent and time dependent properties of protein motion, information regarding the fundamental nature of protein function can be obtained. In what follows we will present several examples to illustrate how our current understanding of the detailed functioning of proteins has evolved.

IV. Protein Structure, Fluctuations and Dynamics.

In this section we will review several results of protein dynamics simulations. We will first focus on the gross features of protein structure and fluctuations, and the role of the surrounding environment in altering these features. Our attention will then shift to more detailed analysis of structure and fluctuations, where the atomic level description of protein and solvent provided by protein dynamics are used to investigate the nature of protein motion and the influence of solvation on structure and catalysis. We conclude the section with a discussion of protein dynamics results for fluorescence anisotropy, a dynamical property, where the emphasis is on the interpretation of experimental results and the development of models for the treatment of experimental data.

To begin a discussion of the fruits of protein dynamics, we will present results which compare directly the crystallographically determined conformation for pancreatic trypsin inhibitor (PTI) and that found from protein dynamics

The results which we present arise from calculations made using different parameterizations of the empirical potential energy function and with a varying degree of sophistication for representing the condensed phase environment surrounding the central molecule of interest. From these comparisons we will be able to assess the ability of the empirical potential energy function to mimic the native protein structure and fluctuations. In addition, the influence of environmental effects, e.g., crystalline versus solvent, on the native structure and fluctuations of protein will be examined.

The backbone structure of PTI is illustrated in figure 2. This protein of 58 amino acids contains typical structural motifs found in many proteins, e.g., the small portion of β-sheet running up the spine of the molecule and small helical units as well as several turns and random coil segments.

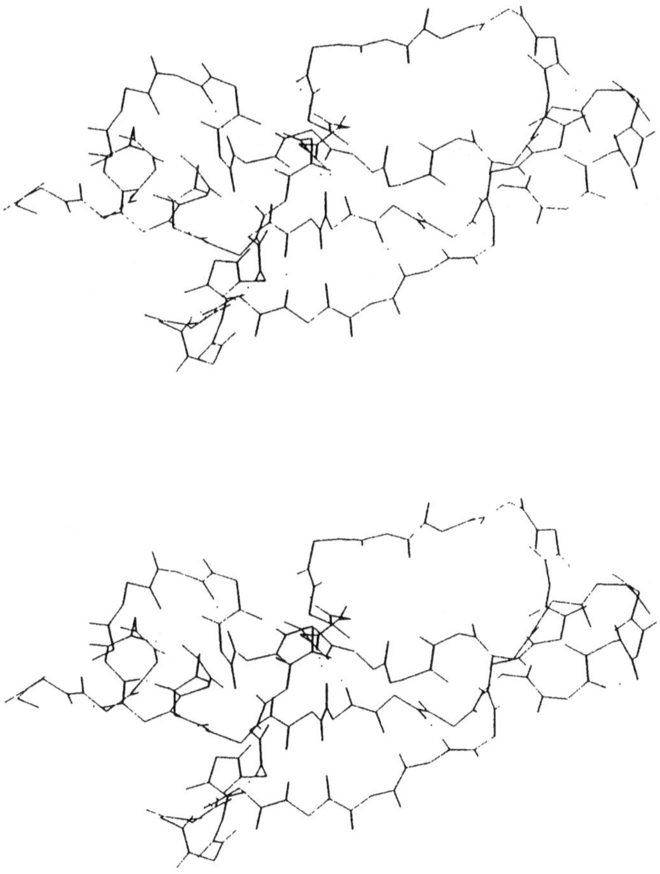

Fig. 2. A stereographic view of the backbone of pancreatic trypsin inhibitor.

Table II. RMS Deviations for PTI.

Simulation Environment	RMS Deviation (A) All Atoms	RMS Deviation (A) C^{α} Atoms
Vacuum[a]	3.02/1.15[b]	2.2/1.01[b]
Static crystal[a]	2.12	1.52
Full crystal[c]		
<8-20ps>[e]	1.62	1.06
<19-20ps>	1.95	1.33
van der Waals solvent[a]	1.94	1.35
Aqueous solvent[d]	2.69	1.81

a. Data from reference 47.
b. Data from recent reparameterization of protein potential, reference 48.
c. Data from reference 49.
d. Data from reference 50.
e. <...> indicate period over which coordinates were averaged.

In Table II we illustrate "typical" results for the comparison of simulated protein structures with the x-ray structure for PTI. The root-mean-square (rms) atomic deviations between the dynamics average structure and the x-ray coordinates is shown for simulations carried out in several different environments - ranging from complete neglect of any environmental effects[47,48] (vacuum) to complete inclusion of crystal images and crystal solvent (note that protein crystals are generally near fifty percent water or solvent by volume), and using three different parameterizations for eqn. (1).[49] From the comparisons made in Table II we find several interesting features:

i) The representation of protein by vacuum potentials generally produces large shifts of the structure from its initial x-ray conformation.[48] Although, as is illustrated by more recent vacuum simulations on PTI, this effect can be reduced by further refinement of the potential.[48]

ii) When a static crystalline environment (representing neighboring protein molecules as fixed and including no crystalline solvent) is included explicitly in the calculation, the resulting average structure deviates from the x-ray less than in the vacuum case. Furthermore, when the full crystalline environment is included the comparison with the x-ray structure is even closer.

iii) Finally, the explicit solvent environment modifies the structure of the protein as measured by the deviation between the x-ray and dynamics averaged structures. However, these changes are argued by van Gunsteren and Berendsen to arise primarily from the absence of crystal neighbors and therefore they conclude the solvent structure is similar to the crystal structure (see figure 3 and discussion below).[50]

From comparisons such as those made in Table II we can proceed with some confidence in the ability of our model to mimic real protein structures. This permits further conclusions to be drawn regarding the nature of differences in structure with varying environment. Thus, we see from a comparison of the results for explicit solvent and crystalline environment calculations that the crystal provides

We may also examine the nature of atomic differences in more detail. In figure 3 the results from calculations by van Gunsteren and Berendsen on PTI in aqueous solution are illustrated.[50] This simulation used the three-site SPC model to represent the solvent and was carried out for a period of 20 ps using periodic boundary conditions. While this represents an extremely short period for computation of average properties, some interesting features emerge. Plotted in the figure is the rms deviation for terminal atoms in the sidechain (figure 3a) or mainchain C^{α} (figure 3b). We see that charged sidechains undergo large shifts in their position in the presence of aqueous solution. It is noted that many of these groups were involved in crystal contacts with other molecules in the crystalline environment but are free to move into solution in the absence of neighboring protein. In addition, we see that the motion of mainchain atoms is much less than that of the more exposed sidechains. This is to be expected on simple geometric grounds since there is intrinsically more freedom for sidechain atoms.

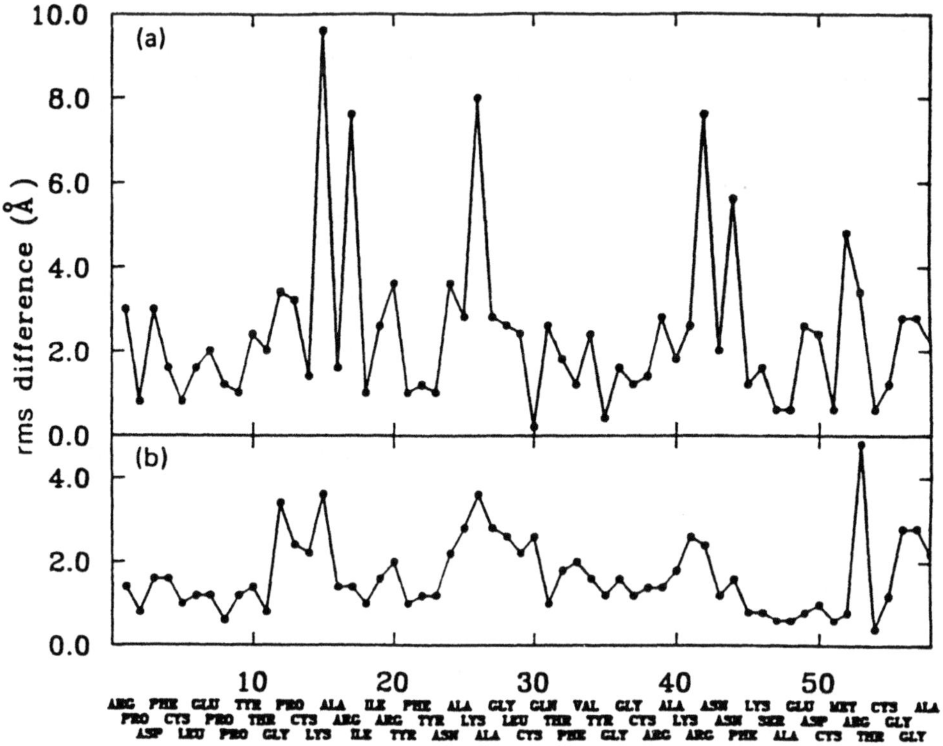

Fig. 3. Atomic rms differences between molecular dynamics and crystal structures of PTI. a) Differences for the terminal atoms in a sidechain and b) the mainchain C^{α} atom.

Before leaving PTI it is of interest to examine the influence of different environments on the "composition" and flexibility of the protein. These points are illustrated in Tables III and IV where the radius of gyration and root-mean-square atomic fluctuations for PTI simulated in the presence of different environments are given.[47]

Table III. Radius gyration for PTI.[a]

Vacuum	10.2 A
Static crystal	10.89 A
Van der waals solvent	10.63 A
x-ray	10.96 A

a. Data from reference 47.

In Table III we see the effect of the crystalline, vacuum and solution environment on the radius of gyration. In the absence of any environmental effects, i.e., in vacuum, the radius of gyration is substantially smaller than that of the x-ray structure. When environmental forces are added the protein expands (from its vacuum configuration) and approaches, in the static crystal environment, the x-ray results. Thus, the overall influence of environment is attractive with the resulting effect of decreasing the density of the protein (and hence increasing the radius of gyration). One might expect that this effect should manifest itself in increased motions of the atoms as well. To some extent this is born out in Table IV, where the atomic rms fluctuations for groups of atoms in PTI are displayed.

Table IV. RMS Atomic Fluctuations in PTI[a]

Atom type	RMS Fluctuation (A		
	Vacuum	vdW Solvent	Static Crystal
All	0.72	0.78	0.67
N	0.52	0.50	0.50
O	0.66	0.65	0.67
C^{α}	0.54	0.54	0.52
C^{β}	0.62	0.69	0.62
C^{δ}	0.78	1.15	0.77

a. Data from reference 47.

In this table we see that the rms fluctuations are largest in the solvent environment and smallest in the static crystalline environment. Furthermore, fluctuations increase as one moves toward the end of the sidechain, e.g., the rms fluctuations for C^{α} atoms are smaller than those for C^{β} and so on. The result that the overall extent of fluctuations is smaller in the crystal than in solution is probably accurate because the former represents a more rigid medium. However, it is most likely that the comparable nature of vacuum and crystal fluctuations is an artifact of the idealized representation used for the crystal in these simulations.

To summarize these findings for protein dynamics simulations on PTI, we see that environmental effects are important in the overall structure and fluctuations of protein. However, the crystalline environment (when explicit solvent and image atoms are included) shows similar character to the solution environment. Moreover, we find that environmental effects generally produce an overall attractive field which reduces the protein density. When atomic details are examined, motion of the atoms increases with increasing distance from the backbone N or C^{α}; an observation which has also been noted in the experimental refinement of motional parameters in crystal structures of protein. Finally, we note that protein dynamics carried out with different potential functions gives similar results for the same protein and these results are consistent with experimental details. Thus the basic features of protein structure and motion are contained in the protein potential function.

We next investigate the relationship between atomic fluctuations computed from protein dynamics and those derived from the refinement of isotropic Debye–Waller factors (B–factors) in protein crystals. In figure 4 we show a comparison of the computed rms fluctuations from a vacuum simulation of ribonuclease A[51] with the same quantity taken from the refinement of the crystal structure B–factors. The rms fluctuations are plotted for the non–hydrogen atoms only, and the fluctuations from the B–factors have been shifted by one unit (the dashed line) for ease of comparison. The rms fluctuations are related to the B–factor through the relationship

$$\Delta r^{rms} = \left\{ \frac{3}{8\pi^2} B_i \right\}^{1/2} \qquad (17)$$

We find from these results that the fluctuations in the vacuum simulation are somewhat larger than those estimated by crystallographic refinement. However, the pattern of fluctuations is similar.

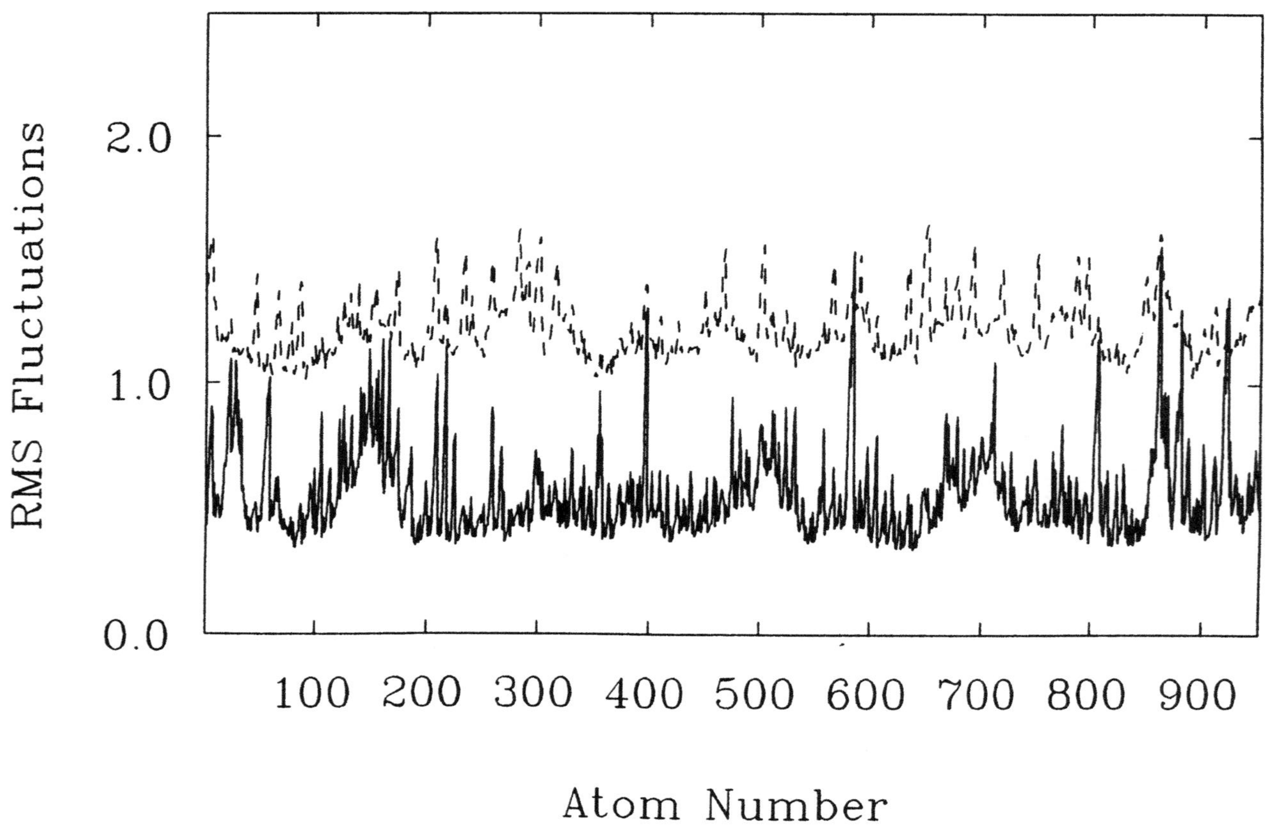

Fig. 4 Atomic rms fluctuations from molecular dynamics and crystallographic B-factors (dashed line) for ribonuclease A.

This result is typical of what is observed in comparison of fluctuations from simulations and refinement, and has led to further investigations of the nature of atomic fluctuations in proteins. The results from two such investigations are illustrated in Table V, where the anisotropy of atomic fluctuations for myoglobin and lysozyme are displayed.[17,52]

Table V. Anisotropy in atomic fluctuations.[a]

Atom type	Myoglobin[b]	Lysozyme[c]
All	0.68	0.85
Backbone	0.52	0.77
Sidechain	0.74	0.93
N	0.55	0.68
C	0.58	0.76
O	0.70	0.93
C^{α}	0.59	0.73
C^{β}	0.67	0.74
γ	0.72	0.90
δ	0.76	0.95
ϵ	0.85	1.03
ζ	0.77	1.14

a. Averages over all atoms at a particular position, except for atoms in valine sidechain.
b. Data from reference 17.
c. Data from reference 52.

The anisotropy is defined by the ratio of the principle moments of fluctuation. If σ_x^2, σ_y^2 and σ_z^2 are the three moments of the atomic fluctuations in an axis system where the fluctuations are decoupled, (i.e. σ_x^2, σ_y^2 and σ_z^2 the elements of the diagonalized fluctuation matrix for atom i given in order of decreasing size $\sigma_x^2 > \sigma_y^2 > \sigma_z^2$) then the anisotropy for atom i is defined as [52]

$$A_i = \left[\frac{\sigma_x^2}{1/2\,(\sigma_y^2 + \sigma_z^2)} - 1 \right] \tag{18}$$

The averages of the anisotropy over similar classes of atoms, as well as the overall anisotropy, indicate that the atomic fluctuations are highly anisotropic (if they were isotropic $A_i = 0$). Hence, the fitting of atomic fluctuations with isotropic Debye-Waller factors (B-factors) will underestimate the amplitude of the motion. This is consistent with the results of figure 4 and suggests that care must be exercised in the direct comparison of fluctuations from protein dynamics and x-ray B-factors.

From these calculations it is also interesting to note that the anisotropy is greater in the protein with the requirement of greater flexibility in function, i.e., lysozyme. Lysozyme is a "hinge" protein with a very open active site which binds polysaccharide molecules; extensive flexibility is necessary for function. Myoglobin contains a heme-proto porphyrin prosthetic group which binds small ligands such as O_2 for storage, and appears not to require large scale motions for its function. It would be of some interest to see where a protein which is intermediate in its requirement for flexibility in function, such as ribonuclease A, lies in this comparison – the suggestion would be that it should lie between myoglobin and lysozyme.

Results of dynamics simulations on proteins can lead to insights on a more detailed level as well. For example, the dynamics and structures which solvated protein simulations yield provide important information on the stabilization of sidechain pairs in configurations which may be relevant for catalytic activity. We examine two examples of this below. The first involves the role of solvation in forming and stabilizing "ion" pairs45 (pairs of like-charged sidechains) and the second is an example from a study of protein–solvent–substrate interactions in the enzymic reaction involving ribonuclease A.[53]

In figure 5 a sequence of "snap-shots" from a protein dynamics study of the solvated active site of lysozyme is shown. The figure illustrates the formation of an "ion" pair between Arg–63 and Arg–71 on the surface of the active site cleft. All solvent within 4.0A of any of the $NH_2^{\delta+}$ moieties in the two argenine sidechains is included, with hydrogen-bonded interactions being represented by dashed lines.[45] The sequence of three snap-shots shows the initial solvation of the vacuum optimized structure (5a), the formation of the solvent stabilized pair (5b), and a structure typical of the fluctuating conformations which exist after formation of the stable pair (5c). What is clear from these figures is that water plays an essential, and molecular, role in stabilizing the close proximity of the two positively charged groups. These findings suggest that qualitatively different results may occur in detailed atomic properties when explicit solvation is included in protein simulations. Furthermore, such effects may be important to protein function.

a

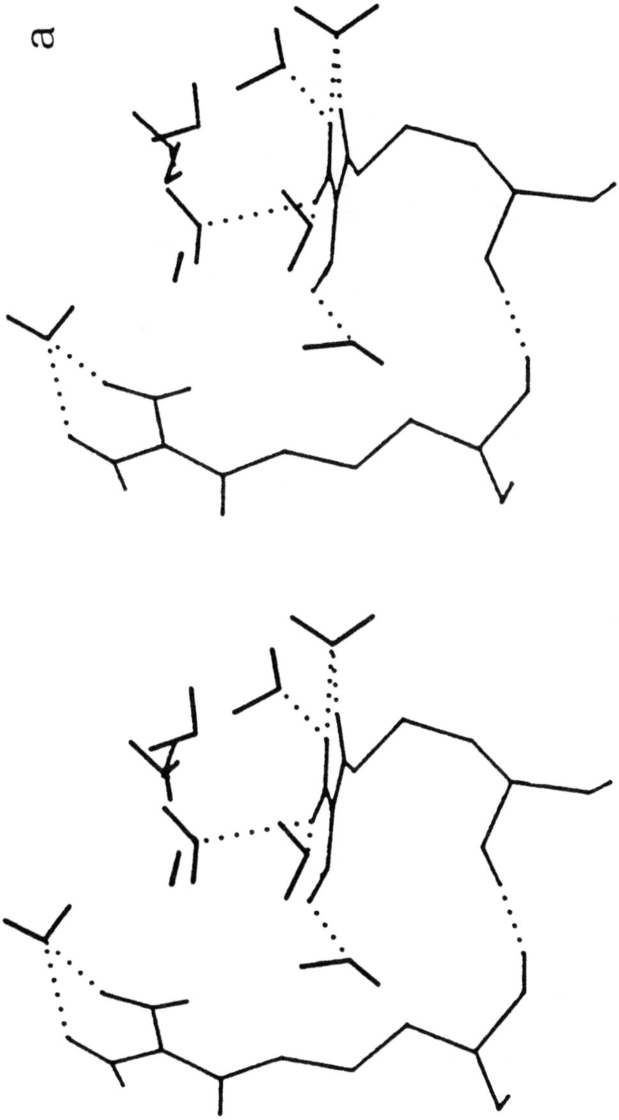

b

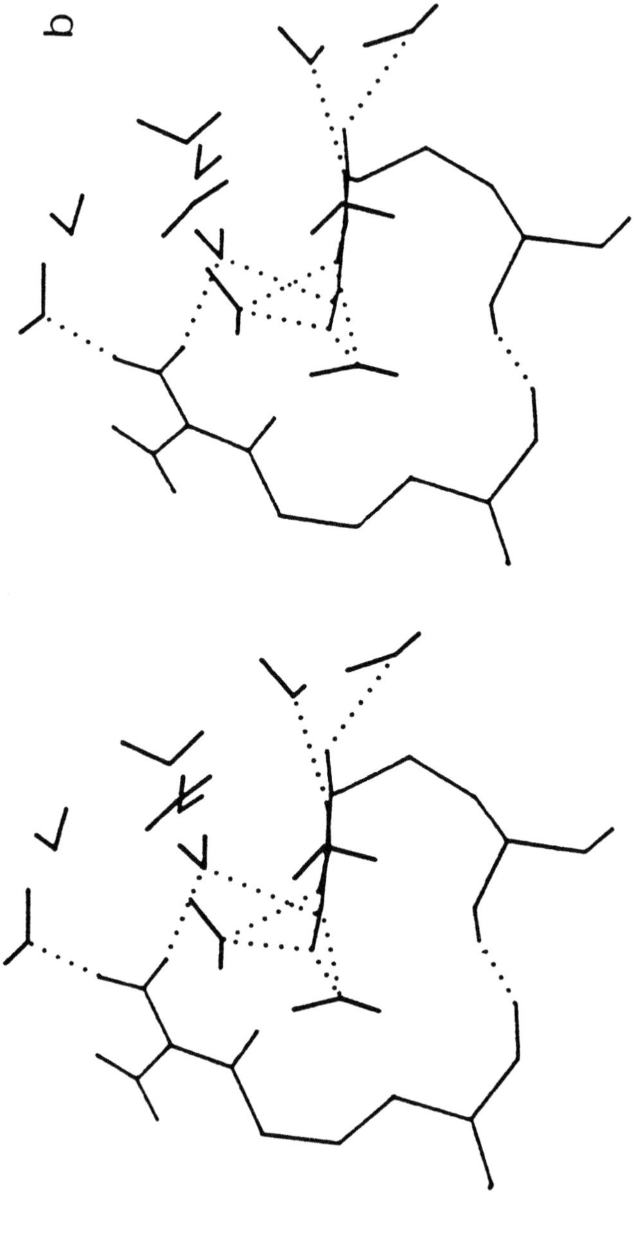

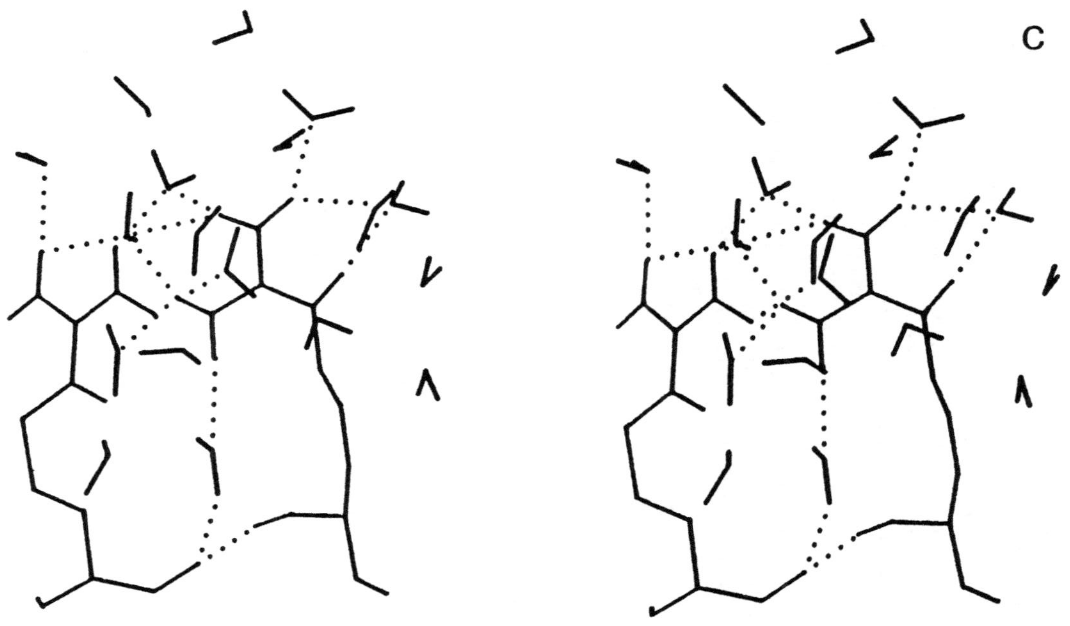

Figure 5. The atomic structure of solvated argenine sidechains in lysozyme. a) An initial "presolvation" structure. b) The formation of solvent linked "close ion" pairs for like-charged sidechains. c) A typical structure from the equilibrium dynamics of the solvated argenines.

The potential importance of solvent in protein function is nicely illustrated with the results of stochastic boundary molecular dynamics simulations carried out on a series of reactive intermediates in the catalytic hydrolysis of small ribonucleotides by ribonuclease A.[53] These calculations represent the first time protein dynamics was used to explore enzyme catalysis. Structures which represent the results of protein dynamics simulations at three key steps in the enzymatic pathway (the native protein active site, figure 6a; the substrate bound complex, figure 6b; and a transition state intermediate with pentacoordinate phosphorous, figure 6c) are shown in Figure 6. From the analysis of structure and fluctuations in these calculations it was apparent that water plays an important role here as well. Specifically, in the native structure (figure 6a) Brunger et al. note that the positively charged residues which are present in the active site region, e.g., Lys–7, Arg–39 and Lys–41, are clustered toward the center of the binding pocket and held in place by water networks. Such a cluster could serve as a "signalling beam" to direct the highly electronegative substrate (cytadine adenine monophosphate) into the binding pocket.[53] When the substrate binds (figure 6b) the water occupying the binding pocket is expelled and the charged cluster breaks up. The catalytically active histidines, His–119 and His–12, now position themselves for the chemistry they are to perform. The substrate is anchored in the binding pocket through hydrogen bonding contacts with residues at the pyrimidine and purine ends of the molecule (hydrogen–bonds are indicated by dotted lines). Once cleavage and expulsion of the purine end of the molecule is accomplished, the chemistry dictates that one of the pentacoordinate O–P bonds must be hydrolyzed to complete the reaction and move to the product state. Preceeding this step the protein dynamics simulations suggest that water again fills the vacant portion of the pocket (figure 6c) and establishes a stabilizing network of hydrogen bonds which brings the charged groups, especially Lys–41, closer to the site of chemical hydrolysis. The presence of such a positive charge greatly enchances the ability of His–119 to protonate the oxygen during hydrolysis, thereby enchancing the reaction rate.[54]

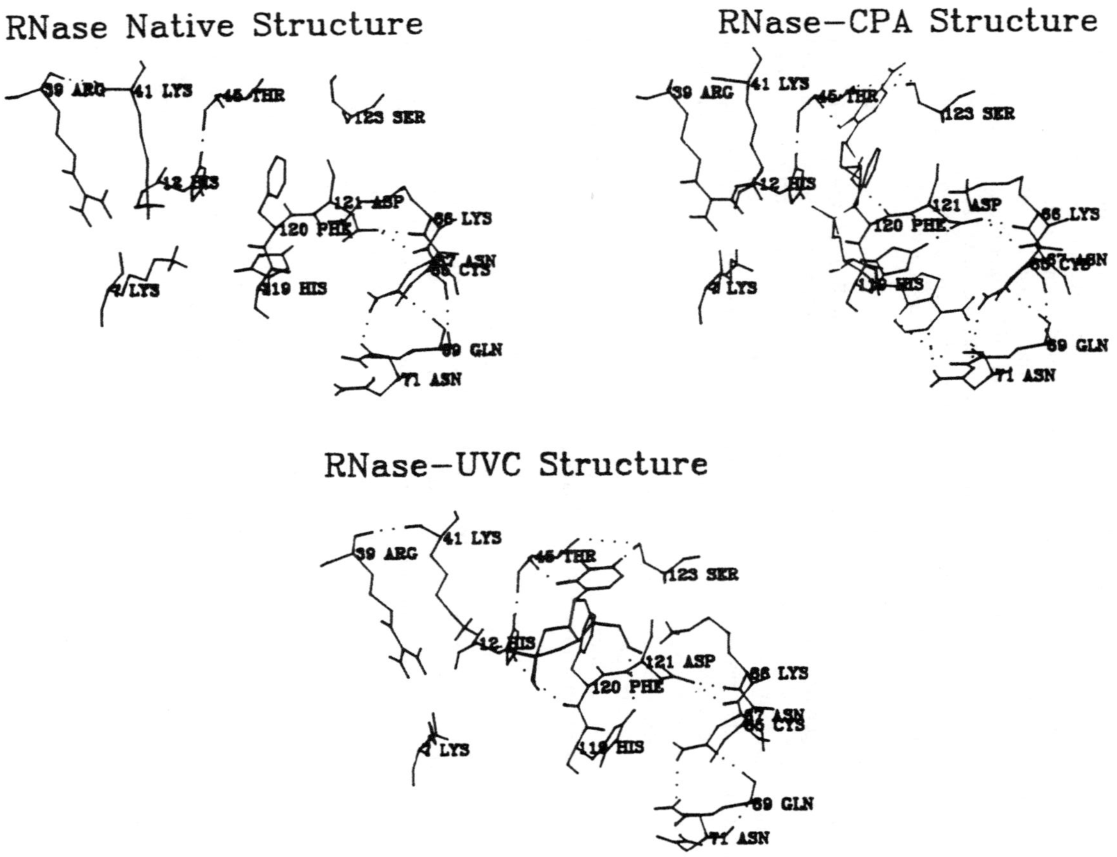

Figure 6. Water and substrate binding in the active site of ribonuclease A: a) the native binding pocket: b) Riboncuclease A – CpA reactant complex: c) Ribonuclease A – uridine vanadate transition–state analog complex.

It is insights from protein dynamics calculations such as those just described which aid and fuel further understanding of protein function. In the examples above it is structure which plays a major role. However, insights into the dynamical character of a protein can also be gleaned from simulations. An example in point can be taken from studies of the dynamics of fluorescence anisotropy depolarization in biological molecules. In this experimental technique fluorescent residues, usually tryptophan or tyrosine, are excited with a short pulse of optical radiation and then the anisotropic components of their emitted fluorescence are monitored, giving rise to the fluorescence anisotropy decay curve, r(t),

$$r(t) = 0.4\, e^{-t/\tau_o} \langle P_2\, [\mu_A(o) \cdot \mu_E(t)] \rangle \quad (19).$$

In eqn. (19) τ_o is the relaxation time for overall tumbling of the protein and P_2 is the second Legendre polynomial of the angle between the absorption and emission dipole vectors. From the decay of the "P_2" correlation function one can obtain information about the magnitude of motion for the dipole vectors, and hence a localized region of the protein. This is extracted through the interpretation of specific models. For example, if the absorption and emission dipoles are coincident at t=0 and their dynamics constitutes restricted diffusional motion (diffusion on the surface of a cone), then the long-time behavior of $P_2(t)$ provides information about the maximum angle of the motion, and hence the degree of local flexibility.[16,54] If, however, the motion is more complicated and multiple conformational substates exist, this simple model interpretation is inadequate for use in extracting motional information.[45,55]

Protein dynamics can be used to examine the details of this motion by explicitly studying the motional character of different fluorescent residues in a protein. Figure 7 illustrates one example of tryptophan residues in hen eggwhite lysozyme, Trp-62 (figure 7a) and Trp-63 (figure 7b). Shown in this figure are results for the P_2 correlation function computed from vacuum (dashed line) and solvated (solid line) simulations.[45] We observe that varying sites in the protein show different motional characteristics in both vacuum and solution (i.e., Trp-62 differs from Trp-63). We also note that solvation has a greater influence on the decay of P_2 for residues which are more solvent accessible, Trp-62 (figure 7a).[45] What is not apparent from simple analysis of the correlation functions, but is quite obvious when the detailed atomic motion of the tryptophans is examined, is the nature of the motions giving rise to the decay. Further analysis of the trajectory for the sidechain dihedral angles indicates that the motion of the dipole absorption/emission vectors is not diffusive but instead consists of local oscillations in a conformational substate with occasional jumps between different substates.[45] From the development of insights like these better models can be (and are being) developed to extract relevant motional information from fluorescence anisotropy decay.

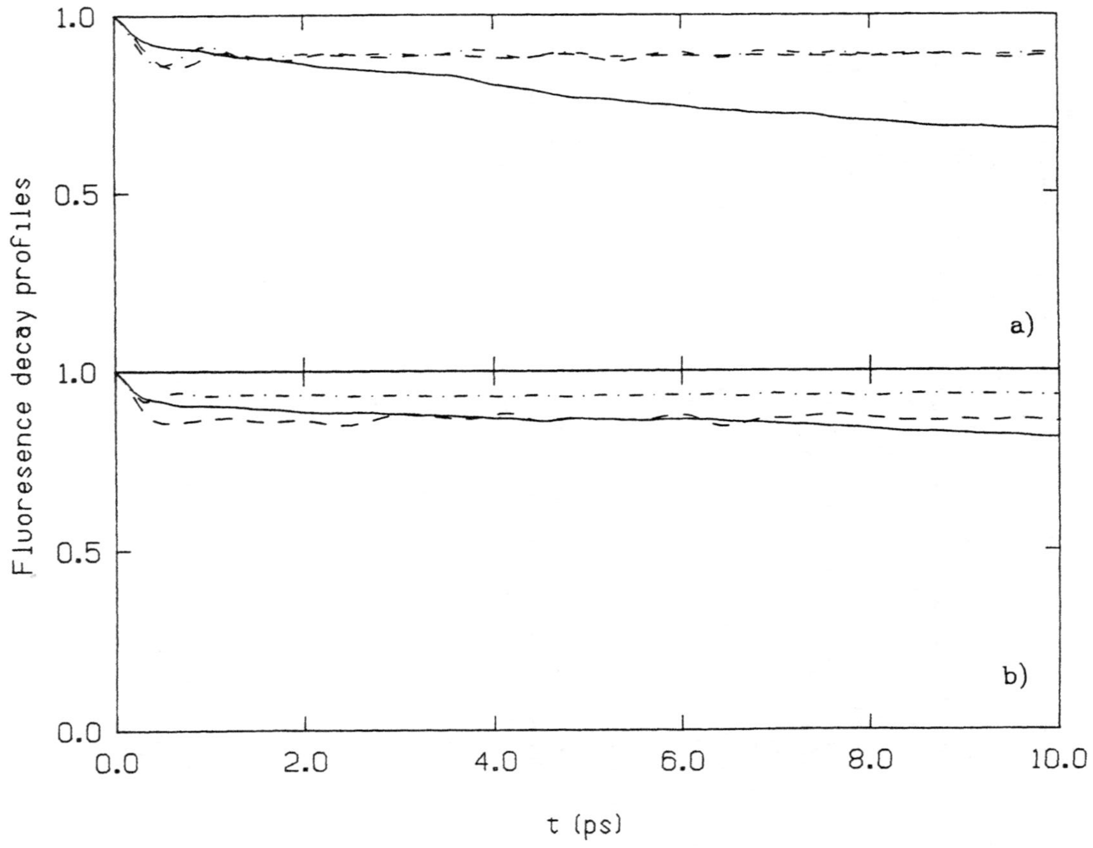

Fig. 7. Decay of the P_2 correlation function for Trp-62 and Trp-63 in lysozyme. a) Trp-62 in vacuum (dashed line) and solvent (solid line). b) Trp-63 in vacuum (dashed line) and solvent (solid line).

The use of protein dynamics simulations to delineate the effects of solution and crystalline environment on protein motion and structure have been illustrated. Similar behavior is found in both environments. More detailed investigations have provided insights into the nature of anisotropy in atomic fluctuations and pointed to specific deficiencies in the interpretation of Debye–Waller factors from crystallographic refinement. From calculations such as these a better understanding of proteins has resulted. Protein dynamics in the presence of solvent can also lead to detailed insights into the explicit role of solvation in stabilizing protein for catalysis and in affecting dynamics as probed by fluorescence depolarization measurements, as illustrated by the preceding examples. These methods are yielding a bountiful harvest of increased understanding of protein structure and dynamics and in the interpretation of experimental measurements.

V. Thermodynamic Simulations: Methods

Protein dynamics has provided detailed atomic information about structure, fluctuations and dynamics of biological molecules, and the potential role of these motions in function.[5] In addition to insights gained into the dynamic behavior of biological macromolecules, simulation techniques are beginning to show promise in evaluating and rationalizing thermodynamic properties, e.g., relative and absolute free energies of solution and transfer.[56] Initial applications of thermodynamic simulation (TS) methods to the calculation of relative free energies of hydration for small solutes,[57,58] relative transfer free energies for amino acids and nucleotides,[59] and the binding properties of small molecules to receptor sites[60,61] (in proteins, DNA or synthetic systems) provide encouraging results which suggest this approach will be useful in the rationalization, prediction and design of drugs and enzymes. In this section we will present the methods used to carry out thermodynamic simulations on proteins. The following section will review a number of recent applications with an emphasis on studies which aim toward deeper understanding of the relationship between structure and thermodynamics.

The general formalism for thermodynamic simulation methods follows from early work by Zwanzig,[62] but takes on several guises in present implementations. The two approaches on which we focus our discussions are termed thermodynamic cycle perturbation theory (TP)[63] and thermodynamic integration (TI).[57] Both of the methods are based on the definition of a hybrid Hamiltonian which represents some admix of the initial state (1) and final state (2) of the system.[56] If we let λ represent the "coordinate" describing the pathway used to interconvert the two systems, then the hybrid Hamiltonian may be defined by,[58,62,65]

$$H(\lambda) = (1-\lambda)^n H_1 + \lambda^n H_2 + H_{env}, \tag{20}$$

where n describes the degree of non-linearity for this thermodynamic pathway. An example may serve to illustrate the nature of the hybrid Hamiltonian. If one is interested in the relative free energy of solvation of methanol (H_1) versus ethane (H_2) in water, the hybrid Hamiltonian is one with which the hydroxy group in methanol is slowly "mutated" (with a dependence on λ as given above) to the methyl moiety in ethane. H_{env} represents the "environment" Hamiltonian which includes all solvent–solvent interactions as well as interations between the solvent and those portions of the solute (hybrid ethane/methanol) common to both states 1 and 2. For a fixed value of λ not equal to 1 or 0 this system is unphysical but specifies a well defined thermodynamic state of the system, with a Hamiltonian given in eqn. (20) above. The phase space for the hybrid system may be sampled using molecular dynamics by integrating the Hamiltonian (or equivalently, the Newtonian) equations of motion with any of the methods described earlier. The equations of motion are given by

$$\dot{p}_\lambda = \frac{\partial H(\lambda)}{\partial q_\lambda} \tag{21a}$$

$$\dot{q}_\lambda = \frac{\partial H(\lambda)}{\partial p_\lambda} \tag{21b}$$

where p_λ and q_λ represent momenta and coordinates for the atoms.

To make connections with thermodynamic properties the phase space configurations are used to compute the average quantity

$$\Delta A(\lambda, \lambda+\Delta\lambda) = -k_B T \ \ln \left\langle \exp\left[\frac{-(H(\lambda+\Delta\lambda) - H(\lambda))}{k_B T} \right]\right\rangle_\lambda, \tag{3}$$

which is the Helmholtz free energy difference between thermodynamic states at λ and $\lambda+\Delta\lambda$. Calculations at several values of λ, spanning $\lambda = 0$ to $\lambda = 1$, are carried out and the averages from these "windows" are then summed to compute the free energy change,

$$\Delta A = \sum_{\lambda=o}^{1-\Delta\lambda} \Delta A(\lambda, \lambda+\Delta\lambda). \tag{23}$$

Alternatively, one may use the methods of thermodynamic integration to compute the free energy change. This technique relies on the connection formula

$$\Delta A(\lambda) = \int_o^\lambda \left\langle \frac{\partial A(\lambda')}{\partial \lambda'} \right\rangle_{\lambda'} d\lambda', \tag{24}$$

where, again, molecular dynamics may be used to compute the average quantities within the broken brackets, $\langle \ldots \rangle_\lambda$. In most cases this average is calculated for a number of windows spanning $\lambda = 0$ to $\lambda = 1$ and numerical quadrature is used to accumulate the free energy change.[57,64] A schematic of the composite calculation of free energy changes using both thermodynamic perturbation (TP) and thermodynamic integration (TI) are illustrated in figure 8.

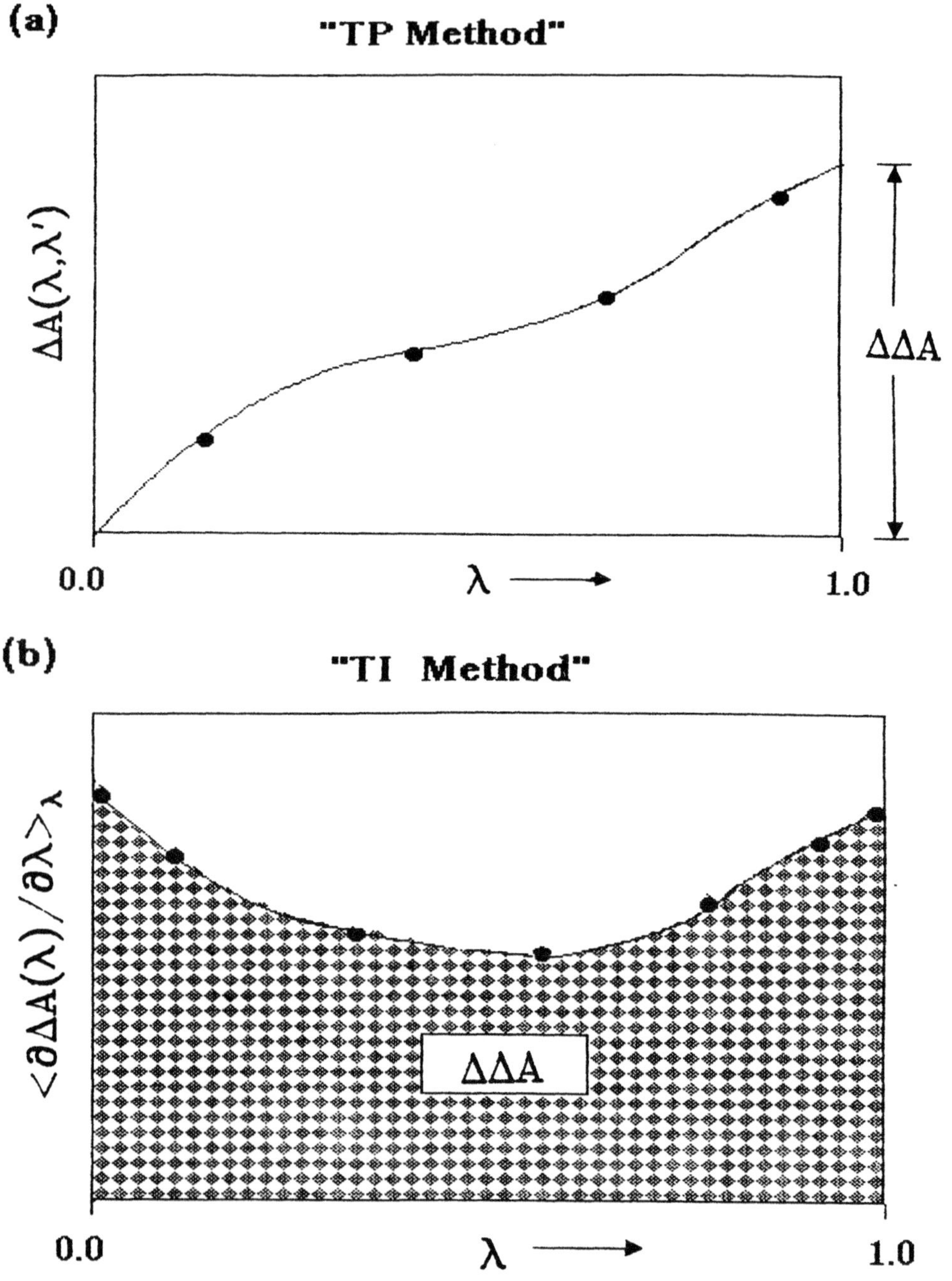

Fig. 8. A schematic diagram of (a) a TP calculation and (b) a TI calculation of the total free energy change in a thermodynamic process.

The thermal decomposition of the free energy into energetic and entropic contributions may be considered as well.[65] These properties are computed from the same set of thermodynamic simulations used to derive the free energy using the connection formula

$$\Delta E(\lambda,\lambda+\Delta\lambda) \cong \frac{k_B T^2}{2\Delta T} \ln \left[\frac{\left\langle \exp\left[\frac{-H(\lambda+\Delta\lambda)}{k_B(T+\Delta T)} + \frac{H(\lambda)}{k_B T} \right] \right\rangle_{\lambda,T}}{\left\langle \exp\left[\frac{-H(\lambda+\Delta\lambda)}{k_B(T-\Delta T)} + \frac{H(\lambda)}{k_B T} \right] \right\rangle_{\lambda,T}} \right.$$

$$\times \left. \frac{\left\langle \exp\left[\frac{-H(\lambda)}{k_B}\left[\frac{1}{T-\Delta T} - \frac{1}{T} \right] \right] \right\rangle_{\lambda,T}}{\left\langle \exp\left[\frac{-H(\lambda)}{k_B}\left[\frac{1}{T+\Delta T} - \frac{1}{T} \right] \right] \right\rangle_{\lambda,T}} \right] \tag{25}$$

and

$$\Delta S(\lambda,\lambda+\Delta\lambda) = k_B \ln \left\langle \exp\left[\frac{-(H(\lambda+\Delta\lambda) - H(\lambda))}{k_B T} \right] \right\rangle_{\lambda,T} + \frac{\Delta E(\lambda,\lambda+\Delta\lambda)}{T} \ . \tag{26}$$

The expressions above are developed by considering the finite difference approximant to the temperature derivative of $\Delta A(\lambda,\lambda+\Delta\lambda)$.[57,64,65] The total energy and entropy changes are computed by summing incremental changes between each window as done for ΔA in eqn. (23) (see also figure 8).

It is also possible to examine the contribution to free energy changes from specific groups of atoms or types of interactions, e.g., electrostatic and van der Waals. This is called thermodynamic component analysis. One approach which has been suggested for separation of van der Waals and electrostatic contributions to the free energy change is to "decouple" the calculations and first make the van der Waals changes, calculating ΔA_{vdw}, and then the electrostatic changes to compute ΔA_{elec}.[59] The total free energy change is the sum of the component parts, $\Delta A_{total} = \Delta A_{vdw} + \Delta A_{elec}$. This method is a well-defined thermodynamic procedure, i.e., it takes place between valid thermodynamic states and represents a thermodynamically consistant decomposition. Its use has provided some insights into the relative importance of van der Waals and electrostatic interactions.[66] However, it does not provide a unique decomposition of the free energy, or its thermal derivatives, into component contributions. For example, if changes in the electrostatic interactions were made first, instead of the van der Waals changes as described above, the total free energy change would be the same (same initial and final states), however the decomposition would not be since the intermediate states differ.

A more rigorous approach to component analysis can be derived by considering the connection formula for thermoynamic integration in eqn. (24) and using the decomposition of the hybrid Hamiltonian, eqn. (20), into component parts,

$$H(\lambda) = \sum_{i=1}^{m} h_i(\lambda) \ . \tag{27}$$

In eqn. (27) $h_i(\lambda)$ is a particular component part of the total hybrid system Hamiltonian (e.g., the electrostatic portion, or the total contribution from solvent molecules, etc.) of which there are m. From here it is simple to show that

$$\Delta A(\lambda) = \sum_{i=1}^{m} \int_{o}^{\lambda} d\lambda' \left\langle \frac{\partial h_i(\lambda')}{\partial \lambda'} \right\rangle_{\lambda'} = \sum_{i=1}^{m} \Delta a_i(\lambda) \tag{28}$$

To verify that this is a thermodynamically consistant decomposition one needs to prove that the thermodynamic energy derived from the thermal derivative of $\Delta A(\lambda)$ is equal to the summation over the thermal derivative of $\Delta a_i(\lambda)$. This follows from simple consideration of eqns. (28) and (24) and yields

$$\Delta E(\lambda) = \sum_{i=1}^{m} \Delta e_i(\lambda) \tag{29a}$$

where

$$\Delta e_i(\lambda) = \frac{-1}{k_B T} \sum_{i=1}^{m} \sum_{j=1}^{m} \int_{o}^{\lambda} \left[\left\langle \frac{\partial h_i(\lambda')}{\partial \lambda'} h_j(\lambda') \right\rangle_{\lambda'} - \left\langle \frac{\partial h_i(\lambda')}{\partial \lambda} \right\rangle_{\lambda'} \langle h_j(\lambda') \rangle_{\lambda'} \right] d\lambda' \tag{29b}$$

We note that this component analysis is valid for any decomposition of the hybrid Hamiltonian obeying eqn. (27) above and is unique, moreover it only follows from consideration of the thermodynamic integration approach to free energy calculations. It cannot be carried out using the thermodynamic cycle perturbation formalism of eqn. (23).

Thermodynamic simulation methods may be used to calculate free energy surfaces using molecular dynamics as well as total or component free energies.[67] A recent approach to this calculation uses a fusion of molecular dynamics and Monte Carlo techniques. The basic connection formula follows from eqn. (22) above, and consideration of the free energy difference between a particular configuration specified at ξ and that at $\xi+\Delta\xi$. This is illustrated in figure 9 for the simplest of condensed phase reaction coordinate, the separation between two spherical particles. If we let $W_i^{\pm}$ represent the weight (Monte Carlo Boltzman factor) for displacing the pair of particles by $\pm\ \Delta\xi$ respectively during the ith molecular dynamics step,

$$W_i^{+} = \exp\left\{\frac{-(U(\xi+\Delta\xi) - U(\xi))}{k_B T}\right\} \tag{30}$$

$$W_i^{-} = \exp\left\{\frac{-(U(\xi-\Delta\xi) - U(\xi))}{k_B T}\right\}, \tag{31}$$

the free energy difference between the configuration at ξ and those at $\xi \pm \Delta\xi$ is simply related to the average of the weights for each step[67a]

$$\Delta A(\xi,\xi \pm \Delta\xi) = A(\xi \pm \Delta\xi) - A(\xi) = \frac{1}{N} \sum_{i=1}^{N} W_i^{\pm} \tag{32}$$

$$= -k_B T \ln\left\langle \exp\left\{ \frac{-(U(\xi \pm \Delta\xi) - U(\xi))}{k_B T} \right\} \right\rangle_{\xi}$$

In eqn. (32) N is the number of dynamics steps used in computing the average and the subscript ξ on the angle brackets indicates this average is over an ensemble constrained at a fixed ξ.

This formalism is implemented by attempting (but never carrying out) Monte Carlo steps $\pm\Delta\xi$ away from the position ξ (see figure 9) and storing the weights for each step in a conventional molecular dynamics simulation. A series of simulations with different ξ spanning the range of reaction coordinate of interest may then be used to compute the reaction free energy surface.[67] Temperature and spatial derivative approximants of $\Delta A(\xi)$ may be computed in analogy to eqns. (25) and (26) above. Thus, this method is ripe for exploitation in understanding condensed phase reactivity and for use in conjunction with TP and TI methods for conformational sampling.[67]

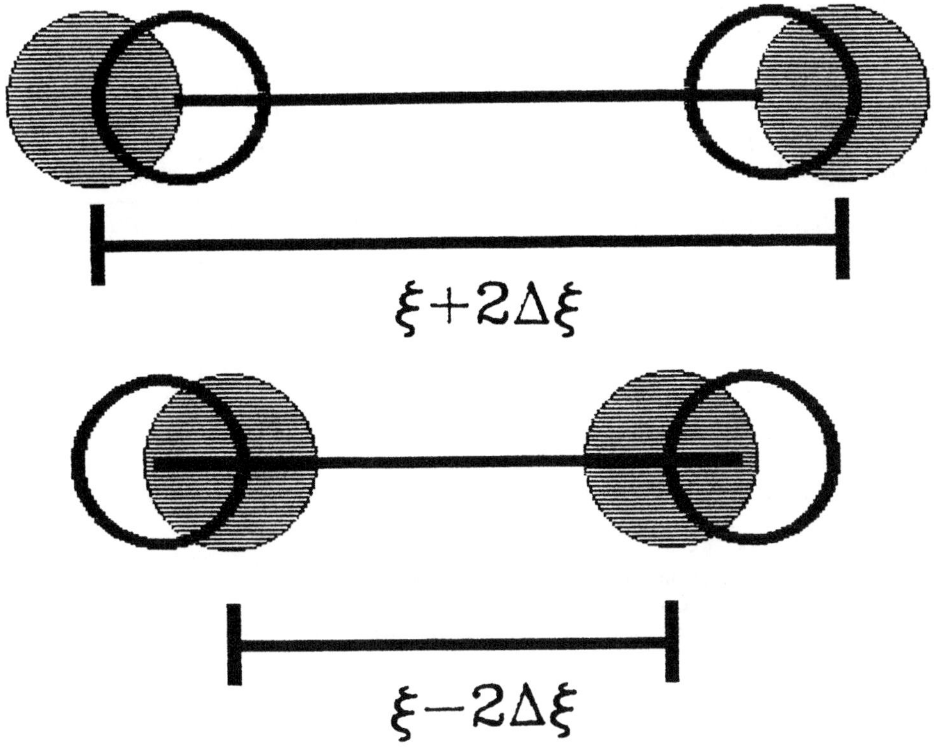

Fig. 9. A diagram illustrating (a) moves out along a reaction coordinate and (b) in along a reaction coordinate for a simple one dimensional reaction.

VI. Thermodynamic Simulations: Results

The methods just described have been used in a number of studies to examine thermodynamic processes ranging from ionic solvation[57] to protein–ligand (drug) binding thermodynamics.[60,61] In this section we will review a number of applications which illustrate the scope of these techniques for use in elucidating basic structure–dynamic information. We begin with a brief description of calculations which probe the accuracy and implementation of the empirical potential functions. Next, examples of the calculation of protein–inhibitor thermodynamics will be presented. We close this section with a review of some applications of new thermodynamic methodologies which permit free energy surfaces to be explored.

The calculation of relative free energies of hydration for small hydrocarbons and alcohols has been used as a benchmark for the development of thermodynamic simulation methods,[57,58] as well as a means of examining current empirical potentials. In recent studies by Fleischman and Brooks the relative thermodynamics of hydration for ethane, propane and butane (hydrophobic solutes), as well as methanol and ethanol (polar solutes) were examined using several different model potential functions. Their results are summarized in Table VI.

Table VI. Aqueous Solvation Thermodynamics of Alcohols and Alkanes.[a]

System and Potential Function	$\Delta\Delta A$ (kcal/mol)[b]			
	aqueous	vacuum	total	experiment[c]
propane → ethane[d]	0.83 ± 0.1	—	0.83 ± 0.1 (2.5 ± 5)	-0.13 (0.65)
propane → ethane[e]	1.63 ± 0.3	0.75	0.87 ± 0.3 (0.7 ± 6)	-0.13 (0.65)
propane → ethane[f]	0.56 ± 0.1	—	0.56 ± 0.1 (0.42 ± 4)	-0.13 (0.65)
butane → propane[f]	-1.43 ± 0.2	-0.40	-1.03 ± 0.2 (7.1 ± 7)	-0.13
methanol → ethane[f]	7.5 ± 0.2	—	7.5 ± 0.2 (7.1 ± 7)	6.9 (6.1)
ethanol → propane[f]	5.9 ± 0.2	—	7.1 ± 0.2 (5.6 ± 6)	7.0 (7.2)

a. Data from reference 68.
b. Numbers in parentheses are energies of hydration, $\Delta\Delta E$.
c. Experimental estimates from reference 69.
d. Potential functions for alkanes are from the CHARMM parameter set, see reference 22.
e. Potential functions from all hydrogen parameter set, see reference 70.
f. Potential functions from Jorgensen OPLS parameter set, see reference 71.

Examining first the results for apolar → apolar transformations, it is interesting to note that all three potential functions examined, two extended hydrocarbon models (CHARMM19[22] and OPLS[71]) and one explicit hydrogen model (a CHARMM parameter set currently under development[70]) give very similar values for the relative free energy of hydration for the propane → ethane transformation. However, we find that this value of $\Delta\Delta A_{solv} \approx 0.75$ kcal/mol is qualitatively incorrect in the sign, $\Delta\Delta G^{exp}_{solv} = -0.13$ kcal/mol.[69] Although the numbers being compared are small, the statistical significance of the difference between the experimental and computed values is at the 90% level.[72] Thus, we must conclude that the three potential models examined in this study do not provide an accurate representation of apolar thermodynamics.

The relative thermodynamics for the transformation of butane → propane were also examined for the OPLS parameter set. The sign now changes for $\Delta\Delta A_{solv}$ and is consistent with the experimental number, however, there is still a significant discrepancy between the two. This suggests a sensitivity of the free energies for these systems on the values of the van der Waals parameters, in particular, the "size" of the hydrophobic group appears to be important.

Free energies of hydration for polar → apolar transformations (methanol → ethane and ethanol → propane) are also included in the table. In both of these cases acceptable agreement with experiment is found. The results are dominated by the change in electrostatic forces, and thus, it is concluded that the potential functions being used are sufficiently accurate in representing the thermodynamics of gross electrostatic changes.

The calculations just described illustrate that the parameterization of current empirical potential functions appears to be inadequate in representing apolar → apolar thermodynamics in aqueous solution. However, polar → apolar transformations, which are governed by the loss of large electrostatic components, are adequately represented. Fleischman and Brooks note that TS methods may be an additional tool for use in parameterization of empirical energy functions. However, to date this has not been successfully applied.

The previous example has illustrated the sensitivity of free energy simulations to the particular values of parameters used in the empirical protein potential. These results provide some guidelines for the interpretation of computed free energies; this suggests that small (1-2 kcal/mol) changes, which might arise from small changes in van der Waals parameters, are near the limit of reliability of present potential function parameterization. We now illustrate another problem which arises in the implementation of these potentials. This is the effect of long-range force truncations on free energies. This point will be demonstrated by examining results from free energy simulations of a protein-inhibitor complex carried out by Wong and McCammon.[73] The system they studied consisted of the inhibitor benzamidine bound to the enzyme trypsin in aqueous solution. Calculations were made to compute the free energy difference for the binding of benzamidine to trypsin versus para-fluorobenzamidine. The binding pocket of trypsin with benzamidine bound is illustrated in figure 10a. The relative binding affinity as a function of the long-range electrostatic cutoff distance used in the calculation of eqn. (22) is shown in figure 10b. A clear dependence on the cutoff radius is obvious from the figure, with the relative free energy changing by a factor of three depending on the cutoff. The results of this work are significant in illustrating another caution which must be applied in free energy simulations and their interpretation. However, with due precautions the methods of thermodynamic simulation should provide an extremely useful tool for understanding thermodynamic aspects of biological function. We illustrate this by reviewing recent calculations which attempt to unravel the underlying principles of protein-inhibitor binding thermodynamics.

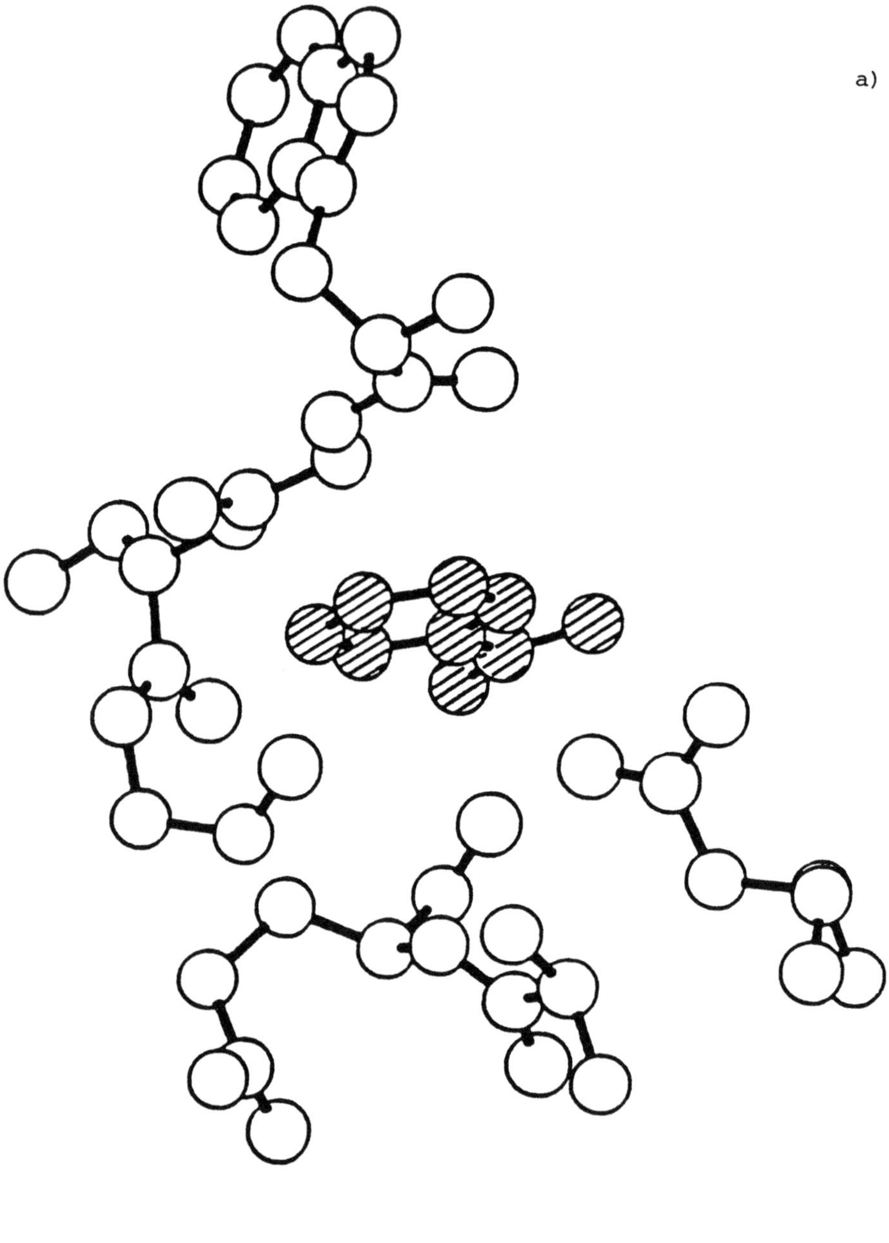

a)

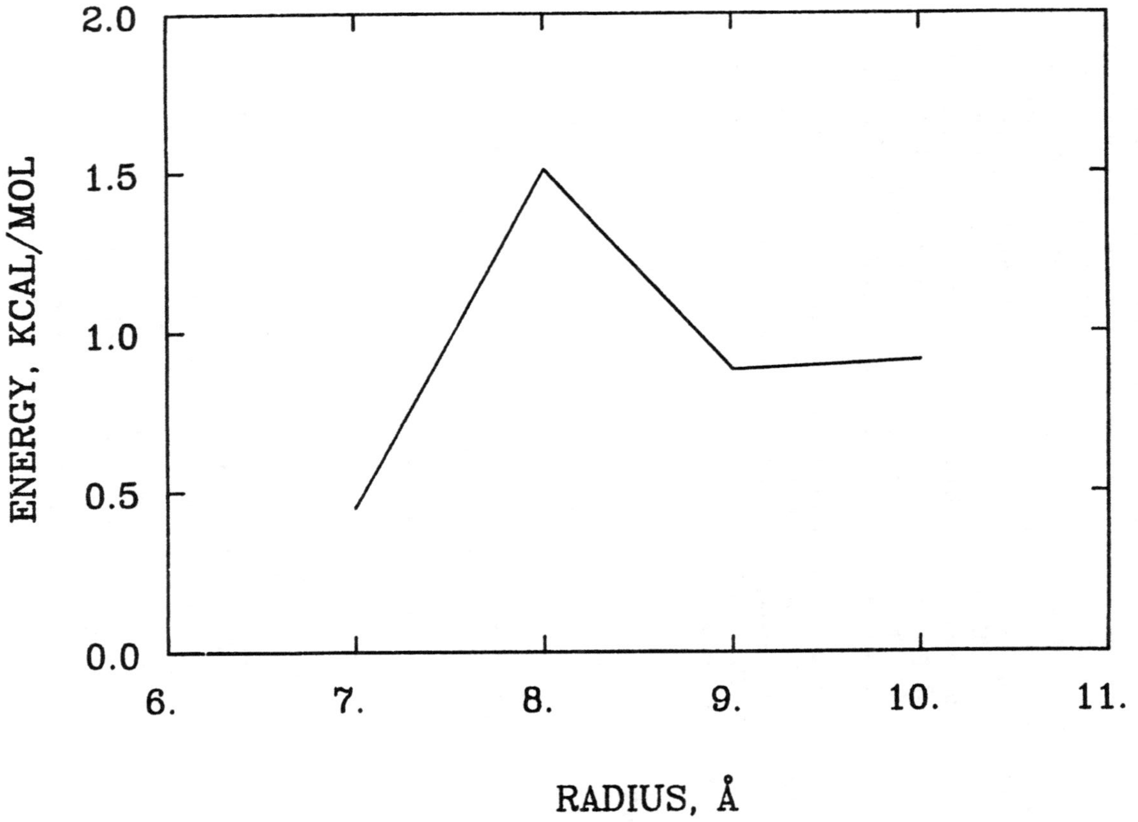

Fig. 10. a) Benzamidine bound to the trypsin active site. b) The influence of long range cutoff on the relative free energy of binding for benzamidine versus para-fluoro-benzamidine.

One of the primary objectives in the development of thermodynamic simulation methods is their use in the rationalization, prediction and design of potentially new compounds (drugs) which possess specific characteristics with respect to their interactions with proteins. To approach this problem two major components must be considered. One is characterization of the desolvation thermodynamics of the drug, which accompanies the overall binding process; the other is the thermodynamic difference between two drugs bound to a common receptor site. It is the combined influence of both which determines the thermodynamic character of protein–inhibitor interactions in aqueous solution. An example which illustrates the use of free energy simulations in the understanding of protein–inhibitor binding thermodynamics is taken from calculations by Fleischman and Brooks.[61]

In Table VII the results of calculations on the relative thermodynamics of interaction for complexes of TMP [2, 4–diamino–5–(3´, 4´, 5´–methoxy–benzyl) pyrimidine] and four of its congeners which substitute ethyl moieties for methoxy at positions 4´(PET), 3´(TOE), 5´(TOE´) and 3´,4´ and 5´(TET), with the NADPH–DHFR complex are listed.[61]

Table VII. Relative Binding Thermodynamics for congeners of TMP with DHFR.[a]

	ΔΔA(kcal/mol)	ΔΔE(kcal/mol)	ΔΔS(e.u.)
(1) PET+DHFR:TMPH$^+$ →TMP + DHFR:PETH$^+$	0.4 ± 0.3	23± 9	76±30
(2) TOE+DHFR:TMPH$^+$ →TMP + DHFR:TOEH$^+$	-2.9 ± 0.1	4± 7	23±23
(3) TOE´+DHFR:TMPH$^+$ →TMP + DHFR:TEO´H$^+$	0.4 ± 0.1	-4± 10	-15±34
(4) TET+DHFR:TMPH$^+$ →TMP + DHFR:TETH$^+$	-1.4 ± 0.7	-6± 10	-15±34

a) Results from reference 61 and Fleischman and Brooks, unpublished results.

For these reactions it is most noteworthy that the compounds PET and TOE´ (reaction 1 and reaction 2) are predicted to bind less tightly than any of the others. In one case the entropic contribution to desolvation is large compared to TMPH$^+$, due to a disruption of the solvent environment at the 3´, 4´, 5´ methoxy end of the drug. This results in the drug moving more deeply into the binding pocket, making additional favorable contacts with the protein. This is illustrated in figure 11 for a comparison of the complexes of TMPH$^+$, PETH$^+$ and TETH$^+$ with DHFR:NADPH. The additional hydrogen–bonds formed between protein and inhibitor are clear for PETH$^+$ (hydrogen bonds are shown as dashed lines). Also, movement of the inhibitor in the binding pocket is evident. Thus, entropy appears to play a passive role in assisting the ligand binding for this system.

It is extremely interesting to notice the thermodynamic decomposition of the relatively small relative free energy changes associated with reactions 1–4; this small free energy change arises from a large energy–entropy compensation – a result usually ignored in other calculations of relative binding thermodynamics. When the results for reaction 4 are compared with experiment,[74] few structural differences are found for the TMP and TET complexes (see figure 11 below). The calculated thermodynamic value for the relative binding affinity, $\Delta\Delta A_{bind}$ = −1.4 kcal/mol, differs from the experimental value of $\Delta\Delta G_{bind}$ = −1.7 kcal/mol by only a small amount.[74] Small relative errors in $\Delta\Delta E_{bind}$ or $\Delta\Delta S_{bind}$ can arise from a number of sources, e.g., potential parameters, sampling, etc., and these errors may result in the overestimation of one component of $\Delta\Delta A_{bind}$.

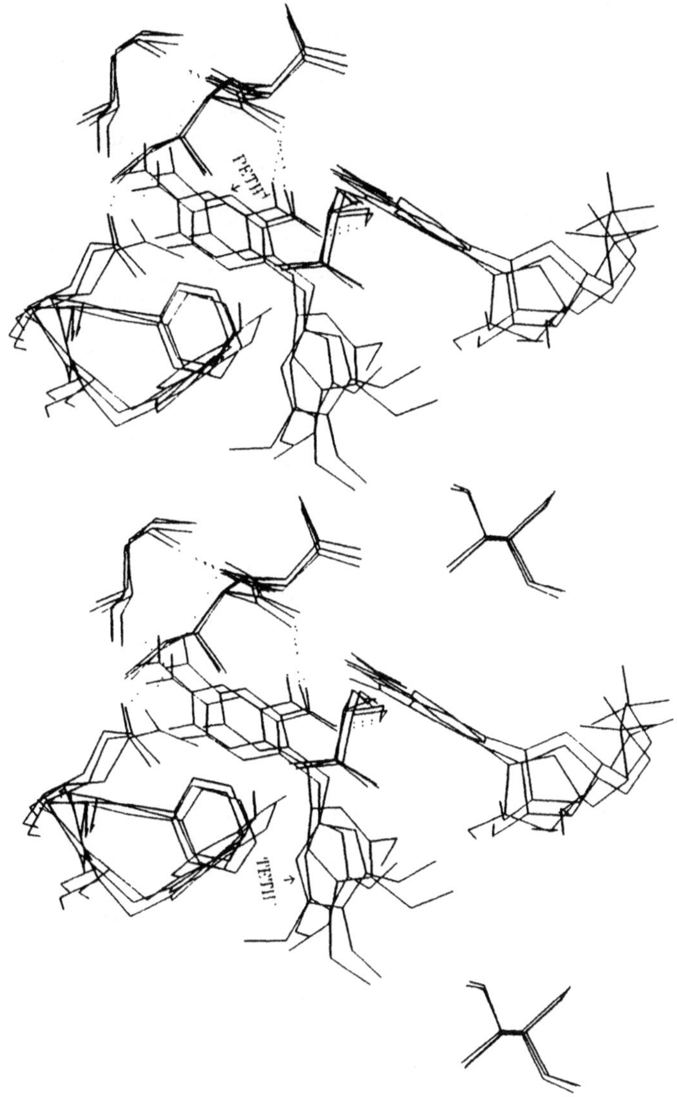

Fig. 11. Stereographic overlay of the dynamics structures of $TMPH^+$ and two of its congeners ($PETH^+$ and $TETH^+$) bound to the binary complex of DHFR:NADPH.

The calculations of Fleischman and Brooks illustrate the potential of these powerful simulation methods. However, they also bring into focus a number of cautionary points. The results of thermodynamic simulations, especially the energetic and entropic components of $\Delta\Delta A$, are quite sensitive the length of the calculation. Small changes in position within the binding pocket can give rise to large changes in the free energy and its attendant energetic and entropic parts. Thus, very long trajectories are necessary to adequately sample the configuration space of the system. In addition, new techniques to more adequately sample conformations of flexible molecules must be sought.[67]

Very recent applications of free energy simulations for the calculation of conformational free energy[67,75] are bringing new insights into our understanding of flexibility and conformational change in ligand binding reactions. To illustrate work from this new and exciting area we describe recent results of Tobias and Brooks on the effects of conformational change in the binding of $TMPH^+$ to a binary complex of DHFR:NADPH.[75] The questions being addressed by these researchers are (i) what role does conformational change in the inhibitor ($TMPH^+$) play in protein–inhibitor binding thermodynamics and (ii) can small conformational changes in the protein control ligand binding.

To examine these questions free energy simulations were carried out to compute the free energy surfaces associated with (i) rotation of the 4′ methoxy group in $TMPH^+$ around the C4′–04′ bond (Φ) and (ii) rotation of the Tyr–31 sidechain about the χ^2 sidechain angle. In figure 12a a schematic of the binding pocket for the DHFR:NADPH system is shown, with the two rotable angles specifed as Φ and χ^2. The results for the associated free energy surfaces are illustrated in figure 12b (Φ) and figure 12c (χ^2), respectively. Focusing first on conformational changes for the drug (figure 12b), we see that the environment dramatically changes the nature of the torsional potential (the gas phase torsion potential is shown as the dashed line). In the complex there is little difference in the binding free energy for ligand binding with the methyl group either above or below the plane of the ring (i.e.,$\Phi \sim \pm 100^{\circ}$). Moreover, there is only a small barrier associated with conversion between these two conformations. This result is not too surprising for the para–methoxy group since it is exposed primarily to solvent and has little direct contact with protein. On the other hand, rotation of the Tyr–31 sidechain about χ^2 does influence the binding. The free energy surface shown in figure 12c indicates that there is a large ($\sim$ 6 kcal/mol) barrier to rotation about χ^2. The origin of this barrier is apparent from examination of figure 11 for the complexes formed with $TMPH^+$, $PETH^+$ and $TETH^+$. In each case the tyrosine sidechain is "nestled" into the space defined by the bridging CH_2 between the phenyl and pyrimidine rings. Rotation away from this position produces a large free energy change, due most likely to collisions with the phenyl ring of $TMPH^+$. These results raise the possibility that the tyrosine sidechain, or the corresponding phenylalanine sidechain in DHFR from other species,[74] may control the binding of ligands in a "lock and key" manner. This suggestion is just speculative at this point, however it suggests that a deeper investigation, both experimentally and theoretically, of the role of conserved aromatic sidechains at this position in DHFR is in order.

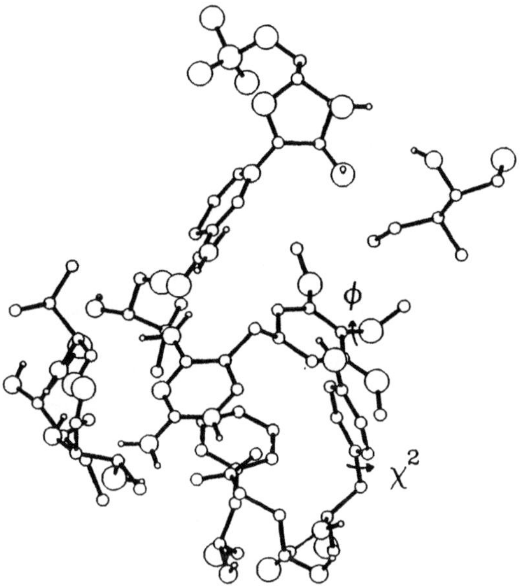

ϕ
χ^2

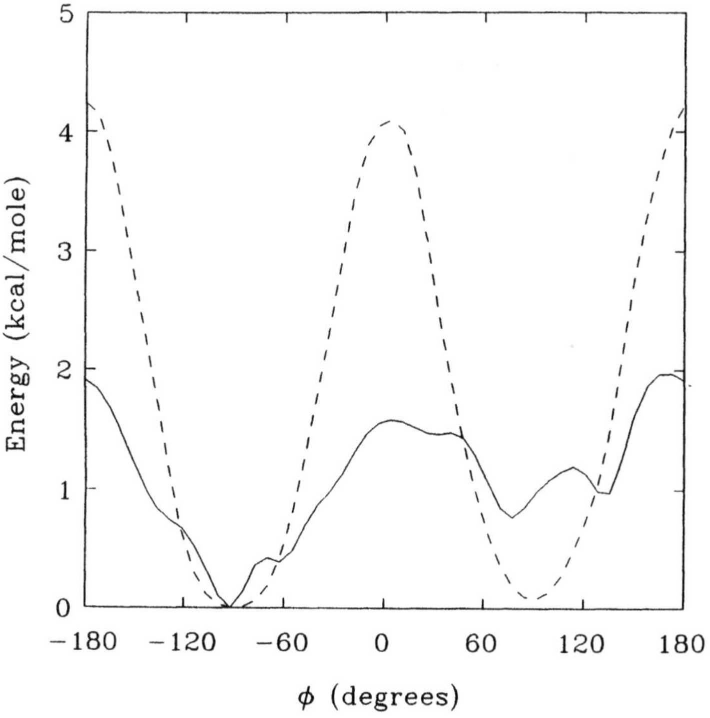

Energy (kcal/mole)
0
1
2
3
4
5
−180
−120
−60
0
60
120
180
ϕ (degrees)

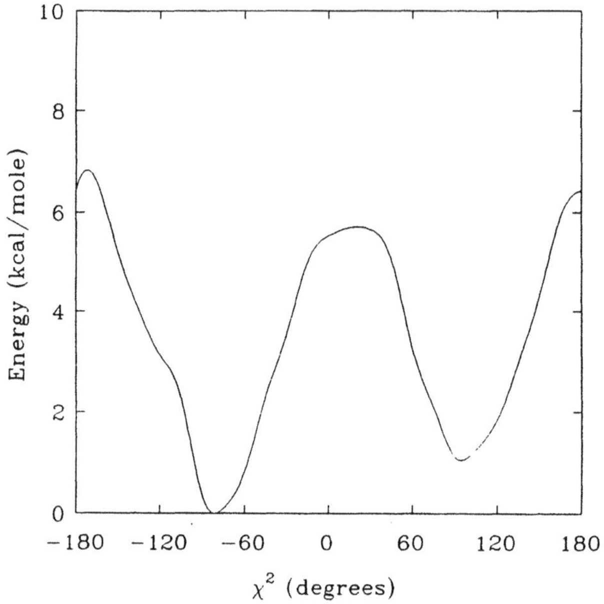

Fig. 12. a) $TMPH^+$ in the DHFR binding pocket with rotable bonds φ and χ^2 shown. b) Free energy surface and vacuum potential (dashed line) for rotation around φ. c) Free energy surface for rotation around χ^2.

The examples just described illustrate the real potential of free energy methods and protein dynamics. Specifically, with free energy simulations structural, dynamic and thermodynamic information may be obtained and direct comparisons with experiment are possible on all levels. In the future we can anticipate the development of closer collaborations between theory and experiment in this area, with the potential for significant advances in our understanding of structure–thermodynamic relations in biological systems.

VII. Summary

In this chapter we have reviewed the methods and techniques of protein dynamics. The application of these methods has been illustrated by results of studies on the dynamics, structure and thermodynamics of proteins in vacuum, crystalline environment, and aqueous solution. From these applications we have learned that the protein empirical potential is adequate to reproduce qualitative (and some instances quantitative) features of structure and fluctuations. It has also proven sufficient in describing dynamics and thermodynamics in many cases. Apart from simply evaluating the potential used in protein calculations, protein dynamics studies have led to insight into the nature of anisotropic atomic motion, dynamics of fluorescence anisotropy and enzyme–inhibitor interactions. These few examples only hint at the full potential of protein dynamics. Its use in method development and model refinement, and its application to a much broader scope of problems than addressed here (e.g., activated processes, enzyme catalytic dynamics, etc.) are happening now or clearly on the horizon.

One of the areas related to protein dynamics, which is of key importance in the future, is potential function development. As was discussed in this ASI (see S. Price in this volume), new approaches to the development and refinement of empirical potentials for protein dynamics are currently under way. These hold the promise of representations of protein structure–energy relationships (i.e., potentials) which are quite accurate, and will fuel a further growth in the area of protein simulations.

In the area of protein dynamics itself, the future points toward the refinement and further development of thermodynamic methods and techniques which combine classical and quantum descriptions to describe reactivity in proteins. The thermodynamic simulation methods described in this chapter are being used to explore fundamental questions of protein stability and protein–ligand interactions. This will continue in the future, and with the refinement of sampling techniques and potential functions these methods will provide quantitative assistance in design and rationalization of medically important compounds.

Acknowledgements

I wish to thank D. J. Tobias and S. H. Fleischman for much of the work presented in these lectures, and S. Sneddon and T. Hoeffel for useful comments and discussion. A special thanks is owed to T. Head–Gordon for careful reading of the manuscript and many useful comments. Acknowledgement to the NIH (GM37554) for financial support and the Pittsburgh Supercomputing Center for computing support is gladly given.

REFERENCES

1. J. A. McCammon, B. R. Gelin and M. Karplus, Nature **267**, 585 (1977).
2. B. J. Alder and T. E. Wainright, J. Chem. Phys. **31**, 459 (1959).
3. A. Rahman, Phys. Rev. A136, 405 (1964).
4(a) J. A. McCammon and M. Karplus, Proc. Natl. Acad. Sci. USA **76**, 3585 (1979).
(b) J. A. McCammon and M. Karplus, Biopolymers **19**, 1375 (1980).
(c) S. H. Northrup, M. R. Pear, C. Y. Lee, J. A. McCammon and M. Karplus, Proc. Natl. Acad. Sci. USA **79**, 4035 (1982).
5. C. L. Brooks III, M. Karplus and B. M. Pettitt, Proteins: A Theoretical Perspective on Structure Dynamics and Thermodynamics, complete volume in Adv. Chem. Phys., **71** (1988).
6. J. A. McCammon and S. Harvey, Dynamics of Proteins and Nucleic Acids, Cambridge University Press (1987).
7. C. L. Brooks III and M. Karplus, Methods in Enzymology **127**, 369 (1986).
8. G. Careri, P. Fasella and E. Gratton, Ann. Rev. Biophys. Bioeng. **8**, 69 (1979).
9. J. A. McCammon and M. Karplus, Ann. Rev. Phys. Chem. **31**, 29 (1980).
10. M. Karplus and J. A. McCammon, Ann. Rev. Biochem. **53**, 263 (1983).
11. M. Levitt, Ann. Rev. Biophys. Bioeng. **11**, 251 (1982).

12. W. van Gunsteren and H. J. C. Berendsen, Biochem. Soc. Trans. **10**, 301 (1982).
13. O. Edholm, L. Nilsson, O. Bery, M. Elnenberg, F. Closens, A. Gräslund, B. Jönsson and O. Teleman, Reviews of Biophysics, **17**, 125 (1984).
14. R. M. Levy, R. P. Sheridan, J. W. Keepers, G. S. Dubey, S. Swaminathan and M. Karplus, Biophys. J. **48**, 509 (1985).
15. R. M. Levy, M. Karplus and J. A. McCammon, J. Am. Chem. Soc. **103**, 994 (1981).
16. T. Ichiye and M. Karplus, Biochemistry **2**, 2884 (1982).
17. J. Kuriyan, G. Petsko, R. M. Levy and M. Karplus, J. Mol. Biol. **190**, 227 (1986).
18. A. T. Brünger, J. Kuriyan and M. Karplus , Science, **235**, 459 (1987).
19. J. A. Pople, B. T. Luke, M. J. Frisch and J. S. Binkley, J. Phys. Chem. **89**, 2198 (1985).
20. See, for example, J–K. Hwang, E. King, S. Creighton and A. Wonshal, J. Am. Chem. Soc., **110**, 5297 (1988).
21. U. Burkhart and N. L. Allinger, Molecular Mechanics (Am. Chem. Soc., Washington, D.C., 1982).
22. B. R. Brooks, R. Bruccoleri, B. Olafson, D. States, S. Swaminathan and M. Karplus, J. Comp. Chem. **4**, 187 (1983).
23. S. J. Weiner, P. Kollman, D. T. Nuguyen and D. Case, J. Comp. Chem. **7**, 230 (1986).
24. A. Dwivedi and S. Krimm, Biopolymers, **23**, 923 (1984).
25. A. T. Hagler, D. J. Osguthorpe, P. Dauber–Osguthorpe and J. C. Hempel, Science **227**, 1309 (1985).
26. K. D. Gibson, H. A. Scheraga, Proc. Nat. Acad. Sci, USA **58**, 420 (1967); L. Dunfield, A. Burgess and H. Scheraga, J. Phys. Chem. **82**, 2609 (1978).
27. A. J. Stone and S. L. Price, J. Phys. Chem., **92**, 3325 (1988).
28. W. Jorgensen, J. Am. Chem. Soc. 103, 335 (1981); ibid. **103**, 341 (1981).
29(a) C. Singh and P. A. Kollman, J. Comp. Chem., **5**, 129 (1984).
(b) L. Chirlian and M. Frank, J. Comp. Chem., **8**, 894 (1987).
30. B.R. Brooks. Supercomputer Research in Chemistry and Chemical Engineering, editors, Kutensen and B. Truhlar, Chap. 8 (1987).
31. C.L. Brooks, B.M. Pettitt and M. Karplus, J. Chem. Phys., **83**, 5897 (1985)
32. A. Brünger, C. L. Brooks III and M. Karplus, Chem. Phys. Lett. **105**, 495 (1984).
33. C. L. Brooks III, A. Brünger and M. Karplus, Biopolymers **24**, 434 (1985).
34. L. Verlet, Phys. Rev. **159**, 98 (1967).
35. R. W. Hockney and J. W. Easfwasel, Computer Simulations Using Particles (McGraw Hill, New York, 1981).
36. D. Beeman, J. Comp. Phys. **20**, 130 (1976).
37. H. C. Andersen, J. Chem. Phys. **72**, 2384 (1980).
38. S. Nose , J. Chem. Phys., 81, 511 (1984); W. G. Hoover, Phys. Rev. A, **31**, 1695 (1985); D. J. Evans and B.L. Hollan, J. Chem.

Phys., **83**, 4069 (1985).
39. D.J. Evans and G.P. Morris, Chem. Phys., **77**, 63 (1983).
40. H. J. C. Berendsen, J. P. M. Postma, W. F. van Gunsteren, A. DiNola and J. R. Haak, J. Chem. Phys. **81**, 3684 (1984).
41. M. Ferrario and J. P. Rychaert, Mol. Phys., **54**, 587 (1985).
42. B. M. Pettitt and M. Karplus, Chem. Phys. Lett. **121**, 194 (1985).
43. S. A. Adelman, Adv. Chem. Phys. **53**, 62 (1984).
44. R. Pastor, B. R. Brooks and A. Szabo, Mol. Phys., to be published.
45. C. L. Brooks III and M. Karplus, J. Mol. Biol., **208**, 159 (1989).
46. C. L. Brooks III and M. Karplus, J. Chem. Phys. **79**, 6312 (1983).
47. W. van Gunsteren and M. Karplus, Biochemistry **21**, 2259 (1982).
48. H. Yu and M. Karplus, unpublished results.
49. H. J. C. Berendesen, W. F. van Gunsteren, H. R. S. Zwindermann and R. G. Geurten, Ann. NY Acad. Sci., **482**, 269 (1986).
50. W. F. van Gunsteren and H. J. C. Berendsen, J. Mol. Biol. **176**, 559 (1984).
51. T. Head–Gordon and C. L. Brooks III, unpublished results.
52. T. Ichiye and M. Karplus, Proteins, **2**, 236 (1987).
53. A. Brünger, C.L. Brooks III and M. Karplus, Proc. Natl. Acad. Sci., **82**, 8456 (1985).
54. C. L. Brooks III, A. T. Brünger, M. France, K. Haydock, L. C. Allen and M. Karplus, International Symposium on Bioorganic Chemistry editor R. Breslow (The New York Academy of Sciences, 1986) p. 295.
54. G. Lipari and A. Szabo, J. Am. Chem. Soc. **104**, 4546 (1982).
55. L. X. Q. Chen, R. A. Eugh, A. T. Bru"nger, D. T. Nguyen, M. Karplus and G. R. Fleming, Biochemistry, **27**, 6908 (1988).
56. M. Mezei and D. L. Beveridge, Annals NY Acad. Sci., **483**, 1(1986).
57. S. H. Fleischman and C. L. Brooks III, J. Chem. Phys., **87**, 3029(1987).
58. W. L. Jorgensen and C. Ravimohan, J. Chem. Phys., **83**, 3050 (1985).
59. a) P. A. Bash, U. C. Singh, F. K. Brown, R. Langridge and P. A. Kollman, Science, **235**, 574 (1987);
    b) U. C. Singh, F. K. Brown, P. A. Bash and P. A. Kollman, J. Am. Chem. Soc., **109**, 1607 (1987).
60. a) C. F. Wong and J. A. McCammon, J. Am. Chem. Soc., **108**, 3830 (1986);
    b) T. P. Lybrand, J. A. McCammon and G. Wipff, Proc. Natl. Acad.Sci. USA, **83**, 833 (1986).
61. S. H. Fleischman and C. L. Brooks III, Proteins, to be published.
62. R. W. Zwanzig, J. Chem. Phys., **22**, 1420 (1954).
63. B. L. Tembe and J. A. McCammon, Comput. Chem., **8**, 281 (1984).
64. S. H. Fleischman, B. Tidor, C. L. Brooks III and M. Karplus, J. Comp.Chem., manuscript in preparation.
65. a) C. L. Brooks III, J. Phys. Chem., **90**, 6680 (1986); b) There were typographical errors in equations 8 and 9 of the above reference, the corrected relationships are given in equations 25 and 26 of the present paper.
66. C. Singh, Proc. Natl. Acad. Sci., **85**, 4280 (1988).

67(a) D. J. Tobias and C. L. Brooks III, Chem. Phys. Lett., **142**, 472(1987); (b)ibid. J. Chem. Phys., 89, 5115 (1988); (c) D. J. Tobias, C. L. Brooks III and S. H. Fleischman, Chem. Phys. Lett. **156**, 256 (1989)

68. C.L. Brooks III, Intl. J. Quant. Chem., proceedings from the Sanibel Symposium, **15**, 221 (1988).

69. a) A. Ben-Naim and Y. Marcus, J. Chem. Phys., **81**, 2016 (1984); b) S. F. Dec and S. J. Gill, J. Solution Chem., **13**, 27 (1984).

70. J. Smith and M. Karplus, J. Comp. Chem., to be submitted.

71. W. L. Jorgensen, J. D. Madura and C. J. Swenson, J. Am. Chem. Soc., **106**, 6638 (1984).

72. W. Mendenhall, Introduction to Probability and Statistics, Duxbury Press, North Scituate, Massachusetts, 1975.

73. C.F. Wong and J.A. McCammon, Isr. J. Chem., **27**, 211 (1986).

74. D. A. Matthews, J. T. Bolin, J. M. Burridge, D. J. Filman, K. W. Volz, B. T. Kaufman, C. R. Beddell, J. M. Champness, D. K. Stammers and J. Kraut, J. Biol. Chem., **280**, 381 (1985); D. A. Matthews, J. T. Bolin, J. M. Burridge, D. J. Filman, K. W. Volz and J. Kraut, J. Biol. Chem., **280**, 392 (1985).

75. D. J. Tobias and C. L. Brooks III, work in progress.

SIMULATION OF PLASTIC CRYSTALS

M. MEYER
LPM CNRS
1 Pl. A. Briand
92195 MEUDON Principal Cedex
France

1. Introduction

Traditionally, the name of "Plastic Crystals" has been reserved, because of their macroscopic properties, to some molecular crystals. These organic compounds have a high temperature phase, just below the melting point where the vapour pressure is relatively high and the deformation is easy. These plastic phases are highly symmetric , generally cubic and at some transition temperature , below the melting point, they transform to another crystalline phase with a lower symmetry. X-Ray diffraction studies of plastic crystals yield only a few diffracted beams and intense diffuse scattering. This is characteristic of an orientational dynamical disorder. The existence of orientational freedom is the property which distinguishes plastic crystal from other molecular solids. This orientational disorder is also observed in the high temperature phases of ionic solids containing molecular ions such as ammonium, sulphate and cyanide. On the microscopic scale the molecular motion is very similar to the one observed in plastic crystals. The essential difference is that the molecular ions are rotating within the monatomic crystal lattice of the second ion species. Consequently on the macroscopic scale they behave quite differently from plastic molecular crystals, i.e. they cannot be as easily deformed and have higher melting entropies and lower vapour pressures. As the orientational disorder is the microscopic property characterizing these high temperature phases it is now common to use the name "Plastic crystals" as a synonym of O.D.I.C. (Orientational Disorder in Crystals).

In most of the papers or reviews e.g. in reference [1], the orientationally disordered state of molecular systems is referred to as a plastic phase even if the plastic behaviour does not exist on the macroscopic scale. The orientational disorder is illustrated in figure 1 with two instantaneous views of a molecular solid (bicyclooctane) in the high temperature cubic phase and in the low temperature trigonal or pseudohexagonal phase. In the (111) plane of the plastic phase one can notice the translational order of the

C.R.A. Catlow et al. (eds.), Computer Modelling of Fluids Polymers and Solids, 335–355.

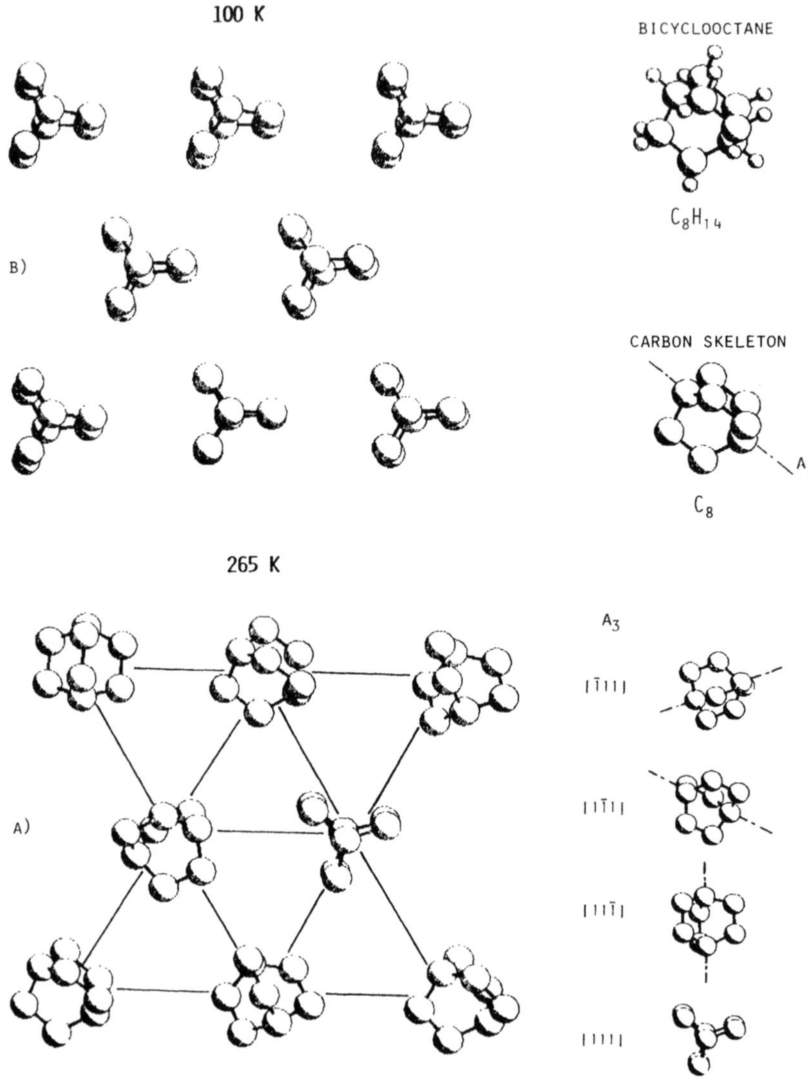

Figure 1. Schematic view of bicyclooctane molecule C_8 H_{14} . The threefold molecular axis A_3 is drawn on the carbon skeleton. The two snapshots correspond respectively to the molecular configurations of the high (A) and low (B) temperature phases.

A) (111) plane of the cubic plastic phase; the orientations and positions of the molecules result from a Molecular Dynamics simulation performed at 265 K. The four preferred orientations of the A_3 molecular axis along the <111> directions are drawn on the right side.

B) (001) plane of the pseudohexagonal phase. Simulation performed at 100 K.

centres of mass while the orientational disorder is obvious. This orientational disorder arises from the existence of four preferred orientations of the threefold molecular axis. The molecules undergo reorientations between these equiprobable positions and they rotate around the threefold molecular axis. The (001) plane of the low temperature phase shows the usual long range translational and orientational order of molecular solids.

Plastic phases occur in substances containing relatively rigid polyatomic units which can be either neutral molecules or molecular ions. The lattice binding energies are quite different for these two molecular species. As a consequence the domain of stability of the ODIC phase is large for the ionic crystals (i.e. several hundreds of K) and more restricted for molecular crystals (some tens of K). There is also some variety of shapes for the molecular units; they can be linear (e.g. N_2) or polycyclic (e.g. bicyclooctane, see figure 1). The geometry of the molecule and its symmetry elements are important parameters for the molecular rotational motion associated with the orientational disorder.

A wide variety of experimental techniques are used to probe the orientational disorder and the rotational motion of the molecules [1]. Structure determinations are achieved with X-Ray and neutron diffraction; phonons are characterized through Brillouin, Raman and neutron scattering. The most powerful technique is neutron scattering since it yields geometrical and dynamical information and allows the determination of structural properties, of molecular reorientation processes and of correlation times [2]. But, because the disorder introduces some limitations, the experimental data are less numerous and less accurate than those obtained in ordered crystals. This is the reason why the interpretation of the experimental results is sometimes difficult.

Other types of experiments are also carried out to characterize the lattice defects, the self-diffusion and the plastic behaviour of various plastic crystals [3,4].

Several models have been proposed to describe the orientational disorder and the rotational motion in plastic molecular crystals [5,6]. They can be schematically classified in three groups :

i) Theories using symmetry properties to calculate some physical quantities, e.g. orientational probability density function [7].

ii) Theories describing the single particle motion : two simple extremes of possible molecular rotational motion are the rotational diffusion and the jump reorientation mechanism between preferred orientations [6].

iii) Theories including collective rotational and translational motion

Numerical simulations have also been performed on plastic phases; since the orientational disorder is essentially a dynamical one, molecular dynamics (M.D.) simulations are well suited to study the properties of ODIC phases. Many simulations are realistic ones and are related to systems which have been studied experimentally such as ionic crystals like NaCN and KCN [8-10] or molecular systems containing diatomic molecules (N_2) [11], tetrahedral molecules (neopentane [12,13], adamantane [14,15], CBr_4 [16,17]) or octahedral molecules (SF_6 [18,19]). They yield important informations on the studied systems and allow some comparisons with experimental results [8,9,14,15] and theoretical approaches. Other simulations are performed on model systems and are mostly aimed at the comparison with theoretical models [20,21]. Recently a new field of interest has been investigated by MD simulations of phase transitions between high temperature plastic phases and low temperature ordered phases [22-26].

2. Simulation techniques

MD simulations require the calculation of intermolecular forces in order to integrate the equations of motion for the molecular units. The potential energy functions used to calculate these forces are generally simplified ones. It is assumed that the potential energy can be separated in two independant parts : an intramolecular one due to chemical bonds and an intermolecular one due to the interactions between molecules. This second part is the only contribution which is taken into account in the force calculations. Moreover, in most of the simulations, the molecules are treated as rigid and many body effects are ignored.

2.1. POTENTIAL ENERGY FUNCTIONS

The simplest procedure to calculate the intermolecular non bonded potential energy $U_{\alpha\beta}$ is to add the interactions of all pairs of non bonded atoms i.e. those which do not belong to the same molecule.

$$U_{\alpha\beta} = \frac{1}{2} \sum_{i,j} v\,(r_{ij}) \tag{1}$$

where r_{ij} is the distance between two atoms i and j belonging respectively to molecules α and β.

The atom-atom potential v is frequently written as the sum of a short range (SR) part typically Lennard-Jones or exp-6 in form and a coulomb interaction between point charges $q_i q_j$ when the molecular unit is an ionic or a polar one.

$$v(r_{ij}) = v_{SR}(r_{ij}) + \frac{q_i\ q_j}{r_{ij}} \tag{2}$$

The intermolecular energy can be calculated in the same way as previously with a potential function v defined as a site-site one. In this case r_{ij} is the distance between groups of atoms (e.g. CH_3, CN^-) When the simulation involves molecular ions or large molecules containing several CH_3 or CH_2 groups the site-site potentials have often been preferred to the atom-atom ones. However in the case of large neutral hydocarbon molecules (e.g. adamantane) the atom-atom potential gives a better description of the structural properties. This implies an increase of computational time but, with vector computers, this is not a real problem. On the other hand for ionic systems site-site potentials are more convenient. An important concept in the choice of potential energy functions is transferability. Potential parameters of the Lennard-Jones or exp-6 short range contribution and coulomb parameters which have been fitted on structural and energetic results can be transferred to other systems containing the same type of sites or atoms. The most commonly used [27,28] transferable potential energy functions have been derived for molecular systems containing C, H, O and N atoms.

2.2. ALGORITHMS

The conventional approach to the solution of the equation of motion of a rigid body treats separately the trajectory of the centre of mass and the orientational parameters which can be expressed in terms of Euler angles. In the simplest form this method is useless in MD calculations because the variables describing the orientations have singularities. This difficulty can be overcome by solving the eulerian equations using unitary quaternions as orientational variables [29]. There is a completely different approach which does not derive the equation of motion in generalized coordinates. It is the method of constraints in Cartesian coordinates [30]. This method was originally derived for partially rigid molecules, but it has also been applied to rigid systems [31]. The n atoms of each molecule are divided in two subsets containing n_b basic atoms and n_s secondary atoms ($n_b + n_s = n$). The n_b basic atoms are held together by bond constraints σ_{ij} and the secondary atoms are linked to the basic ones by linear constraints τ_α.

$$\sigma_{ij} \equiv (\mathbf{R}_i - \mathbf{R}_j)^2 - d_{ij}^2 = 0 , \qquad i > j = 1, n_b \tag{3}$$

$$\tau_\alpha \equiv \sum_{i=1}^{n_b} C_{\alpha i} \mathbf{R}_i - \mathbf{r}_\alpha = 0 , \qquad \alpha = 1, n_s \tag{4}$$

$\mathbf{R}_i$ and $\mathbf{r}_\alpha$ are, in cartesian coordinates, the vectors corresponding to the positions of the basic and secondary atoms respectively. d_{ij} are the bond lengths . The $C_{\alpha i}$ values depend on the geometry of the molecule. The equations of motion using the Lagrange multipliers λ_{jk}, μ_α associated to the constraints σ_{jk}, τ_α are :

$$M_i \ddot{R}_i = F_i - \sum_{k=1}^{n_b-1} \sum_{j=k+1}^{n_b} 2R_{jk} (\delta_{ij} - \delta_{ik}) \lambda_{jk} - \sum_{\alpha=1}^{n_s} C_{\alpha i} \mu_\alpha \tag{5}$$

$$m_\alpha \ddot{r}_\alpha = f_\alpha + \mu_\alpha \tag{6}$$

All we need to know is the time evolution of the basic coordinates R_i ($i=1, n_b$); the time evolution of the secondary coordinates follows from the constraint relation (4). Equation (5) is reformulated by using μ_α given in relation (6) together with the following equation : $\sum_{i=1}^{n_s} C_{\alpha i} \ddot{R}_i = \ddot{r}_\alpha$ derived from the constraint relation (4). This results in a transfer to the basic particles of the forces acting on secondary ones and in a rescaling of the masses in order to take into account the contribution of the secondary particles. The equation of motion of the basic atoms becomes :

$$\ddot{R}_i(t) = \mathcal{F}_i(t) - \sum_{k-1}^{n_b-1} \sum_{j=k+1}^{n_b} \mathfrak{R}^i_{jk}(t) \lambda_{jk}(t) \tag{7}$$

where $\mathcal{F}_i(t)$ is a function of F_i, f_α, M_i, m_α, $C_{\alpha i}$ and $\mathfrak{R}^i_{jk}$ a function of R_{jk}, $C_{\alpha i}$, M_i, m_α. The equations of motion are then integrated using Verlet's algorithm to predict the coordinates at time t + h.

$$R_i(t+h) = 2R_i(t) - R_i(t-h) + h^2\mathcal{F}_i(t) - h^2 \sum_{k=1}^{n_b-1} \sum_{j=k+1}^{n_b} \mathfrak{R}^i_{jk}(t) \lambda_{jk} \tag{8}$$

In the first stage, the coordinates $R'_i(t+h)$ of all the basic atoms are predicted in the absence of bond constraints using relation (9).

$$R'_i(t+h) = 2R_i(t) - R_i(t+h) + h^2\mathcal{F}_i(t) \tag{9}$$

Then the coordinates are reset in order to follow the bond constraints. The unknown quantities λ_{jk} are calculated using relations (8) and (3), in an iterative way in order to obtain the required bond lengths with a tolerance Δd . Less than five iterations are necessary when Δd has a value of 10^{-7}Å. This is not a time consuming procedure since it requires only loops on the n_b basic atoms and when a tridimensional molecule is considered , $n_b = 4$.

2.3. EXTENDED-SYSTEMS

The method of constraints in Cartesian coordinates, described previously, has been derived for MD simulations performed at fixed energy, volume and number of molecules. The methods developped in other statistical ensembles for atomic systems allow us to carry out MD calculations at constant pressure and/or constant temperature. To achieve a constant pressure MD simulation, the volume is allowed to fluctuate [32] and an extension of this method gives to the MD cell the possibility of changing its shape [33]. The volume evolution is controlled by the balance between internal stresses and an externally set pressure. In constant temperature MD simulations, the velocities are scaled in order to maintain the kinetic energy at a constant value [34].

These methods require the introduction of additional degrees of freedom in an "extended" system [35], where the coordinates and momenta are related to space and time in the system of interest by dynamical scaling. They have been generalized to molecular systems. This extension requires an adaptation to maintain the geometry of the rigid molecules. The internal constraints are decoupled from space scaling which is only applied to the centres of mass. Rigid molecular systems can be described in generalized coordinates [36] where minor extensions are required to modify the basic theory derived for atomic systems. The generalization of the method of constraints in Cartesian coordinates to extended systems is obtained by introducing additional "constrained" degrees of freedom [37].

3. Applications

3.1. ORIENTATIONAL DISORDER

3.1.1. *Orientational probability density function.* X-Ray and neutron scattering are used to probe experimentally the orientational disorder. This result is not directly obtained from the experimental data and models are required to fit the Bragg intensities. It is assumed that translational and rotational motions can be decoupled. The amplitude of scattering contains the usual Debye Waller factor which accounts for the centre of mass displacement and a rotational form factor. This rotational form factor can be expanded in terms of symmetry adapted functions by introducing an orientational probability density function $\mathcal{P}(\theta,\varphi)$ where θ and φ are the polar angles defining the orientation of a given bond or axis of the molecule. In the case of cubic systems, the relevant symmetry adapted functions are the cubic harmonics [29] ${}_iK_\ell$ which are polynomials of degree ℓ in terms of Cartesian coordinates. They are listed up to $\ell = 10$ in the Appendix. The choice of the ${}_iK_\ell$ functions necessary to obtain a complete basis of symmetry adapted functions depends on the point group symmetry of the molecule and of the lattice site [7], [38]. The orientational probability density function for a tetrahedral molecule in a cubic site is then given by the following relation :

$$P(\theta,\varphi) = 1/4\pi \left(1 + \sum_{\ell,i} \langle {}_iK_\ell \rangle \; {}_iK_\ell(\theta,\varphi)\right) \tag{10}$$

with ℓ values equal to 4, 6, 8, 9, 10 when the expansion is limited to $\ell = 10$.

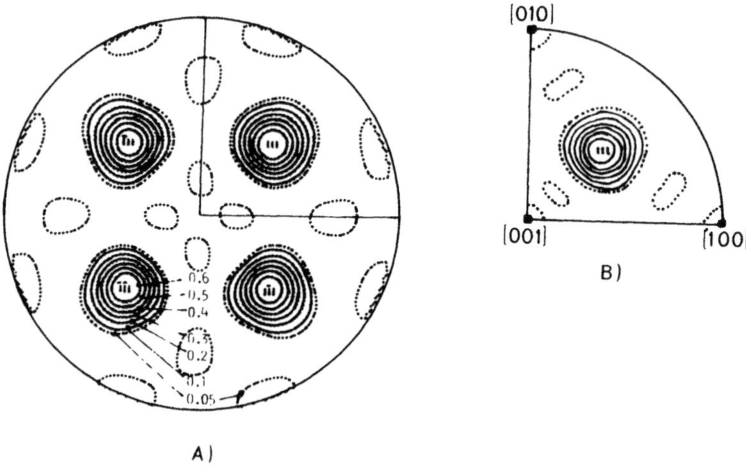

Figure 2. A) Stereographic plot of the orientational probability density function calculated for adamantane at 309 K . The contour lines correspond to equal values of $P(\theta,\varphi)$ ranging between 0.6 to 0.05. Half of the orientational space is shown here.
B) Experimental orientational probability density function obtained for adamantane by fitting X-Ray data.

This orientational probability density function has been calculated for a tetrahedral molecule in a MD simulation of the plastic phase of adamantane [14]. The coefficients $\langle {}_iK_\ell \rangle$ of the expansion are calculated for the four threefold axes of the molecules by averaging the coordinates of four carbon atoms located on these axes. The coordinates are relative to the centre of mass of the molecule and normalized since they must define a unit vector. This orientational probability density function is plotted in figure 2a on a stereographic projection. The contour lines corresponding to equal values of P (θ,φ) are plotted. One can notice that the maxima of this function are obtained when the ternary molecular axes coincide with the <111> crystalline directions; this is characteristic of a T_d orientation as will be shown in § 3.1.2. Figure 2b shows the results obtained experimentally by fitting the X-Ray diffraction Bragg intensities [39]. The calculated orientational function compares very well with the experimental one; the maxima have the same location and the contour lines are quite similar for the values of $P(\theta,\varphi)$ ranging between 0.6 and 0.1. The discrepancies occuring for $P(\theta,\varphi)$=0.05 can be explained by the difference in the choice of ${}_iK_\ell$: ${}_1K_9$ is not taken into account in the experimental fitting.

3.1.2. *Order parameters*. The expansion coefficients $<{}_iK_\ell>$ of $P(\theta,\varphi)$ can be used as order parameters as is done in experimental work. Other symmetry adapted functions which are, as the cubic harmonics, linear combinations of spherical harmonics can give information on the orientational order. Using the formalism of group theory they are classified in sets belonging to the irreducible representations of the point groups. The orientational behaviour of CN^- ions in KCN and NaCN has been investigated with symmetry adapted functions belonging to E_g and T_{2g} representations of the cubic point group [40].

$$Y_1^{(2)} = (5/16\pi)^{1/2}(3Z^2 - 1) \;,\; Y_2^{(2)} = (5/16\pi)^{1/2}(x^2 - y^2) \qquad E_g \qquad (11)$$

$$Y_3^{(2)} = (15/4\pi)^{1/2}xy \;,\; Y_3^{(2)} = (15/4\pi)^{1/2}yz \;,\; Y_5^{(2)} = (15/4\pi)^{1/2}xz \qquad T_{2g} \qquad (12)$$

They allow us to investigate the orientations of CN^- ions which correspond either to <100> symmetry (i.e. E_g) or to <111> symmetry (i.e. T_{2g}).

The orientation of a tetrahedral molecule in a cubic lattice can be characterized with cubic harmonics. The tetrahedral rotator functions $M_n(\Omega)$, n = 1,7 are also interesting for this configuration. They are functions of the Euler angles (Ω) giving the orientation of the tetrahedral molecule in the crystal lattice reference frame [41]. They are classified into three independant sets and only one expression per set is given in the following list.

$$M_1 = \frac{3\sqrt{3}}{4}\sum_{i=1}^{4} x_i y_i z_i \qquad (13)$$

$$M_2 = \frac{3\sqrt{5}}{40}\sum_{i=1}^{4} x_i\,(5x_i^2 - 3) \qquad (14)$$

$$M_5 = \frac{3\sqrt{3}}{8} \sum_{i=1}^{4} x_i (y_i^2 - z_i^2) \qquad (15)$$

where $r_i = (x_i, y_i, z_i)$ is a unit vector directed along the direction joining the centre of the tetrahedron to one of the four vertices. To understand how these rotator functions can be used as order parameters, one has to consider the four ideal orientations T_d, D_{2d}, C_{3v} and C_{2v} of a tetrahedron in a cubic site as shown in figure 3 [42].

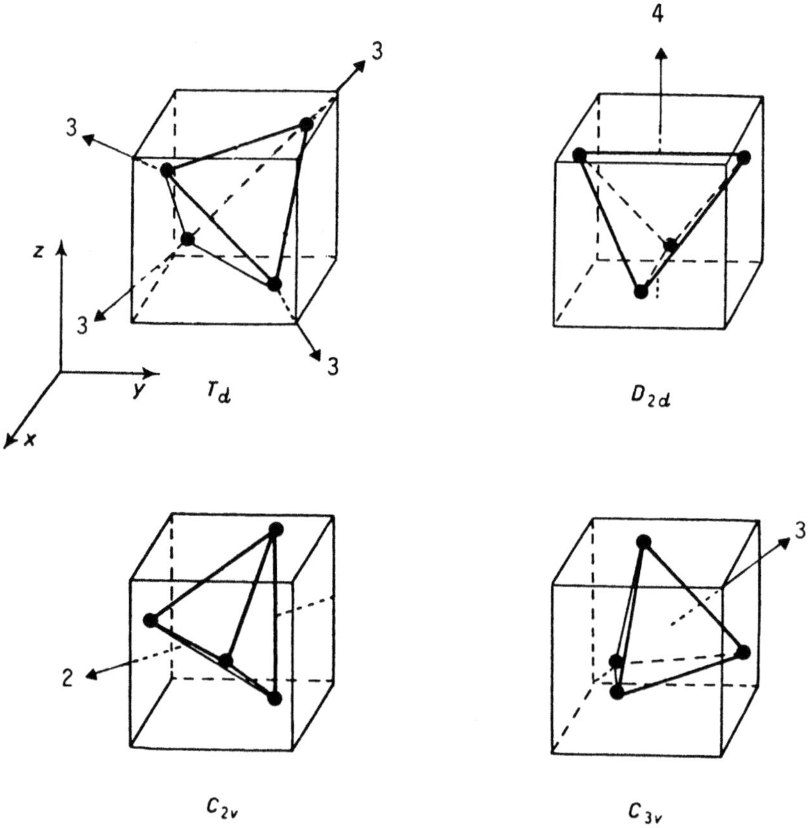

Figure 3. Schematic view of some ideal orientations of a tetrahedral molecule on a cubic site (see text)

In T_d orientations all the symmetry elements of the tetrahedron (24) coincide with those of the cubic site. The number of common symmetry elements decreases from D_{2d} (8) to C_{3v} (6) and then to C_{2v} (4) [14]. In the T_d orientation shown in figure 3, M_1=-1 ; if the tetrahedron is rotated of $\pi/2$ around a fourfold cubic axis we obtain a second T_d orientation for which M_1=+1. There are only two distinct T_d orientations in a cubic field. The values of M_1 are equally distributed between +1 and -1 when the fourfold rotations of the cubic site are performed. In a plastic phase the two T_d orientations are equally distributed and in an ideal configuration $\langle M_1 \rangle$ would be equal to zero but $\langle M_1^2 \rangle = 1$ while $\langle M_2^2 \rangle = \langle M_5^2 \rangle = 0$. Similar considerations can be applied to the other ideal orientations and one can conclude that $\langle M_1^2 \rangle$, $\langle M_2^2 \rangle$, $\langle M_5^2 \rangle$ are order parameters characterizing T_d, D_{2d}, C_{3v} and C_{2v} orientations; their values are given in table I.

The cubic harmonic values are also given in table I. The existence of a T_d order has been found in adamantane where $\langle M_1^2 \rangle$, $\langle M_2^2 \rangle$ and $\langle M_5^2 \rangle$ are respectively equal to 0.69, 0.041 and 0.05 for a temperature T = 309K [14]. The α phase of NH_4Br shows a significant degree of C_{3v} orientations since the values of the M_n order parameters are 0.02, 0.25 and 0.08 [42].

Table I. Order parameters for the ideal orientations of a tetrahedron.

	$\langle M_1^2 \rangle$	$\langle M_2^2 \rangle$	$\langle M_5^2 \rangle$	$\langle {}_1K_4 \rangle$	$\langle {}_1K_6 \rangle$
T_d	1	0	0	- 1.528	2.266
D_{2d}	0	0	0.333	- 0.255	- 1.7
C_{3v}	0.012	0.329	0	0.735	0.961
C_{2v}	0	0.312	0.021	0.7	0.779

3.2. ROTATIONAL MOTION

The evidence of rotational motion can be seen directly by plotting the time evolution of some orientational parameters. On the stereographic plot of figure 4, the orientations of the four threefold axes of an adamantane molecule are followed during a trajectory of 17 ps. This plot shows the librations of the molecule around the ideal T_d orientations when the molecular axes are close to the <111> directions of the crystal. During this trajectory the molecule undergoes a C_4 rotation (i.e. a rotation of $\pi/2$ around a fourfold crystal axis). The initial and final ideal T_d orientations corresponding to this C_4 rotation are illustrated in figure 6 together with the description of the C_3 rotation (i.e. a rotation of $2\pi/3$ around a threefold crystal axis). Of course this approach can only give qualitative results but it is useful to find evidence for the type of motion occuring in a plastic phase.

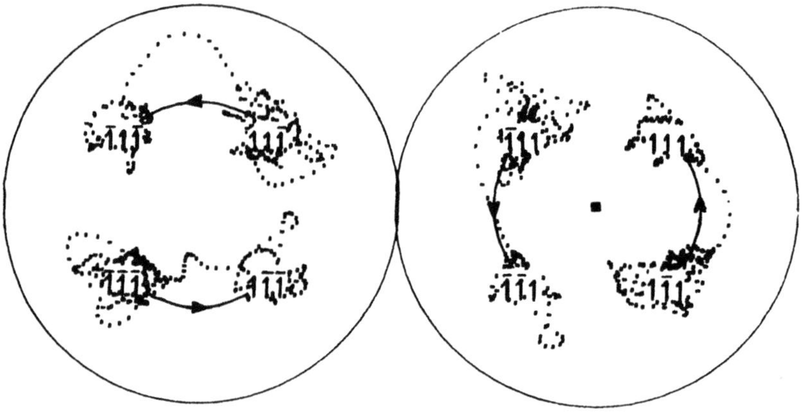

Figure 4. Stereographic plot of the trajectories of the threefold axes of an adamantane molecule. They result from a MD simulation performed at T=400K in the high temperature cubic phase. The molecule undergoes a C_4 rotational jump around the [001] direction (■) of the crystal. The solid line show the ideal path for the reorientation between 2 T_d orientations and the arrows show the direction of rotation.

3.2.1. *Decay of the order parameters*. The order parameters described in § 3.1.2 were used to identify the possible preferred orientations of adamantane, CN^- and NH_4^+ ions.

These parameters do not distinguish between static and dynamical disorder; this is why the study of the autocorrelation functions (e.g. $< M_n(t) (M_n (0)>$) of the order parameters provides useful information on the time scale of the persistence of the orientational order. The relaxation of the order parameters characterizing the orientations of the NH_4^+ ions has been studied in the two high temperature disordered cubic phases α and β of NH_4 Br [42]. The correlation functions associated with the M_1, M_2 and M_5 rotator functions have been computed and are plotted in figure 5. In the α phase the function (n = 2) associated with the C_{3v} order is the longest lived. In the β phase the C_{3v} and D_{2d} (n = 5) curves are almost unchanged when compared to the corresponding ones in the α phase ; the T_d type order (n = 1) is strongly persistent.

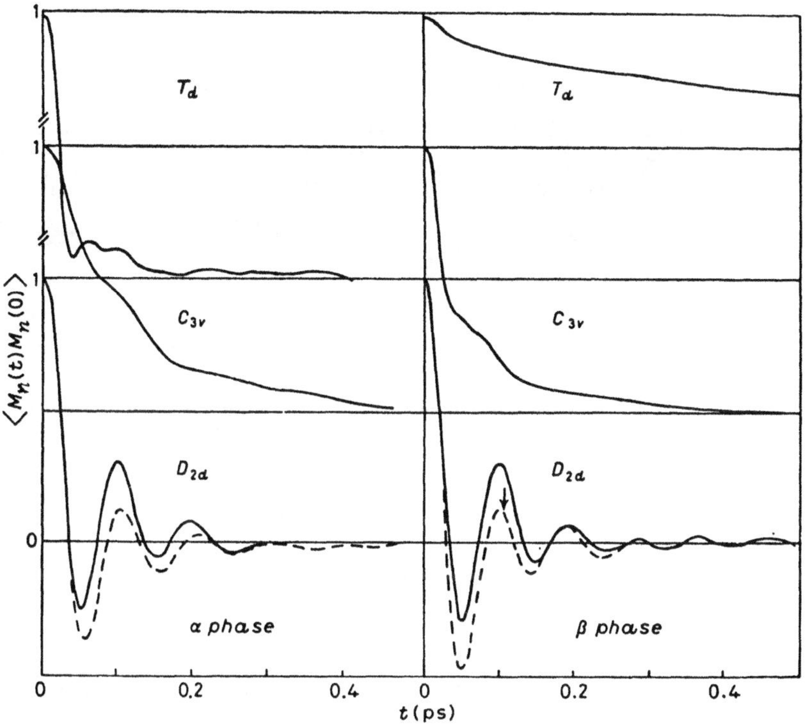

Figure 5. Decay of order parameters [42]. The continuous lines are the relaxation of the order parameters for NH_4^+ in the ODIC phases of NH_4Br . The dashed curves are the velocity autocorrelation functions of the protons. The arrow indicates the experimental value of the oscillation period of the ammonium ions.

When the rotational motion of the molecules is governed by reorientational jumps between preferred orientations, the direct analysis of the time evolution of an order parameter allows us to identify the jumps. In figure 6 two different jumps corresponding to C_3 and C_4 rotations of an adamantane molecule are indicated on a graph where the rotator function M_1 is plotted as a function of time. The analysis of these curves allows us to count the number of rotational jumps and to calculate the corresponding jump frequencies. The results thus obtained for adamantane are compared to experimental ones. The model used to analyse the neutron scattering experiments involves only a C_4 rotational jump mechanism, while MD simulations show that the C_3 and C_4 rotations occur with comparable frequencies [15]. In order to

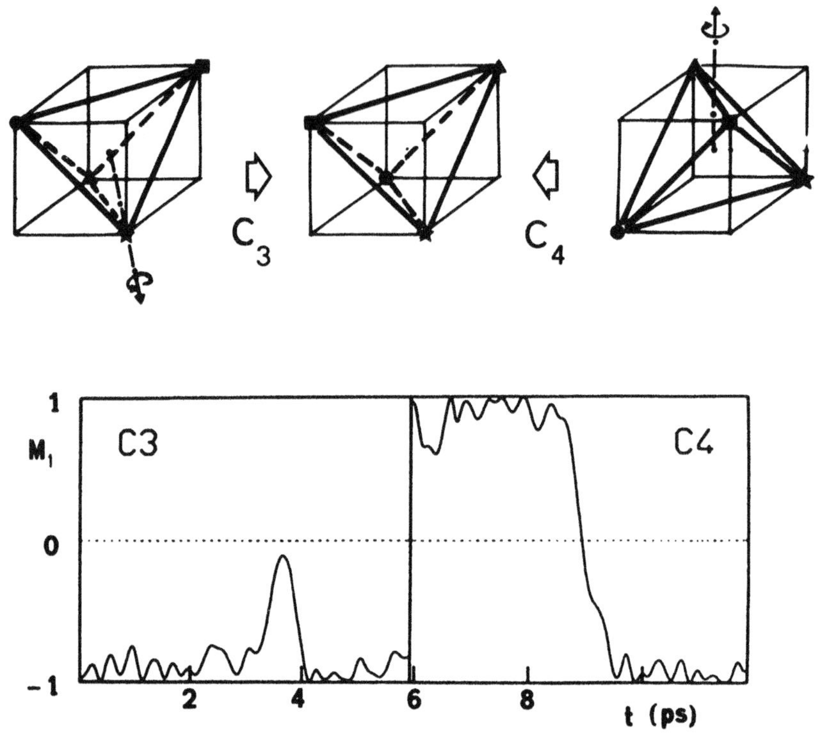

Figure 6. Upper part : schematic view of C_4 and C_3 rotations (see text § 3.2.) for a tetrahedral molecule in a cubic field.

Lower part : Time evolution of the rotator function M_1 showing two reorientional jumps of an adamantane molecule at T=309 K

allow the comparison with the experimental results, accumulated jump frequencies have been calculated at different temperatures. They include the different types of jumps identified during the MD simulations. These frequencies are plotted in figure 7 together with the experimental ones. If one takes into account the restrictive assumptions introduced during the analysis of experimental data, there is a good agreement between calculated and experimental data. Nevertheless it would be interesting to reanalyse the the experimental data using the frequencies calculated for each type of jump.

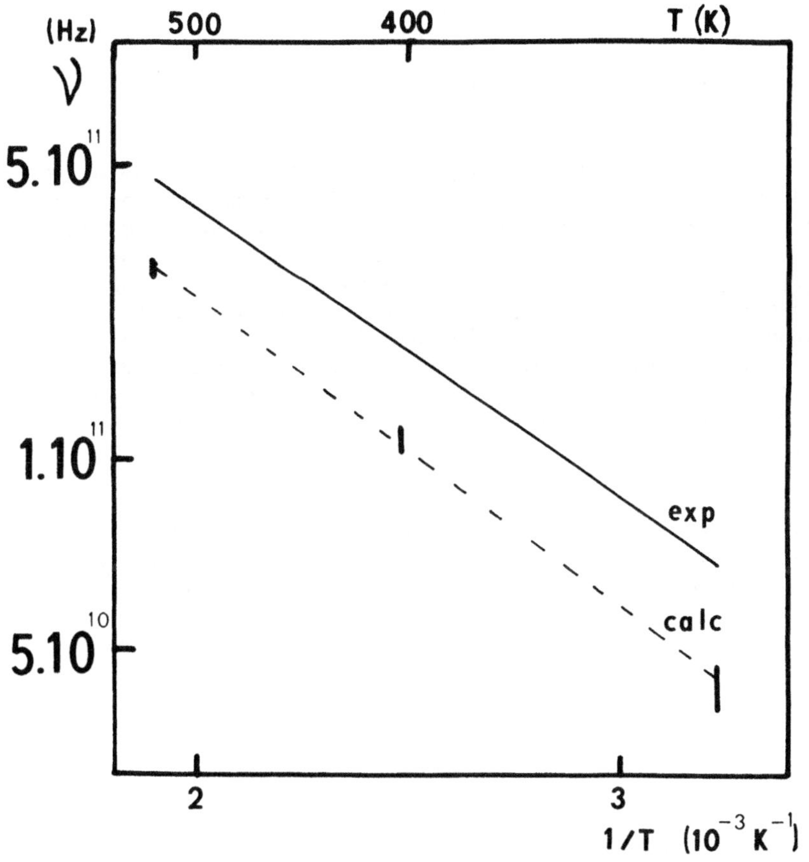

Figure 7. The bars indicate the rotational jump frequencies ν of adamantane molecules calculated at T = 309, 400 and 520 K. The experimental values are reported by plotting $\nu = \nu_0 \exp(-E_R/T)$ (continuous line).

3.3 COLLECTIVE MODES

Neutron scattering techniques are often used to study collective modes in plastic crystals. The quantity measured in neutron scattering experiments is the dynamical structure factor $S(Q, \omega)$ where Q is the scattering vector and $\hbar\omega$ is the neutron energy transfer. $S(Q,\omega)$ is the time Fourier transform of the intermediate scattering function $F(Q,t)$ which is the autocorrelation function of the density operator $\rho(Q,t)$ defined as follows :

$$F(Q,t) = \langle \rho(Q,t)\rho(-Q,t) \rangle \tag{16}$$

$$\rho(Q,t) = \sum_{j\alpha} b_{j\alpha} \exp\,[iQ.r_{j\alpha}(t)] \tag{17}$$

where $b_{j\alpha}$ is the scattering length of the j-th atom of the molecule α and $r_{j\alpha}$ is the vector defining the instantaneous position of the atom $j\alpha$. The intermediate scattering function and the dynamical structure factor have been computed in many MD simulation of plastic crystals [43]. They can be used to construct complete dispersion curves for phonons. A direct comparison between experimental data and simulated values has been realized for SF_6 [44]. The dynamical structure factor is obtained by calculating the Fourier transform of the intermediate scattering function F(Q,t). This function is truncated before the Fourier transform operation in order to avoid spurious peaks due to the oscillatory noise. To allow the comparison with the experimental results, the calculated $S(Q,\omega)$ is convoluted with an experimental resolution function. The values of $S(Q,\omega)$ calculated for several scattering vectors Q_{calc} are plotted in figure 8 together with the experimental data obtained for values of Q_{obs} very close to Q_{calc}. The agreement is qualitatively good. There are some discrepancies in

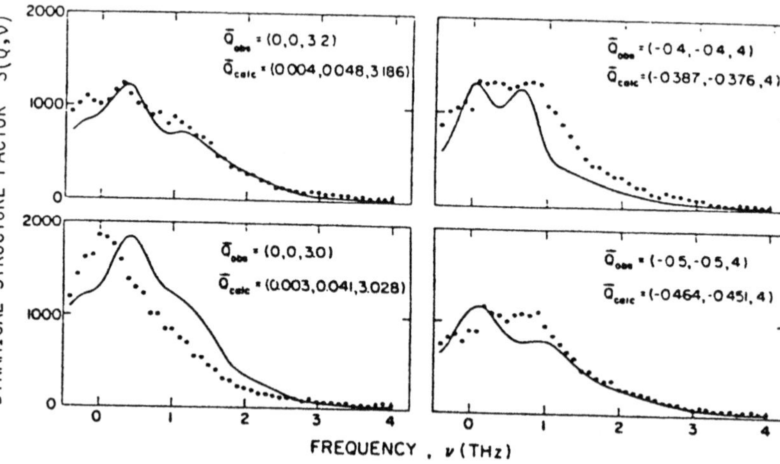

Figure 8. Comparison of calculated (line) and observed dynamical structure factors for SF_6.

peak intensities and frequencies; they may arise from the technical details of the calculations: the resolution function is a simplified one and some artificial structure may result from the transformation of F(Q,t).

3.4. PHASE TRANSITIONS

Plastic crystals undergo on cooling a transition to a low temperature phase which is generally a less symmetric ordered crystal or sometimes a glassy system. The rearrangement of the molecules leads to tetragonal, orthorhombic or even less symmetric structures. The simulation of these phase transitions requires techniques in which the cell can spontaneously transform its shape in response to change of temperature and/or pressure. The extended system approach described in § 2.3 has been successfully employed to study structural transformations in plastic crystals.

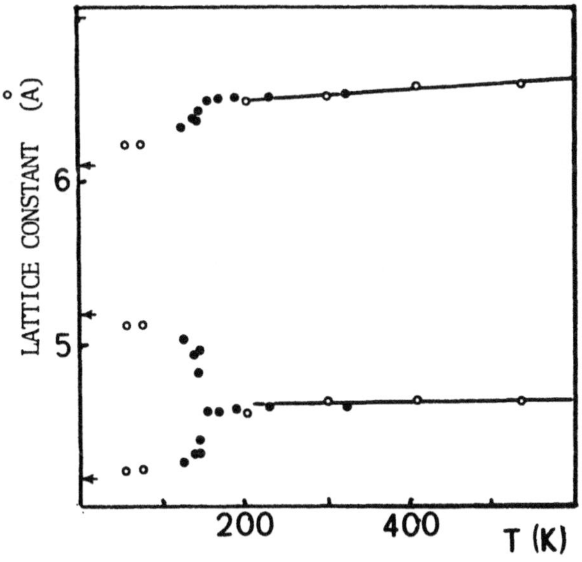

Figure 9. Temperature dependance of the lattice constants of KCN expressed in a body centered tetragonal reference frame. The splitting indicates the transition from cubic to orthorhombic phase. The arrows refer to the experimental values for the orthorhombic lattice parameters.

3.4.1. *Potassium cyanide.*

A detailed study was carried out of the cubic to orthorhombic transition in KCN at constant pressure [10]. The results of the calculation are shown in figure 9. The cooling of the orientationally disordered cubic phase at low pressure yields an orthorhombic crystal with the correct lattice parameters. The CN^- ions align along the [110] direction of the initial cubic lattice with a head to tail disorder. The transition to a fully antiferro ordered structure was not obtained; this is due to the slow reorientation rate in the orthorhombic phase. The transition temperature is about 150K and compares favourably with the experimental one of 168 K.

3.4.2. *Adamantane*. This molecular solid undergoes at 208.6K a transition to a body-centered-tetragonal phase [45]. The simulation performed at constant temperature and pressure by slowly cooling a configuration equilibrated at 300 K failed to reproduce the phase

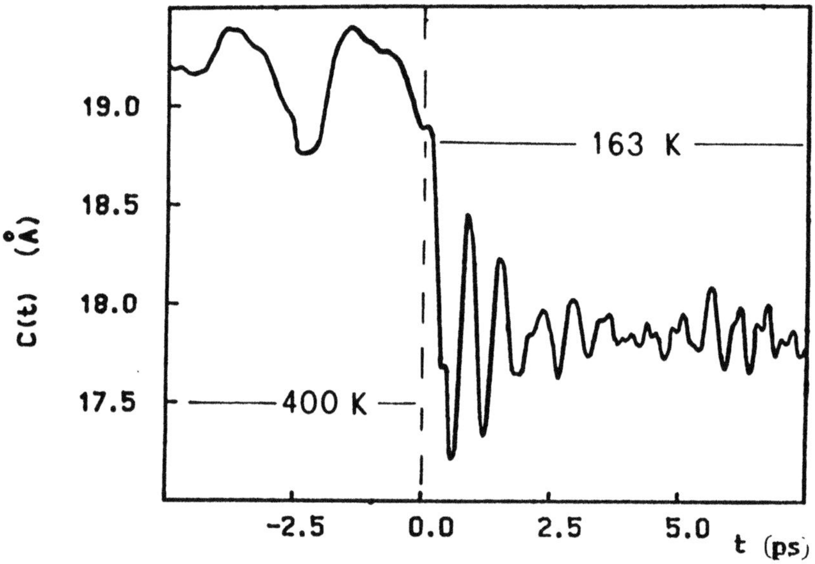

Figure 10. MD Simulation of phase transition in adamantane. Time evolution of the cell parameter C on cooling after reordering ($C=2c_T$ in the tetragonal phase and C=2a in the cubic phase). For T=400 K : cubic phase and T=163 K : tetragonal phase.

transition. The reason for this failure is related to the slow reorientation rate of the molecules at low temperature. Any rearrangement of the molecules is therefore a slow process. Another attempt was then realized: a suitable initial state was prepared by extending to the whole system the short-range orientational order which appears in the low temperature range (i.e. 250-300K) of the plastic phase. The reordered system was prepared from an equilibrium configuration at 400K, quenched at 163K and then allowed to evolve under constant temperature and pressure [25]. The evolution of the cell parameter C (figure 10) during this rapid quenching shows that the transition to the tetragonal phase occurs almost immediately.

The average value of the tetragonal lattice parameters c_T and a_T obtained at 163K are respectively equal to 8.91 and 6.64 Å: this is in rather good agreement with the experimental value of 8.81 and 6.60 Å [46]. The molecules are rotated through angles of ± 7° around the c axis with respect to the initial T_d orientations ; this is also in good agreement with the experimental value of 9°.

4. Conclusions

This brief survey has not intended to give a complete description of the various topics related to the ODIC phases. Some examples of the results obtained by simulation have been reviewed together with their comparison with experimental data when they are available. This is one aspect of the use of simulation techniques. Moreover simulation experiments give a very accurate description of the structural and dynamical properties on the macroscopic scale.

Aknowledgments : It is pleasure to thank O. Hardouin Duparc for kindly providing unpublished results. I am also grateful to M.C. Dolou, O. Dumon, C. Marhic and J.P. Plaut for their technical help.

Appendix

Cubic harmonics [39]

$$_1K_0 = 1,$$

$$_1K_3 = \sqrt{105}\ xyz$$

$$_1K_4 = \frac{5\sqrt{21}}{4}\,[x^4 + y^4 + z^4 - 0.6]$$

$$_1K_6 = \frac{231\sqrt{26}}{8}\,[x^2y^2z^2 + [_1K_4]/22 - \frac{1}{105}]$$

$$_1K_7 = \frac{11\sqrt{1365}}{4}\,xyz\,[x^4 + y^4 + z^4 - \frac{5}{11}]$$

$$_1K_8 = \frac{65\sqrt{561}}{16}\,[x^8 + y^8 + z^8 - 5.6\,[_1K_4] - \frac{210}{143}\,[_1K_6] - \frac{1}{3}]$$

$$_1K_9 = {}_1N_9\,xyz\,[x^2y^2z^2 + \frac{3}{34}\,(x^4 + y^4 + z^4) - \frac{27}{442}]$$

$$_1K_{10} = \frac{11 \times 17 \times 19\sqrt{227.5}}{80}\,[x^{10} + y^{10} + z^{10} - \frac{45}{19}\,[_1K_8] - \frac{126}{17}\,[_1K_6] - \frac{210}{143}\,[_1K_4] - \frac{3}{11}\,]$$

with

$$ {}_1N_9 = 2187.44 , \qquad x^2 + y^2 + z^2 = 1 \qquad \text{and} \qquad [\,{}_1K_\ell\,] = \frac{{}_iK_\ell}{{}_iN_\ell} $$

References

[1] *The Plastically Crystalline State* edited by Sherwood J.N., John Wiley & Sons (1979).
[2] Leadbetter A.J. and Lechner R.E., in Ref. [1] p.285.
[3] Sherwood J.N., in Ref. [1] p.39.
[4] Chadwick A.V. in *Mass Transport in Solids* edited by Bénière F. and Catlow C.R.A., NATO ASI Series B 97, 285 (1983).
[5] Pethrick R.A., in Ref. [1] p.334; Huller A. and Press W. in Ref. [1] p.345.
[6] Press W. in *Single-Particle Rotations in Molecular Crystals* Springer Verlag (1981).
[7] Yvinec Y. and Pick R.M., *J. Physique* 41, 1045 (1980).
[8] Bounds D.G., Klein M.L., Mc Donald I.R., *Phys. Rev. Letters* 46 1682 (1981).
[9] Lynden-Bell R.M., Mc Donald I.R. and Klein M.L., *Mol. Phys.* 48 1093 (1983).
[10] Impey R.W., Sprik M. and Klein M.L., *J. Chem. Phys.* 83 3638 (1985).
[11] Klein M.L., Lévesque D. and Weis J.J., *J. Chem. Phys.* 74 2566 (1981).
[12] Mountain R.D. and Brown A.C., *J. Chem. Phys.* 82 4236 (1985).
[13] Breymann W. Thèse de doctorat de l'Université de Paris 6 (1987).
[14] Meyer M. and Ciccotti G., *Mol. Phys.* 56 1235 (1985).
[15] Meyer M., Marhic C. and Ciccotti G., *Mol. Phys.* 58 723 (1986).
[16] Dove M.T., *J. Phys. C* 19 3325 (1986).
[17] Dove M.T. and Lynden-Bell R.M., *J. Phys. C* 19 3343 (1986).
[18] Dove M.T. and Pawley G.S., *J. Phys. C* 16 5969 (1983).
[19] Dove M.T. and Pawley G.S., *J. Phys. C* 17 6581 (1984).
[20] Yvinec M., *Mol. Cryst. Liq. Cryst.* 109 303 (1984).
[21] Buchet P., Pick R.M., *J. Physique* 48 821 (1987).
[22] Impey R.W., Klein M.L. and Mc Donald I.R., *J. Chem. Phys.* **82** 4690 (1985).
[23] Nose S. and Klein M.L., *Phys. Rev. Letters* 50 1207 (1983).
[24] Ferrario M., Klein M.L. and Mc Donald I.R., *J. Chem. Phys.* 87 4823 (1987).
[25] Ciccotti G., Ferrario M., Memeo E. and Meyer M., *Phys. Rev. Letters* 59 2574 (1987).
[26] Lewis L.J. and Klein M.L., *J. Phys. Chem.* 91 4990 (1987).
[27] Williams D.E., *J. Chem. Phys.* 47 4680 (1968).
[28] Jorgensen W.L., *J. Am. Chem. Soc.* 103 335 (1981); *J. Chem. Phys.* 77 4156 (1982).
[29] Evans D.J. and Murad S., *Mol. Phys.* 34 327 (1977).
[30] Ryckaert J.P., Ciccotti G. and Berendsen H.J.C., *J. Comput.*

Phys. 23 327 (1977).
[31] Ciccotti G., Ferrario M. and Ryckaert J.P., *Mol. Phys.* 47 1253 (1982).
[32] Andersen H.C., *J. Chem. Phys.* 72 2384 (1980).
[33] Parrinello M. and Rahman A., *Phys. Rev. Letters* 45 1196 (1980).
[34] Nosé S., *Mol. Phys.* 52 255 (1984).
[35] Nosé S., *J. Chem. Phys.* 81 511 (1984).
[36] Nosé S. and Klein M.L., *Mol. Phys.* 50 1055 (1983)
[37] Ferrario M. and Ryckaert J.P., *Mol. Phys.* 54 587 (1985)
[38] Meyer M., in *Molecular Dynamics Simulation of Statistical Mechanical systems* ed. Ciccotti G. and Hoover W. G., North Holland p. 477 (1986)
[39] Amoureux J.P. and Bee M., *Acta Cryst.* B26 2636 (1980)
[40] Klein M.L. and Mc Donald I. R., *J. Chem. Phys.* 79 233 (1983)
[41] James H. and Keenan T.A., *J.Chem. Phys.* 31 12 (1959)
[42] Klein M.L., Mc Donald I.R. and Ozaki Y., *J. Chem. Phys.* 79 5579 (1983)
[43] Klein M.L., in *Molecular Dynamics Simulation of Statistical Mechanical Systems* ed. Ciccotti G. and Hoover W.G., North Holland p. 449 (1986)
[44] Dove M.T., Pawley G.S., Dolling G. and Powell M.B., *Mol. Phys.* 57 864 (1986)
[45] Chang S.S. and Westrum E.F., *J. Phys. Chem.* 64 1547 (1960)
[46] Nordmann C.E. and Schmittkons D.L., *Acta Cryst.* 18 764 (1965)

MOLECULAR DYNAMICS SIMULATIONS OF AQUEOUS SYSTEMS

K. Heinzinger

Max-Planck-Institut für Chemie, D-6500 Mainz, Federal Republic of Germany

1. INTRODUCTION

Aqueous electrolyte solutions are especially suited for investigation by computer simulations. Huge amounts of experimental data on such system have been collected over the last hundred years because of their importance in chemistry and biology, but they are still too complicated for treatment by an analytical theory. Therefore, it is no surprise that shortly after the first successful simulation of pure water by Rahman and Stillinger [1] a first paper on an MD simulation of an electrolyte solution was published [2].

It is demonstrated in this review that MD simulations can predict structural and dynamical properties of electrolyte solutions which cannot directly be measured and that they can explain macroscopic properties on a molecular level. The decisive factor for the reliability of the simulation results are the potentials employed in the simulation. We stress that the best test for the reliability of the potentials is the agreement with experimental results as far as they can be deduced unambiguously from measurements.

In the following two sections the structural properties of alkali halide and alkaline earth chloride solution are discussed on the basis of various radial distribution functions, the orientation of the water molecules and their geometrical arrangement in the hydration shells of the ions. The water models employed in these simulations are presented together with the ion-water pair potentials. In subsequent sections the dependence of the hydration numbers on the ion-oxygen distances are reported. The changes in the structure of the solutions with concentration, temperature and pressure are discussed. Dynamical properties of the solutions – e.g. self-diffusion coefficients, spectral densities of hindered translations, librations, and internal vibrations – are derived from the simulations with the help of various autocorrelation functions. They are calculated separately for the three water subsystems – bulk water, hydration water of the cation and of the anion – in order to investigate the single ion effect. Finally, results of simulations of water/metal interfaces are reported.

2. STRUCTURE OF ALKALI HALIDE SOLUTIONS

At the time the simulations of electrolyte solutions were started there was only the ST2 water model available in the literature. It had been proposed by Rahman and Stillinger [1]. They had shown that this model was able to describe many structural and dynamical properties of water sufficiently accurately. It is a four point charge model with the charges arranged tetrahedrally around the oxygen atom (Fig. 1). The positive

C.R.A. Catlow et al. (eds.), Computer Modelling of Fluids Polymers and Solids, 357–394.

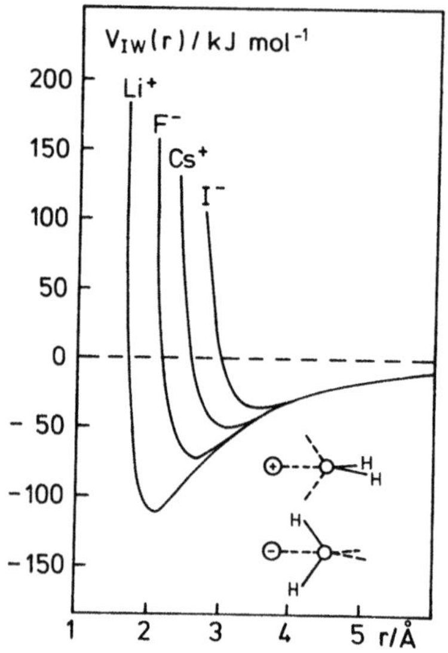

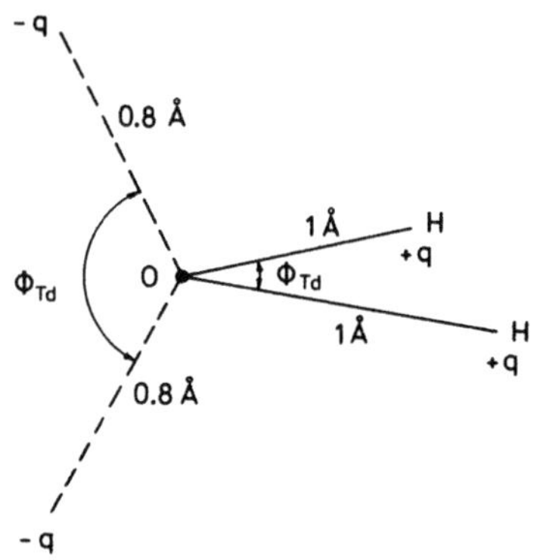

Figure 1: The ST2 point charge model for water. The tetrahedral angle $\Phi_{Td} = 109°28'$ and $q=0.23|e|$.

Figure 2: Ion-water pair potentials as a function of ion-oxygen distance for selected alkali and halide ions and water molecule orientations as shown in the insertion.

charges are located at the hydrogen atom positions at a distance of 1 Å from the oxygen atom, nearly the real distance in the water molecule. The negative charges are located at the other two vertices of the tetrahedron but at a distance of only 0.8 Å from the oxygen atom. The charges were chosen to be 0.23 elementary charges leading to roughly the correct dipole moment of the water molecule. The tetrahedrally arranged point charges render possible the formation of hydrogen bonds in the right directions. The ST2 model is completed by adding a (12;6) Lennard-Jones (LJ) potential, the centre of which is located at the oxygen atom.

In the simulations where the ST2 model is employed, the alkali and halide ions are modelled as LJ spheres with a point charge at the center [3]. With these models for the two kinds of particles – water and ions – it is easy to formulate the effective pair potentials for the six different kinds of interactions: cation-cation, anion-anion, cation-anion, cation-water, anion-water, and water-water. All six pair potentials consist of an LJ term:

$$V_{ij}^{LJ}(r) = 4\varepsilon_{ij}\left[(\sigma_{ij}/r)^{12} - (\sigma_{ij}/r)^{6}\right] \tag{1}$$

where i and j refer either to ions or water molecules, and a Coulomb term, different for water-water, ion-water, and ion-ion interactions, given by:

$$V_{ww}^{C}(r, d_{11}, d_{12}, \ldots) = S_{ww}(r) \cdot q^2 \sum_{\alpha,\beta=1}^{4} (-1)^{\alpha+\beta}/d_{\alpha\beta} \tag{2a}$$

$$V^{C}_{\substack{+w\\(-w)}}(d_{\substack{+1\\(-1)}}, d_{\substack{+2\\(-2)}}, \ldots) = \underset{(+)}{-}\sum_{\alpha=1}^{4}(-1)^{\alpha} q \cdot e/d_{\substack{+\alpha\\(-\alpha)}} \tag{2b}$$

$$V^{C}_{\substack{\pm\pm\\(+-)}}(r) = \underset{(-)}{+} e^2/r. \tag{2c}$$

The switching function, $S_{WW}(r)$, in the Coulomb term of the water pair potential has been introduced by Rahman and Stillinger in order to reduce unrealistic Coulomb forces between very close water molecules. d and r denote distances between point charges and LJ centers, respectively, q the charge in the ST2 model. The sign of the Coulomb term is correct if α and β are chosen to be odd for positive and even for negative charges.

The LJ parameters for the cations are taken from the isoelectronic noble gases [4]. Comparing e.g. Pauling radii, it is obvious that halide ions have a larger ionic radius than the isoelectronic alkali ions. In order to describe all interactions consistently, new LJ parameters had to be determined for the halide ions on the basis of the Pauling radii. The procedure employed is given in Ref. [5]. Knowing the parameters for cation-cation and anion-anion interactions, the parameters for cation-water and anion-water interactions have been determined by applying Kong's combination rules [6]. The results of this procedure are given in Table 1.

TABLE 1: Lennard-Jones parameters in the pair potentials for cation-cation, anion-anion, cation-water, and anion-water interactions. In the ST2 model: $\sigma = 3.10\,\text{Å}$ and $\varepsilon = 0.317\,\text{kJ/mol}$.

Ion	Pauling radius [Å]	σ_{II} [Å]	ε_{II} [kJ mol^{-1}]	σ_{IW} [Å]	ε_{IW} [kJ mol^{-1}]
Li^+	0.60	2.37	0.149	2.77	0.224
Na^+	0.95	2.73	0.358	2.92	0.330
F^-	1.36	4.00	0.050	3.53	0.123
K^+	1.33	3.36	1.120	3.25	0.568
Cl^-	1.81	4.86	0.168	4.02	0.185
Rb^+	1.48	3.57	1.602	3.39	0.641
Br^-	1.95	5.04	0.270	4.16	0.215
Cs^+	1.69	3.92	2.132	3.61	0.662
I^-	2.16	5.40	0.408	4.41	0.228

In Fig. 2 the ion-water pair potentials according to Eq.(2b) and based on the LJ values given in Table 1 are shown for various alkali and halide ions. The depth and the shift in the position of the potential minima show the expected changes with increasing ion size. The curves coincide beyond about 4 Å where only Coulomb interactions remain.

In the MD simulations of the aqueous alkali halide solutions the basic periodic cube contained 200 water molecules, 8 anions, and 8 cations equivalent to 2.2 molal solutions. The sidelength of the cube is determined from the experimental densities and amounts to about 20 Å for all solutions. The classical equations of motion are integrated in time steps of about $2 \cdot 10^{-16}$ s. The simulations extended over about 10 ps. The Ewald summation is employed for the ion-ion interactions, while the ion-water and water-water interactions are cut-off at a distance of about 10 Å, half of the sidelength of the basic cube. This simple cut-off procedure leads to jumps in the potential energy and the forces each time a particle crosses the cut-off sphere and results in trends in the total energy of the system. To overcome this problem, the so-called "shifted force potential"

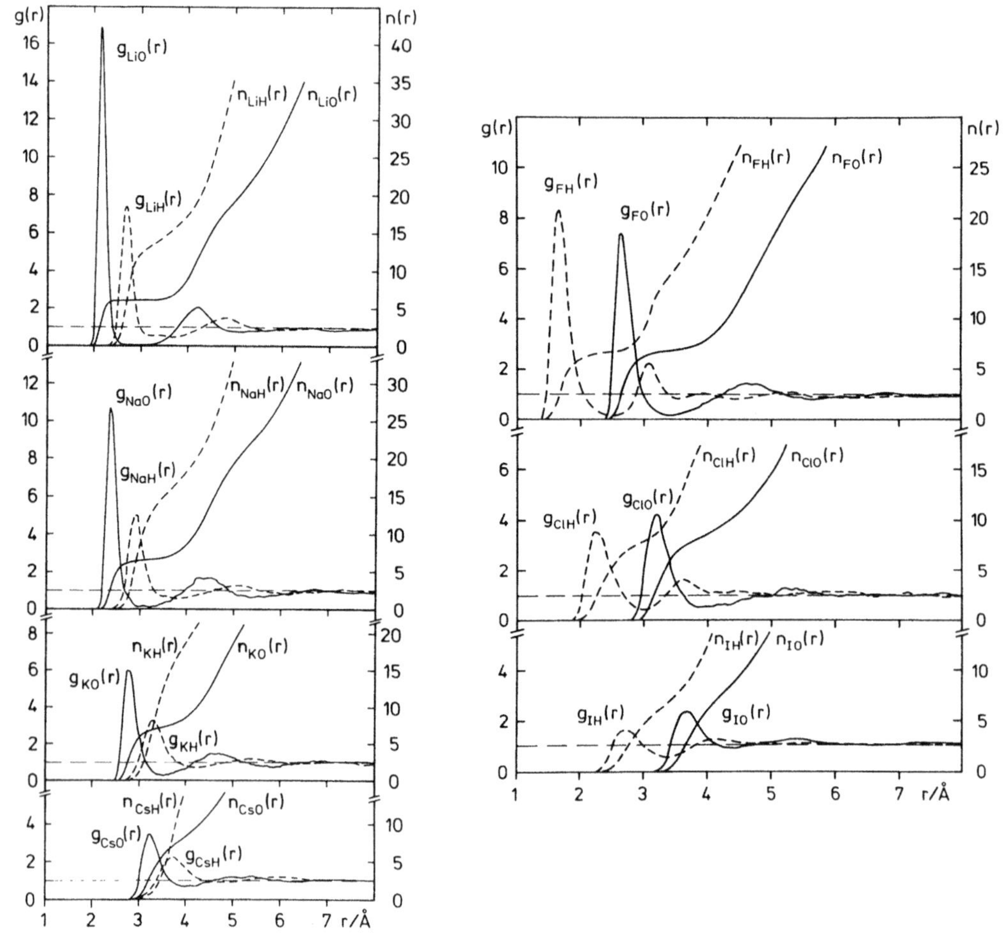

Figure 3: Ion-oxygen (full) and ion-hydrogen (dashed) radial distribution functions and running integration numbers from MD simulations of 2.2 molal LiI [8], $NaClO_4$ [9], KCl [10], NH_4Cl [11], and CsF [12] solutions.

as proposed by Streett *et al.* [7] has been employed. With this procedure, the energy change $\Delta E/E$ during the total simulation was smaller than $5 \cdot 10^{-5}$ in all cases and the average temperature remained constant without rescaling, which is very important for the reliability of the dynamical properties calculated from velocity autocorrelation functions. This modification of the pair potentials seems to be acceptable in view of all the other uncertainties in the choice of the potentials.

The formation of hydration shells is the most obvious effect which occurs when ions are added to water. The first properties derived from MD simulations as far as the hydration shells are concerned are the various radial distribution functions (RDFs), $g_{xy}(r)$. In Fig. 3 the ion-oxygen and ion-hydrogen RDFs are shown for various alkali

TABLE 2: Comparison of characteristic values of the radial distribution functions for various alkali and halide ions. $r_{IO}^{(M)}$ and $r_{IH}^{(M)}$ denote the positions of the first maxima in the ion-oxygen and ion-hydrogen RDF, respectively. The hydration number is defined – according to Eq.(3) – as n at the position of the first minimum, $r_{IO}^{(m)}$.

Ion	$r_{IO}^{(M)}$			$n(r_{IO}^{(m)})$			$r_{IH}^{(M)}$		
	MD[a]	MC[b]	MD[c]	MD[a]	MC[b]	MD[c]	MD[a]	MC[b]	MD[c]
Li^+	2.13	2.10	1.98	6.1	6.0	5.3	2.68	2.70	2.57
Na^+	2.36	2.35	2.29	6.5	6.0	6.0	2.90	2.89	2.95
K^+	2.80	2.71	2.76	7.8	6.3	7.5	3.52	3.19	3.35
Cs^+	3.22	–	–	7.9	–	–	3.72	–	–
F^-	2.64	2.60	2.67	6.8	4.1	5.8	1.65	1.68	1.73
Cl^-	3.22	3.25	3.29	8.2	8.4	7.2	2.24	2.25	2.35
I^-	3.68	–	–	8.7	–	–	3.40	–	–

[a] 2.2 molal solutions of LiI [8], $NaClO_4$ [9], KCl [10], CsF [12], and NH_4Cl [11]
[b] One ion surrounded by 215 water molecules [14]
[c] One ion surrounded by 64 or 125 water molecules [15].

and halide ions. In addition, the corresponding running integration numbers, $n_{xy}(r)$, are drawn. They are defined as:

$$n_{xy}(r) = 4\pi\rho_0 \int_0^r g_{xy}(r')r'^2 dr', \qquad (3)$$

where ρ_0 is the number density of the water molecules.

Fig. 3 shows that with increasing ion size the first hydration shell becomes less pronounced, as expected. The height of the first peak in the ion-oxygen RDF decreases and the first minimum gets filled up. Accordingly, the plateau in n(r) disappears and the end of the first hydration shell becomes less well defined. The existence of a second hydration shell around Li^+ with about twelve water molecules is well established and has been confirmed by X-ray diffraction studies [13]. Even in the case of Na^+ and F^-, the formation of a second shell is indicated in Fig. 3. The positions of the first maxima and first minima in the ion-oxygen RDFs together with the hydration number for various alkali and halide ions are compared in Table 2 with the results of Monte Carlo calculations of Mezei and Beveridge [14] and MD simulations of Impey *et al.* [15] and found to be in good agreement although in these simulations the MCY model for water [16] is employed and the ion-water interactions are derived from ab initio calculations. It is also satisfying to note that in most cases there is good agreement with the neutron diffraction studies with isotopic substitution by Enderby and Neilson [17]. The positions of the first maxima in the ion-oxygen RDFs of the alkali and halide ions almost coincide with the minima of the ion-water pair potentials for the energetically most favorable orientations. This means that the ion-oxygen distances are not influenced by the water-water interactions, but are solely determined by the ion-water interactions.

The best test for the reliability of the potentials as far as the structural properties discussed so far are concerned is the comparison with X-ray and neutron diffraction measurements. Immediately the difficulty arises that in normal diffraction measurements only one structure factor results but in an electrolyte solution there are ten different $g_{xy}(r)$. The decomposition into the partial X-ray structure factors can only be achieved by fitting a model structure factor to the measured one. As in this fitting procedure a large number of parameters are involved it usually does not lead to an unambiguous result. For a detailed discussion see e.g. Ref. [5]. Therefore, the comparison between simulation and X-ray data can only be performed reliably for the total structure factor or the total radial distribution function which is simply its Fourier transform. By such a comparison practically only the first neighbour distances can be checked and these only as far as they are sufficiently different from the O–O first neighbour distance. It can be stated that generally the agreement between simulation and X-ray measurements for such distances is better than ±0.1 Å. The results reported from different laboratories for the same solutions scatter in about the same range.

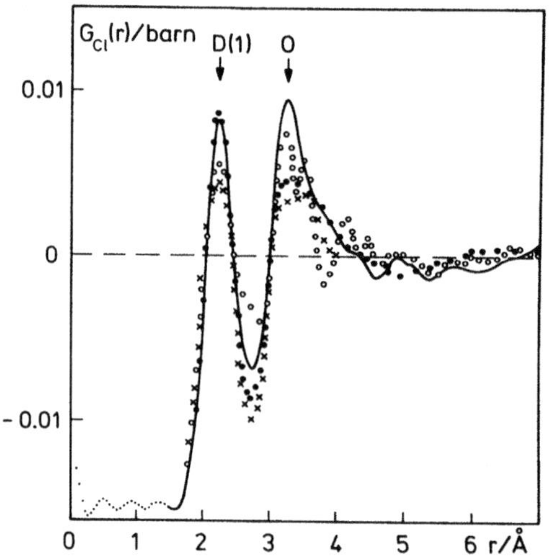

Figure 4: Comparison of the weighted Cl^--water radial distribution function from an MD simulation of a 1.1 molal $MgCl_2$ solution with results from neutron diffraction studies of a 5.32 molal NaCl(∘), a 3.0 molal $NiCl_2$(×), and a 9.95 molal LiCl(•) solution [21].

Important progress in the experimental evaluation of ion-water RDFs was achieved through the neutron diffraction measurements with isotopic substitution of Soper *et al.* [18]. This kind of experiments renders possible the direct determination of separate cation-water and anion-water RDFs. In Fig. 4 the Cl^--water RDF – composed of the weighted $g_{ClO}(r)$ and $g_{ClH}(r)$ – predicted from an MD simulation of a 1.1 molal $MgCl_2$ solution [19] is compared with results from such neutron diffraction measurements of a 5.32 molal NaCl, a 3.0 molal $NiCl_2$ and a 9.95 molal LiCl solution [20]. The experimental curves have been rescaled for concentration while the influence of counterions and concentrations on $g_{ClO}(r)$ and $g_{ClH}(r)$ is not known. Part of the deviations from the overall good agreement has, therefore, to be attributed to this effect [21].

The second peak in Fig. 4 comprises the oxygen and the second hydrogen atom. It is rather broad and its end is not well defined. Unlike the usual definition of the hydration numbers as the integral over the ion-oxygen RDF up to the first minimum, it seems, therefore, to be more appropriate to employ for the evaluation of the experimental hydration number of Cl^- the first peak in the ion-hydrogen RDF. Of course, this will be true for all halide ions. In the comparison of the hydration numbers of Cl^- from experiment and simulation this difference in the definition was not recognized immediately and caused some confusion in early work. Unlike the values given in Table 2 a value of six was deduced from the neutron diffraction studies. The rather soft hydration shell of Cl^- causes the difference resulting from the two definitions. There are water molecules in the first hydration shell which happen to have orientations in such a way that their dipole moment vector points away from Cl^-, reducing in this way the number of nearest hydrogen atoms. From the series of simulations of alkaline earth chloride solutions reported in the next section the number of nearest neighbor hydrogen atoms to Cl^- ranges from 6.0 to 6.7. This number as well as its independence of the counterions is in good agreement with the neutron diffraction measurements in the limits of statistical uncertainty estimated to be about ± 0.5 [20].

A good example of a successful prediction by an MD simulation is the hydration number of Li^+. Up to the end of the 1970's a hydration number of 4 for Li^+ was generally accepted. Then, from simulations of LiCl and LiI solutions a value of 6 was predicted, which was later on confirmed by neutron diffraction studies with isotopic substitution [22].

The orientation of the water molecules in the hydration shells of the ions could be calculated from the distances of the first maxima in the ion-oxygen and ion-hydrogen RDFs. Only in this way can information on the orientation of the water molecules be deduced from diffraction studies. The difficulties connected with this approach have been discussed in detail in Ref. [21]. For the MD simulations these difficulties do not exist as the orientations can be calculated immediately from the data produced.

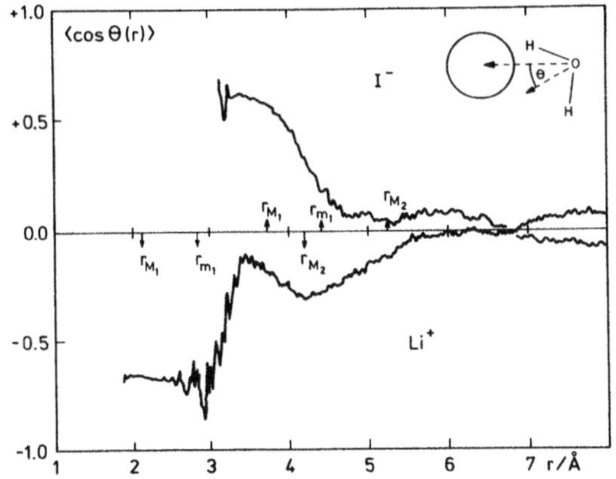

Figure 5: Average value of $\cos\theta$ as a function of distance from the iodide and the lithium ion from an MD simulation of a 2.2 molal LiI solution. r_{M1}, r_{M2} and r_{m1} indicate the position of the first and second maximum and the first minimum in the ion-oxygen RDF, respectively [8].

In Fig. 5 the average value of $\cos\theta$ – where θ is defined in the insertion – is shown as a function of distance from the ion for Li^+ and I^- as example, calculated from an MD simulation of a 2.2 molal LiI solution [8]. In the case of Li^+ $\langle\cos\theta\rangle$ is constant over the whole range of the first peak in $g_{LiO}(r)$ indicating a strong preference for a lone pair orientation of the water molecules in the first hydration shell towards the lithium ion. This result is in agreement with neutron diffraction measurements by Enderby and Neilson [17]. For I^- the average value of$\cos\theta$ decreases over the range of the first peak in $g_{IO}(r)$ significantly, showing that only the water molecules very near to I^- form linear hydrogen bonds with the ion. The preferential orientation of the water molecules decreases beyond the first hydration shell rapidly for both ions, except for Li^+ in the range of the second hydration shell.

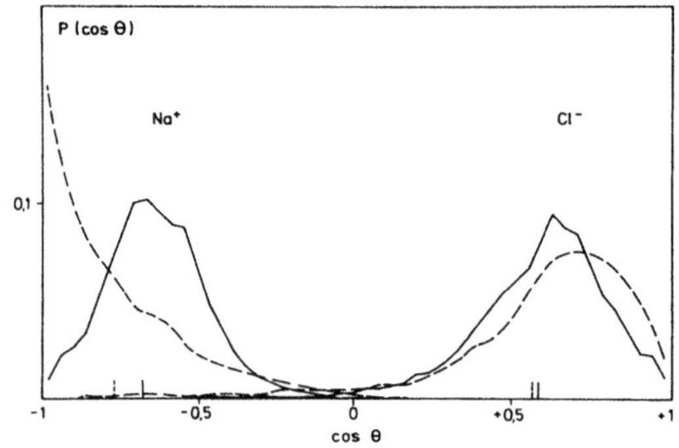

Figure 6: Distribution of $\cos\theta$ for the water molecules in the first hydration shells of Na^+ and Cl^- from simulations of 2.2 molal NaCl solutions with the ST2 model (full) and the CF model (dashed) for water [23].

The orientation of the water molecules in the first hydration shells of the ions is one of the rare cases where the results of the simulations depend upon the water model employed. This is demonstrated in Fig. 6 where the distribution of $\cos\theta$ is shown for the water molecules in the first hydration shells of Na^+ and Cl^- [23]. The result of the simulation with the ST2 model of water and the ion-water interactions according to Table 1 are presented as full lines, while the dashed lines result from simulations with the BJH model (see below) and ion-water pair potentials derived from *ab initio* calculations. In both cases 2.2 molal solutions have been simulated at room temperature. In the case of the chloride ion, both simulations agree in showing a preference for a linear hydrogen bond formation with the anion. For Na^+ the simulation with the BJH model results in a preferentially trigonal orientation of the water molecules while in the ST2 case a preference exists for a lone pair orbital directed towards the cation. The reason for this discrepancy might be that in the ST2 model the directionality of the lone pair orbitals is exaggerated by the negative point charges.

The geometrical arrangement of the water molecules in the hydration shells of the ions is determined by the ion-water as well as by the water-water interactions. It can be calculated from the knowledge of the positions of all particles as a function of time

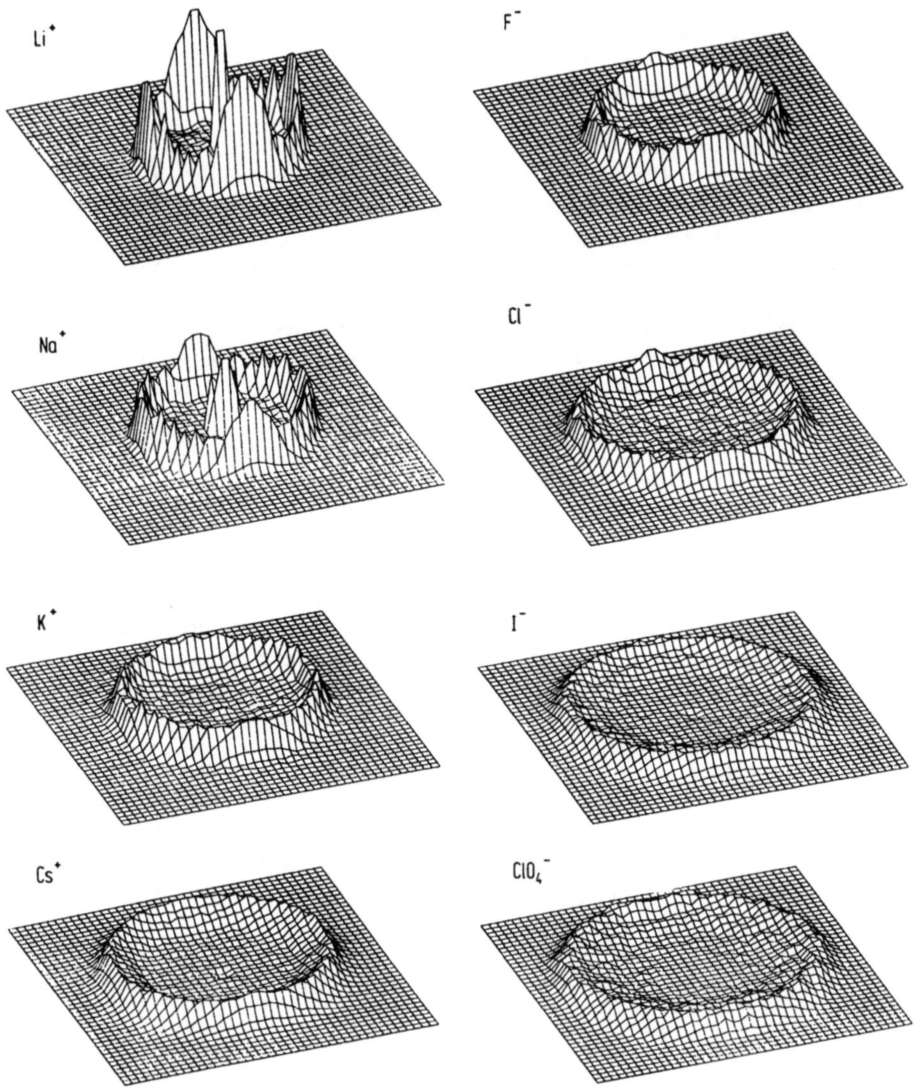

Figure 7: Densities of the projections of the oxygen atom positions of the six and eight nearest neighbor water molecules around a Li^+, Na^+, F^- and Cl^- and a K^+, Cs^+, I^- and ClO_4^-, respectively, onto the xy-plane of a coordinate system as defined in the text. The drawings are calculated from MD simulations of 2.2 molal CsF [12], LiI [8], KCl [10], and $NaClO_4$ [9] solutions.

provided by the MD simulations. In order to achieve this aim, a coordinate system has to be introduced where the ion defines the origin, one oxygen atom of the hydration shell water molecules the z-axis, and a second one the xz-plane. The registration of the water positions (oxygen atoms) in this ion-centered coordinate system at several hundred times spread over the whole simulation run provides the ensemble and time averaged geometrical arrangement.

In Fig. 7 the densities of the projections onto the xy-plane of the oxygen atom positions of the water molecules in the first hydration shells of various singly charged ions are shown in the form of axonometric drawings. The figure shows unambiguously that the six water molecules in the first hydration shell of Li^+ strongly prefer an octahedral arrangement with some distortions around the octahedral positions. With increasing alkali ion size the distributions around the octahedral positions broaden resulting in a uniform distribution of the eight water molecules in the first hydration shells of K^+ and Cs^+. With the exception of F^-, where a small preference for the occupation of octahedral sites is indicated, for all the other anions investigated a uniform distribution results.

3. STRUCTURE OF ALKALINE EARTH CHLORIDE SOLUTIONS

Unlike the simulations of the alkali halide solutions where the ST2 model of water was employed, in the simulations reported in this section the BJH model of water is used [24]. It is based on the latest version of the CF model by Stillinger and Rahman [25]. This model consists of oxygen and hydrogen atoms, bearing partial charges, where the water molecule geometry is solely preserved by an appropriate set of oxygen-hydrogen and hydrogen-hydrogen pair potentials. Thus, the CF model has the advantage that the influence of ions on the intramolecular properties of water can be studied. Unfortunately, it is unable to describe correctly the gas-liquid frequency shifts of the intramolecular vibrations.

Therefore, in the BJH model the total potential is separated into an intermolecular and an intramolecular part. The intermolecular pair potential is an only slightly modified version of the CF model and is given by:

$$\begin{aligned} V_{OO}(r) = {} & 604.6/r + 111889/r^{8.86} - 1.045\{\exp[-4(r-3.4)^2] \\ & + \exp[-1.5(r-4.5)^2]\} \end{aligned} \tag{4a}$$

$$\begin{aligned} V_{OH}(r) = {} & -302.2/r + 26.07/r^{9.2} - 41.79/\{1+\exp[40(r-1.05)]\} \\ & - 16.74/\{1-\exp[5.439(r-2.2)]\} \end{aligned} \tag{4b}$$

$$V_{HH}(r) = 151.1/r + 418.33/\{1+\exp[29.9(r-1.968)]\} \tag{4c}$$

The intramolecular part is based on the water potential in the formulation of Carney, Curtiss and Langhoff [26]:

$$V_{intra} = \sum L_{ij}\rho_i\rho_j + \sum L_{ijk}\rho_i\rho_j\rho_k + \sum L_{ijkl}\rho_i\rho_j\rho_k\rho_l, \tag{5}$$

with $\rho_1 = (r_1 - r_e)/r_1$, $\rho_2 = (r_2 - r_e)/r_2$ and $\rho_3 = \alpha - \alpha_e = \Delta\alpha$, where r_1, r_2 and α are the instantaneous O–H bond lengths and H–O–H angle; the quantities r_e and α_e are the corresponding equilibrium values ($r_e = 0.9572\,Å$, $\alpha_e = 104.52°$). The finally adopted parameter set is given in Table 3.

TABLE 3: Potential constants used for the intramolecular part of the BJH model for water in units of kJ/mol [24]. The notations are according to Eq. (5).

$\rho_1\rho_2(\rho_1+\rho_2)$	−55.7272	$(\rho_1^2+\rho_2^2)$	2332.27
$\rho_1^2+\rho_2^2)\Delta\alpha$	237.696	$\rho_1\rho_2$	−55.7272
$(\rho_1^4+\rho_2^4)$	5383.67	$(\rho_1+\rho_2)\Delta\alpha$	126.242
$\rho_1\rho_2(\rho_1^2+\rho_2^2)$	−55.7272	$(\Delta\alpha)^2$	209.860
$(\rho_1^3+\rho_2^3)\Delta\alpha$	349.151	$(\rho_1^3+\rho_2^3)$	−4522.52

In the simulations of the 1.1 molal alkaline earth chloride solutions reported in this section the ion-oxygen and ion-hydrogen pair potentials (as well as the ion-ion ones which are not very important in the case of the dilute solution) are derived from *ab initio* calculations. They are given in Table 4. The basic periodic cube contained again 200 water molecules, 4 cations and 8 chloride ions. All other details of the simulations are quite similar to the ones given in the preceding section and can be found also in the original communications cited.

TABLE 4: Ion-oxygen and ion-hydrogen pair potentials employed in the simulations of the alkaline earth chloride solutions. Energies are given in kJ/mol and distances in Å.

$$V_{BeO}(r) = -1832/r - 1387/r^2 + 10.35 \cdot 10^4 \exp(-3.78r)$$
$$V_{BeH}(r) = 916.5/r + 62.45/r^2 + 29.81 \exp(-0.0859r)$$
$$V_{MgO}(r) = -1832/r - 890.7/r^2 + 26.95 \cdot 10^4 \exp(-4.08r)$$
$$V_{MgH}(r) = 916.5/r + 82.02/r^2 + 73.83 \exp(-0.349r)$$
$$V_{CaO}(r) = -1832/r - 1572/r^2 + 25.97 \cdot 10^4 \exp(-3.49r)$$
$$V_{CaH}(r) = 916.5/r + 626/r^2 + 12.02 \cdot 10^4 \exp(-6.79r)$$
$$V_{SrO}(r) = -1832/r - 1192/r^2 + 17.64 \cdot 10^4 \exp(-3.11r)$$
$$V_{SrH}(r) = 916.5/r + 352.3/r^2 + 14.45 \exp(-0.158r)$$
$$V_{ClO}(r) = 916.5/r - 111.3/r^2 + 37.96 \cdot 10^4 \exp(-3.21r)$$
$$V_{ClH}(r) = -458.2/r + 18.90 \cdot 10^{25} \exp(-34r)$$

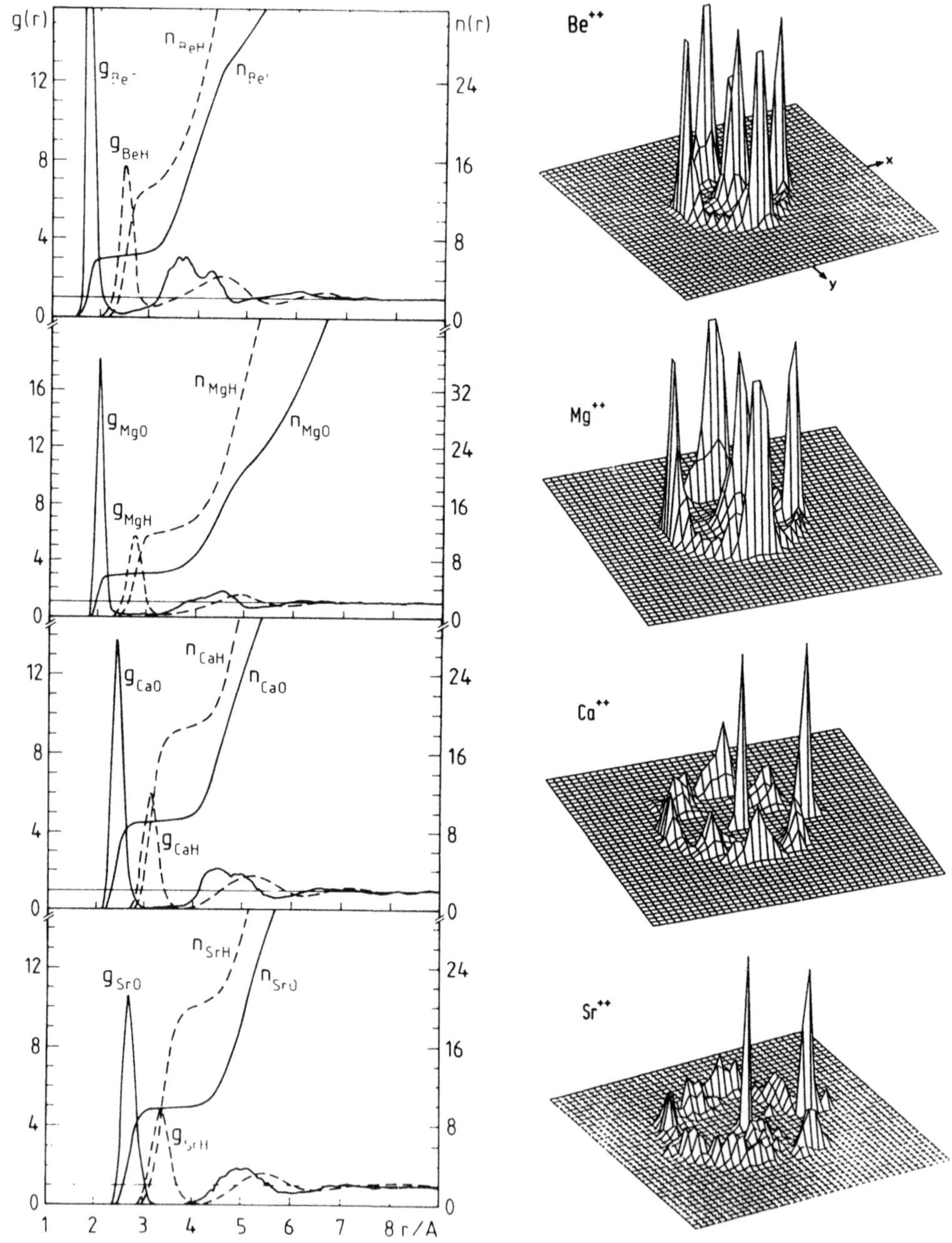

Figure 8: Ion-oxygen and ion-hydrogen radial distribution functions together with the running integration numbers(left) and projection densities of the oxygen atom positions of the water molecules in the first hydration shell of the cations onto the xy-plane of a coordinate system as defined in the text (right) from MD simulations of 1.1 molal $BeCl_2$ [27], $MgCl_2$ [19], $CaCl_2$ [28], and $SrCl_2$ [29] solutions.

The ion-oxygen and ion-hydrogen RDFs together with the running integration numbers as defined by Eq. (3) are shown in the left column of Fig. 8. Unlike the alkali ion series, for all four alkaline earth ions the first hydration shells are well defined ($g_{IonO}(r)$ becomes practically zero beyond the first peak) and a second shell exists. Because of the double charge the first peaks are higher and sharper than for the alkali ions. The positions of the first maxima and first minima of the ion-oxygen RDFs together with the hydration numbers are given in Table 5.

TABLE 5: Positions of the first maxima and the first minima of the ion-oxygen radial distribution functions for various alkaline earth ions together with their hydration numbers, which are defined as $n(r^{min}_{IonO})$.

	Be^{2+}	Mg^{2+}	Ca^{2+}	Sr^{2+}
r^{max}_{IonO}	1.75	2.00	2.39	2.63
r^{min}_{IonO}	2.3	2.3	3.0	3.2
$n_{IonO}(r_{min})$	(6.0)	6.0	9.2	9.8
Reference	[27]	[19]	[28]	[29]

The structural properties of the solutions as calculated from the simulation are compared in detail with diffraction measurements in the original papers. An overall good agreement has been found in the limits of uncertainty except as far as the hydration number of Be^{2+} is concerned. Here the X-ray diffraction measurements seem to indicate a value of 4 although this result is not too reliable at the low concentration because of the small scattering power of Be^{2+}. On the other hand the simulation was performed under the assumption of pair additivity of the ion-water interaction potentials, an assumption which seems to be questionable for the small Be^{2+} in the light of more recent simulations. The Ca^{2+}-O first neighbor distance and the hydration number of Ca^{2+} are also in good agreement with neutron diffraction studies with isotopic substitution [30].

In the right column of Fig. 8 the projection densities are depicted for the alkaline earth ion series. They show a quite different behavior from the alkali ions (Fig. 7). The six water molecules in the first hydration shells of Be^{2+} and Mg^{2+} are arranged octahedrally with practically no distortion and only a narrow distribution around the octahedral sites. This structure is in keeping with the results for Li^+ considering that Li^+ is of similar size but has only half of the charge of Mg^{2+}. The hydration number of Ca^{2+} has been found to be about 9.5. This 50% increase is a consequence of the larger Ca^{2+}-O nearest neighbor distance (2.40 Å) compared with the Mg^{2+}-O one (2.00 Å). Unlike the alkali ion series, where with increasing ion size a preference for octahedral arrangement persists until a unfiform distribution is reached, the change in the structure of the hydration shell on going from Mg^{2+} to Ca^{2+} is dramatic. Although a well-defined structure exists for Ca^{2-} the hydration shell shows no regular symmetry [31]. The hydration shell structure of Sr^{2+} is similar to that of Ca^{2+} but with a strong tendency to a more uniform distribution of the water molecules. The significantly stronger ion-water interactions in the case of the alkaline earth ions must be responsible for the different size dependence of the hydration shell structure of the two ion series.

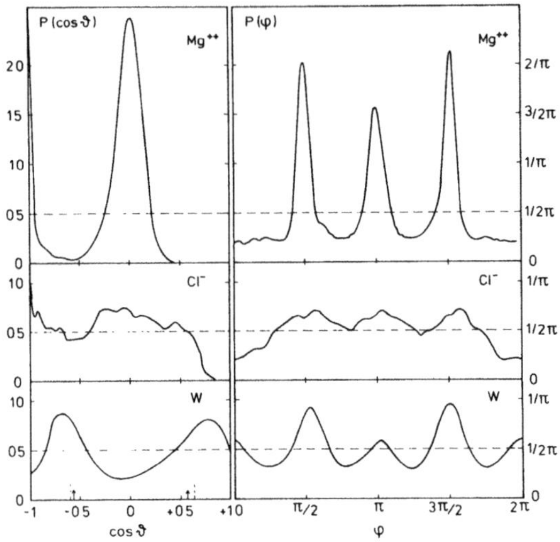

Figure 9: Probability densities for the polar angle ϑ and the azimuth angle φ of the oxygen atom positions of the six nearest neighbour water molecules around an Mg^{2+} and a Cl^- and for the four nearest neighbour water molecules around a central one in the coordinate systems defined in the text. The dashed lines indicate uniform distributions [32].

The qualitative picture of the geometrical arrangement of the water molecules in the first hydration shells of the ions as presented in Figs. 7 and 8 and around a central water molecule can be quantified by the calculation of the probability densities of the polar angle ϑ and the azimuth angle φ – denoted by $P(\cos\vartheta)$ and $P(\varphi)$ – in the coordinate systems as defined above. As an example the results for Mg^{2+}, Cl^- and bulk water are presented in Fig. 9. The dashed lines indicate uniform distributions [32]. In the case of the ions, $\vartheta = 0$ by definition for the oxygen atom which defines the positive z-axis. Therefore, this water molecule is not included in $P(\cos\vartheta)$. The φ-value of this water molecule is not defined and the second molecule defining the xz-plane has $\varphi = 0$ by definition. Therefore, these two water molecules are not included in $P(\varphi)$. In the case of water these exclusions are not necessary because of the inherent coordinate system employed. Fig. 9 confirms quantitatively the conclusions drawn above. There exists a well defined octahedral arrangement of the six water molecules in the first hydration shell of Mg^{2+}. There is only a very slight indication for a preference of the occupation of octahedral sites in the hydration shell of Cl^-. There is a strong preference for the occupation of tetrahedral sites by the four nearest neighbours around a central water molecule with a narrower distribution in the hydrogen atom direction when compared with the lone pair directions.

The distributions of $\cos\theta$ for the water molecules in the first hydration shells of Sr^{2+} and Cl^- are shown in Fig. 10, where θ is defined in the insertion. For Sr^{2+} a strong preference exists for a trigonal orientation while the water molecules in the first hydration

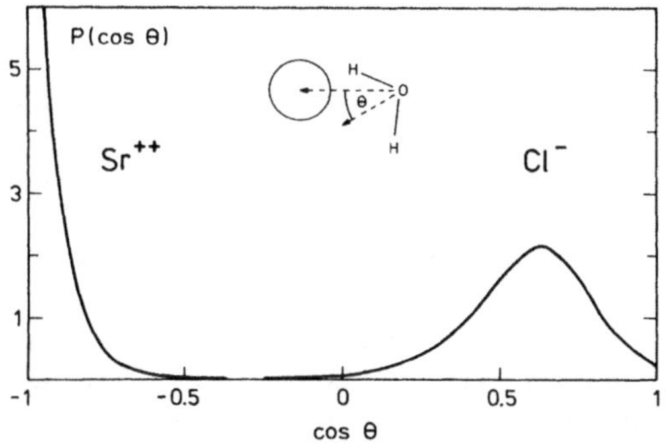

Figure 10: Distributions of $\cos\theta$ for the water molecules in the first hydration shells of Sr^{2+} and Cl^-. θ is defined in the insertion [29].

shell of Cl^- form preferentially linear hydrogen bonds. The $\cos\theta$ distributions are only slightly sharper for the hydration shells of the other (smaller) alkaline earth ions. The distribution for Cl^- does not significantly depend on the counterions.

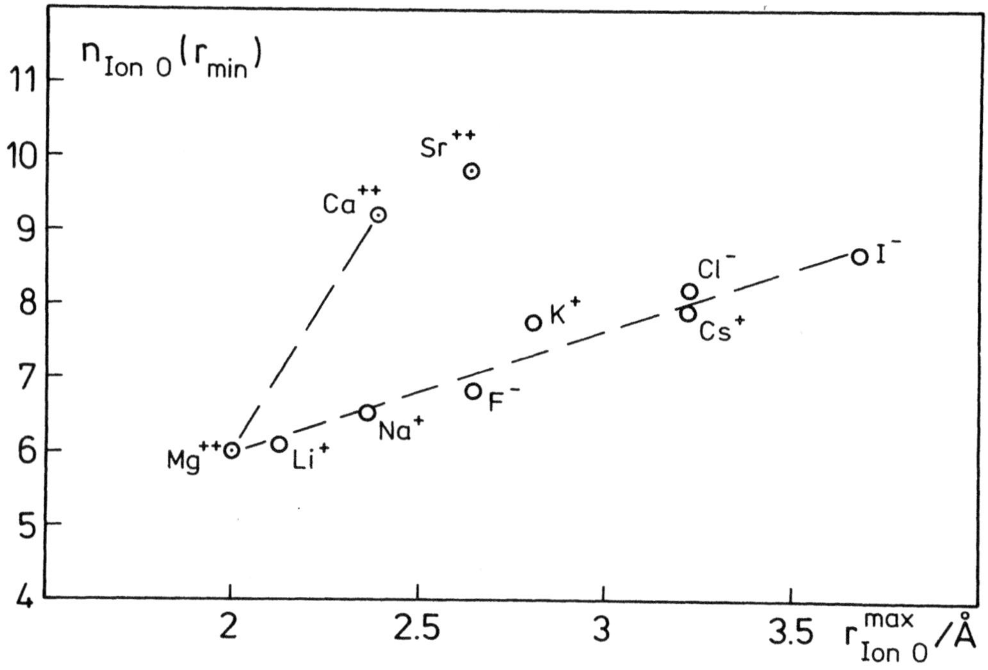

Figure 11: Hydration numbers for various ions as a function of the position of the first maximum in the ion-oxygen radial distribution function. The dashed lines are drawn to guide the eye.

4. HYDRATION NUMBERS

The hydration numbers as defined by the running integration numbers at the first minimum of the ion-oxygen RDF, $n_{IonO}(r_{min})$, are drawn in Fig. 11 as a function of the positions of the first maxima in the ion-oxygen RDFs, r^{max}_{IonO}. A linear dependence results for all singly charged ions. It demonstrates that water-water interactions mainly determine the hydration numbers as for hard spheres an increase with the square of r^{max}_{IonO} would be expected. In the case of the alkaline earth ions the hydration numbers for Mg^{2+}, Ca^{2+} and Sr^{2+} have been established rather reliably from MD simulations. A conclusion on their distance dependence can only be reached after the hydration number of Be^{2+} is known. Even a value of 4 for Be^{2+} would lead to linear relationship again only if quite large errors are assumed for the hydration numbers of Ca^{2+} and Sr^{2+}.

5. THE EFFECT OF PRESSURE AND TEMPERATURE ON STRUCTURAL PROPERTIES

The effect of pressure on the hydration shells of Na^+ and Cl^- has been studied with simulations of a 2.2 molal NaCl solution at densities of $1.0792\,g/cm^3$ and $1.3067\,g/cm^3$ [33], corresponding to normal pressure and a pressure of 10 kbar at room temperature, respectively. The relevant RDFs are presented in Fig. 12. The position and width of the first peak in $g_{NaO}(r)$ remain unchanged, while its height decreases slightly with increasing pressure. The coordination number increases by about 0.5. This small increase in the coordination number (the density is 21% higher) as well as the slight decrease in peak height seem to indicate a certain loss of structure of the Na-hydration sphere with increasing pressure. The chloride hydration seems to undergo larger changes with pressure increase, suggesting that the hydration layer of this ion is

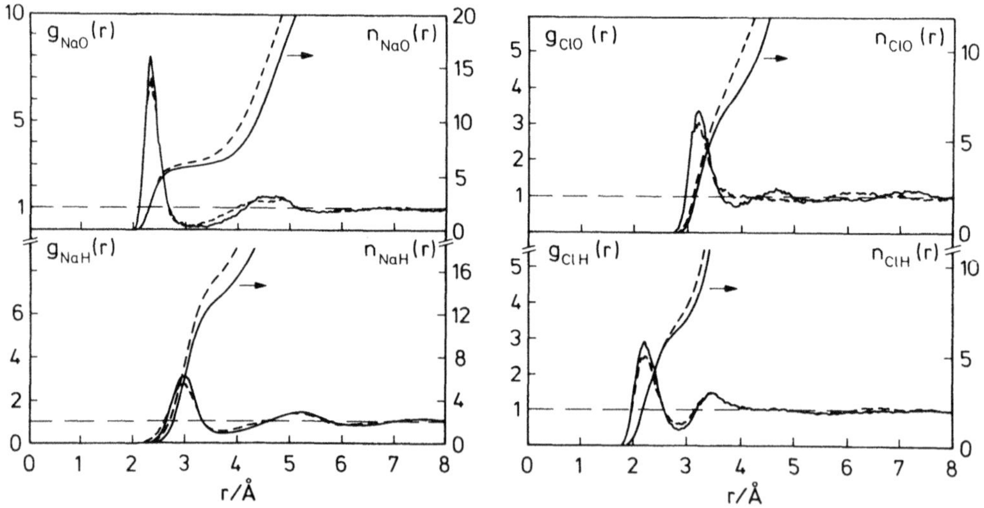

Figure 12: Ion-oxygen and ion-hydrogen radial distribution functions and running integration numbers for high pressure (dashed) and normal pressure (full) 2.2 molal NaCl solutions [33].

even less developed under pressure than it is normally, though the preference for forming linear hydrogen bonds between Cl^- and water seems not to be affected. Unlike Na^+ the hydration number of Cl^- increases by almost two.

The effect of elevated temperature and pressure on the ion-oxygen RDFs has been investigated for a 0.55 molal LiI solution by MD simulations at 308 K and 508 K and constant density, where the higher temperature corresponds to a pressure of about 3 kbar [34]. It can be seen from Fig. 13 that the increase in temperature (and pressure) reduces significantly the height of the first peak in $g_{LiO}(r)$ and broadens it (full line). At the same time the gap between first and second hydration shell begins to get filled up. In this way the number of water molecules (six) in the first hydration shell of Li^+ remains constant, as can be seen from $n_{LiO}(r)$ at $r = 3$ Å. The second hydration shell of Li^+ almost disappears at high temperature and pressure. Accordingly, there is a significant difference between the two curves for $n_{LiO}(r)$ at $r = 4.5$ Å which slowly disappears with increasing distance. In the case of the iodide ion the first hydration shell is, even at normal temperature and pressure, not well pronounced (lower part of Fig. 13). Consequently not much change can be observed. Only a slight smearing out of the first peak results from the temperature (and pressure) increase.

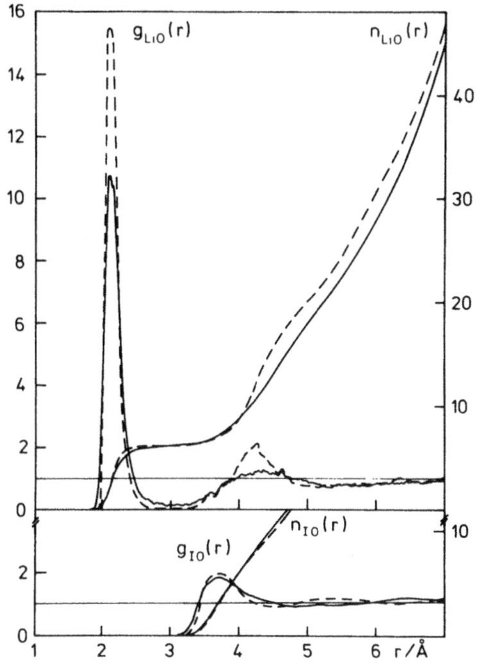

Figure 13: Ion-oxygen radial distribution functions and running integration numbers for a 0.55 molal LiI solution at 508 K (full) and 308 K (dashed) with the same density of 1.05 g/cm^3. At the high temperature the pressure is about 3 kbar [34].

While the pressure increase does not significantly influence the hydration shells, it has a strong effect on the hydrogen bond structure of water. The projection densities for the eight nearest neighbour water molecules around a central one and separately for the first four and the second four onto the xy-plane of a coordinate system as defined in the insertion are shown in Fig. 14, calculated from MD simulations of pure BJH water at 1 bar and 22 kbar [35]. Eight is about the number of nearest neighbours in the high pressure (HP) case. It can be clearly observed that there is a decrease in the preference for the occupation of tetrahedral positions in the HP water. The neighbours five to eight in HP water are seen to be closer to the central molecule than the second neighbours in the normal pressure (NP) water, but they do not show any preference for occupying tetrahedral positions. Accordingly, it seems reasonable to conclude that the first coordination sphere of HP water consists of neighbours of two different types.

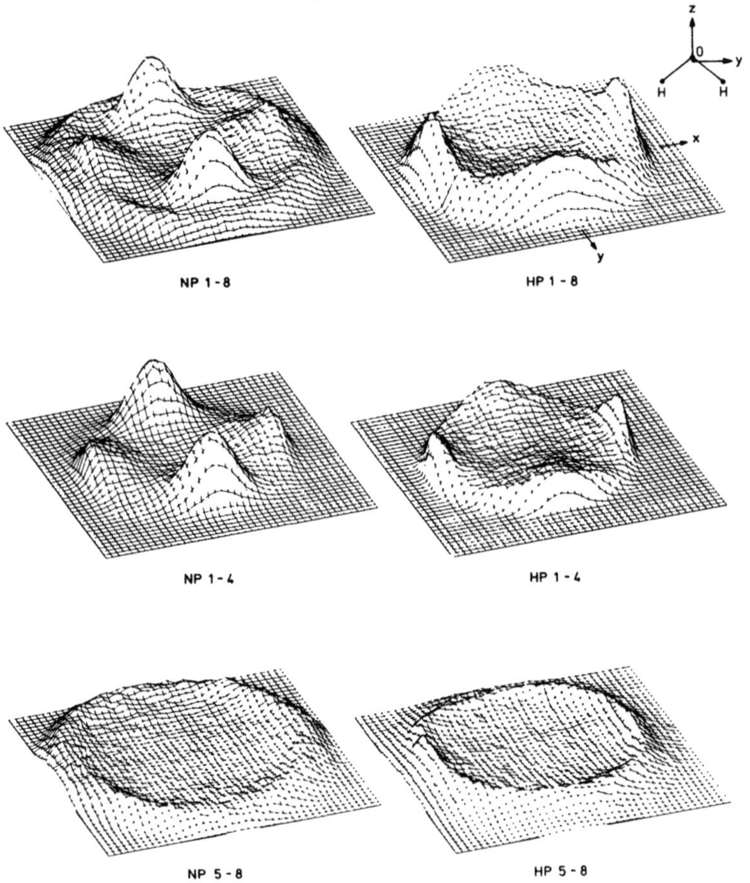

Figure 14: Densities of the projections of the oxygen atom positions of the eight – and separately for the first and second four – nearest neighbor water molecules around a central one onto the xy-plane of a coordinate system as defined in the insertion, calculated from MD simulation of pure BJH water at 1 bar (NP, left) and 22 kbar (HP, right) [35].

Similar conclusions have been drawn from the results of two others MD simulations in which an interpenetration of the subsections of the random hydrogen bond network has been proposed [36,37].

6. ION-ION POTENTIALS AND SOLUTION STRUCTURE AT HIGH CONCENTRATION

In the simulations of the 2.2 molal alkali halide solutions the ion-ion interactions were described by (12;6) LJ potentials while they were derived from *ab initio* calculations in the case of the 1.1 molal alkaline earth chloride solutions. At these relatively low concentrations the choice of the ion-ion potentials does not seem to have any significant influence on the results of the simulations. At high concentrations this statement is no longer correct [38]. Two simulations have been performed of a 18.5 molal LiCl solution both with the BJH model for water and the ion-water potential as derived from *ab initio* calculations. In one simulation the (12;6) LJ potential with the parameters as given in Table 1 was employed for ion-ion interactions [39] while in the second one a Born-Mayer-Huggins (BMH) type of potential was used which is usually written in the form:

$$V_{ij} = z_i z_j e^2 r^{-1} + A_{ij} b \exp\left[(\sigma_i + \sigma_j - r)/\rho\right] - C_{ij} r^{-6} - D_{ij} r^{-8}, \qquad (6)$$

with the parameters taken from Tosi and Fumi [40] and Okada *et al.* [41]. The BMH potential had proved to lead to a good description of the structural and dynamical properties of various molten salts.

Both ion-ion pair potentials are shown in Fig. 15. The differences between the two kinds of potential are very large at small distances and disappear beyond 4 Å, where even in $V_{ClCl}(r)$ only the Coulomb interaction remains. The changes in the ion-ion potential from LJ to BMH do not only cause changes in the ion-ion RDFs but also in the ion-water RDFs as can be seen from Fig. 16. The shift of the first maximum in $g_{ClO}(r)$ by about 0.1 Å to larger distances is the most important change and leads to a much better agreement with X-ray measurements [38]. The hydration number of Li^+ decreases by about 10% to 3.6 and the number of nearest neighbor hydrogen atoms around Cl^- by about 1 to a value of 4.

The smallest change of the ion-ion RDFs occurs in $g_{LiLi}(r)$ (Fig. 16). The increase of contact ion pairs can be read from $n_{LiCl}(r)$ at the position of the first minimum in $g_{LiCl}(r)$ at around 3.5 Å and amounts to an average of about 1.5 Li^+ in contact with a Cl^-. The shift of the position of the first peak in $g_{LiCl}(r)$ by about 0.5 Å follows directly from the difference in the position of the potential minima for the BMH and the LJ potential as shown in Fig. 15. As a consequence of the much softer BMH potential relative to the LJ one the first peak in $g_{ClCl}(r)$ between 4.5–5 Å practically disappears and the Cl^-–Cl^- RDF extends down to 3 Å, the distance of closest approach of two Cl^-. The dotted line results from the application of the second order difference method to neutron diffraction studies of a 14.9 molal aqueous LiCl solution with chloride isotope substitution by Copestake et al. [42]. With the change of the ion-ion potential from LJ to BMH type the discrepancy between simulation and experiment is reduced but still remains significant. It seems very surprising that the experiment gives such a far ranging pronounced order of the chloride ions in the solution. It is very difficult to imagine a

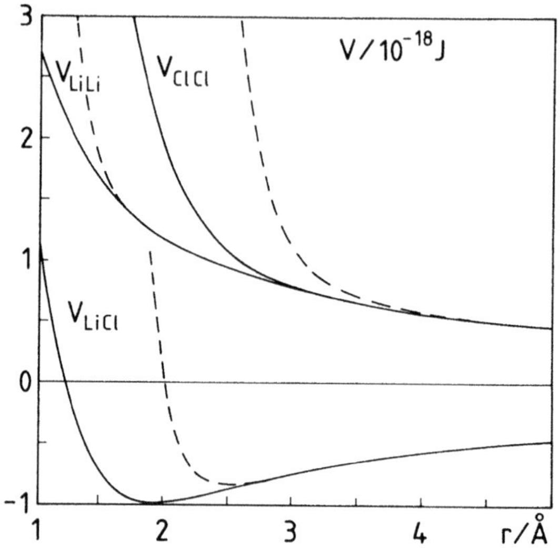

Figure 15: Ion-pair potentials of Born-Mayer-Huggins type with parameters taken from [40,41] (full) and of (12;6) Lennard-Jones type with an elementary charge in the centre and parameters taken from Table 1 (dashed).

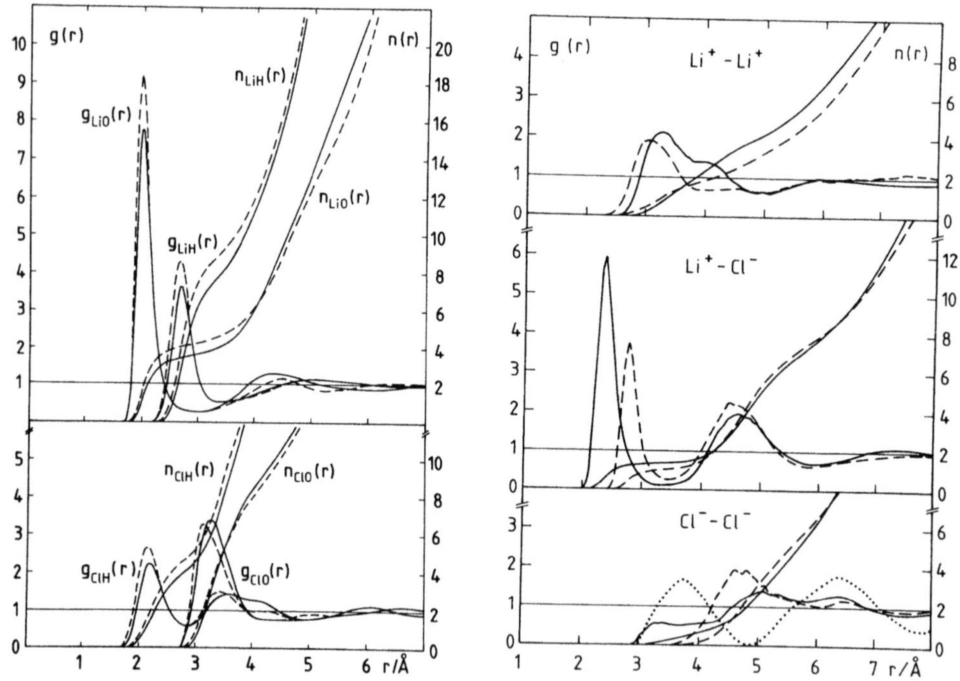

Figure 16: Ion-oxygen and ion-hydrogen (left) and ion-ion (right) radial distribution functions and running integration numbers from MD simulations of an 18.5 molal LiCl solution with a Born-Mayer-Huggins (full) and a Lennard-Jones (dashed) potential for the ion-ion interactions. The dotted line gives the result of a neutron diffraction study of a 14.9 molal LiCl solution with chloride isotope substitution [42].

consistent set of ion-ion potentials which would lead to agreement with the experimental Cl^-–Cl^- RDF.

7. SELF-DIFFUSION COEFFICIENTS

The self-diffusion coefficients can be determined from the simulation either by the mean square displacement:

$$D = \lim_{t\to\infty} \langle [\vec{R}(t) - \vec{R}(0)]^2 \rangle / 6t \tag{7}$$

or – as done here – through the velocity autocorrelation functions with the help of the Green-Kubo relation:

$$D = \lim_{t\to\infty} \frac{1}{3} \int_0^t \langle \vec{v}(0) \cdot \vec{v}(t') \rangle dt'. \tag{8}$$

The averages have been calculated according to:

$$\langle \vec{v}(0) \cdot \vec{v}(t) \rangle = \frac{1}{N_T N} \sum_{i=1}^{N_T} \sum_{j=1}^{N} \vec{v}_j(t_i) \cdot \vec{v}_j(t_i + t) \tag{9}$$

where N denotes the number of particles, N_T the number of time averages and $\vec{v}_j(t)$ the velocity of particle j at time t.

As an example, the normalized velocity autocorrelation functions for the ions and the water molecules in the LiI solution are shown in Fig. 17 [43]. For a better comparison of their different time dependences they are drawn only up to 0.5 ps. The strongly different curves for I^- and Li^+ reflect the translational motions of the heavy iodide ion, slowly changing its velocity, and the almost vibrational motion of the lithium ion in the cage of the six firmly attached water molecules of its first hydration shell. Obviously, the decay times are different for the water molecules and the two ions. Therefore, the upper limit of integral (8) has been chosen differently to be 1.5 ps for water, 0.6 ps for Li^+, and 2.2 ps for I^-, times above which the correlation functions vanish, apart from statistical noise. The statistical quality of the results depends on the length of the correlation time. The shorter the correlation time the more time averages can be employed for a given simulation length. The results are compared in Table 6 with the experimentally determined self-diffusion coefficients [44]. The errors are estimated from the variations of the integrals at the upper integration limit, while the uncertainties of the experimental values should be about 10%. There is good agreement within the limits of error.

TABLE 6: Self-diffusion coefficients of the ions and solvent water (D_w) in a 2.2 molal LiI solution from the MD simulation and from experiments at 305 K in units of 10^{-5} cm^2/s [43].

	D_w	D_{Li^+}	D_{I^-}
MD	2.48 ± 0.06	0.7 ± 0.3	1.40 ± 0.15
Exp.	2.35	1.0	1.47

Having shown that the pair potentials employed in the simulation lead to good agreement with the experimental results, it can be expected also that the other dynamical properties which are not directly accessible by experiment can be calculated from the simulation with a high degree of reliability. In order to investigate how the single ions contribute to the difference in the self-diffusion coefficients between pure water and solvent water, the velocity autocorrelation functions have been calculated separately for the three water subsystems – water in the first hydration layer of Li^+, I^-, and not in any first hydration layer (called "bulk" water) – and are shown also in Fig. 17. The resulting self-diffusion coefficients are given in Table 7. The following conclusions can be drawn: i) In all three water subsystems the diffusional motion is reduced relative to pure water. ii) The difference in the self-diffusion coefficients between bulk water and pure water contributes more than half to the change which occurs on going from pure water to the 2.2 molal LiI solution. iii) The self-diffusion coefficient of Li^+ and its hydration water is smaller by more than a factor of two than that of bulk water. iiii) The self-diffusion coefficient of I^- is much smaller than that of its hydration water and of bulk water, which are approximately equal.

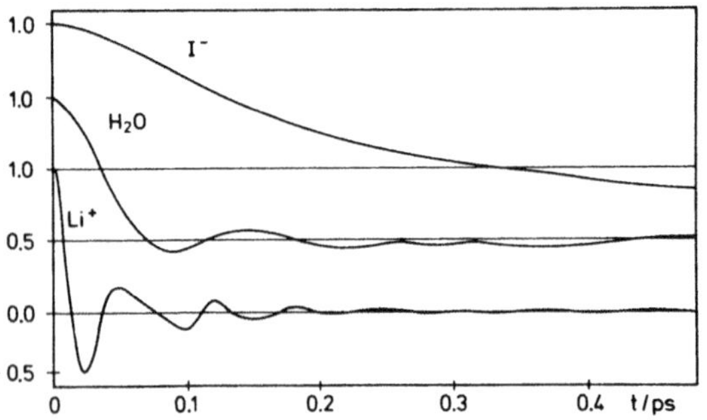

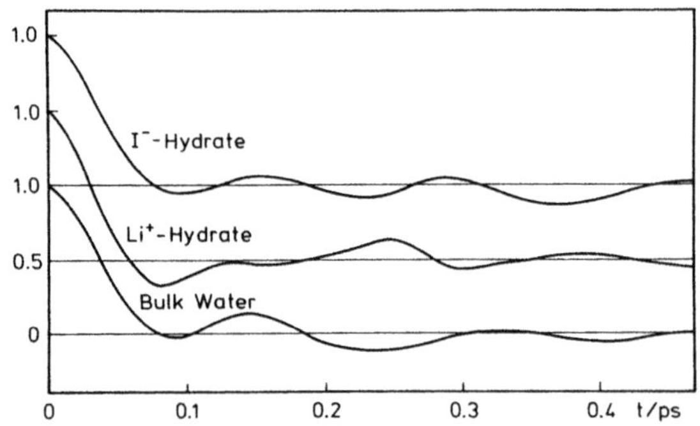

Figure 17: Normalized velocity autocorrelation functions $\langle \vec{v}(0)\cdot\vec{v}(t)\rangle \;/\; \langle \vec{v}(0)^2\rangle$ for water molecules, lithium ions, and iodide ions (top) and separately for bulk water, hydration water of Li^+ and I^- (bottom) in a 2.2 molal LiI solution [43].

The difference in the self-diffusion coefficients between bulk water in the 2.2 molal LiI solution and pure water can easily be explained on the basis of the structure of the solution as discussed above. All water molecules outside the first hydration

TABLE 7: Self-diffusion coefficients in units of 10^{-5} cm^2/s for bulk water (D^b), hydration water of Li^+(D^+) and of I^-(D^-) from an MD simulation of a 2.2 molal LiI solution at 305 K. D^0 denotes pure water [43]

i	D^i	D^i/D^0
b	2.85 ± 0.08	0.84
+	1.33 ± 0.10	0.39
−	2.67 ± 0.10	0.78

shells ($n^+ = 6$; $n^- = 8$) are counted as bulk water but it can be seen from Fig. 3 that twelve additional water molecules form a second hydration shell around Li^+. Therefore, the dynamical properties of all water molecules in the solution are influenced by the ions and are expected to differ from pure water significantly even at this concentration. Occasionally, models are developed to explain experimental data which use the assumption that bulk water has the same properties as pure water [45]. This example shows, that even at moderate concentrations, this models may lead to erroneous conclusions.

One would expect in the hydration shell of I^- a larger self-diffusion coefficient than in pure water. However, just the opposite has been found (Table 7). As some of these water molecules belong to the second and some even to the first hydration shell of Li^+, it is easy to understand that the simulation leads to very similar results for the water molecules in the hydration shell of I^- and bulk water. Actually, bulk water should be written with quotation marks, as most of the water molecules belong to the second hydration sphere of Li^+. It is important to note that the ST2 model employed in the simulation is not responsible for the above reversed effect. From an MD simulation where a single ion is surrounded by 215 ST2 water molecules, it has been shown by Geiger [46] that for the hydration sphere of a negatively charged ion the self-diffusion coefficient is increased.

The self-diffusion coefficient for the water molecules in the first hydration shell of Li^+ is somewhat larger than for the ion itself but smaller than for pure water by a factor of 2.5 (Table 7). This results from the long residence time of the water molecules in the hydration shell. The simulation shows that more than 50% of the water molecules remain in the first hydration sphere of Li^+ over the whole simulation time of 10 ps.

The pressure dependence of the self-diffusion coefficients of the ions and of solvent water have been calculated from the two simulations of the 2.2 molal NaCl solution at normal pressure (NP) and a pressure of 10 kbar (HP), the structural properties of which are reported above [33]. The results are compared with the available experimental data in Table 8. Measurements of the pressure dependence of the self-diffusion coefficients of the ions in aqueous solutions are not known.

The self-diffusion coefficient of the solvent water has been corrected for the different temperatures of the NP- and HP-MD simulations by using the temperature dependence of D of pure water [47]. The comparison of the self-diffusion coefficients obtained from the MD simulations with the experimental results shows that the simulation leads to smaller values in all cases. Considering the uncertainties in the experimental results for D_{Na^+}

TABLE 8: Self-diffusion coefficients of the ions (D_{Na^+}, D_{Cl^-}) and water (D_W) from MD simulations at normal pressure (NP, 1 bar) and high pressure (HP, 10 kbar) and from experiments at room temperature in units of 10^{-5} cm^2/s.

		D_W	D_{Na^+}	D_{Cl^-}
MD	NP(299 K)	1.60(1.83[a])	0.82	1.08
	HP(303 K)	1.51	0.55	1.25
Exp.	NP(298 K)	1.92[48]	1.12[49]	1.60[49]
			1.30[50]	1.85[50]

[a] Extrapolated to 303 K according to [47].

and D_{Cl^-} as indicated by the values reported from different laboratories, the differences between simulation and experiments seem to be small enough to justify the conclusion that the changes in the self-diffusion coefficients for water and the ions with increasing pressure as calculated from the simulations are reliable – at least qualitatively. Table 8 shows that the 21% density increase is accompanied by about 20% decrease in the self-diffusion coefficient of solvent water D_W. (Separate calculations lead to similar changes for all three water subsystems – bulk water hydration water of Na^+ and of Cl^-). In pure water at 77°C a 38% density increase yielded an about 35% decrease in the self-diffusion coefficient [51], which suggests that at high pressures the increased steric hindrances between the molecules plays an important role in controlling the diffusional motion of water molecules both in the pure water and in the solution. In the NP solution D_{Na^+} is found to be significantly smaller than D_{Cl^-}, in good agreement with the experimental results [49, 50]. This can be rationalized as in the case of the LiI solution [52] by assuming that the Na^+ is moving through the solution with its hydration shell attached while the Cl^- diffuses essentially without its hydration shell. It is also remarkable that D_{Na^+} decreases while D_{Cl^-} increases with increasing pressure. It is interesting to note that in a recent study of the effect of pressure on the conductance of KCl in heavy water Nakahara *et al.* [53] found that the residual friction coefficient for K^+ slightly increases with pressure while that for Cl^- decreases.

The self-diffusion coefficients of Li^+, Cl^- and H_2O have been calculated from the simulation of the 18.5 molal LiCl solution – introduced above – to be 0.20(0.21), 0.17(0.21), and 0.35(0.41) in units of 10^{-5} cm^2/s, respectively. They are by up to one order of magnitude smaller than those at low concentration and compare favorably with the experimental results as given in parentheses. The result indicates that the pair potentials employed are able to describe also the dynamical properties of highly concentrated solutions reliably.

8. HINDERED TRANSLATIONAL MOTIONS

The spectral densities $f(\omega)$ of the hindered translational motions will be discussed here again for a 2.2 molal LiI solution. They have been calculated by Fourier transformation:

$$f(\omega) = \int_0^\infty \frac{\langle \vec{v}(0) \cdot \vec{v}(t) \rangle}{\langle \vec{v}(0)^2 \rangle} \cos(\omega t) dt \qquad (10)$$

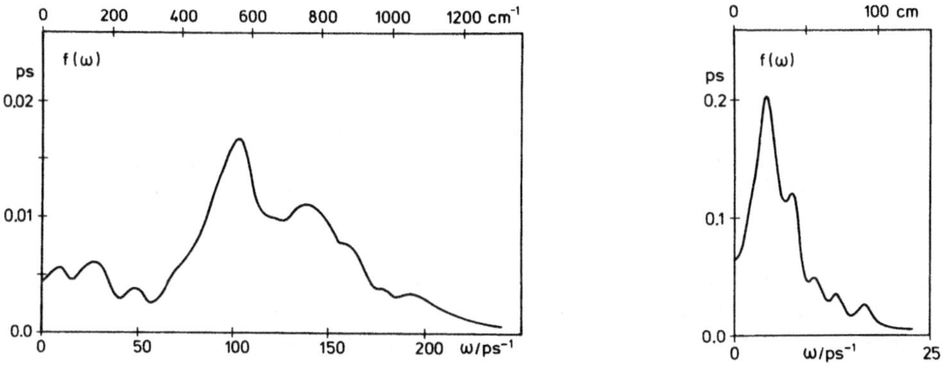

Figure 18: Spectral densities of the hindered translations of the lithium ion (left) and the iodide ion (right) from an MD simulation of a 2.2 molal LiI solution [52].

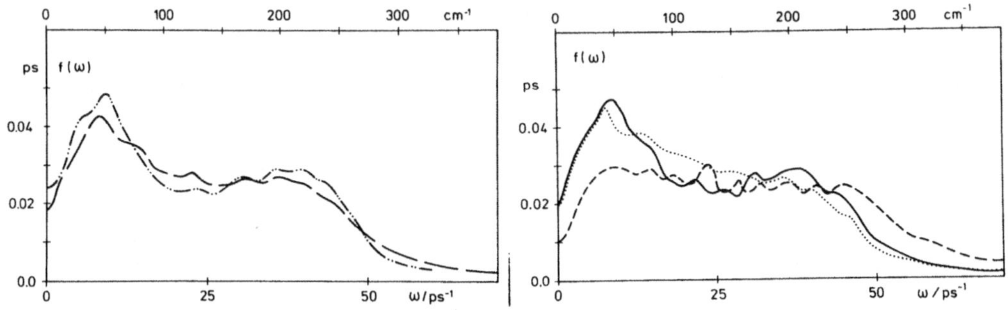

Figure 19: Spectral densities of the hindered translations of the water molecules from MD simulations of pure water and of a 2.2 molal LiI solution. Left: pure water (—··—) total water of the solution (— —). Right: bulk water (———) hydration water of Li^+ (— — —) and of I^- (·······) [52].

from the normalized velocity autocorrelation functions for the two ions and the total water of the solution as well as separately for the three water subsystems as shown in Fig. 17. The results are presented in Figs. 18 and 19.

In Fig. 18(left) $f(\omega)$ is shown for the lithium ion. There is a main peak at 560 cm^{-1} and a second one at 760 cm^{-1}. The high frequencies of the hindered translations are expected from the velocity autocorrelation function (Fig. 17) and fall in the range of the librational band of the water molecules in the first hydration shell of Li^+. As the spectral densities of the librations around the dipole moment direction are similar to those of the hindered translations it can be inferred that the translational motion of the Li^+ is coupled with the librational motions of the strongly attached six water molecules in its first hydration shell.

The spectral densities of the hindered translations of I^-, shown in Fig. 18 (right), are very different from those of Li^+. There is a main peak at 23 cm^{-1} followed by a strong decrease, with $f(\omega)$ being practically zero beyond 100 cm^{-1}. This is in accordance with the picture of the heavy iodide ion moving without its hydration shell and being hindered occasionally by collisions with neighboring water molecules. This picture of the translational motions of Li^+ and I^- is supported by the self-diffusion coefficients for the two ions (Table 6). The bare I^- translates significantly faster than the Li^+ with its six water molecules in the first hydration shell. The movement of Li^+ is further hindered by strong hydrogen bonds between the water molecules in its first and second hydration shell [13].

The hindered translational motions of the water molecules are shown in Fig. 19 for pure water and the total water in the 2.2 molal LiI solution. The curve for pure water roughly resembles that obtained from spectroscopic investigations, where the first peak at 50 cm^{-1} has been assigned to hydrogen bond bending (O–O–O flexing) motions and the broad peak around 175 cm^{-1} to O–O stretching motions. The curve for the total water of the solution shows a slight decrease around 50 and 200 cm^{-1} and a slight increase around 100 cm^{-1}, a tendency similar to the changes observed by Stillinger and Rahman for an increase in temperature of pure water [1].

In order to understand the surprisingly small difference between pure water and total water of the solution in more detail, the spectral densities of the hindered translations have been calculated separately for hydration water of Li^+ and I^- and for bulk water and are shown in Fig. 19. The result for bulk water is very similar to that for pure water and needs no further discussion. The effect of the Li^+ on its hydration shell water molecules is twofold. The water molecules in the first hydration shell are strongly attached to the ion and the O–O distances between the water molecules in the first and second hydration shell (2.72 Å [13]) are strongly reduced relative to pure water (2.85 Å). Both effects lead to higher frequencies for the hydrogen bond bending motion (strong reduction of the peak around 50 cm^{-1}) and the O–O stretching frequencies (extension beyond 250 cm^{-1}) and result in an overall shift in $f(\omega)$ to higher frequencies relative to pure water, as can be seen from Fig. 19. The figure shows also that the hydration water of I^- is responsible for the enhancement of $f(\omega)$ in the range 75–150 cm^{-1}. The interaction of I^- with its neighboring water molecules leads to a loosening of their hydrogen bonds and results, therefore, in a displacement of the O–O stretching motions toward the low frequency range.

9. LIBRATIONAL MOTIONS AND INTRAMOLECULAR VIBRATIONS

In order to separate the various librational and vibrational modes of the water molecules the following scheme has been employed [29]: The instantaneous velocities of the two hydrogen atoms in the center-of-mass system are projected onto the instantaneous unit vectors: i) in direction of the corresponding O–H bond ($\vec{u}_1$ and $\vec{U}_2$), ii) perpendicular to the O–H bonds in the plane of the molecule ($\vec{v}_1$ and $\vec{v}_2$), and iii) perpendicular to the plane of the molecule ($\vec{p}_1$, and $\vec{p}_2$).

Using capital letters to denote the projections of the hydrogen velocities onto the corresponding unit vectors, the following quantities are defined:

$$Q_1 = U_1 + U_2 \qquad R_x = V_1 - V_2$$

$$Q_2 = V_1 + V_2 \qquad R_y = P_1 + P_2$$

$$Q_3 = U_1 - U_2 \qquad R_z = P_1 - P_2$$

where Q_1, Q_2, and Q_3 describe approximately the three normal mode vibrations usually referred to as symmetric stretch, bend, and asymmetric stretch, respectively. R_x, R_y, and R_z approximate the instantaneous rotations around the three principal axes of the water molecule, as defined in the insertion of Fig. 20. The normalized velocity autocorrelation functions calculated in this way for the water molecules in an 18.5 molal LiCl solution are shown as examples in Fig. 20 [38]. The differences between x, y, and z reflect the different moments of inertia around the three axes. Q_2 denotes the bending mode, the frequency of which is about half of that of the stretching modes.

The Fourier transforms of these velocity autocorrelation functions result in the spectral densities of the corresponding modes. They are shown in Fig. 21 for a 1.1 molal $SrCl_2$ solution as an example, calculated separately for the three water subsystems. The frequencies of the peak maxima are given in Table 9.

TABLE 9: Frequencies, in cm^{-1}, of the peak maxima of the spectral densities of the librations and vibrations for all water molecules in the solution and separately for the three water subsystems. In the case of the vibrations the widths at half height, $\Delta\tilde{\nu}$, are given additionally.

		total water	bulk water	hydration water of cation	anion
R_x	$\tilde{\nu}_{max}$	445 ± 10	410 ± 10	500 ± 10	465 ± 10
R_y	$\tilde{\nu}_{max}$	620 ± 10	595 ± 10	640 ± 15	650 ± 15
R_z	$\tilde{\nu}_{max}$	405 ± 10	405 ± 10	390 ± 10	420 ± 20
Q_1	$\tilde{\nu}_{max}$	3450 ± 10	3475 ± 10	3335 ± 10	3420 ± 10
	$\Delta\tilde{\nu}$	–	230 ± 10	230 ± 10	360 ± 10
Q_2	$\tilde{\nu}_{max}$	1705 ± 5	1705 ± 5	1690 ± 10	1710 ± 10
	$\Delta\tilde{\nu}$	–	165 ± 10	150 ± 10	180 ± 10
Q_3	$\tilde{\nu}_{max}$	3580 ± 10	3595 ± 10	3460 ± 10	3540 ± 20
	$\Delta\tilde{\nu}$	–	260 ± 10	270 ± 10	330 ± 10

Both ions cause a significant blueshift of the librational frequencies around the x- and y-axes relative to bulk water while the maxima in the spectral densities for the rotations around the dipole moment axis show only small changes, hardly outside the limits of statisical uncertainty (Fig. 21 and Table 9). The effect of anion and cation on the peak maximum of the librations around the y-axis is quite similar while the shift caused by Sr^{2+} in the case of the x-axis is about twice as large as that of Cl^- [29].

The strong Sr^{2+}-water interactions lead to a well defined first hydration shell where the oxygen atom density goes up to about 10 times the average one (Fig. 8) and results in a strong preference for an orientation where the dipole moment of the water

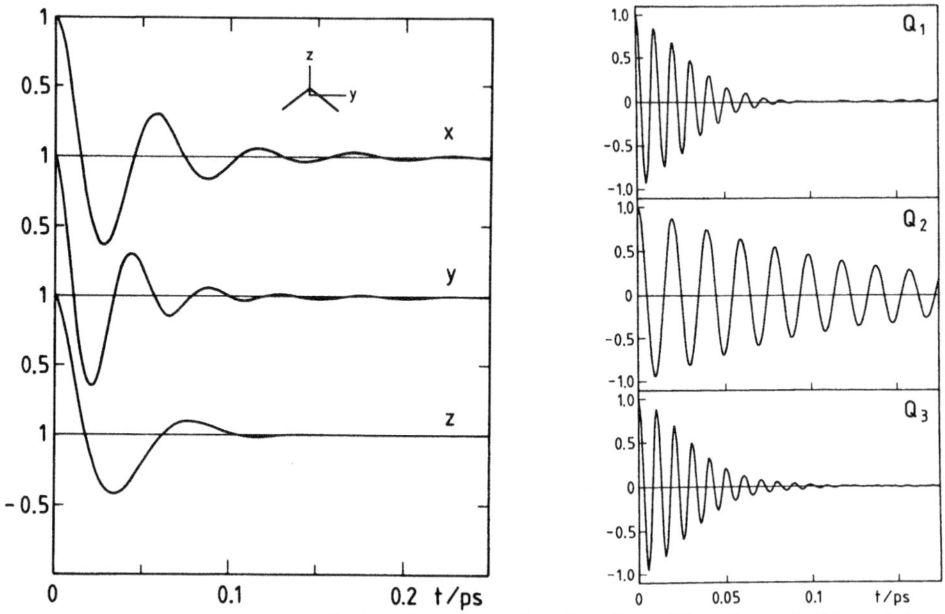

Figure 20: Velocity autocorrelation functions for the three librations (left) and for the three intramolecular vibrations (right) of the water molecules in an 18.5 molal LiCl solution. The librations are calculated separately for the three components in a molecule fixed coordinate system as defined in the insertion. Q_1, Q_2, and Q_3 denote the symmetric stretching, bending, and asymmetric stretching modes, respectively [38].

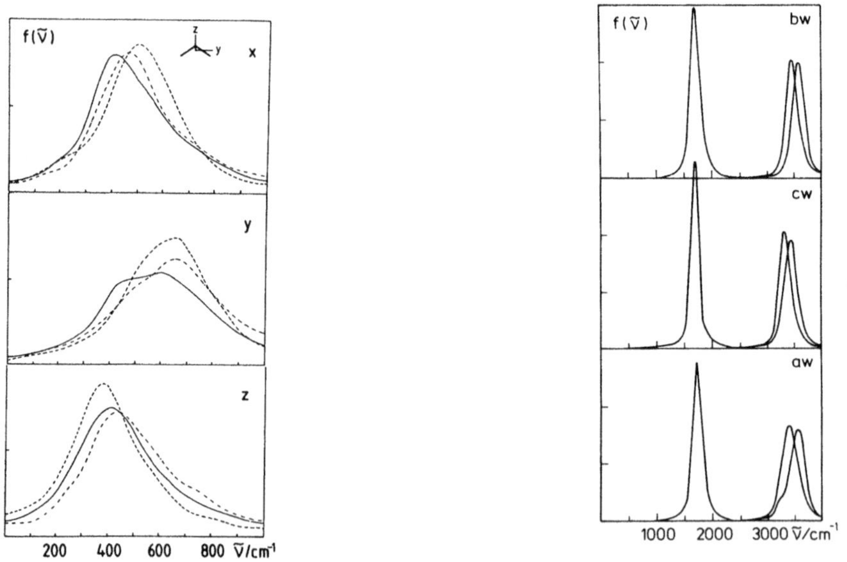

Figure 21: Spectral densities of the librations (left) and the intramolecular vibrations (right) of the water molecules from a simulation of a 1.1 molal $SrCl_2$ solution, calculated separately for the hydration water of Sr^{2+} (cw, — — —), Cl^- (aw, —·—), and bulk water (bw, ———) and given in arbitrary units [29].

molecule points away from the centre of the ion (Fig. 10). Consequently these ion-water interactions cause an additional hindrance for the rotations around the x- and y-axes which explains the blueshift for the hydration shell molecules while the rotation around the z-axis remains rather unaffected. The small blueshift found for the librational bands of the hydration shell water molecules of Cl^- relative to bulk water might result from a stronger hydrogen bond formed between Cl^- and water than between water and water, which is evident from the corresponding redshift of the O–H stretching vibration (see below).

The ion-water interactions lead to an increase of the intramolecular O–H distance and a decrease of the H–O–H angle for the water molecules in their hydration shells resulting in average values of 0.9753, 0.9768, and 0.9815 Å as well as 100.7°, 100.3°, and 99.2° for bulk water, hydration water of Cl^- and Sr^{2+}, respectively. These changes in geometry together with the partial charges on the O and H atoms lead to an increase of the dipole moment for the hydration shell water molecules of Sr^{2+} by about 2% compared with bulk water for which a value of 2.00 D has been calculated. From the well known empirical relationship between the intramolecular O–H distances and the frequencies of the O–H stretching modes of about 20.000 cm^{-1}/Å [54] redshifts relative to bulk water of about 120 cm^{-1} and 30 cm^{-1} are expected for the hydration shell molecules of Sr^{2+} and Cl^-, respectively. The values given in Table 9 are in accordance with these expectations. The effect of both ions on the bending vibration is hardly outside the limits of error [29].

10. WATER AND ELECTROLYTE SOLUTIONS AT INTERFACES

The investigation of the influence of non-polar walls on structural and dynamical properties of water by computer simulation started several years ago. The walls were either assumed to be hard and the water-wall interactions were described by a simple exponential or an LJ potential. In the only simulation of an electrolyte solution reported so far LJ walls were employed. Polar interfaces were investigated by simulations of water near an insoluble NaCl crystal and at mineral and lecithin surfaces where the atoms bear elementary and the polar groups partial charges. The simulations of water-metal interfaces started just recently. In all three studies published so far the metallic character of the uncharged wall was described by image charges. For further details the reader is referred to a review article which was published recently [55].

In all three water-metal interface simulations a strong preference for an orientation of the dipole moment of the water molecules parallel to the surface resulted in disagreement with measured work function changes. Additionally, in the study where a corrugated surface was employed the water-wall potential used has led to a preferential adsorption on the hollow sites of the surface, which is also in contradiction to several quantum mechanical calculations which indicate that an adsorption on top of the metal atom is the preferred site. Therefore, it seemed to be appropriate to try to find a water-metal potential which leads to better agreement between simulation on the one hand and experiments and MO calculations on the other.

The development of the new water-metal interaction potential is based on extended Hückel molecular orbital calculations of a water molecule on top of a five-atomic platinum cluster [56]. Pair potential functions have been adjusted to the cluster energies in such a way that the analytical form of the metal-water potential is given by:

$$V_{H_2O-Pt} = V_{O-Pt}(r_{O-Pt}, \rho_{O-Pt}) + V_{H-Pt}(r_{H_1-Pt}) + V_{H-Pt}(r_{H_2-Pt}) \tag{12}$$

where

$$\begin{aligned} V_{O-Pt} = &\left[1894.2 \exp(-1.1004r) - 1886.3 \exp(-1.0966r)\right] f(\rho) \\ &+ 10^6 \exp(-5.3568r)\left[1 - f(\rho)\right] \end{aligned} \tag{13}$$

and

$$V_{H-Pt} = 1.7142 \exp(-1.2777r) \tag{14}$$

with

$$f(\rho) = \exp(-0.5208\rho^2). \tag{15}$$

The site-site distances are denoted by r. ρ is the length of the projection of the distance vector onto the surface plane. The energy is given in units of 10^{-19} J if r and ρ are given in Å. The pair potential V_{O-Pt} is attractive for larger Pt–O distances only if ρ is small, leading to the preferential adsorption site on top of a platinum atom. The energeticaly most favourable orientation is the one where the dipole moment vector points away from the surface in agreement with the Hückel calculations [56]. The total interaction energy between one water molecule and all Pt atoms is given as a function of the x- and y-coordinates in Fig. 22 for a water-surface distance Δz chosen such that the potential has its minimum. The figure shows clearly that the potential leads to a surface corrugation with deep energy minima at the platinum atom positions. The water-water interactions are described again by the BJH model [24] and the platinum-platinum interactions are simply modeled by a harmonic nearest-neighbor potential [57].

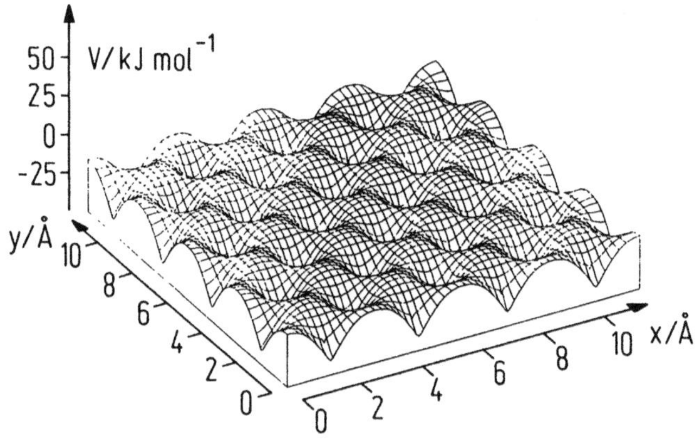

Figure 22: Water-metal interaction potential as a function of x and y for a distance Δz chosen in such a way that the potential has its minimum value. The dipole moment vector points away from the surface.

With these potentials two kinds of simulations have been performed. In one case the interval $-12.7 < z < 12.7$ Å of a rectangular basic box with sidelengths $L_x = L_y = 19.6$ Å and $L_z = 45$ Å as shown in Fig. 23 is occupied by 305 water molecules resulting in a water

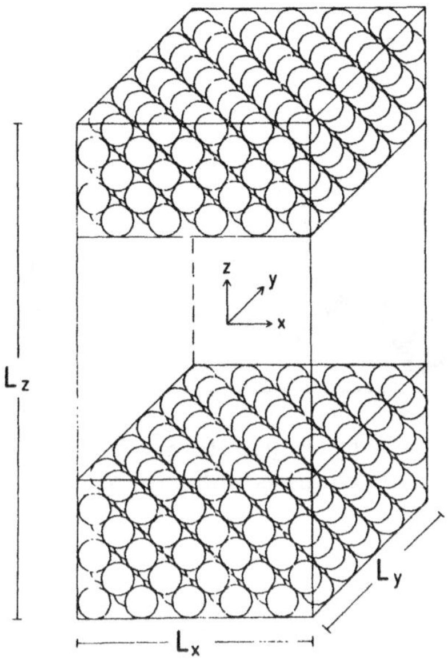

Figure 23: Sketch of the basic tetragonal cell used in a simulation of a water lamina between two (100) faces of a platinum crystal. The water molecules are located in the centre of the box and the platinum atoms are represented by the circles [55].

density of 1 g/cm^3 in the inner region of the lamella. In the regions $12.7 \leq | z | \leq 22.5$ Å 550 platinum atoms are located. As a consequence of the introduction of periodic boundary conditions in all directions the platinum subsystem can be regarded as one infinitely extended slab with a thickness of 19.6 Å which interacts with water molecules on both surfaces. The values of L_x and L_y have been chosen in accordance with the periodicity of the face-centered cubic platinum lattice with lattice constant $a = 3.92$ Å. The crystal slab consists of $5 \times 5 \times 5$ face-centered cubic unit cells, each containing four atoms and the (100) plane being parallel to the xy-plane of the periodic box [58]. The second kind of simulation differs from the first one only in respect to the number of water molecules involved. Several simulations have been performed with the number of water molecules reduced in such a way that the dependence of various properties on the degree of coverage θ could be investigated. Θ values of 0.2, 0.4, 0.6, 0.8, and 1.4 have been chosen [59].

The Ewald summation method for two-dimensional systems [60] has been used to calculate the long-range Coulomb interactions. For the non-Coulomb interactions the shifted force method [7] was used with a cut-off distance of $L_x/2$. After an extensive equilibration period of 15 ps the simulation of the lamella extended over 35000 steps of 0.25 fs at an average temperature of 349 K without rescaling the velocities. This is equivalent to a total elapsed time of 8.75 ps. The simulations of the monolayers of various coverages extended between 8 and 30 ps.

The oxygen and hydrogen atom density profiles are shown in Fig. 24. The first maxima of both profiles coincide at a distance of about 2.5 Å from the Pt surface and the reduced

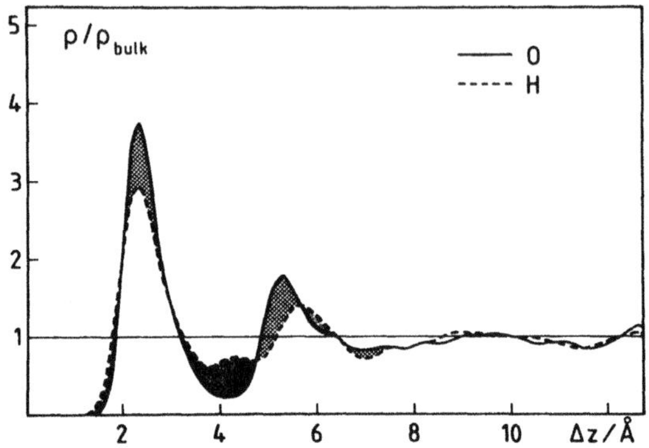

Figure 24: Normalized oxygen and hydrogen atom densities as a function of distance from the Pt(100) surface. The shaded areas indicate an excess of positive or negative charges [58].

oxygen atom density is by about 30% higher than that of the hydrogen atoms. After a much less pronounced second peak in both density profiles the distributions become uniform except for statistical noise. The coincidence of the positions of the first peaks indicates a strong preference for a water molecule orientation parallel to the Pt surface. The shaded areas in Fig. 24 indicate a non-vanishing dipole moment in the first two layers. The dipole moment density is very large in the first layer compared with the second one. Accordingly, the solvent contribution to the interfacial potential drop which can be calculated from the dipole density ρ_μ through:

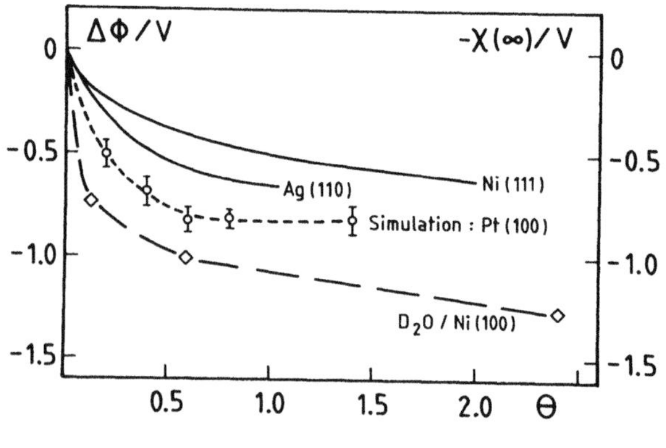

Figure 25: Comparison of experimental and calculated values of the work function change. The data are taken from Ref. [61] for H_2O on Ni(111), from Ref. [62] for H_2O on Ag(110), and from Ref. [63] for D_2O on Ni(100) (diamonds). Dashed lines are drawn to guide the eyes while full lines are taken from the papers cited. Error bars are not available for experimental results [59].

$$\chi(\Delta z) = \frac{1}{\epsilon_0} \int_0^{\Delta z} \rho_\mu(z')dz', \tag{16}$$

where ϵ_0 is the dielectric permittivity of the vacuum, results mainly from the first layer water molecules.

In Fig. 25 the simulated values of $-\chi(\infty)$ as a function of the degree of coverage Θ, calculated according to Eq. (16), are compared with experimental work function changes for water adsorption on several metal surfaces. The simulation results reproduce the experimental trend very well. After a steep initial decrease $-\chi(\infty)$ remains constant for the coverages above about 0.5. The experimental results show a further decrease which is, however, significantly smaller than that for small Θ. Quantitative agreement can be expected only from simulations which take explicitly into account the electronic structure of water and especially that of the metal.

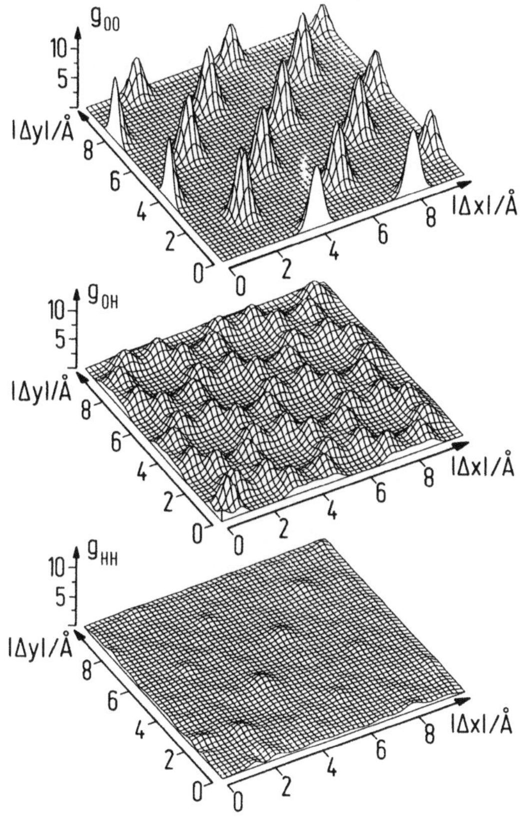

Figure 26: Oxygen-oxygen, oxygen-hydrogen and hydrogen-hydrogen pair correlation functions between the water molecules in the adsorbate layer ($\Delta z_1, \Delta z_2 < 4.2Å$) [58].

The atom-atom pair correlation functions (PCFs) have been calculated according to the defining relation:

$$g(\vec{r}_1, \vec{r}_2) = \frac{\rho^{(2)}(\vec{r}_1, \vec{r}_2)}{\rho^{(1)}(\vec{r}_1) \cdot \rho^{(1)}(\vec{r}_2)}, \tag{17}$$

with $\rho^{(i)}$ the local i-particle density and $\vec{r}_1$ and $\vec{r}_2$ the cartesian coordinates of the two particles. Averaging over only the water molecules in the surface layer leads to the PCFs depicted in Fig. 26. It can be seen that a far ranging symmetrical arrangement of high maxima and relatively low minima in the oxygen-oxygen correlation exists which is obviously produced by crystal-mediated interactions which force the molecules (via interactions with the oxygen atoms) onto the adsorption sites and thus into a rigid mutual arrangement. Due to the water-surface potential employed the hydrogen atoms experience the surface corrugation only mediated through the oxygen-crystal interactions. Consequently, g_{OH} shows a much smaller degree of correlation. There are groups of four maxima in g_{OH}, each corresponding to one of the four different directions of an intramolecular O–H bond along the two diagonals $x = y$, $x = -y$ which are the preferential directions for hydrogen bonds as a consequence of the arrangement of adsorption sites. The hydrogen-hydrogen PCF exhibits only very small correlations. The PCFs of the adsorbate layer indicate that water forms a planar quadratic overlayer with hydrogen bonds to its neighbours in the plane. The hydrogen atoms are disordered in a similar manner as in proton disordered ice.

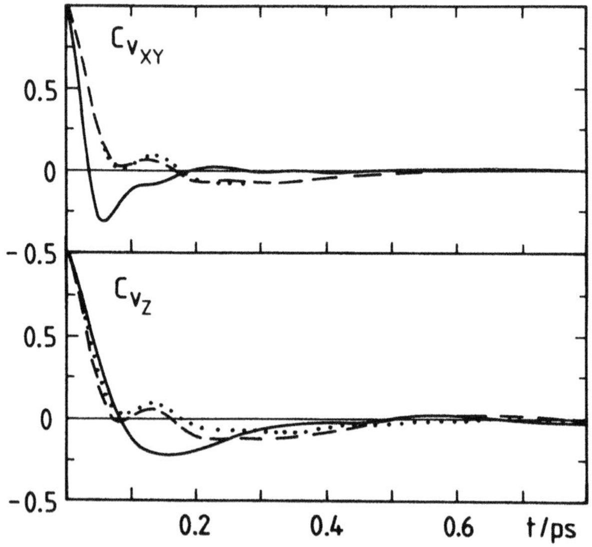

Figure 27: Center-of-mass velocity autocorrelation functions for the motions parallel (top) and perpendicular (bottom) to the surface for three different layers, defined by $\Delta z \leq 4.2$Å (full), $4.2 < \Delta z \leq 6.2$Å (dashed), and $6.2 < \Delta z$ (dotted) at 349 K where Δz denotes the distance of the oxygen atom from the surface [58].

The velocity autocorrelation functions for the water molecules have been calculated separately for three different layers and for the motions parallel and perpendicular to the surface. They are depicted in Fig. 27. The curves are very similar for the second and third layer and do not differ significantly from that of pure water. In the comparison with Fig. 17 it has to be kept in mind that the temperature here is about 50° higher. The autocorrelation functions for both components of the velocity – parallel and perpendicular to the surface – of the water molecules in the adsorbate layer are strongly different from that of pure water.

The self-diffusion coefficients have been calculated according to Eq. (8) from the six functions drawn in Fig. 27. In the adsorbate layer they are too small ($< 0,1 \cdot 10^{-5}$ cm^2/s) to be calculated accurately from the simulation which extended over about 9 ps. In the third layer the self-diffusion coefficients are the same for both directions and do not differ from that of pure water in the limits of error. An interesting feature is found forthe second layer. The self-diffusion perpendicular to the surface is strongly reduced for

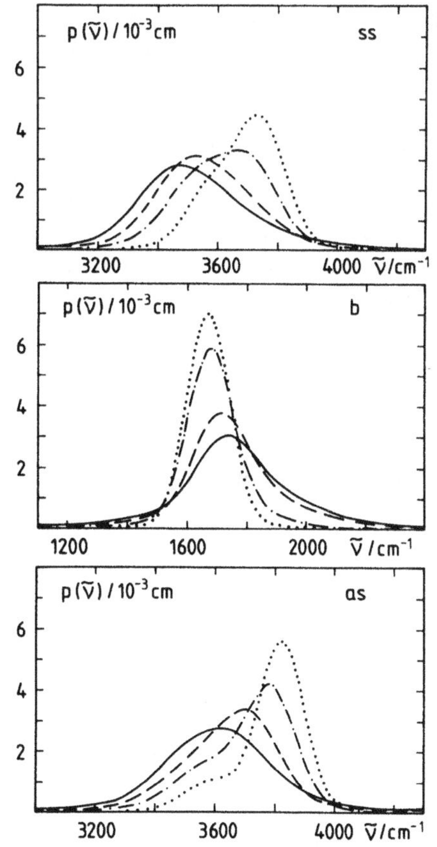

Figure 28: Spectral densities of the three intramolecular vibrations of the water molecules for different degrees of coverage Θ. The full line gives the results for the adsorbate layer of the complete system while dashed, dash-dotted, and dotted lines are for surface coverages of 0.8, 0.4, and 0.2, respectively [58].

these water molecules as the strongly attached adsorbate layer allows a mobility only in one direction. On the contrary, the self-diffusion parallel to the surface is increased as the number of hydrogen bonds between first and second layer is smaller than in pure water.

As a second example of the effect of the walls on dynamical properties the intramolecular vibrations of the water molecules will be discussed. The maxima of their spectral densities, namely symmetric stretch, bend, and asymmetric stretch, are found at 3475, 1710, and 3610 cm^{-1}, respectively. The differences between the three layers are hardly outside the limits of statistical uncertainty which are estimated to be $\pm 20\,cm^{-1}$. As in the case of the librations, the vibrations are not influenced significantly by the surface. But it can be seen from Fig. 28 that the frequency shifts occuring with the change in the coverage extend over the whole range of frequencies between gas phase and liquid water. The bending frequency is found experimentally at 1630 cm^{-1} [64] which is below the range of frequencies shown in Fig. 28. The bands at 2870, 3380, and 3670 cm^{-1} found in the electron energy loss spectra have been assigned by Ibach and Lehwald to O–H stretching vibrations. While the latter two are in the frequency range found from the simulations it remains unclear what should cause the strong redshift resulting in the band at $2870\,cm^{-1}$ [64].

The results presented here seem to indicate that computer simulations can contribute significantly to a better understanding of structural and dynamical properties of water/metal interfaces. There is undoubtedly a strong need for an improvement of the potentials employed in the simulations. More unambiguous experimental data would be very helpful in finding more reliable potentials. The successful simulation of an electrochemically important electrolyte solution/metal interface appears to be possible in the near future.

ACKNOWLEDGEMENT

Financial support by Deutsche Forschungsgemeinschaft is gratefully acknowledged.

REFERENCES

[1] F.H. Stillinger and A. Rahman, *J. Chem. Phys.* **60**, 1545 (1974)
[2] K. Heinzinger and P.C. Vogel, *Z. Naturforsch.* **29a**, 1164 (1974)
[3] K. Heinzinger and P.C. Vogel, *Z. Naturforsch.* **31a**, 463 (1976)
[4] W. Hogervorst, *Physica* **5**, 2464 (1973)
[5] G. Pálinkás, W.O. Riede and K. Heinzinger, *Z. Naturforsch.* **32a**, 1137 (1977)
[6] C.L. Kong, *J. Chem. Phys.* **59**, 2464 (1973)
[7] W.B. Streett, D.J. Tildesly and G. Saville, *AGS Symp. Ser.* **86**, 144 (1978)
[8] Gy.I. Szász, K. Heinzinger and W.O. Riede, *Z. Naturforsch.* **36a**, 1067 (1981)
[9] G. Heinje, W.A.P. Luck and K. Heinzinger, *J. Phys. Chem.* **91**, 331 (1987)
[10] M. Migliore, S.L. Fornili, E. Spohr, G. Pálinkás and K. Heinzinger, *Z. Naturforsch.* **41a**, 826 (1986)
[11] Gy.I. Szász and K. Heinzinger, *Z. Naturforsch.*. **34a**, 840 (1979)
[12] Gy.I. Szász and K. Heinzinger, *Z. Naturforsch.* **38a**, 214 (1983)

[13] T. Radnai, G. Pálinkás, Gy.I. Szász and K. Heinzinger, *Z. Naturforsch.* **36a**, 1076 (1981)
[14] M. Mezei and D.L. Beveridge, *J. Chem. Phys.* **74**, 6902 (1981)
[15] R.W. Impey, P.A. Madden and I.R. McDonald, *J. Phys. Chem.* **87**, 5071 (1983)
[16] O. Matsuoka, E. Clementi and M. Yoshimine, *J. Chem. Phys.* **64**, 1351 (1976).
[17] J.E. Enderby, S. Cummings, G.J. Herdman, G.W. Neilson, P.S. Salomon and N. Skipper, *J. Phys. Chem.* **91**, 5851 (1987)
[18] A.K. Soper, G.W. Neilson, J.E. Enderby and R.A. Howe, *J. Phys. C* **10**, 1793 (1977)
[19] W. Dietz, W.O. Riede and K. Heinzinger, *Z. Naturforsch.* **37a**, 1038 (1982)
[20] S. Cummings, J.E. Enderby, G.W. Neilson, J.R. Newsome, R.A. Howe and A.K. Soper, *Nature* **287**, 714 (1980)
[21] Gy.I. Szász, W. Dietz, K. Heinzinger, G. Pálinkás and T. Radnai, *Chem. Phys. Lett.* **92**, 388 (1982)
[22] Gy.I. Szász, K. Heinzinger and G. Pálinkás, *Chem. Phys. Lett.* **78**, 194 (1981)
[23] P. Bopp, W. Dietz and K. Heinzinger, *Z. Naturforsch.* **34a**, 1424 (1979)
[24] P. Bopp, G. Jancsó and K. Heinzinger, *Chem. Phys. Lett.* **98**, 129 (1983)
[25] F.H. Stillinger and A. Rahman, *J. Chem. Phys.* **68**, 666 (1978)
[26] G.D Carney, L.A. Curtiss and S.R. Langhoff, *J. Mol. Spectr.* **61**, 371 (1976)
[27] T. Yamaguchi, H. Ohtaki, E. Spohr, G. Pálinkás, K. Heinzinger and M.M. Probst, *Z. Naturforsch.* **41a**, 1175 (1986)
[28] M.M. Probst, T. Radnai, K. Heinzinger, P. Bopp and B.M. Rode, *J. Phys. Chem.* **89**, 753 (1985)
[29] E. Spohr, G. Pálinkás, K. Heinzinger, P. Bopp and M.M. Probst, *J. Phys. Chem.* **92**, 6754 (1988)
[30] N.A. Hewish, G.W. Neislon and J.E. Enderby, *Nature* **297**, 138 (1982)
[31] G. Pálinkás and K. Heinzinger, *Chem. Phys. Letters* **126**, 251 (1986)
[32] G. Pálinkás, T. Radnai, W. Dietz, Gy.I. Szász and K. Heinzinger, *Z. Naturforsch.* **37a**, 1049 (1982)
[33] G. Jancsó, K. Heinzinger and P. Bopp, *Z. Naturforsch.* **40a**, 1235 (1985)
[34] Gy.I. Szász and K. Heinzinger, *Earth Planet. Sci. Lett.* **64**, 163 (1983)
[35] G. Pálinkás, P. Bopp, G. Jancsó and K. Heinzinger, *Z. Naturforsch.* **39a**, 179 (1984)
[36] F.H. Stillinger and A. Rahman, *J. Chem. Phys.* **61**, 4973 (1974)
[37] R.W. Impey, M.L. Klein and I.R. McDonald, *J. Chem. Phys.* **74**, 647 (1981)
[38] Y. Tamura, K. Tanaka, E. Spohr and K. Heinzinger, *Z. Naturforsch.*, in press
[39] K. Tanaka, N. Ogita, Y. Tamura, I. Okada, H. Ohtaki, G. Pálinkás, E. Spohr and K. Heinzinger, *Z. Naturforsch.* **42a**, 24 (1987)
[40] M.P. Tosi and F.G. Fumi, *J. Phys. Chem. Solids* **25**, 31 (1964)
[41] I. Okada, H. Okano, H. Ohtaki and R. Takagi, *Chem. Phys. Lett.* **100**, 436 (1983)
[42] A.P. Copestake, G.W. Neilson and J.E. Enderby, *J. Phys. C: Solid State Phys.* **18**, 4211 (1985)
[43] Gy.I. Szász, K. Heinzinger and W.O. Riede, *Ber. Bunsenges. Phys. Chem.* **85**, 1056 (1981)
[44] L. Endom, H.G. Hertz, B. Thül and M.D. Zeidler, *Ber. Bunsenges. Phys. Chem.* **71**, 1008 (1967); M.I. Emelyanov, *Zhurn. Struct. Khim.* **6**, 295 (1965); D.W. McCall and D.C. Douglas, *J. Phys. Chem.* **69**, 2001 (1965)
[45] H.G. Hertz, in: *Water – A Comprehensive Treatise*, Vol.3 (F. Franks, ed.), Plenum Press, New York and London 1973
[46] A. Geiger, *Ber. Bunsenges. Phys. Chem.* **85**, 52 (1981)

[47] H. Weingärtner, *Z. Phys. Chem.* **132**, 129 (1982)
[48] K. Tanaka, *J. Chem. Soc. Faraday Trans.I* **71**, 1127 (1975)
[49] R.A. Robinson and R.H. Stokes, *Electrolyte Solutions*, Butterworths, London 1955
[50] E. Hawlicka, private communication
[51] G. Jancsó, P. Bopp and K. Heinzinger, *Chem. Phys.* **85**, 377 (1984)
[52] Gy.I. Szász and K. Heinzinger, *J. Chem. Phys.* **79**, 3467 (1983)
[53] M. Nakahara, M. Zenke, M. Ueno and K. Shimizu, *J. Chem. Phys.* **83**, 280 (1985)
[54] S.I. LaPlaca, W.C. Hamilton, B. Kramb and A. Prakash, *J. Chem. Phys.* **58**, 567 (1973)
[55] E. Spohr and K. Heinzinger, *Electrochim. Acta* **33**, 1211 (1988)
[56] S. Holloway and K.H. Bennemann, *Surf. Sci.* **101**, 327 (1980)
[57] J.E. Black and P. Bopp, *Surf. Sci.* **140**, 275 (1984)
[58] E. Spohr, to be published
[59] E. Spohr and K. Heinzinger, *Ber. Bunsenges. Phys. Chem.* **92**, 1358 (1988)
[60] D.M. Heyes, *J. Chem. Phys.* **79**, 4010 (1983)
[61] C. Nöbl and C. Benndorf, *Surf. Sci.* **182**, 499 (1987)
[62] K. Bange, T.E. Madey, J.K. Sass and E.M. Stuve, *Surf. Sci.* **183**, 334 (1987)
[63] D.E. Peebles and J.M. White, *Surf. Sci.* **144**, 512 (1984)
[64] H. Ibach and S. Lehwald, *Surf. Sci.* **91**, 187 (1980)

COMPUTER SIMULATION OF INORGANIC MATERIALS

R.A. JACKSON

Department of Chemistry, University of Keele, Keele, Staffs ST5 5BG

Abstract

This chapter begins with a description of the potential models commonly used in computer simulations of inorganic materials. Methods of parameterising these potentials are discussed, and an example of potential derivation is given. This leads into the second section of the chapter which gives some recent examples of simulations, based on energy minimisation methods.

1. INTRODUCTION

The reliability of any computer simulation depends crucially upon the potentials used to describe the interatomic interactions. For inorganic materials, reliable potentials have been becoming available for a number of years, ranging from potentials for largely ionic materials, to potentials capable of simulating materials with a degree of covalency. The basic potential model has, however, remained largely unchanged.

In the first part of this chapter, the fundamental details of potential models for inorganic materials are described. This is followed by a discussion of the different ways potentials can be parameterised. Finally, an example is given of the deviation of a potential for α–quartz, and the extent to which this potential may be transferred to related materials is discussed. The second part of the chapter gives examples of recent simulations.

Fundamentals of Potential Models

In considering inorganic materials, we are invariably dealing with systems that are ionic to a greater or lesser extent, and therefore ion–ion interactions are of great importance.

The interaction energy of a pair of ions is given by Coulomb's Law:

$$V = \frac{q_i q_j}{r_{ij}} \qquad (1)$$

where q_i and q_j are the charges on ions i and j and r_{ij} is the distance between them.

The choice of charges on the ions is a matter of debate, but the only unambiguous choice is to use the formal charges. This approach has invariably been used with great success.

The next question which occurs is that of ionic polarisability. The rigid ion model treats ions as point charges, neglecting this effect. However, potential models assuming rigid ions can model structures accurately [1]. Ionic polarisability can be included via the shell model [2], which describes polarisable ions as a core and shell, linked by a harmonic spring. The mass of the ion lies at the core, and the charge

C.R.A. Catlow et al. (eds.), Computer Modelling of Fluids Polymers and Solids, 395–404.

is divided between the core and the shell. The polarisability α is related to the shell charge Y and spring constant k as follows:

$$\alpha = Y^2/k \qquad (2)$$

The shell model has been widely and successfully used, and its use has proved to be essential in simulations of defects in inorganic materials, since the presence of defects can cause strong polarisation of surrounding ions.

We note however, in passing, that at present molecular dynamics simulations are usually still restricted to rigid ion potentials, for reasons of cpu time. This restriction is expected to disappear in the future.

The total electrostatic energy of the system is therefore obtained by summing terms of form (1). Such a sum is slowly convergent in real space, and convergence may be achieved via the Ewald method, which expresses a slowly convergent sum in real space as a sum of two rapidly convergent sums, one in real space and one in reciprocal space. Details of the method are not given here; the reader is referred to one of two alternative explanations [1,3].

In addition to the Coulombic or electrostatic interaction, other interactions, generally referred to as "short range interactions", need to be included in the potential. These include repulsion due to electron charge cloud overlap, and attractive interactions such as the van der Waals or dispersive interactions.

A potential form frequently used to describe these interactions is the Buckingham potential:

$$V(r) = A \exp(-r/\rho) - Cr^{-6} \qquad (3)$$

The parameters A, ρ, and C need to be obtained; this can be done by empirical fitting or direct calculation. Further details of these two approaches are given in the next section.

Other potential forms have been used to describe short range interactions, including the Lennard-Jones potential

$$V(r) = Ar^{-12} - Cr^{-6} \qquad (4)$$

Bonded interactions sometimes have to be included in potentials for inorganic materials. Potentials used include descriptions based on simple harmonic models, and Morse potentials [4].

So far, only two body interactions have been represented in the potentials discussed. Three body potentials have been used in situations where, for specific reasons, two body potentials cannot represent all the interactions of importance.

For example, quartz and the very wide range of silicate and aluminosilicate minerals contain $(SiO_4)^{4-}$ units, in which there is a strong directionality of bonding of the four oxygen ions with respect to the silicon ion. This effect can be included in the potential by a "bond bending potential" which provides an energetic constraint to the angle between silicon and each pair of oxygens. It has the form

$$V_{(3\ \mathrm{body})} = \tfrac{1}{2} k (\theta - \theta_0)^2 \qquad (5)$$

where θ_0 is the equilibrium bond angle, in this case the tetrahedral angle (109.47°), and k is the harmonic bond force constant, empirically fitted to the structure of interest.

It should be noted that, while the bond bending potential is a three body potential in the sense that it involves the interaction of three ions, it is only one of the possible terms in the total three body potential. It can, however, successfully account for some effects omitted by two-body potentials, and has been used in studies of α-quartz [5], zeolites [1], silicate minerals [6,7], rutile [8] and La_2CuO_4 [9].

Other forms of three body potential have been used. For example, in recent studies of the silver halides, triple-dipole potentials of the form originally suggested by Axilrod and Teller [10] are being used [11].

Potential parameterisation

Potential parameters may be obtained with one of two main approaches – empirical fitting or direct calculation. These approaches will be considered in turn.

(a) empirical fitting

The procedure is to fit a potential to crystal structure and lattice properties – i.e., starting with some initial guess of the parameters, to adjust them systematically until the difference between calculated and experimental structure and properties is minimised. The procedure is summarised in figure 1.

The technique depends upon the availability of crystal structures and also lattice properties, such as elastic and dielectric constants, as it is generally insufficient to fit to structures alone. However, as will be shown in the next section, empirically determined potentials can often be transferred to other related structures, i.e. a potential is fitted to a structure for which experimental data is available, and then transferred to a less-studied system.

FIGURE 1

Fitting procedure to obtain empirical potentials
(THBFIT, M. Leslie)

A. Initial guess of parameters in specified expression for short-range interaction

B. Calculate, at the observed structure:
   1. Crystal Properties
   2. Lattice (bulk) and internal (ion) strains

C. Minimise:
   1. Difference between observed and calculated properties
   2. Strains
   By adjusting potential parameters (numerical derivative taken to calculate size of adjustment)

D. Improved potential mode

E. Stop at minimum

nb. Weights may be applied to observables at stage B

(b) direct calculation – ab initio methods

Two approaches have been used to calculate potentials directly. The first employs the "electron gas approximation" which is based on methods developed by Wedepohl [12] and applied by Gordon and Kim [13]. More recently, Post and Burnham [14] have used this technique to calculate potentials for a range of interactions in minerals. The technique has the advantage of being computationally inexpensive.

The second "direct calculation" approach involves the use of ab initio molecular orbital methods, usually Hartree–Fock SCF techniques. The procedure is to represent the interacting ions, and possibly some of their immediate neighbours, quantum mechanically, and then to employ an array of point charges to represent the remainder of the lattice. The point charges are placed at lattice sites, and their charges chosen so that the correct Madelung field and energy is obtained at the position of the interacting ions. The relative geometry of interacting ions is then varied, and the energy obtained for each configuration. An energy surface is thus obtained, which can be fitted to one of the functional forms mentioned in the previous section.

(c) relative strengths of the two approaches

Both the empirical method and the ab initio method have their strengths and weaknesses. The empirical method will calculate potentials that are only applicable with certainty to the range of interatomic distances found in the crystal used in the fitting. This can lead to problems, e.g. if the potential is used in a molecular dynamics simulation, which might access interatomic distances not used in the fitting. Ab initio potentials have the problem of being too specific for particular systems if point charge arrays are used to represent a particular lattice environment. However, with advances in available computer power, ab initio potentials are becoming more reliable, and applications for which they are essential are increasing. An example is the need to calculate potentials for the interaction of hydrocarbons with zeolite lattices, for which ab initio methods are almost the only available approach.

(d) Example: the calculation of an empirical potential for α–quartz, and its transfer to zeolite structures.

This example will show how a potential was empirically fitted to the α–quartz structure and lattice properties.

The potential parameters fitted were the Si–O A, ρ and C parameters (equation 3) and a bond bending term about all SiO_2 bonds (equation 5). In addition, the shell model parameters and shell displacements (for the oxygen ions) were included in the fit. The O–O parameters were the same as those used in many previous calculations, and were those obtained by Catlow [15]. The properties fitted to, apart from the crystal structure, were the elastic constants, static and high frequency dielectric constants. Table 1 shows the calculated and experimental lattice properties.

Having fitted the potential, it was put to a number of additional tests to establish its general validity. These included (i) the calculation of phonon dispersion curves, (ii) the simulation of the structural properties of α–quartz under high applied pressures and (iii) the simulation of the structural and cohesive properties of other polymorphs of SiO_2 (α–cristobalite, coesite and α–tridymite). In each case good agreement with experiment was obtained, and further details will be found in reference 5.

In the discussion of empirical potentials it was noted that in cases where there is insufficient information available to fit a potential, it is sometimes possible to use a potential fitted to a related material. This procedure was tried in the case of zeolites, aluminosilicates of major industrial importance, but for which only structural data is available.

Na–zeolite A was chosen as an example of a well–studied zeolite, and the experimental structure determined by Pluth and Smith [16] was used to compare the results of simulations with experiment. Details are given in a later section.

TABLE 1
Observed and calculated properties of α–quartz

	Experimental	Bond-bending potential	Two-body potential
Elastic constants (10^{11} dyn cm^{-2})			
c_{11}	8.683	8.815	6.204
c_{33}	10.498	10.605	7.466
c_{44}	5.826	5.296	3.301
c_{66}	3.987	4.269	2.737
c_{14}	-1.8064	-1.666	-1.012
c_{13}	1.193	1.151	1.629
Static dielectric constants			
ϵ_{11}	4.520	4.452	5.513
ϵ_{33}	4.640	4.812	6.086
High-frequency dielectric constants			
	2.4	2.04	2.069

Lattice Energy Minimisation

This technique is used to obtain structural and energetic information on crystal structures. It requires as input interatomic potentials and a starting structure. In calculating the structure corresponding to the lattice energy minimum, either atomic positions only are adjusted (constant volume minimisation) or atomic positions and lattice parameters are adjusted (constant pressure minimisation). An example of the application of this technique to zeolites is given later in the chapter.

Defect Energy Calculations

This technique has been thoroughly discussed elsewhere [17] and will only be described briefly here. Atomistic defects are modelled by explicit consideration of the region immediately surrounding the defect – i.e. atoms interact through interatomic potentials, and outside this region, a dielectric continuum approximation, due to Mott and Littleton [18] is employed. An example – the calculation of solution energies for dopant ions in $SrTiO_3$ – is given later.

Examples

1. Lattice energy minimisation studies of Na zeolite A using potentials fitted to the structure and lattice properties of α–quartz.

In this example, the potential derived for α–quartz was taken and used in a simulation of Na^+ zeolite A. The other potentials needed, i.e. Al–O and Na–O were transferred from other sources. The aim of the calculations was to establish whether the α–quartz potentials could be used in a transferable manner, and also to see how accurately zeolite structures can be reproduced by lattice energy minimisation.

To further test the sensitivity of the calculations to choice of potential, a rigid ion version of the α–quartz potential was obtained for purposes of comparison.

In all, five different potential parameterisations were used, and the results of constant volume energy minimisations compared with the experimental results of Pluth and Smith [16]. Table 2 gives the potential parameters used, and table 3 compares experimental and calculated bond lengths and bond angles.

The following conclusions may be drawn

(i) all potentials reproduce the experimental structures to much the same accuracy; note that the shell model potentials do not do appreciably better then the rigid ion potentials.

(ii) all potentials tend to slightly underestimate the Si–O bond lengths.

(iii) the most difficult angles to reproduce are those involving two tetrahedral atoms and a bridging oxygen atom, but the most detailed model, model (e) does reasonably well here.

(iv) the worst agreement in bond lengths concerns the Na(2) atoms located in the 8–rings. This is mainly due to the fact that there are 4 possible positions within the 8–ring that may be occupied by Na(2) atoms, one of which has to be chosen for the simulation.

(v) It is clear that the criteria of potential transferability applies here; all the potentials were transferred from other sources, and good agreement with experiment is obtained. Some improvement (eg. in the Si–O bond lengths) might result from refitting to the zeolite A structure.

In conclusion, this example shows that it is possible to derive an empirical potential for one material and to transfer it successfully to a related material. It also shows that zeolite structures can be modelled successfully by lattice energy minimisation. This has important consequences, given the industrial importance of zeolites, and these are fully discussed in reference 1.

TABLE 2

(i) Short range potentials

interaction		A(eV)	ρ(Å)	C(eVÅ^6)	model(s)
$Si^{4+}...O^{2-}$	rigid ion	1584.167	0.32962	52.64511	(a),(b)
$Si^{4+}...O^{2-}$	shell model	1283.907	0.32052	10.66158	(c)-(e)
$Al^{3+}...O^{2-}$	potential(1)	1460.3	0.29912	0.0	(a)-(d)
$Al^{3+}...O^{2-}$	potential (2)	498.602	0.38528	0.0	(e)
$O^{2-}...O^{2-}$		22764.0	0.149	27.88	(a)-(e)
$Na^{+}...O^{2-}$	potential (1)	1226.84	0.3065	0.0	(a)-(e)
$Na^{+}...O^{2-}$	potential (2)	5836.84	0.2387	0.0	(b),(c),(e)

(ii) harmonic and 3 body interactions

O^{2-} harmonic potential Y = 2.86902 1e1, k=74.92 eVÅ^{-2} (c)-(e)

SiO_2 3 body potential (rigid ion O^{2-}) k=4.5815 eV rad^{-2} (a),(b)
θ_o=109.47°

SiO_2 3 body potential (shell model O^{2-}) k=2.09724 eVrad^{-2} (c)-(e)
θ_o=109.47°

AlO_2^- 3 body potential (shell model O^{2-}) k=2.075 eV rad^{-2} (e)
[nb only used in model (e)] θ_o=109.47

note: Except in model (e), Al^{3+} is treated the same as Si^{4+} as far as 3 body interactions are concerned.

2. CALCULATION OF SOLUTION ENERGIES FOR TRIVALENT DOPANT IONS IN $SrTiO_3$

This example shows how the technique of defect energy minimisation may be applied to the problem of the incorporation of trivalent ions in $SrTiO_3$. $SrTiO_3$ is of interest as a potential electronic ceramic, and its well-studied neighbour $BaTiO_3$ is used in thermistor devices. The semiconductor properties of both materials are dependent on doping by trivalent impurities, but the resulting properties will depend on where the impurity ions are incorporated in the structure. The work of Lewis and Catlow [19] showed that, depending on the dopant ion, either the Ba or the Ti sites may be occupied, and same ions preferentially occupy both sites. The present work is intended to establish whether $SrTiO_3$ behaves in the same way as $BaTiO_3$ as far as doping by trivalent cations is concerned.

The defect energy minimisation code can be used to calculate the energy involved in substitution of the impurity ion at the Sr or Ti site in $SrTiO_3$, but in each case there is a resulting charge imbalance which must be compensated for. For substitution at Sr, Ti sites, two ways in which this can be achieved are considered, and the self-compensation mechanism is also considered. Table 4 gives the defect reactions involved. Solution energies are calculated bearing in mind all the processes involved in the solution process; the expression for the solution energy for mechanism 1 is given as a footnote to table 4. The other terms needed to calculate the solution energies are either obtained by calculation (lattice energies, substitution energies for electron/hole states) or from tables (ionisation energies, electron affinities, bond dissociation energies). Table 5 gives the calculated solution energies for a range of trivalent ions, for each of the five mechanisms.

From table 5 it is concluded that, except for Al^{3+}, all ions preferentially self-compensate. This differs from $BaTiO_3$, which shows roughly equal numbers of ions in each category. This suggests an interesting difference between the defect chemistry of $BaTiO_3$ and $SrTiO_3$, which may be relevant to the interpretation of experimental data. Electrical conductivity measurements on La-doped $SrTiO_3$ were interpreted using a model which assumed that La substitutes at the Sr site (in the same way as it substitutes at the Ba site in $BaTiO_3$) [20]. Poor agreement was obtained between the calculated curves and the experimental data points. The calculations reported here suggest that there is a preference by La for self-compensation, and suggestion that a re-interpretation of the conductivity data in terms of this mechanism would be worthwhile.

Summary

This chapter has described the potential models used in simulations of inorganic materials, and has described the main ways of obtaining potential parameters. Potential transferability is illustrated, and examples are given of the techniques of lattice energy minimisation, and defect energy calculation.

TABLE 3
Experimental and calculated bond lengths and angles for Na zeolite-A. Details of models (a)-(e) in table 2

bond length or angle	expt [16]	(a)	(b)	(c)	(d)	(e)
mean Si-O	1.595	1.537	1.493	1.525	1.523	1.531
O(1)-Si-O(2)	108.8	109.5	110.4	107.7	106.3	106.2
mean Al-O	1.731	1.728	1.731	1.759	1.764	1.757
O(1)-AlO(2)	108.1	109.5	107.1	114.1	117.0	118.0
T-O(1)-T	142.2	142.4	152.1	137.1	141.3	142.2
Na(1)-O(3)	2.323	2.471	2.359	2.477	2.374	2.378

TABLE 4
Reaction mechanisms for M^{3+} solution in $SrTiO_3$

(1) subs. at Ti^{4+} site, O^{2-} vacancy compensation

$$M_2O_3 + 2Ti_{Ti} + O_o \Longrightarrow 2M'_{Ti} + V_o^{\cdot\cdot} + 2TiO_2$$

(2) subs. at Ti^{4+} site, hole compensation

$$M_2O_3 + 2Ti_{ti} + 2O_o + \tfrac{1}{2}O_2 \Longrightarrow 2M'_{Ti} - 2O_o^{\cdot} + 2TiO_2$$

(3) subs. at Sr^{2+} site, electron compensation

$$M_2O_3 + 2Sr_{sr} + 2Ti_{Ti} \Longrightarrow 2M_{Sr}^{\cdot} + 2Ti'_{Ti} + 2SrO + \tfrac{1}{2}O_2$$

(4) subs. at Sr^{2+} site, Ti^{4+} vacancy compensation

$$M_2O_3 + 2Sr_{Sr} \Longrightarrow 2M^{\cdot} + \tfrac{1}{2}V_{Ti}'''' + \tfrac{1}{2}SrTiO_3 + {}^3/_2SrO$$

(5) self compensation

$$M_2O_3 + Sr_{Sr} + Ti_{Ti} \Longrightarrow M_{Sr}^{\cdot} + M'_{Ti} + SrTiO_3$$

Calculation of solution energies

e.g. case 1: $E_{SOL} = -E_{lat}(M_2O_3) + 2E_{sub}(M^{3+}/Ti^{4+}) + E(O^{2-}\text{ vac}) - 2E_{lat}(TiO_2)$

TABLE 5
Solution energies for M^{3+} substitution in $SrTiO_3$

	solution energies /eV				
dopant	(1)	(2)	(3)	(4)	(5)
Al^{3+}	4.57	8.79	13.65	15.84	4.89
Sc^{3+}	4.47	8.69	8.45	10.64	2.25
Cr^{3+}	6.24	10.46	5.76	7.95	1.78
Mn^{3+}	4.12	8.34	11.92	14.11	3.8
Y^{3+}	6.54	10.76	4.78	6.97	1.44
La^{3+}	10.34	14.56	3.44	5.63	2.67
Nd^{3+}	8.82	13.04	5.72	7.91	3.05
Eu^{3+}	7.76	11.98	4.06	6.25	1.69
Gd^{3+}	7.48	11.7	4.22	6.41	1.63
Ho^{3+}	6.53	10.75	4.75	6.94	1.42
Yb^{3+}	5.94	10.16	5.4	7.59	1.45
Lu^{3+}	5.75	9.97	5.55	7.74	1.43
Pu^{3+}	8.63	12.85	3.77	5.96	1.98

Lowest energy mechanisms underlined

REFERENCES

1. R.A. Jackson and C.R.A. Catlow, Molecular Simulation, 1, 207–244 (1988).
2. B.G. Dick and A.W. Overhauser, Phys. Rev. B, 112, 90 (1958).
3. M.P. Tosi, Solid State Physics, 16, 1 (1964).
4. P. Saul, C.R.A. Catlow and J. Kendrick, Phil. Mag. B, 51, 107 (1985).
5. M.J. Sanders, M. Leslie and C.R.A. Catlow, J. Chem. Soc. Chem. Comm., 1273 (1984).
6. G.D. Price and S.C. Parker, in "Physical Properties and Thermodynamic Behaviour of Minerals" (ed E. Salje), 591 (1988).
7. G.D. Price, S.C. Parker and M. Leslie, Phys. Chem. Miner., 15, 181 (1988).
8. S.M. Tomlinson, C.M. Freeman, C.R.A. Catlow, H. Donnerburg and M. Leslie, J. Chem. Soc. Far. Trans. II in press.
9. M.S. Islam, M. Leslie, S.M. Tomlinson and C.R.A. Catlow, J. Phys. C: Solid State Physics, 21, L109 (1988).
10. B.M. Axilrod and E. Teller, J. Chem. Phys., 11, 299 (1943).
11. R. Baetzold, C.R.A. Catlow, J. Corish, F.M. Healy, P.W.M. Jacobs and

Y. Tan, accepted for publication in J. Phys. Chem. Solids.
12. P.T. Wedepohl, Proc. Phys. Soc., 92, 79 (1967).
13. R.G. Gordon and Y.S. Kim, J. Chem. Phys. 56, 3122 (1972).
14. J.E. Post and C.W. Burnham, Amer. Miner., 71, 142 (1986).
15. C.R.A. Catlow, Proc. Roy. Soc. A353, 533 (1977).
16. J.J. Pluth and J.V. Smith, J. Am. Chem. Soc. 102, 4704 (1980).
17. C.R.A. Catlow, in "Defects in Solids: Modern Techniques", ed. A.V. Chadwick and M. Terenzi (NATO ASI Series B: Physics vol. 147) pub. Plenum, 1987.
18. N.F. Mott and M.J. Littleton, Trans. Far. Soc. 34, 485 (1938).
19. G.V. Lewis and C.R.A. Catlow, J. Phys. Chem. Solids, 47, 89 (1986).
20. B.K. Flandermeyer, M.M. Nasrallah, D.H. Sparlin and H.U. Anderson, in "Transport in Nonstoichiometric Compounds", ed. G. Simkovich and V.S. Stubican (NATO ASI Series B: Physics vol. 129) pub. Plenum, 1985.

COMPUTER MODELLING OF THE STRUCTURE AND THERMODYNAMIC PROPERTIES OF SILICATE MINERALS

S. C. PARKER
School of Chemistry, University of Bath, Claverton Down, Bath, U.K.

G. D. PRICE
Department of Geological Sciences, University College London, Gower Street, London, U.K.

1. INTRODUCTION

One of the major aims of mineralogical research is to determine which minerals exist under given pressure, temperature and compositional conditions. This is of particular importance as mass transport deep in the Earth's Mantle, and ultimately plate motion, will be governed by the behaviour of the component minerals. Hence a detailed understanding of their thermodynamic and kinetic properties under extreme pressures and temperatures will aid in our understanding of Mantle dynamics. Much of our current knowledge of the composition and properties of the Mantle comes from seismic studies, which show that the Earth has a layered structure. For example the Mantle is divided into three distinct layers, with the Upper Mantle comprising principally of olivine-structured magnesium silicate (forsterite), the Transition Zone of the same composition but comprised of minerals with the spinel structure, and finally another magnesium silicate mineral which is believed to have the perovskite structure forms the bulk of the Lower Mantle. These assignments are not unambiguous however but recent work using X-ray diffraction techniques with high pressure diamond anvil cells [1] is making progress in elucidating the crystal structure and relative stability of these phases under realistic conditions. As these experimental studies are often difficult, an attractive alternative is the use of computer simulation methods, which can now be used to model both the pressures and temperatures existing in the Mantle. Two recent developments in the field of atomistic simulation have made it possible to model silicate minerals under Mantle conditions. These are the development of a reliable potential model for silica [2], and the inclusion of entropic effects in the calculation of free-energies via a new computer code, PARAPOCS [3]. The new potential for silica includes three-body terms to simulate the covalent effects which are so important in determining the

C.R.A. Catlow et al. (eds.), Computer Modelling of Fluids Polymers and Solids, 405–429.

geometry and properties of the silicate (SiO_4^{4-}) tetrahedron. The calculation of free-energies enables us to predict crystal structures at elevated temperatures unlike earlier simulations where only zero Kelvin properties could be evaluated.

In this article we focus on some recent work which has made progress in the simulation of structural and thermodynamic properties of these important mantle-forming minerals to a high degree of reliability, and in determining which structures will be most stable under the Mantle conditions.

2. ATOMISTIC SIMULATION OF SILICATE MINERALS

To date the most reliable approach for predicting structural and thermodynamic data of inorganic solids without recourse to experiment is atomistic simulation. Here, the interactions between pairs or groups of atoms are described by potential functions, the sum of which yields the lattice energy of the crystal. This lattice energy is given by:

$$U(r_{ij}) = \sum_{ij} \frac{q_i q_j}{r_{ij}} + \sum_{ij} \phi_{ij} + \sum_{ijk} \Theta_{ijk} \qquad (1)$$

where the first term is the long range, Coulombic interaction corresponding to the interaction between the effective charges, q_i and q_j, on each atom. The second term, ϕ_{ij}, includes the repulsion between neighbouring charge clouds, van der Waals attraction, polarisability, etc. The electronic polarisability is modelled using the shell model [4], where an ion is represented by a mass-less shell attached to a core by a harmonic spring. The polarisability of the free-ion is approximated by the expression:

$$\alpha_i = \frac{Y_i^2 e^2}{k_i} \tag{2}$$

where $Y_i e$ is the charge of the shell and k_i the spring constant. The final term in equation (1) represents explicit three body interaction energy terms. In the case of silicates the expression usually chosen for the three body component is a simple harmonic about the equilibrium tetrahedral bond angle, given by:

$$\Theta_{ijk} = \frac{k_{ijk}}{2}(\Theta - \Theta_0)^2 \tag{3}$$

where Θ_0 is the equilibrium bond angle such as the O-Si-O bond angle, and k_{ijk} is the bond bending force constant.

In general, the expressions chosen for the interaction energies in the potential model are parameterised analytical expressions which readily yield analytical expressions for the derivatives of the potential. The reliability of the simulation in modelling a particular solid depends on the accuracy of the chosen parameters. Indeed it is usually the reliability of the potential model rather than the approximations in the simulation methods which determines the success of the simulation. Thus the inclusion of all the important interatomic interactions, and then the subsequent derivation of the appropriate parameters is critical in determining the outcome of the simulation. To date, simulation studies of silicates usually employ empirically derived potentials , where the parameters describing the interatomic forces are obtained by least squares fitting to experimental data, such as the crystal structure and the elastic constants, see Jackson this volume [4] and ref. [5]. The difficulty in fitting parameters to the silicate mineral properties is that many parameters need to be derived and often there are insufficient experimental data with which to make a comparison. Hence, the approach of assuming transferability of potential parameters is adopted. Thus parameters derived from the component oxides are transferred to the minerals in question, i.e. the parameters for

Mg_2SiO_4 are transferred from MgO and SiO_2 [2] (which we denote as THB1).

This approach for obtaining potentials is preferred when considering minerals over the modified electron gas method or *ab initio* methods because the latter do not include important effects such as dispersion and electronic polarisability, which are important for modelling defect properties and can be included in the empirical potentials. However, we recognise that with the increased reliability of the non-empirical potentials the position is likely to change. Indeed, considerable success has been enjoyed by the theoretical approach developed by Wedepohl [8] and Gordon and Kim [9] for ionic solids where the interaction energy is modelled using a modified electron gas approach, and recent work by Hemley et al. [10] has shown that such potentials can reproduce many properties of selected silicates.

As noted above the olivine phase of Mg_2SiO_4 (forsterite) was investigated because of its geological importance and because of the abundance of experimental data available on the phase with which to verify our potential models. The potential was tested by comparing the calculated structural properties of the polymorphs of Mg_2SiO_4 at zero Kelvin with those determined experimentally. Forsterite is the most stable phase and has an approximately hexagonal close-packed oxygen sublattice with magnesium and silicon residing in octahedral and tetrahedral holes respectively. The packing is such that the silicate tetrahedra are isolated from each other as can be seen from Figure 1.

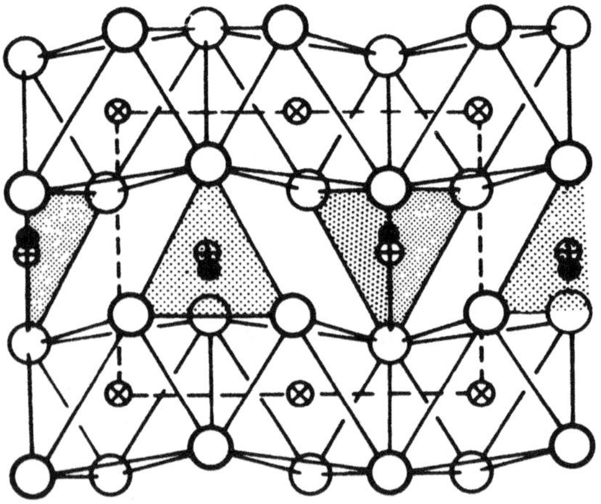

Figure 1. Structure of Mg_2SiO_4, olivine, where the open circles represent

oxide ions, the crossed are silicons and the filled circles are magnesium ions.

In table 1 we give the structural and elastic properties of forsterite as calculated by the empirical potentials.

	Observed[6,7]	Calculated
a(A)	4.754	4.784
b(A)	10.194	10.261
c(A)	5.981	5.991
Volume(cm^3mol^{-1})	43.67	44.26
Av. bond distance		
Si - O	1.61 - 1.64	1.61 - 1.66
Mg - O	2.05 - 2.13	2.02 - 2.26
K(Mbar)	1.33	1.54
μ (Mbar)	0.84	0.77

Table 1. Observed and predicted values for the properties of olivine Mg_2SiO_4.

The structure and elasticity data were chosen for detailed study as these properties depend on the first and second derivatives of the lattice energy and hence provide a check on the functional form of the potential model. The results demonstrate the success of the calculations in predicting the crystal structure as the calculated structural parameters are in general correct to within 1-2%. The largest discrepancy occurred for the MgO separation. This may be due to the neglect of three body interactions in the magnesium sublattice, which from the magnitude of the Cauchy violations on MgO (the c_{44} and c_{12} elastic constants are not equal) may be significant.

A further test of the model's validity is whether it can reproduce the properties of the two other Mg_2SiO_4 polymorphs which have the spinel and beta-phase structures. These phases are also geologically important as olivine in the Earth's Mantle is believed to transform first to beta-phase and then to spinel below the seismic discontinuity occurring at approximately 400 km depth. The structures differ from the olivine structure in the packing of the oxygen sublattice, which is cubic in spinel and beta-phase. Spinel and beta-phase are both spinelloid phases and can be thought to be built from the same structural units (see Figure 2a).

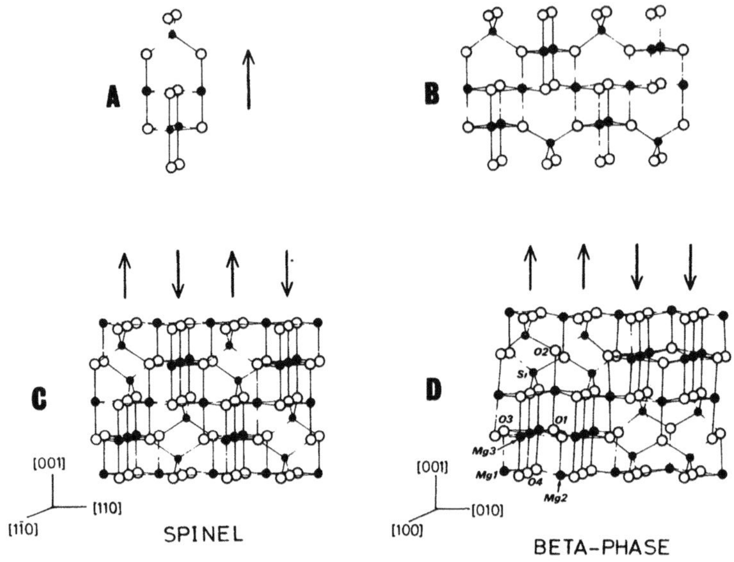

Figure 2. The structures of the spinel and beta-phase polymorphs.

The major structural difference is that the silicate (SiO_4^{4-}) groups found in olivine and spinel structures condense to form $Si_2O_7^{6-}$ groups in the beta-phase. As a result, one oxygen in the interface region between the parallel units in the beta-phase is bonded to five magnesium ions, and another to two silicon and one magnesium, in contrast to the usual coordination in spinel in which oxygen ions are bonded to one silicon and three magnesium ions. The calculated and experimentally determined structural and elastic data are summarised in Table 2.

The structural and elastic data are generally in good agreement, the largest deviation coming from the bulk modulus of spinel; the simulations predict spinel to be stiffer than is observed experimentally. However, the agreement is sufficiently good to give us the confidence to investigate the dynamic properties which should provide a still greater test for the potential model.

SPINEL	Observed [11]	Calculated
a(A)	8.065	8.020
Volume (cm^3mol^{-1})	39.65	38.82
Av. bond distance		
Si - O	1.66	1.66
Mg - O	2.07	2.05
K [12]	1.84	2.50
μ	1.19	1.26
BETA		
a(A)	5.698	5.658
b(A)	11.438	11.371
c(A)	8.257	8.291
Volume	40.54	40.18
Av. bond distance		
Si - O	1.63	1.63
Si - O (bridging)	1.70	1.73
Mg - O	2.02 - 2.13	2.02 - 2.12
K(Mbar) [14]	1.74	2.21
μ (Mbar)	1.37	1.17

Table 2. Observed and predicted properties for spinel and beta-phase.

3. THE LATTICE DYNAMICS OF MAGNESIUM SILICATE POLYMORPHS

The method described in this article for calculating the free-energy and thermodynamic properties of silicate minerals is based on lattice dynamics. This is an approach for calculating vibrational frequencies in periodic structures. The major limitation of calculations using lattice dynamics is the quasi-harmonic approximation, in which the vibrational motions in the solid are assumed to comprise independent quantised harmonic oscillators whose frequencies vary with cell volume. An alternative method for calculating the vibrational properties of solids is to use molecular dynamics where Newton's Laws of Motion are solved explicitly [15-17]. The major advantage of molecular dynamics is that the full anharmonicity of the potential energy surface can be calculated. This is possible because the atoms are in motion and hence after each time-step the atoms will sample a different part of the potential energy surface. In contrast, the atoms in lattice dynamics are held fixed when evaluating vibrational frequencies, and thus only sample the energy at the specific lattice sites where the shape of the potential energy-well is assumed to be harmonic. The disadvantage of molecular dynamics is that the calculation of thermodynamic properties is intensive in computer cpu time, and hence certain terms in the potential model are usually neglected, particularly the electronic polarisability, the inclusion of which would considerably increases the computer time. This may result in a simulation incorrectly predicting which phase is most stable as the energy differences caused by neglecting polarisability are often greater than the relative free-energies between different polymorphs. Thus lattice dynamics currently provides the most efficient method for determining thermodynamic properties and phase stabilities, provided anharmonic effects are not significant as is the case when the temperature of the mineral approaches its melting point or if it undergoes a displacive or soft-mode phase transition.

The first step in using the lattice dynamics approach to evaluate crystal properties is to determine the phonon frequencies. The method was elucidated by Cochran(18) for a shell model potential. The equations of motion to be solved are:

$$\frac{m\,d^2u}{dt^2} = -\frac{dV}{du} \qquad (4)$$

in which u is the displacement of the core, m its corresponding mass and V is the potential energy. In the case of the mass-less shells we have the condition:

$$0 = \frac{dV}{dw} \qquad (5)$$

where w is the shell displacement, and hence this corresponds to a zero net force on the each shell. The potential energy is again expanded to second order with respect to ion displacements i.e. the harmonic approximation:

$$V(r')-V(r) = (W_{uu}.u.u + W_{uw}.u.w + W_{wu}.w.u + W_{ww}.w.w) \qquad (6)$$

This assumes the unit cell is at equilibrium, W_{uu} represents the second derivative matrix between cores, W_{uw} between core and shell and W_{ww} between shells. Finally before solving the equations of motion the periodic nature of the solid must be considered by including the dependence of atomic displacements on the wavevector, q i.e.

$$u = u \exp(i(q.r - \varpi t)) \quad (7)$$

and

$$w = w \exp(i(q.r - \varpi t)) \quad (8)$$

where r is the atom position and ϖ the vibrational frequency. The second derivative or force constant matrix is similarly dependent on the wavevector. For example the force constant matrix for the interactions involving only shells is given by:

$$S = W \exp(iq(r_a - r_b)) \quad (9)$$

which is summed over all ions r_b. Similar expressions are obtained for the core-shell (T) and the core-core (R) matrices. Hence we can now solve the equations of motion (4) and (5) to get:

$$\varpi^2 m.u = R.u + T.w \quad (10)$$

and

$$0 = T.u + S.w \quad (11)$$

where m is the diagonal matrix of core masses. Removing the shell displacements w by substituting 10 into 11 gives:

$$\varpi^2 m.u = (R + T.S^{-1}.T).u \quad (12)$$

Finally by defining the dynamical matrix D as:

$$D = m.(R - T.S^{-1}.T).m \quad (13)$$

and

$$u = m.n \tag{14}$$

to give

$$\varpi^2 n = D.n \tag{15}$$

which is now the eigenvector problem and hence the frequencies can be calculated by diagonalising the dynamical matrix. Before describing the calculation of thermodynamic properties the predicted frequencies can be compared with those determined by spectroscopy.

3.1 SPECTROSCOPIC PROPERTIES

The infrared and Raman spectra of minerals, particularly those of geological importance, e.g. forsterite, have been extensively studied, by for example Servoin and Piriou [19], Paques-Ledent and Tarte [20] and Iishi [21]. We will illustrate the application to spectroscopy by considering two mantle-forming magnesium silicates, first forsterite which was described above and secondly, a $MgSiO_3$ phase with the ilmenite structure, which differs from the silicates discussed previously as silicon ions reside in octahedrally coordinated sites.

Forsterite has four formula units, and consequently 28 atoms, per unit cell, which give rise to 84 normal modes of vibration. Detailed analysis of the symmetry of these modes has been presented in a variety of works (see McMillan [22]) and has the point group symmetry of mmm. At the centre of the Brillouin zone ($q = 0$), three of these 84 modes are purely translational. The remaining 81 modes, however, are truly vibrational in nature. Of these, those which are symmetric with respect to inversion (grade or g) are Raman active, 10 with A_u symmetry are spectroscopically inactive, and the remaining 35 modes with B_{1u}, B_{2u} and B_{3u} symmetry are infrared active. These vibration frequencies can be modelled by simply solving equation (15), i.e. by diagonalising the dynamical matrix.

There is general agreement on the frequencies of the characteristic vibrations. Most infrared and Raman bands have been unambiguously identified, but there is still some controversy over some of the weaker peaks (see McMillan [22]). The spectroscopically active vibrational modes have frequencies that lie in the range between 1100 cm^{-1} and 150 cm^{-1}, but modes with frequencies greater than 800 cm^{-1} are separated by a 150 cm^{-1} wide gap. It is generally agreed that the high frequency modes are the result of Si-O stretching motions, while modes with frequencies below 650 cm^{-1} are the result of vibrations involving more complex displacements of Si, O and Mg. This is a characteristic of silicate minerals which the simulation should model. In addition, there is considerable uncertainty concerning the detailed nature of the atomic motions in many vibrational modes, despite the efforts which have been made to elucidate them by using isotopic substitution techniques (e.g. Paques-Ledent and Tarte [20]). Atomistic simulation provides a powerful tool for assigning the vibrational modes to particular atomic displacements, because the solution of equation (15) yields the eigenvectors which are directly related to the atomic displacements, from equation (14). Indeed, in the future this technique, if used in conjunction with experiment, could be part of a routine approach for analysing the infrared and Raman spectra of solids.

Detailed prediction of the vibrational modes of forsterite is difficult to assess as there appears to be some experimental uncertainty over the exact frequencies of some of the infrared active modes. The average error in the predicted Raman frequencies is only 20 cm^{-1}, and the predicted magnitudes of the TO-LO splittings of the infrared active modes are generally comparable with those observed. The success of the potential model can be judged by comparison with two other calculations, which were developed by using force constant models with parameters fitted directly to the observed vibrational frequencies. The two theoretical approaches differed in the number of parameters fitted: Devarajan and Funck [23] used twenty-four parameters, while Iishi [21] used fourteen. The differences between experiment and those calculated using the empirical potential are small with only 12 frequencies in error by more than 30 cm^{-1} [24], which compares favourably to the two force constant models where 11 and 16 frequencies are in error by more than 30 cm^{-1} [23].

Recent work by Wall and Price [25] has demonstrated that this approach can also be applied to high pressure silicates, such as the ilmenite-structred magnesium silicate. The structure is based on the corundum structure in which all cations occupy six-fold coordinated sites in a hexagonal close packed oxygen sublattice, as shown in Figure 3.

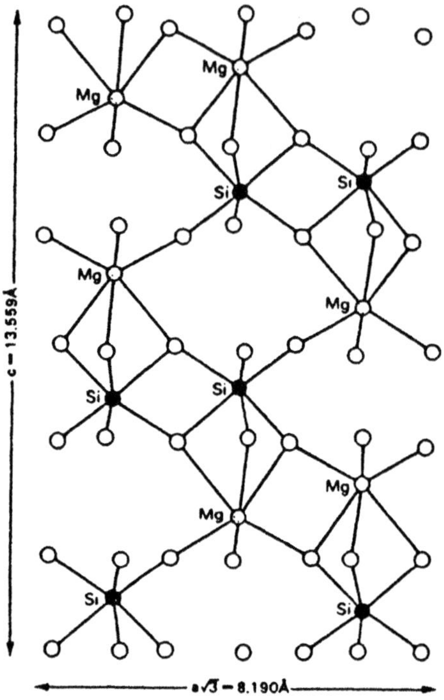

Figure 3. The structure of $MgSiO_3$ ilmenite.

The ilmenite phase has received attention as it may be a constituent of the Transition Zone. Wall and Price followed a similar procedure as that described above. As there are ten atoms in the primitive unit cell, this will give rise to twenty-seven optic modes of which fifteen will be Raman active, giving ten spectral bands and twelve will be infrared active giving eight absorption bands. The calculated Raman frequencies are compared with experiment [26], in Figure 4 and show good agreement with the largest difference of 40 cm^{-1}.

The power of this technique is clearly demonstrated as the experimental method could only identify nine of the ten bands, and it was left to simulation to predict that the unidentified band is at approximately 721 cm^{-1}. An important further advantage of this technique is that in addition to modelling spectroscopic data thermodynamic properties can be evaluated.

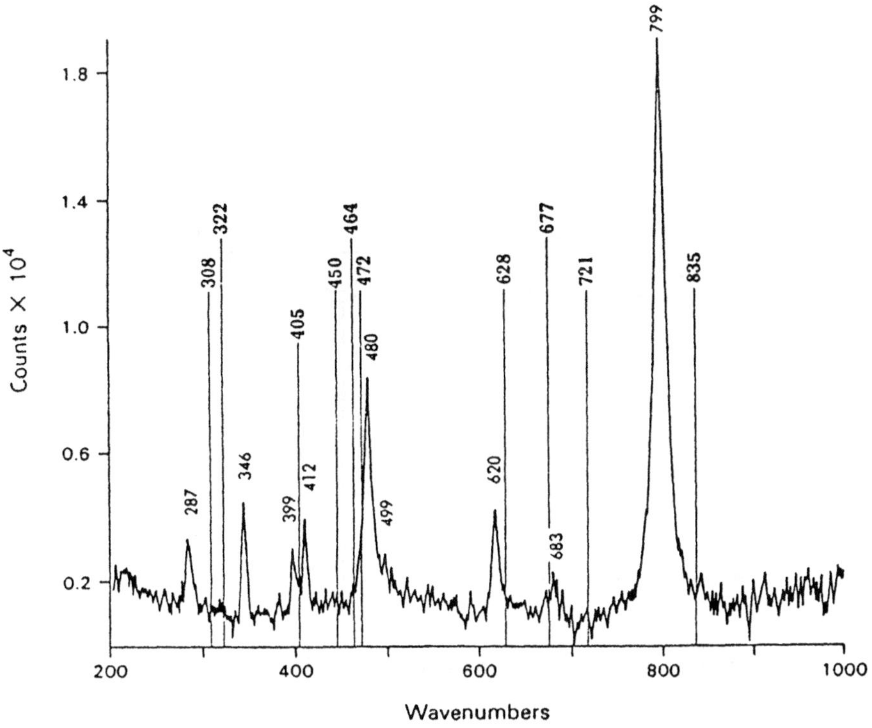

Figure 4. The calculated (bold figures) and experimental Raman frequencies for $MgSiO_3$-ilmenite

3.2 THERMODYNAMICS PROPERTIES

The calculation of thermodynamic properties is complicated by the fact that phonon frequencies depend on the wavevector. This dependence of vibrational frequency on q (inversely related to the wavelength of the lattice vibration) gives rise to the widely described phonon dispersion relationships, and reflects the effect that the phase of the vibrational mode in adjacent unit cells has upon the frequency of the mode. Because of the periodic nature of crystals, phonon dispersion can be fully described by simply considering the behaviour of phonons within the first Brillouin zone (the Wigner-Seitz reciprocal lattice unit cell, figure 5). Phonons with a wavelength large in comparison with the unit cell size (such as those stimulated by infrared radiation) have a q-vector which approaches zero, and so appear close to the Brillouin zone centre (the Γ point, figure 5), while those with a wavelength close to that of the unit cell, will fall near to the edge of the zone (e.g. points X,Y and Z in Figure 5).

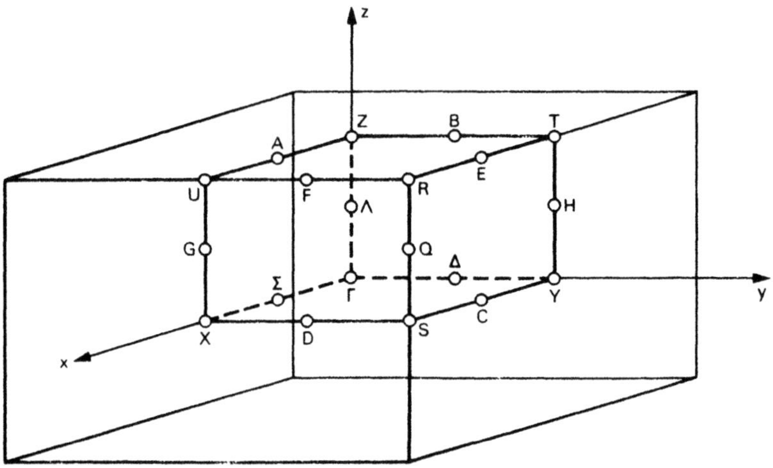

Figure 5. The Wigner-Seitz reciprocal lattice unit cell.

Thus when calculating the spectroscopic properties only one value of q near the zone centre needs to be considered. In contrast the thermodynamic functions E, S, F and C_V i.e. the vibrational energy, entropy, free energy and heat capacity respectively, are integrated over all values of q and are determined from the following expressions:

$$E = k_B T \ \Sigma_i \ (x/2 \ + x/(e^x - 1)) \tag{16}$$

$$S = k_B \ \Sigma_i \ (-\ln(1-e^{-x}) + x/(e^x - 1)) \tag{17}$$

$$F = k_B T \ \Sigma_i \ (x/2 \ + \ln(1-e^{-x})) \tag{18}$$

$$C_V = k_B \ \Sigma_i \ (x^2 e^x)/(e^x - 1) \tag{19}$$

where each property is summed over the total number of phonon frequencies, M, and where $x = h\varpi_i/kT$. The zero point energy $\Sigma_i h\varpi_i/2$ is also included in the vibrational energy and free-energy.

The need to calculate the frequencies for all wave vectors is overcome in part by choosing only those frequencies on a three-dimensional mesh of points within the Brillouin zone and using an appropriate weighting factor. The weighting will depend on the number of times a point appears in the Brillouin zone, thus the zone centre q=(0,0,0) which appears only once will have a weighting factor of one, while q=(0.5,0.5,0.5) for an orthorhombic crystal will have a weighting factor of eight. Fortunately the thermodynamic properties converge rapidly with the size of the mesh for temperatures above 50K. In Table 3 we give the entropy and heat capacity of forsterite which show good agreement with experiment [27,28].

	OBS	CALC.		OBS	CALC.	
T(K)	C_P	C_P^*	C_P^{**}	S	S^*	S^{**}
300	119.1	115.8	116.9	94.85	94.86	94.16
400	137.7	135.1	136.3	132.0	129.0	130.7
500	148.3	147.0	148.3	163.9	160.6	162.5
600	155.8	154.7	156.2	191.6	188.1	190.4
.						
.						
1000	175.1	168.9	171.1	276.2	272.9	274.6

* Calculated using the O K structure

** Calculated using the structure equilibrated at T K

Table 3. Observed and predicted values for C_P and S ($J.mol^{-1}K^{-1}$).

This also demonstrates the rapid convergence of the predicted thermodynamic properties with an increasing number of grid points in reciprocal space. Figure 6 shows that one such property, namely C_V has converged by considering eight points and is typical of other thermodynamic properties.

The failure at very low temperatures is because acoustic phonons with wave vectors close to the zone centre are the only thermally excited modes, therefore we need a fine mesh in this region to determine correctly the phonon frequencies and hence the thermodynamic properties. This may be overcome by using 'uneven sampling' techniques [29,30] which give an appropriate weighting to wave vectors near the zone centre.

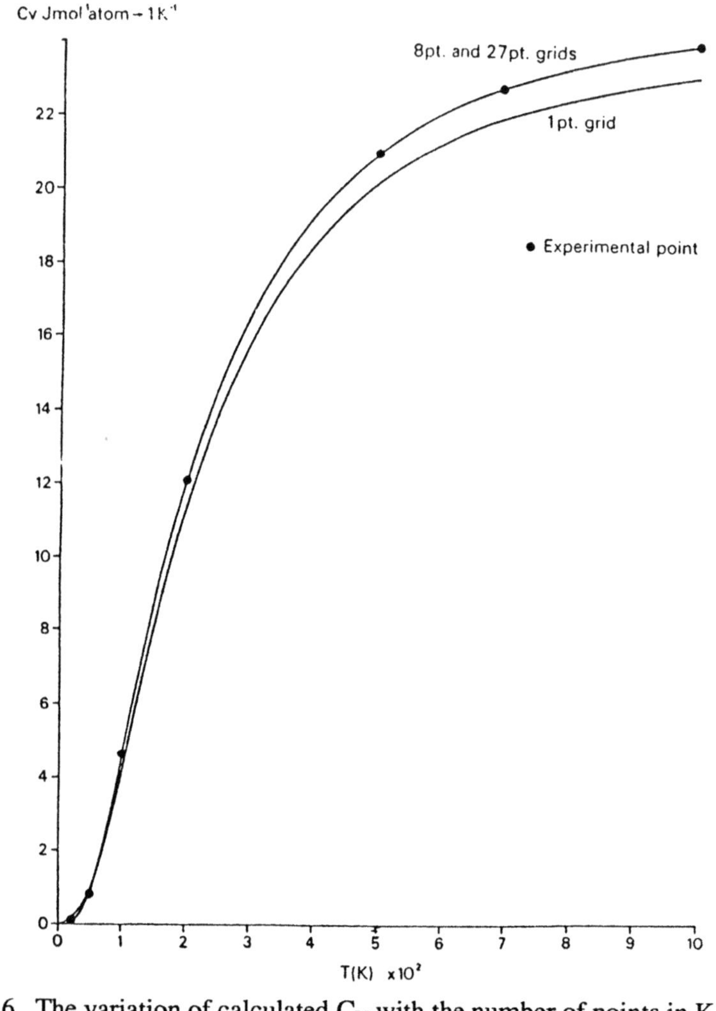

Figure 6. The variation of calculated C_V with the number of points in K-space compared to that determined experimentally.

3.3 THERMAL EXPANSION

One measure of the temperature dependence of the structure is the Gruneisen parameter, as it links the microscopic properties, namely the change in frequencies with the macroscopic properties, such as thermal expansion coefficient. The thermal Gruneisen parameter can be determined from the change in frequencies with volume directly, by first calculating the mode Gruneisen parameters for each frequency:

$$\gamma_i = \frac{d\ln \varpi_i}{d\ln V} \tag{20}$$

The bulk or thermal Gruneisen parameter is evaluated by taking the average of the heat capacity weighted mode Gruneisen parameters:

$$\gamma_T = \frac{1}{C_V} \sum C_i \gamma \tag{21}$$

where C_i is the contribution to the total heat capacity of mode i. The alternative approach is to use standard thermodynamic functions requiring the thermal expansion coefficient α, the isothermal compressibility, K_T or the adiabatic compressibility, K_S along with the heat capacities.

$$\alpha = \frac{\gamma_T C_V}{K_T V} = \frac{\gamma_T C_P}{K_S V} \tag{22}$$

The results are given in Figure 7 and show that although there is not exact agreement with experiment the same trends are observed. This approach has an important inconsistency which is that the mode Gruneisen parameters are obtained from the phonon frequencies calculated using the 0K structure and then extrapolated to the appropriate temperature.

Preferably the simulation should determine the mode Gruneisen parameters calculated from the structure equilibrated at the appropriate temperature. The predicted high temperature Gruneisen parameter was 0.90 using the frequencies derived from the 0K structure and the standard thermodynamic relations. This is compared to 1.14 determined by White et al. [31]. The simulated value represents a severe under-estimate, demonstrating the shortcomings of not modifying the cell volume with increased temperature. Thus we needed to determine the crystal structure at a given temperature and then determine its corresponding mode Gruneisen parameters.

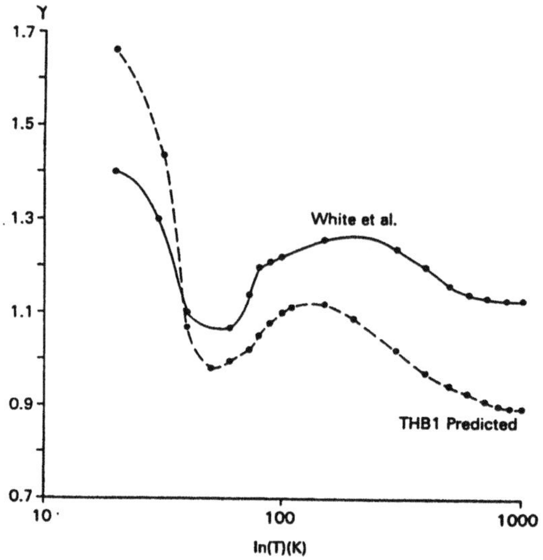

Figure 7. The calculated and predicted Gruneisen parameter as a function of temperature.

3.4 FREE ENERGY MINIMISATION

The crystal structure at a given temperature can be predicted by minimising the free-energy, which is the basis of the computer code PARAPOCS. The approach is to adjust the cell volume and coordinate positions until the net pressure or stress is zero. The pressure is simply the derivative of free energy with respect to volume. Thus for a cubic material:

$$P = \frac{dF}{dV} \tag{23}$$

The pressure is determined by calculating the free-energy at a given volume and then recalculating it after making a small adjustment to the cell volume dV. The problem becomes slightly more complicated for non-cubic materials as the volume will not expand isotropically. For these systems we have to consider six different strain components, ϵ_j. However, the same approach is used except that a small strain is applied in each of the six directions and hence the pressure becomes the difference in free-energy for each component [33]:

$$P_j = \frac{1}{V}\frac{dF}{d\epsilon_j} \qquad (24)$$

During the minimisation each atom position is adjusted so that it remains at a potential energy minimum, by performing a constant volume energy minimisation each time the cell volume is modified, see Jackson in this volume. This is carried out to reduce the possibility that an atom will be moved to a position where the harmonic approximation breaks down, such as at a potential energy maximum, leading to imaginary frequencies. Thus by minimising to constant pressure and including the vibrational component of the free-energy the crystal structure at a given temperature can be predicted.

The effect of adjusting the cell volume on the crystal properties can be shown by again considering the magnesium silicate polymorphs by minimising their free-energies for given temperatures. The effect on the thermodynamic properties is small. In Table 3 we give the entropy and heat capacities at a range of temperatures for forsterite which show a small improvement over the properties calculated using the 0 Kelvin structure.

The major improvements in the agreement with experiment arise when considering thermal expansion, see for example the thermal expansion coefficients which show an improvement of about 10%, as as can be seen in Figure 8.

The predicted high temperature Gruneisen parameters of the spinel and beta-phase obtained from the 0 Kelvin structures were 0.90 and 0.88 for the beta-phase and spinel respectively. On minimising to constant pressure at 1500K the values are also much improved increasing to 1.03 and 1.01 but are still less than the experimental high temperature values of 1.25 and 1.37. Unfortunately the small remaining differences can still have important consequences for the determination of the phase diagram. This can be demonstrated by considering the free-energy of the olivine and spinel polymorphs of Mg_2SiO_4 at 1000K over a range of pressures. The plot of free-energy versus pressure gives a straight line for both phases as shown in Figure 9.

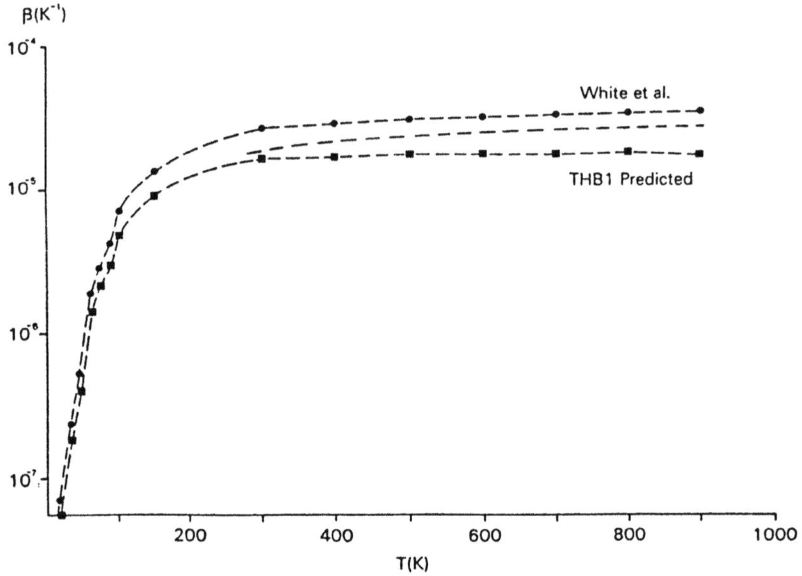

Figure 8. The experimental[31](black circles) thermal expansion of olivine compared with the predicted values, first using the 0 K structure (black squares), and second from the fully relaxed structure.

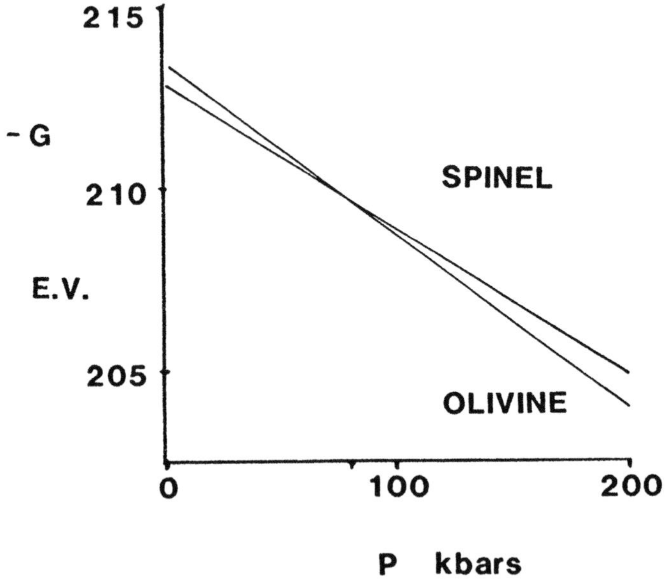

Figure 9. The variation of free-energy of the olivine and spinel structured magnesium silicate, versus pressure at 1000K

The gradients are dependent on the cell volumes and as the slopes for olivine and spinel are almost equal a small deviation in the gradient will have a marked effect on the point of intersection of the lines. However the point of intersection represents the pressure at which spinel is more stable than olivine. Thus a small error in the potential model, and hence cell volume can result in a large error in the predicted pressure for the phase transition for a given temperature. Taking due regard of this possible problem with the potential model the crystal structures and their free-energies of the three phases of Mg_2SiO_4 can be calculated for a range of pressures at different temperatures.

The variation of free-energy of the phases with pressure is shown in Figure 9, with the points of intersection representing the pressure at which one phase becomes more stable than another. The phase diagram can be obtained when these points of intersection are plotted as a function of temperature and pressure. In Figure 10 we give the experimental determination of the phase diagram by Kawada [36], which shows the three distinct regions of stability with no triple-point. In the high pressure region the spinel phase is the most stable and the middle region the beta-phase is most stable. This is then compared with a phase diagram inferred by Akoagi et al. [37], Sawamoto [38] and our own calculation [39]. The work by Akoagi and coworkers found a distinct triple point with a very small stability field for the beta-phase. Our calculation, in contrast, agrees more with Kawada showing a large beta-phase stability field. We also note that the calculated gradients are smaller than those derived by Kawada and agree more closely with the results determined by Akoagi which we believe to be reliable. There is however, still considerable uncertainty in our calculated stability field of the beta-phase because of the small free-energy differences between the phases involved.

A further benefit of the simulation techniques is that with reliable potentials the approach can be extended to pressures and temperatures not easily accessible to experiment. An important example of this is the phase transition from beta-phase and spinel to $MgSiO_3$ and MgO where the former has the perovskite structure and MgO has the rock-salt structure. This phase change is believed to account for the seismic discontinuity which separates the Transition Zone from the Lower Mantle. The difficulty with using our potential for perovskite is that the silicon is in six-fold coordination and the magnesium is in eight-fold coordination with cation-oxygen bond distances greater than found in the component oxides from which the potential was originally fitted. Despite this, we evaluated the phase relations which are given in Table

4 and shown in Figure 11 which display the characteristic negative slope in good agreement with available experimental data [37].

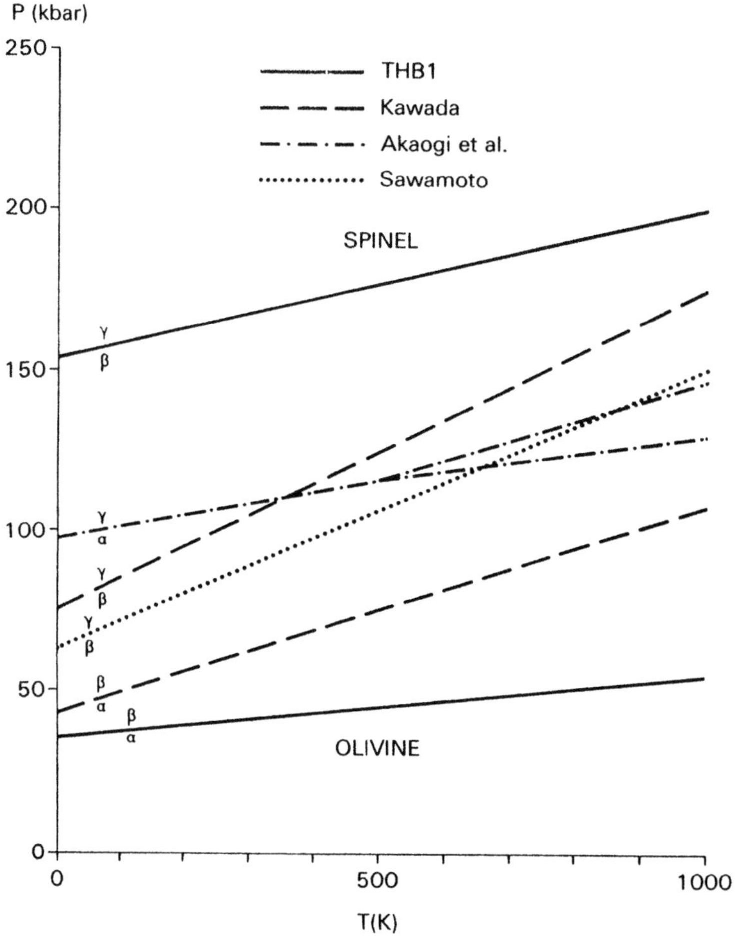

Figure 10. A comparison of the experimentally derived phase diagrams with the simulated phase diagram using the potential THB1

	Experimental [36-38]	Calculated
olivine - beta phase	98 + 0.035T 116 + 0.025T 57 + 0.058T	51 + 0.018T
beta phase - spinel	100 + 0.055T 117 + 0.048T 70 + 0.058T	113 + 0.056T
spinel - perovskite	273 - 0.020T	279 - 0.033T

Table 4. The phase relations for Mg_2SiO_4 in kbars.

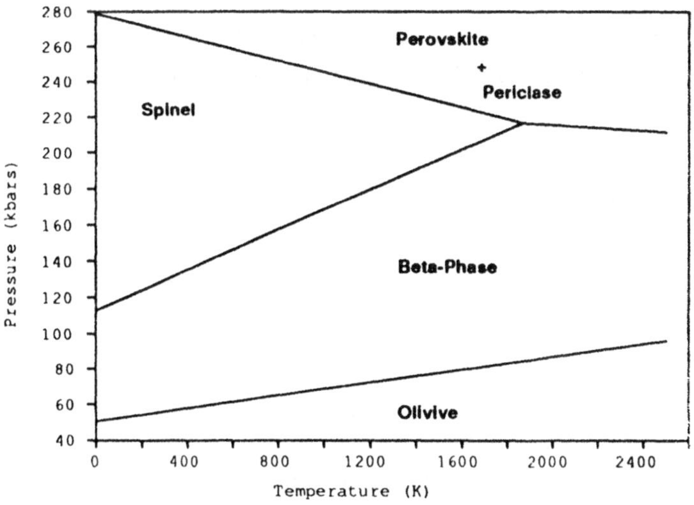

Figure 11. The predicted phase diagram for Mg_2SiO_4.

The simulation can be extended further to consider impurities, for example iron which is believed to occur in significant quantities in the Mantle and similar calculations could establish the effect of iron and other impurities on the phase diagram. The comparable experiments in which the influence of each possible impurity is investigated would be very difficult.

4. CONCLUSIONS

This work demonstrates the potential of computer simulation techniques in predicting the structures and phase diagrams of geologically important minerals. The agreement with experiment is encouraging and will increase as improved potential models become available. The improvements in the potential models for silicate minerals should come from two sources, first from the non-empirical techniques such as the modified electron gas method or more likely Hartree-Fock molecular orbital methods. Secondly, from the inclusion of potential parameters that are dependent on the Madelung fields generated by the component ions. Future developments in the simulation methods must come from the inclusion of anharmonic effects which play an important role in the calculation of high quality thermodynamic data. Methods for including anharmonic effects in lattice dynamics are already well-documented [29,30] but will greatly increase the computational requirements. The inclusion of such effects should ensure that lattice dynamics remains an important tool in the study of minerals.

ACKNOWLEDGEMENTS

We thank A. Wall for helpful comments and criticism and acknowledge CCP5 for supporting this collaboration.

REFERENCES

1. A. Navrotsky, Prog. Solid St. Chem. **17**, (1987), 53.
2. M.J. Sanders, M. Leslie and C.R.A. Catlow, J. Chem. Soc. Chem. Comm. (1984) 1271.
3. S.C. Parker and G.D. Price, Advances in Solid State Chemistry **1** (1989) 295.
4. R.A. Jackson - this volume.
5. G. Lewis and C.R.A. Catlow, J. Phys C **18** (1985) 1149.
6. R.M. Hazen, Amer. Mineral. **61**, (1976) 1280.
7. E.K. Graham and G.R. Barsch, J. Geophys. Res. **74** (1969) 5949.
8. P.T. Wedepohl, Proc, Phys. Soc. **92**, (1967) 79.
9. R.G. Gordon and Y.S. Kim, Phys. Rev. **12**, (1972), 431.
10. R.J. Hemley, M.D. Jackson and R.G. Gordon, Phys. Chem. Minerals, **14**, (1987) 2.

11. S. Sasaki, C.T. Prewitt, Y. Sato and E. Ito, J. Geophys. Res. **87** (1982) 7829.
12. D.J. Weidner, H. Sawamoto, S. Sasaki, J. Geophys. Res. **87** (1984) 7852.
13. H. Horiuchi and H. Sawamoto, **66**, (1981) 568.
14. H. Sawamoto, D.J. Weidner, S. Sasaki and M. Kumazama, Science -in press.
15. M.J. Sangster and M. Dixon, Adv. Phys **25** (1976) 247.
16. M.J. Gillan, Physica **B157** (1985).
17. A. Ladd - this volume.
18. W. Cochran, Crit. Rev. Sol. St. Sci. **2**, (1977), 1
19. J.L. Servoin and B. Piriou, Phys. Status Solidi **B55**, (1973) 677
20. M.T. Paques-Ledent, P. Tarte, Spectrochim. Acta **A29**, (1973) 1007.
21. K. Iishi, Amer. Mineral **63**, (1978) 1198
22. P. McMillan In Microscopic to Macroscopic. Reviews in mineralogy, **14**, (1985) (eds S.W. Kieffer and A. Navrotsky) Mineral. Soc. Amer, Washington, 9.
23. V. Devarajan and E. Funck, J. Chem. Phys. **62** (1974) 3406.
24. G.D. Price, S.C. Parker and M.Leslie, Phys. Chem. Minerals **15** (1987), 181.
25. A. Wall and G.D. Price, Amer. Mineral. **73**, (1988) 224.
26. N.L. Ross and P. McMillan, Amer. Mineral. **69**, (1984) 719.
27. R.A. Robie, B.S. Hemingway and H. Takei, Amer. Mineral. **67** (1982) 470.
28. R.G. Berman and T.H. Brown, Cont Mineral Petrol. **89**, (1985) 168.
29. G. Filippini, C.M. Gramacciolli, M. Simonetta and G.B. Suffritti, Acta Cryst **A32**, (1976), 259.
30. D.J. Chadi and M.L. Cohen Phys. Rev. **B8** (1973), 5747.
31. G.K. White, R.B. Roberts and J.G. Collins High Temp High Press **17**, (1985), 61.
32. G.D. Price and S.C. Parker In 'Physical properties and thermodynamic behaviour of minerals' (ed E.K.H. Salje) Kluwer Academic Publishers (1988).
33. T.H.K. Barron, J.G. Collins and G.K. White, Adv. in Physics **20** (1980) 609.
34. H. Watanabe In 'High-temperature research applications to geophysics (eds M.H. Manghnani and S. Akimoto) Centre for Academic Publications, Tokyo (1982)
35. Q. Williams, R. Jeanloz, M. Akaogi Phys. Chem. Minerals **13**, (1986), 141.
36. K. Kawada PhD Thesis, University of Tokyo (1977).
37. M. Akaogi, N.L. Ross, P. McMillan and A. Navrotsky Amer. Mineral. **69**, (1984), 499.
38. H. Sawamoto Phys. Chem. Minerals **13**, (1986), 1.
39. I.A. Marshall, S.C. Parker and G.D. Price - to be published.

COMPUTER SIMULATION EXERCISES

M. P. ALLEN
H. H. Wills Physics Laboratory
Royal Fort, Tyndall Avenue, Bristol BS8 1TL
D. M. HEYES
Department of Chemistry
Royal Holloway and Bedford New College, Egham, Surrey
M. LESLIE
TCS Division
SERC Daresbury Laboratory, Daresbury, Warrington WA4 4AD
S. L. PRICE
University Chemical Laboratory
Lensfield Road, Cambridge CB2 1EW
W. SMITH
TCS Division
SERC Daresbury Laboratory, Daresbury, Warrington WA4 4AD
D. J. TILDESLEY
Department of Chemistry
The University, Southampton SO9 5NH

Abstract

We present an overview of methods and applications of computer simulations in the field of fluids, polymers, and solids, through a set of exercises and projects. These exercises were actually undertaken by students attending the Advanced Study Institute: most involve access to a computer, but some require pen and paper only.

1 Introduction

A set of exercises, intended to give the students 'hands-on' experience of computer modelling techniques, formed an integral part of the Advanced Study Institute. Instruction sheets were made available to the participants, who were each allocated a

C.R.A. Catlow et al. (eds.), Computer Modelling of Fluids Polymers and Solids, 431–536.

user identifier on the Bath University mainframe computer. A brief introduction to the essentials of the operating system was given, together with information regarding the compiling and running of FORTRAN programs, the linking of numerical libraries, plotting facilities, etc. The problems classes themselves were conducted in the afternoons, with the authors and local experts on hand to give assistance and guidance where required, but access to the computer was fairly free, and some keen students worked well into the evenings!

Here we reproduce the instructions, modified only slightly from the originals. Some problems involved accessing files on the computer provided by the authors. We have had to omit these for reasons of space; nonetheless, the essentials of the problem should be clear in each case. Each problem was given a 'star' rating according to degree of difficulty: introductory exercises were rated '⋆' while small projects (perhaps suitable for a team) were rated '⋆⋆⋆⋆⋆'.

Here is a complete list of problems, grouped together by general subject area, with their authors identified by initials, and the degree of difficulty indicated. The numbers indicate sections within this chapter.

The first set of introductory exercises do not involve extensive programming; in some cases, they merely require pen and paper.

2 Long-range corrections and power-law potentials ⋆ (dmh)

3 Permanent electrostatic interactions ⋆ (djt)

4 To demonstrate the differences between simple electrostatic models ⋆⋆ (slp)

5 Lennard-Jones lattice energies ⋆⋆ (mpa)

6 Ionic crystal energy calculations. Direct summation ⋆⋆ (ml)

7 Lennard-Jones Monte Carlo ⋆⋆ (mpa)

8 Lennard-Jones molecular dynamics ⋆⋆ (djt)

9 Monte Carlo by hand ⋆ (mpa)

These are followed by some tasks involving the modification, writing, and running, of simulation programs.

10 Hard sphere Monte Carlo in one dimension ⋆⋆⋆ (djt)

11 Hard sphere dynamics in one dimension ⋆⋆⋆ (mpa)

12 A Monte Carlo simulation of Lennard-Jones atoms in one dimension ⋆⋆⋆ (djt)

13 Ising model simulations ⋆⋆⋆⋆ (mpa)

The third group of exercises concern the development of realistic intermolecular potentials.

14 The effects of small non-sphericities in the model atom-atom intermolecular potential on the predicted crystal structure of an A_2 molecule ⋆⋆ (slp)

15 Deriving a model potential for CS_2 by fitting to the low temperature crystal structure ⋆⋆ (slp)

This is followed by a series of problems highlighting a variety of interesting applications of simulation.

16 The Lennard-Jones fluid: a hard-sphere fluid in disguise? ⋆⋆⋆ (dmh)

17 Time correlation functions ⋆⋆ (djt)

18 Polymer reptation Monte Carlo ⋆⋆⋆ (mpa)

19 Simulations of crystal surfaces ⋆⋆⋆ (mpa)

20 Irreversible aggregation by Brownian dynamics simulation ⋆⋆⋆ (dmh)

21 Diffusion-limited aggregation ⋆⋆⋆⋆ (djt)

22 Shear flow in 2d Lennard-Jones fluid by non-equilibrium molecular dynamics ⋆⋆⋆⋆ (dmh)

The next four exercises deal with applications of the Fourier transform: they are preceded by an introductory section.

23 The Fourier transform and applications to molecular dynamics (ws)

24 Harmonic analysis ⋆ (ws)

25 Correlation functions ⋆⋆ (ws)

26 Particle density calculations ⋆⋆ (ws)

27 Quantum simulations ⋆⋆⋆ (ws)

Finally, there is a set of rather more substantial projects.

28 Percolation cluster statistics of 1D fluids ⋆⋆⋆⋆ (dmh)

29 Quaternions and constraints ⋆⋆⋆⋆⋆ (mpa)

30 The chemical potential of liquid butane ⋆⋆⋆⋆⋆ (djt)

31 Molecular dynamics on the Distributed Array Processor ⋆⋆⋆⋆⋆ (mpa)

2 Long-range corrections and power-law potentials

2.1 Summary

The long-range correction to the internal energy and the virial can have important consequences in many areas of simulation. This is a pen-and-paper exercise to give you practice at working out some of these expressions.

2.2 Background

Let us consider a monatomic fluid or crystal in which the particles interact through a simple pair potential of the form

$$\phi(r) = \epsilon(\sigma/r)^n, \tag{2.1}$$

where r is the separation between the two particles, n is usually an integer (positive), ϵ is a characteristic energy and σ is a characteristic "molecular" diameter. The internal energy, U, and pressure, P, in a simulation are defined by

$$U = \sum_{i=1}^{N-1} \sum_{j>i}^{N} \phi(r_{ij}) \tag{2.2}$$

and

$$P = \frac{1}{3V}(\sum_{i=1}^{N} m_i \dot{r}_i^2 - \Psi), \tag{2.3}$$

where Ψ is the so-called *virial*,

$$\Psi = \sum_{i=1}^{N-1} \sum_{j>i}^{N} r_{ij} \frac{d\phi(r_{ij})}{dr}. \tag{2.4}$$

In these expressions $\rho = N\sigma^3/V$, where N is the number of particles in a simulation cell of volume V. The mass of particle i is m_i. $\mathbf{r}_i$ is the position of particle i; $\dot{\mathbf{r}}_i = d\mathbf{r}_i/dt$, $\mathbf{r}_{ij} = \mathbf{r}_i - \mathbf{r}_j$.

In any simulation it is necessary to truncate pair-interactions beyond some specified pair separation distance, which we will denote by r_c. Short range potentials, $n > 1$, are normally truncated at several σ and certainly $r < V^{1/3}/2$ for a cubic unit cell, or half the minimum sidelength for a non-cubic simulation cell - so that the particle only interacts with one image of another particle. Therefore the contributions to eqns (2.2) - (2.4) for those interactions for which $r_{ij} > r_c$ are set to zero. An estimate of the effect of these neglected interactions can be made by assuming that there is a "smeared-out" distribution of molecules for $r_{ij} > r_c$ of density ρ. These *long-range corrections* to the internal energy and virial *in three dimensions* are:

$$U_{lrc} = 2\pi\rho \int_{r_c}^{\infty} r^2\phi(r)dr. \tag{2.5}$$

and

$$\Psi_{lrc} = 2\pi\rho \int_{r_c}^{\infty} r^3 \frac{d\phi(r)}{dr} dr. \tag{2.6}$$

In eqns (2.5) and (2.6) we have assumed that the pair radial distribution function, $g(r) = 1, r > r_c$.

2.3 The problem

The object of this exercise is to evaluate expressions for the long range corrections to the energy and pressure for potentials $\phi(r) = \epsilon(\sigma/r)^n$ in 1, 2 and 3 dimensions. Take the truncation distance to be r_c. Assume that the pair distribution function is unity for pair separations $r > r_c$.

2.4 Further tasks

- Is the long-range correction larger for the internal energy or the pressure?
- By inspection discover the smallest value of n one can use in each dimension, without having summation convergence problems.
- Are eqns (2.5) and (2.6) also valid for a crystal simulation?

3 Permanent electrostatic interactions

3.1 Summary

This problem offers some exercises on modelling permanent electrostatic interactions for use in simulations. In the main these problems can be tackled using pencil and paper, but you will need to resort to the computer from time to time.

3.2 Background

There are three types of long-range intermolecular interactions which bind molecules together in condensed phases. All molecules interact through an attractive dispersion potential. This arises from the coupling between an induced-dipole on one molecule and an induced-dipole on a neighbour. The interaction falls off with the inverse sixth power of the intermolecular separation. It is approximately pairwise additive, generally anisotropic, and increases with increasing molecular polarizability. It is the dominant attractive interaction in many condensed phases e.g. liquid N_2 and solid CS_2 . A less important interaction is that due to induction. For instance, the electrostatic dipole in HCl creates a field at a neighbouring molecule which couples with the polarizability to produce a temporary dipole in a neighbour. The induced and permanent dipole attract. This interaction is not pairwise additive and is normally small if the molecule is in a reasonably symmetrical environment such as provided by the solid or liquid. The permanent electrostatic interactions arise from the coupling of the charge distributions. The lowest non-zero electrostatic moment depends on the symmetry of the molecule: for a heteronuclear diatomic such as CO, the lowest non-zero moment is a dipole; for a homonuclear diatomic such as O_2 it is a quadrupole, and for a tetrahedral molecule such as CH_4 it is an octopole. These interactions are usually anisotropic, exactly pairwise additive, and can be attractive or repulsive. In computer simulations of solids and liquids, the electrostatic interactions are normally added to the core of the molecule. This core might be the three Lennard-Jones sites in a model of CO_2 which provide the anisotropic dispersion and repulsion between molecules. The electrostatic potential is added to the core in one of three ways. We can represent the charge distribution as a set of point electrostatic moments located at the centre of mass of each molecule. For two CO molecules the first term in the series will be the dipole-dipole interaction which falls off as r^{-3} . The orientational dependence is a sum of first order spherical harmonics, and the strength of the interaction will depend on the square of the magnitude of the dipole moment. CO also has a sizeable quadrupole and the next terms in the series will be the dipole-quadrupole and the corresponding quadrupole-dipole interactions. The quadrupole-quadrupole interaction falls off as r^{-5} , depends on second order spherical harmonics of the relative molecular orientations and the square of the magnitude of the quadrupole moment. There are many higher order terms and the series may not be rapidly convergent. It is certainly not convergent for intermolecular separations which are less than the length of the molecules. In an alternative description of the charge distribution the moments can always be represented by distributing partial charges inside the core of the molecule. These charges should produce the lowest non-zero moment and will also produce higher moments automatically. Even if the magnitude of the charges are chosen to fit the lowest moment, there is no certainty that these higher moments be the same size or sign as the experimentally measured values. A dipole requires a minimum of two charges and a quadrupole a minimum of three. The lowest moments can of course be represented by more charges and this gives some flexibility for fitting some of the higher moments, i.e. a useful model for water involves four charges distributed tetrahedrally to produce a dipole and to mimic the lone pairs of electrons. These charge distributions can be unrealistic. For example, N_2 has a negative quadrupole moment which can be modelled as two negative charges, one on each atom, and a double positive charge at the centre of mass of the molecule. However, chemical intuition tells us that there should be a build up of negative charge in the centre of the triple bond. Perhaps the most attractive

Figure 3.1: A configuration of CS_2 molecules

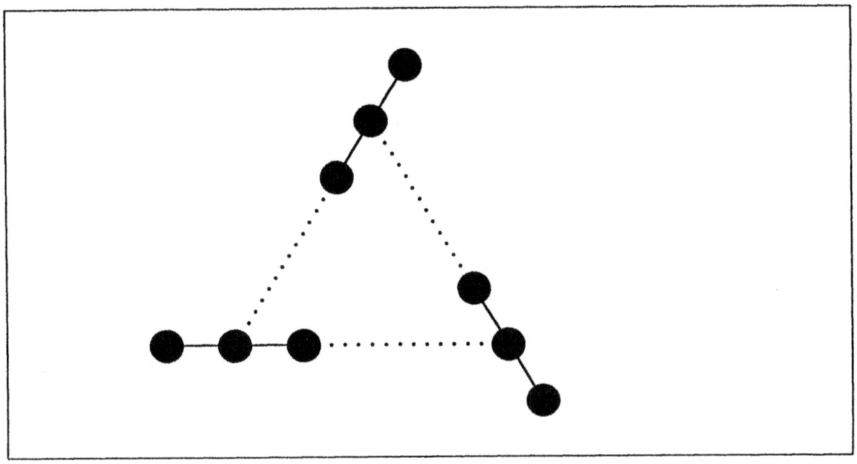

scheme is the use of a number of multipole moments distributed at sites within the hard core. This approach has been pioneered by Stone and co-workers and can give an accurate and simple description of the complete charge distribution, which is more rapidly convergent than the traditional multipole expansion. An interesting example is chlorine where the overall charge distribution can be represented by a dipole and quadrupole situated on each of the atoms. There is not sufficient space in these notes to give any sort of description of the detail of these three techniques. However there are many useful references which give explicit formulae and a more detailed discussion of the problem [1,2,3,4]. The reader is also referred to appendix D of [2], which has a complete discussion of the units used to describe multipoles and charges, and a comprehensive list of experimentally and theoretically measured multipoles.

3.3 Task

The following problems are roughly in order of difficulty. It is not necessary to do them all. You can join in when the bar is at the right height.

- The quadrupole moment of CS_2 is $7 \times 10^{-40} \mathrm{Cm}^2$. The separation of the sulphur atoms is 0.267nm. What charges (in units of the electronic charge) should be placed at the three nuclei to represent the quadrupole? Three CS_2 molecules are arranged in a plane with their centres on the vertices of an equilateral triangle, as shown in the figure. If the centres of the molecules are separated by 0.4nm, what is the quadrupolar energy of the trimer?
- In a recent MD study of Cl_2 , the charge distribution was described by placing a dipole and quadrupole on each of the chlorine atoms. The bondlength of the

molecule is 0.1994nm, the dipole is -0.1449 ea_0, and the quadrupole is 1.8902 ea_0^2 . (These moments are given in the usual atomic units). What are the quadrupole and hexadecapole moments of this molecule referred to its centre of mass. In a simulation of chlorine all interactions were cutoff at centre of mass separation of 0.35nm. What is the long-range correction to the electrostatic energy in this model?

- The lowest non-zero moment of CF_4 is the octopole moment. The value obtained from collision induced adsorption studies is 4.0×10^{-3}esu cm^3. Develop a five charge model for the electrostatic interaction between a pair of CF_4 molecules. What is the most favourable orientation of the pair?

- The first four non-zero moments of nitrogen have been estimated using ab-initio methods: $M_2 = -1.041ea_0^2; M_4 = -6.600ea_0^4; M_6 = -15.44ea_0^6; M_8 = -29.40ea_0^8$. Develop a five charge model which describes this charge distribution.

- The dipole moment of OCS is 0.7152×10^{-18}esu cm. The quadrupole moment is -2.0×10^{-26}esu cm^2. If the higher order moments are zero, what is the favoured orientation of a pair of OCS molecules?

4 To demonstrate the differences between simple electrostatic models

4.1 Task

Model potentials for small molecules often include an electrostatic term which is designed to give the correct electrostatic energy at long range (molecular separation $R \gg$ molecular dimensions) by ensuring that the model reproduces the experimental value for the first multipole moment. This can be done by either using the first central multipole moment in a one site model, or by using a set of point multipoles on every atomic site, or other choices of multiple interaction sites. The different choices can produce very different estimates of the electrostatic energy for smaller separations of the molecules, as they correspond to different implicit values of the higher multipole moments.

To demonstrate this, use the formulae below to derive (a) a point charge (Q_{00}) model with sites on each C and H atom, and (b) a distributed quadrupole (Q_{20}) model with sites on just the carbon atoms, for acetylene (HCCH) and benzene (C_6H_6) which correspond to their central multipole moments $Q_{20}^T = 5.46ea_0^2$ and $Q_{20}^T = -7.22ea_0^2$ respectively (SCF values). What assumptions are these two models making about the form of the charge distributions of these molecules? Acetylene has the z axis along the molecule and the bondlengths are C-C 2.274 a_0 and C-H 2.004 a_0. Benzene has the z axis along the sixfold axis and the C-C and C-H bondlengths are 2.634 a_0 and 2.033 a_0 respectively.

Write a small FORTRAN program for each molecule to compare electrostatic energy calculated from these two multi-site models with the energy calculated from the central quadrupole moment, for various orientations. For acetylene, consider a T-shaped, crossed (X) and staggered parallel structure. For benzene, consider a parallel plate sandwich structure, and a T shaped dimer, both with a side and a vertex pointing into the ring. The seperation of the molecules should be such that there is no significant overlap of the molecular charge distributions, so the C···C intermolecular separations should not be much smaller than about 4Å.

It is also instructive to assess the importance of the electrostatic term relative to the other contributions to the potential. This can be done very approximately by adding an isotropic atom-atom repulsion-dispersion potential such as the one given below. Use this to predict the dimer geometry for these molecules when there is no electrostatic term, and with the different simplified electrostatic models above. (There is a discussion of the importance of the electrostatic model in predicting the dimer structure of benzene by Price and Stone, [5]).

4.2 Notes

Atomic units $1a_0 = 0.529$Å; $1E_h = 2.6255 \times 10^3\,\mathrm{kJ\,mol^{-1}}$

Formulae for total multipole moments [4]:

$$Q_{20} = \sum_{sites} Q^i_{00}\tfrac{1}{2}(3z_i^2 - r_i^2) \quad = \sum_{sites} Q^i_{20} \tag{4.1}$$

Formulae for electrostatic energy for charge and quadrupole models [4]:

$$U_{estatic} = (4\pi\epsilon_0)^{-1} Q^1_{00} Q^2_{00} R^{-1} \tag{4.2}$$

$$U_{estatic} = (4\pi\epsilon_0)^{-1} Q^1_{20} Q^2_{20} \tfrac{3}{4}[35\mathbf{z}_1.\hat{\mathbf{R}}^2\mathbf{z}_2.\hat{\mathbf{R}}^2 - 5\mathbf{z}_1.\hat{\mathbf{R}}^2 - 5\mathbf{z}_2.\hat{\mathbf{R}}^2 \\ -20\mathbf{z}_1.\hat{\mathbf{R}}\ \mathbf{z}_2.\hat{\mathbf{R}}\ \mathbf{z}_1.\mathbf{z}_2 + 2\mathbf{z}_1.\mathbf{z}_2{}^2 + 1]R^{-5}. \tag{4.3}$$

The orientation dependence of the quadrupole-quadrupole interaction is given in terms of the unit vectors along the quadrupole axes ($\mathbf{z}_1$ and $\mathbf{z}_2$) and a unit vector $\hat{\mathbf{R}}$ from site 1 to 2.

Repulsion-dispersion potential (from [6]) has atom-atom form:

$$U_{atom-atom} = B\exp(-CR) - AR^{-6} \tag{4.4}$$

with parameters given in the Table.

Table 4.1: Williams potential parameters [6]

	$A/\,\mathrm{kJ\,mol^{-1}\,\AA^{6}}$	$B/\,\mathrm{kJ\,mol^{-1}}$	$C/\mathrm{\AA}^{-1}$
C···C	2439.8	369743	3.60
H···H	136.4	11971	3.74
C···H	$\sqrt{A_{CC}A_{HH}}$	$\sqrt{B_{CC}B_{HH}}$	$\frac{1}{2}(C_{CC}+C_{HH})$

5 Lennard-Jones lattice energies

5.1 Summary

Calculate the lattice energy for the Lennard-Jones fcc crystal and obtain the zero-pressure equilibrium density and nearest-neighbour separation. Estimate the relative contributions to the lattice energy due to close neighbours and those at large distances, and compare with the approximate long-range correction formula used for isotropic liquids.

5.2 Background

The energy of a perfect crystal at zero temperature can be calculated, given the crystal structure, the unit cell dimensions or nearest-neighbour separation, and the form of the interaction potential. Let us assume pairwise additivity for simplicity:

$$\mathcal{V} = \frac{1}{2}\sum_i \sum_{j\neq i} v(r_{ij}). \tag{5.1}$$

Each sum goes from 1 to N, the total number of atoms, and the term $i = j$ is omitted. The factor of $\frac{1}{2}$ compensates for double-counting of interactions. In many cases, we take a simple Lennard-Jones pair potential

$$v(r) = 4\epsilon\left(\left(\frac{\sigma}{r}\right)^{12} - \left(\frac{\sigma}{r}\right)^{6}\right) \tag{5.2}$$

or more generally a sum of algebraic terms

$$v(r) = \sum_n v_n(r) \tag{5.3}$$

with

$$v_n(r) = C_n r^{-n}. \tag{5.4}$$

In these equations, ϵ and σ are, respectively, energy and length parameters in the potential. In eqn (5.4) the constant C_n has the units of $\epsilon\sigma^n$.

Table 5.1: Lattice sums A_n

n	simple cubic	face-centred cubic	body-centred cubic
4	16.5323	25.3383	22.6387
5	10.3775	16.9675	14.7585
6	8.4019	14.4539	12.2533
7	7.4670	13.3593	11.0542
8	6.9458	12.8019	10.3552
9	6.6288	12.4925	9.8945
10	6.4261	12.3112	9.5644
11	6.2923	12.2009	9.3133
12	6.2021	12.1318	9.1142

Writing, in the same way, $\mathcal{V} = \sum_n \mathcal{V}_n$, using the fact that each atom i in a simple lattice is equivalent to any other, and placing such an atom at the origin, we can write eqn (5.1), for each term $\mathcal{V}_n$, as

$$\frac{\mathcal{V}_n}{N} = \frac{1}{2} C_n \sum_{\mathbf{r}_j \neq \mathbf{0}} r_j^{-n}. \tag{5.5}$$

The work of summing the right-hand side over all lattice sites $\mathbf{r}_j$ need only be performed once for each crystal structure. Distances are conveniently scaled by r_{nn}, the nearest-neighbour separation, so we define $\tilde{\mathbf{r}}_j = \mathbf{r}_j / r_{nn}$. Then,

$$\frac{\mathcal{V}_n}{N} = \frac{1}{2} C_n r_{nn}^{-n} \left(\sum_{\tilde{\mathbf{r}}_j \neq \mathbf{0}} \tilde{r}_j^{-n} \right) = \frac{1}{2} C_n r_{nn}^{-n} A_n. \tag{5.6}$$

In the case of the Lennard-Jones potential, this becomes

$$\frac{\mathcal{V}}{N} = 2\epsilon \left(A_{12} \left(\frac{\sigma}{r_{nn}} \right)^{12} - A_6 \left(\frac{\sigma}{r_{nn}} \right)^{6} \right). \tag{5.7}$$

The lattice sums A_n are just constants; they appear in the table, for the common cubic Bravais lattices [7]. $\mathcal{V}/N$ depends on density through r_{nn}. The equilibrium, zero-pressure, value may be obtained by differentiating the above expression with respect to r_{nn} and setting the derivative to zero. Further details may be found in standard texts on solid-state physics [8].

The contributions of nearest-neighbours to the total energy can be worked out directly, given r_{nn}, and a little more work gives the next-nearest-neighbour contributions and so on. The residue is due to long-range interactions. In liquid-state simulations [3], it is common practice to truncate the interaction beyond a certain cutoff range r_c, and to approximate the long-range part by the equation

$$\frac{\mathcal{V}^{LRC}}{N} = 2\pi\rho \int_{r_c}^{\infty} r^2 v(r) dr. \tag{5.8}$$

For the Lennard-Jones potential, this gives

$$\frac{\mathcal{V}^{LRC}}{N} = \frac{8}{9}\pi\rho\sigma^3 \left(\frac{\sigma}{r_c}\right)^9 - \frac{8}{3}\pi\rho\sigma^3 \left(\frac{\sigma}{r_c}\right)^3. \tag{5.9}$$

These equations assume that there is no structure ($g(r) = 1$) beyond $r = r_c$, a reasonable assumption in liquids provided r_c is not too small.

5.3 Task

Use eqn (5.7) to determine the equilibrium (minimum energy) nearest-neighbour distance in the fcc Lennard-Jones solid at zero temperature. Calculate the lattice energy $\mathcal{V}/N$. How much of this is due to atomic neighbours at distance $r > 1.25\sigma$? How much due to neighbours at $r > 2.5\sigma$? Compare these two values with the approximation of eqn (5.9).

5.4 Further work

It is instructive to try and calculate the sums A_n for a simple lattice. A crude approach is to sum over neighbouring atoms, out to some large distance beyond which, to the required accuracy, the values of A_n no longer change. Write a computer program to calculate $A_n, n = 4, 6, 8, 12$ for the hexagonal close-packed lattice; you will need to look up the lattice vectors for this structure [8]. Expect the answers to be very close to those for the fcc lattice. In fact, you will need more significant figures than are given in the Table, in order to distinguish these structures. You may wish to repeat the calculation for fcc, to obtain greater accuracy, and so convince yourself that there is a difference! Particularly for low n, the direct summation method does not converge very rapidly, and more sophisticated methods are preferable [7].

6 Ionic crystal energy calculations. Direct summation

6.1 Summary

Write a program to compute the electrostatic energy of an ionic crystal using brute force methods. A number of input datasets will be provided.

6.2 Background

Ionic crystal energies are usually calculated using the Ewald method [9]. This problem investigates brute force summation of small clusters of ions to highlight the difficulties inherent in this method and show the need for the Ewald sum. Clusters will be set up with the same parallelepiped shape as the unit cell. The examples all have the calcium fluoride structure. Let the lattice vectors be $\mathbf{V}(1)$, $\mathbf{V}(2)$ and $\mathbf{V}(3)$, the NBAS basis atoms have coordinate vectors $\mathbf{R}(I)$ and charges Q(I). The energy required to remove an ion from the central cell of a cluster of ions made by repeating the central cell from -M to +M in all 3 directions is

$$E(I) = \sum_{L1=-M}^{M} \sum_{L2=-M}^{M} \sum_{L3=-M}^{M} \sum_{J=1}^{NBAS} \frac{Q(I) * Q(J)}{|\mathbf{R(I)} - (\mathbf{R(J)} + L1 \times \mathbf{V}(1) + L2 \times \mathbf{V}(2) + L3 \times \mathbf{V}(3))|} \tag{6.1}$$

The denominator in this sum is the distance between ion I and ion J translated by L1 times lattice vector 1 + L2 times lattice vector 2 + L3 times lattice vector 3. The interaction when I = J and I and J are both in the central cell is of course omitted. Compute also the total energy of the unit cell,

$$ETOT = \frac{1}{2} \sum_{I=1}^{NBAS} E(I) \tag{6.2}$$

It is also of interest to compute two more quantities. The dipole moment of the central cell

$$\mathbf{D} = \sum_{I=1}^{NBAS} Q(I) \times \mathbf{R}(I) \tag{6.3}$$

and the second moment of the charge distribution of the central cell

$$S = \sum_{I=1}^{NBAS} Q(I) \times \left|\mathbf{R}(I)^2\right| \tag{6.4}$$

6.3 Task

A number of input datasets are provided: `clus1.dat`, `clus2.dat`, `clus3.dat`, `clus4.dat` The first three lines give the Cartesian components of the unit cell lattice vectors. `clus1.dat` and `clus2.dat` are primitive unit cells with three atoms per unit cell and lattice vectors $(\frac{1}{2}\frac{1}{2}0)$, $(\frac{1}{2}0\frac{1}{2})$, $(0\frac{1}{2}\frac{1}{2})$. `clus3.dat` and `clus4.dat` are non-primitive unit cells with 12 atoms per unit cell and non-primitive lattice vectors (100), (010), (001). The next line in the input dataset gives an integer number of ions to be read in NBAS. This is followed by NBAS lines giving the three Cartesian coordinates of the ion position $\mathbf{R}(I)$ and the ion charge $Q(I)$. Write a program to compute the electrostatic energy of a unit cell of these crystals. Compute also the individual ion energies, the unit cell dipole moment and the second moment. It is suggested that you set up clusters of ions with $M = 1, 3, 5$ and 7. Examine the results to see:-

- Does the total energy of the unit cell and the energy required to remove a single ion converge as M increases, and does the convergence depend on the central cell dipole moment and second moment?
- If the total energy and single ion energies do converge, are the results the same for the four different input datasets?

There is no need to worry about the units the result is computed in; only the relative values matter for this exercise.

7 Lennard-Jones Monte Carlo

7.1 Summary

Run a Monte Carlo program, which simulates a system of Lennard-Jones atoms. Starting from a face-centred cubic lattice, monitor the various indicators of equilibration (potential energy, pressure, translational order parameter, mean-square displacement) as the crystal melts to form a liquid. Observe the effects of changing the maximum atomic displacement parameter in the Monte Carlo algorithm. Compare your results with the known properties of the Lennard-Jones fluid.

7.2 Background

A system of atoms interacting via the pairwise Lennard-Jones potential

$$v(r) = 4\epsilon\left((\sigma/r)^{12} - (\sigma/r)^{6}\right) \tag{7.1}$$

is a very common test-bed for simulation programs. In this exercise, a Monte Carlo program to simulate a system of 108 Lennard-Jones atoms is provided. It can be found in the file `mclj`, and it may be helpful to have a copy in front of you (on the screen or on paper) as you read this.

The program starts from an initial configuration, which can either be read in from a file or set up on a face-centred cubic (fcc) lattice. It then conducts a conventional Metropolis Monte Carlo simulation, for a predetermined number of attempted moves. The program calculates various quantities during the simulation: by monitoring these it is possible to see how quickly the system equilibrates. Following equilibration, a production run gives proper simulation averages, which can be compared with the known properties of the Lennard-Jones system, and with the output from other simulations (for example, molecular dynamics, as treated in another exercise).

The basic Monte Carlo method is described in the standard references [3,10,11,12]. Atoms are selected one at a time, either in sequence or randomly, for an attempted (trial) move. Each trial move involves displacing an atom by an amount $(\delta x, \delta y, \delta z)$ from its current position. The three Cartesian components are typically chosen, at random, from a uniform distribution between $\pm\delta r_{max}$, where the maximum displacement parameter δr_{max} is chosen at the start of the run. The change in potential energy $\delta\mathcal{V}$ that would result from this move is then calculated, and the Metropolis prescription used to decide whether or not to accept it. Downhill moves $(\delta\mathcal{V} \leq 0)$ are accepted unconditionally; uphill moves $(\delta\mathcal{V} > 0)$ are only accepted with probability $\exp(-\delta\mathcal{V}/k_B T)$ where T is the temperature. For small values of δr_{max}, large atomic overlaps are unlikely to result, and the acceptance probability is high; for larger trial moves, the acceptance probability decreases. Often, δr_{max} is chosen to give an overall acceptance rate of about 50%, but the most rapid progress through configuration space may well result from accepting a smaller proportion of larger moves. Part of this exercise is to investigate this point.

A simulation may be started from an initial configuration that is atypical of the state point of interest. It is common practice, for example, to start a series of liquid-state runs from a low-density metastable lattice configuration. The 'equilibration' of the system should then be followed carefully, before a 'production run' (giving true equilibrium averages) is undertaken.

Several properties indicate the progress of equilibration. Both the potential energy and the pressure will change significantly as the metastable solid 'melts'. Translational order parameters may be defined in terms of Fourier components of the single-particle density

$$\rho(k) = \frac{1}{N}\sum_{i=1}^{N}\exp(i\mathbf{k}.\mathbf{r}_i) \tag{7.2}$$

where $\mathbf{k}$ is a reciprocal vector characterizing the initial lattice. For example, a suitable choice for the fcc lattice would be $\mathbf{k} = ((2N)^{\frac{1}{3}}\pi/L)(1,1,-1)$ where L is the (cubic) box size. In particular, the real part of ρ

$$\mathcal{O}_1 = \Re\rho(k) = \frac{1}{N}\sum_{i=1}^{N}\cos(\mathbf{k}.\mathbf{r}_i) \tag{7.3}$$

should be of order unity in the solid (provided the coordinate origin is chosen appropriately) and it should vary about zero, with amplitude of order $N^{-1/2}$ in the liquid. A better choice, perhaps, is the modulus

$$\mathcal{O}_2 = |\rho(k)| = (\rho(k)\rho(-k))^{\frac{1}{2}} = \left[\left(\frac{1}{N}\sum_{i=1}^{N}\cos(\mathbf{k}.\mathbf{r}_i)\right)^2 + \left(\frac{1}{N}\sum_{i=1}^{N}\sin(\mathbf{k}.\mathbf{r}_i)\right)^2\right]^{\frac{1}{2}} \tag{7.4}$$

since this is origin-independent: it is essentially a component of the instantaneous structure factor. This quantity is unity in the perfect lattice, and small but positive, of order $N^{-1/2}$, in a liquid. Finally, another interesting quantity is the mean-square displacement from the initial lattice positions, at a given time t during the simulation:

$$< \delta r^2(t) >= \frac{1}{N} \sum_{i=1}^{N} |\mathbf{r}_i(t) - \mathbf{r}_i(0)|^2. \tag{7.5}$$

This rises to a constant in a stable crystal; it continues to rise, linearly with time, in a liquid as the atoms diffuse around.

In the foregoing, we introduced the notion of 'time' t. Although the concept of time has no physical meaning in Monte Carlo simulations, it is useful to measure the progress of the run either in terms of the total number of moves attempted or, more sensibly for an N-atom system, in terms of the number of 'cycles' of N such trials. The computer time involved in a single MC cycle is comparable with that needed for a single MD timestep. In this exercise, we take the Monte Carlo 'cycle' as our unit of time.

7.3 Task

Examine the program, and satisfy yourself that it should perform as indicated above. Compile and run it: it is written for interactive use, but later you will be able to submit it as a batch job. You will be prompted for several parameters, some in Lennard-Jones reduced units ($\epsilon = 1, \sigma = 1$).

- A run title (80 characters maximum).
- The number of MC cycles.
- The number of cycles between lines of output.
- An option controlling the start configuration: 0 for fcc lattice start, 1 to read the configuration in from a file.
- The configuration file name. This file is optionally read in, and at the end of the run the file is created, or overwritten if it already exists.
- The density and temperature.
- The potential cutoff distance.
- The maximum atomic displacement parameter.

At the specified intervals, the program prints out the number of cycles, the number of trial moves so far, the ratio of successful moves to trials (the acceptance ratio), and the instantaneous values of the potential energy, pressure, mean-square displacement and translational order parameter.

Choose a point in the liquid region of the Lennard-Jones phase diagram, and start from a lattice. Experiment with the displacement parameter δr_{max}, observing the effect on the acceptance ratio. Choose a value that leads to an acceptance ratio of about 50%. Then, see how long the system takes to equilibrate. This run will probably be of a few hundred cycles, perhaps more than a thousand, and it is best submitted as a batch job. Plot, as a function of the number of cycles, the four instantaneous values mentioned above. Do these various indicators of 'melting' agree with one another? Repeat the above procedure for higher and lower values of δr_{max}. What choice gives the most rapid equilibration in terms of numbers of cycles? Which represents the most efficient use of computer time? After equilibration, perform a production run, and compare the simulation averages (printed out at the end) with the known properties of the Lennard-Jones fluid. These can be obtained by running the program `ljeq`. For argon, σ = 0.34nm and ϵ = 120K. Convert your chosen density, temperature and measured pressure into SI units: are your results sensible?

7.4 Further work

This program can be used to explore further questions of simulation efficiency, at different state points.

- The same system is investigated using molecular dynamics, in a separate exercise (see section 8). You may have attempted this yourself; if not, find someone who has. Compare the maximum efficiency of MC, as seen here, with that of MD, in terms of the rate of lattice melting per unit computer time.

- Perhaps the results will be dramatically different as the state point is changed. Choose two other densities and temperatures: one in the moderately dilute gas regime, and one in the solid state. What differences in behaviour do you see, as δr_{max} is varied?

8 Lennard-Jones molecular dynamics

8.1 Summary

Run a molecular dynamics program, which simulates a system of Lennard-Jones atoms in three dimensions. Starting from a face-centred cubic lattice, monitor the various indicators of equilibrium (total, kinetic, and potential energy, pressure, temperature, translational order parameter, mean-square displacement) as the crystal melts to form a liquid. Explore the effect of changing the initial temperature. Plot your results as a function of time, and compare with the known properties of the Lennard-Jones fluid.

8.2 Background

This problem is to familiarize you with the computer available during the Summer School. In particular we would like you to compile and run a molecular dynamics simulation program and to plot the output using the plotting routines available. The program supplied performs a molecular dynamics simulation of atoms interacting through the normal Lennard-Jones pair potential

$$v(r) = 4\epsilon[(\sigma/r)^{12} - (\sigma/r)^{6}]. \tag{8.1}$$

The program can be found in the file `mdlj`, and it may be helpful to have a copy in front of you (on screen or on paper). The program is a molecular dynamics simulation of 108 atoms [3,13,14]. The initial configuration can be set up as a face-centred cubic (fcc) lattice or read in from a file. The size of the simulation box is chosen to give the required density and velocities are chosen from an appropriate Maxwell-Boltzmann distribution at an initial temperature. The program use the leap-frog algorithm [3] to move the atoms

$$\mathbf{v}(t + \frac{1}{2}\delta t) = \mathbf{v}(t - \frac{1}{2}\delta t) + \delta t \mathbf{a}(t) \tag{8.2}$$

$$\mathbf{r}(t + \delta t) = \mathbf{r}(t) + \delta t \mathbf{v}(t + \frac{1}{2}\delta t). \tag{8.3}$$

In the first equation the velocity is advanced from half a timestep behind the positions to half a timestep in front of the positions. The new velocity at $t + \frac{1}{2}\delta t$ is then used to advance the positions forward a full timestep. In this simulation the total energy is conserved, but the potential energy and the kinetic energy fluctuate around their equilibrium values once they are attained. We can also monitor the progress towards equilibrium by calculating a translational order parameter,

$$\mathcal{O} = \left\langle \left[\left(\frac{1}{N} \sum_{i=1}^{N} \cos(\mathbf{k}.\mathbf{r}_i) \right)^2 + \left(\frac{1}{N} \sum_{i=1}^{N} \sin(\mathbf{k}.\mathbf{r}_i) \right)^2 \right]^{1/2} \right\rangle \tag{8.4}$$

where $\mathbf{k}$ is a reciprocal lattice vector of the fcc structure. $\mathcal{O}$ is unity for the perfect lattice and is of order $N^{-1/2}$ in the fluid. The program also calculates the mean-square displacement of the atoms from their initial lattice positions as a function of t.

$$< \delta r^2(t) >=< \frac{1}{N} \sum_{i=1}^{N} |\mathbf{r}_i(t) - \mathbf{r}_i(0)|^2 > . \tag{8.5}$$

This rises to a constant value in a crystal, but increases linearly in a liquid.

8.3 Task

Examine the program, compile it and run it. The program runs interactively, and begins by asking for the following parameters:

- a title of not more than eighty characters;
- the number of steps (say 1000);
- the number of steps between output lines (say 10);
- a flag to determine whether the configuration is from a lattice or file, 0 for a lattice start, 1 for a file start;
- the name of a file for the output configuration, not more than 30 characters;
- the reduced density ($\rho\sigma^3$);
- the reduced starting temperature (kT/ϵ);
- the reduced cut-off distance for the Lennard-Jones potential (say 2.5σ);
- the reduced timestep (say 0.005).

The routine then converts LJ units to simulation units which are based on a box length of one. The long-range corrections appropriate to the potential and virial are calculated, and a reciprocal lattice vector of the starting lattice is calculated. The accumulators for the properties and their fluctuations are set to zero. The program then begins the main loop over time steps. In this loop the atoms are moved using the second leapfrog equation, the forces and accelerations at the new positions are calculated and the velocities are advanced using the first leapfrog equation. The order parameter and the temperature are calculated. Finally the accumulators are incremented with information from the current time step. When the main loop is completed, final averages are computed and displayed. There are eight subroutines:

- `SUBROUTINE FCC` sets up an initial configuration in an fcc lattice;
- `SUBROUTINE COMVEL` chooses the initial velocities at a particular temperature;
- `SUBROUTINE READCN` reads in a starting configuration;
- `SUBROUTINE FORCE` calculates the forces, accelerations, energy, and virial;
- `SUBROUTINE MOVE` advances the velocities and positions of all the atoms;
- `SUBROUTINE KINET` calculates the kinetic energy;
- `SUBROUTINE ORDER` calculates the translational order parameter, which is one for a perfect fcc lattice and close to zero for a liquid;
- `SUBROUTINE WRITCN` writes out a configuration to file.

In [15] you will find a phase diagram of the three-dimensional Lennard-Jones fluid. Try to simulate at a point in the liquid region i.e. where the translational order parameter is close to zero. Remember that if you start from an expanded lattice, the potential energy will rise as the lattice melts and the temperature (kinetic energy) will fall. Be careful that it does not fall so far from your starting temperature that your simulation freezes. Plot the following as a function of time:

- the total energy, corrected for the cut-off (`CUTENERGY`);
- the kinetic energy (`KINETIC`);
- the potential energy (`POTENT`);
- the pressure (`PRESSURE`);
- the temperature (`TEMPER`);
- the order parameter (`OPARAM`);
- the mean-square dispalcement (`MDISP`).

Repeat these plots starting from a disordered configuration prepared at the temperature you require. Make sure that you are completely familiar with compiling and running `FORTRAN` programs, using the editor to manipulate data and plotting the data. You will find a parameterized equation of state for the Lennard-Jones fluid on the computer. This equation was fitted to a large amount of previous simulation data [16]. The equation is accessed by running the job `ljeq`. You feed in the reduced temperature and density and it returns an estimate of the thermodynamic properties. Compare these results to your average values for the density and final temperature of your run.

8.4 Further work

- Experiment with the time-step, and the length of the simulation run.
- The same system is investigated using Monte Carlo in section 7. You may have attempted this yourself; if not, find someone who has. Compare the efficiency of molecular dynamics and Monte Carlo in terms of the rate of lattice melting per unit computer time.
- Run the simulation for state points typical of a dilute gas and a crystal. Plot the mean square displacement as a function of time in each case.

9 Monte Carlo by hand

9.1 Summary

In this exercise you will try some simple Monte Carlo simulations by hand, using a six-sided die to generate random numbers. The idea is to study a small, one-dimensional, Ising spin system, employing both asymmetric and symmetric transition probabilities, to get a feel for the way the method works. This leads on, in a later project, to computer simulations of the same system.

9.2 Background

The Ising model is one of the most fundamental models in statistical physics. At each site of a lattice is a spin s_i which may point up or down: $s_i = \pm 1$. There are interactions of the form

$$E_{ij} = -J s_i s_j \tag{9.1}$$

between nearest-neighbour spins i and j, where J (here assumed positive) is a coupling constant. This system can be thought of as a representing a ferromagnetic metal. We can add an external magnetic field (a term of the form $E_i = H s_i$) if we wish, but here we consider the field-free case for simplicity. An isomorphism with the lattice gas model (each site either occupied or unoccupied) means that the system can also represent, in a highly idealized way, an atomic liquid or gas. The phase transitions in these types of model reflect those of real systems, thanks to universality, although for the simple one-dimensional system studied here there is no phase transition.

In this exercise we use small Ising systems to test out the basic simulation methods. Useful background material on the simulation of spin systems can be found in the standard references [10,11,17]. The basic Monte Carlo method consists of repetitions of the following steps:

- select a spin (sequentially or at random);
- calculate a transition probability for flipping this spin;
- choose to flip the spin or not, according to this probability.

There are many different ways of choosing the transition probabilities, and we investigate two here.

For this task you require pen and paper, and a six-sided die (provided). Consider a one-dimensional system of 8 Ising spins, as shown in Figure 9.1. Each spin has two nearest neighbours, with interaction energies given by eqn (9.1). As usual, periodic boundary conditions apply.

Consider flipping a spin. This will, in general, result in an energy change ΔE. The Metropolis formula [18] for the probability of accepting this flip is

$$P(\Delta E) = \min(1, \exp(-\Delta E/kT)) \tag{9.2}$$

Figure 9.1: A one-dimensional configuration of spins with periodic boundary conditions

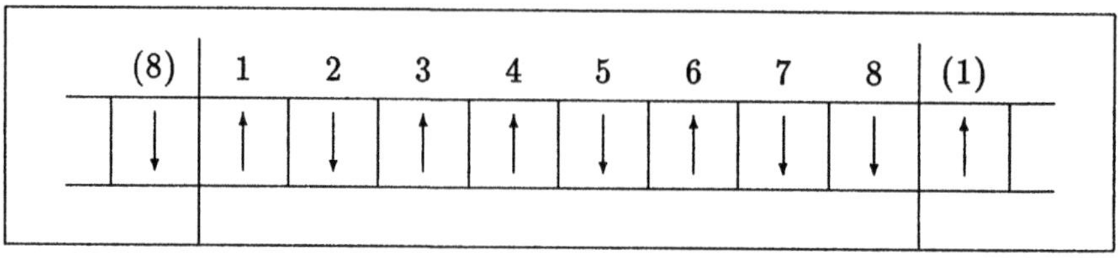

Figure 9.2: A low-energy one-dimensional configuration of spins

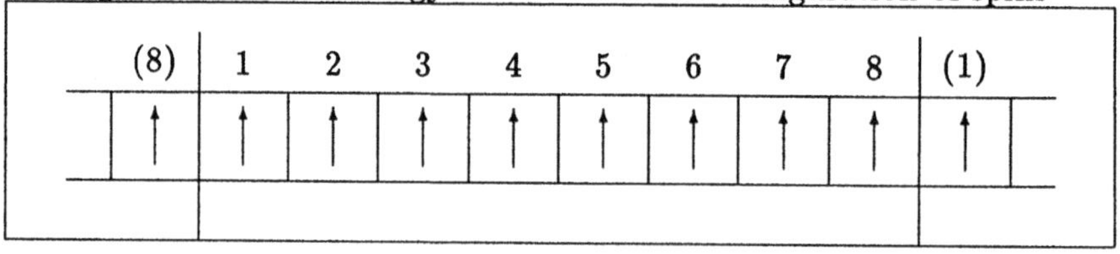

where T is the temperature and k Boltzmann's constant. In other words, if ΔE is negative (downhill), accept the flip unconditionally; if ΔE is positive (uphill), accept with probability $\exp(-\Delta E/kT)$. An alternative, the Glauber formula [19] is more symmetrical with respect to uphill and downhill moves:

$$P(\Delta E) = \frac{\exp(-\Delta E/kT)}{1 + \exp(-\Delta E/kT)}. \tag{9.3}$$

Both these prescriptions satisfy the basic equation of detailed balance (or microscopic reversibility), i. e. that $P(\Delta E)/P(-\Delta E) = \exp(-\Delta E/kT)$. This ensures that the simulation, in principle, generates proper canonical ensemble averages. (Here we are assuming equal underlying probabilities for *attempting* a forward move and the reverse move.) Accepting a flip 'with a given probability' is simply a matter of choosing a random number uniformly from a given range, typically (0,1). Suppose we wish to accept a flip with probability 30%; if the random number is 0.3 or less, we flip the spin, if it is more than 0.3 we do not. In special cases, when only a small set of acceptance probabilities are possible, the decision can be made by tossing a coin or throwing a die.

9.3 Task

Consider the one-dimensional system of Figure 9.1, with a temperature chosen such that $J/kT = \frac{1}{4}\ln 2$. What are the possible changes $\Delta E/kT$ involved in flipping a spin in this system? Calculate the flip probabilities for all possible arrangements of neighbouring spins, for each of the two prescriptions above, eqns (9.2),(9.3). For these probabilities, it should be easy to carry out a Monte Carlo simulation of the system,

Figure 9.3: A high-energy one-dimensional configuration of spins

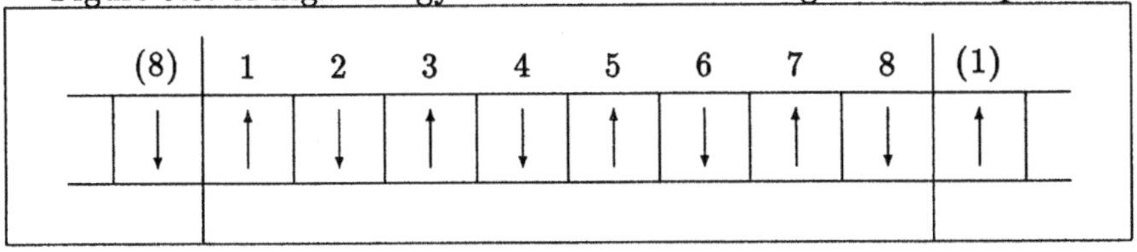

using a six-sided die as a random number generator. Starting from the configuration of Figure 9.1, carry out a few sweeps (one sweep is one attempted Monte Carlo flip per spin, i. e. 8 flips here), using each prescription. For simplicity, select each spin for flipping sequentially. Keep a record of the flip probabilities, whether or not you need to throw the die, and the outcome, for each attempt. Try to get a feel for the relative rate at which flips actually occur: which algorithm moves the system through configuration space the fastest?

Consider what would happen to an initial configuration with the lowest possible energy, and to one with the highest possible energy, as shown in Figures 9.2, 9.3, under these algorithms. What is the evolution likely to be at very high temperatures and at very low temperatures?

10 Hard sphere Monte Carlo in one dimension

10.1 Summary

Write a Monte Carlo program for a one dimensional system of hard spheres. Using a modest number of particles (e.g. $N = 50$), and choosing some suitable density, run the program for a few thousand Monte Carlo cycles and calculate the pressure. Compare with the exact result in the thermodynamic limit ($N \to \infty$).

10.2 Background

Simulation studies of model liquids such as hard spheres in three dimensions and hard disks in two dimensions have played a key rôle in understanding the structure of simple liquids and in the development of theories of liquids [15]. In this exercise you are going to write a Monte Carlo program for hard disks whose centres are confined to a line.

A simulation box containing five atoms is shown below. We imagine that the line is actually on a circle so that the points A and B are identical. In other words, if an atom leaves the cell to the right by passing through B it re-enters from the left by passing through A. This technique of periodic boundary conditions removes surface effects which would occur if we reflected the atoms at A and B.

Figure 10.1: A one-dimensional system of $N = 5$ hard spheres in periodic boundaries (with two periodic images shown)

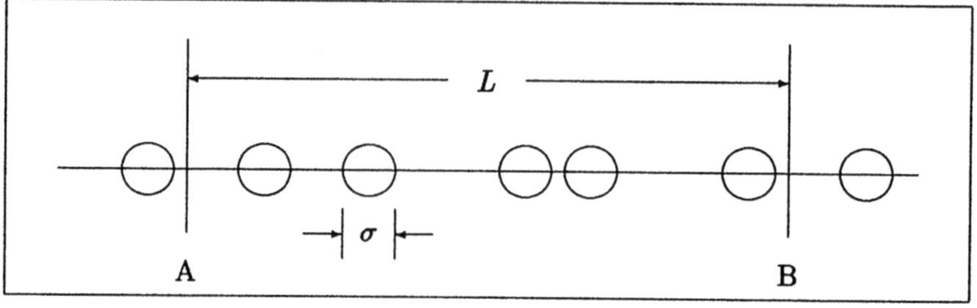

The properties of our one dimensional fluid depend only on the density or packing fraction of atoms on the line. This packing fraction, η, is an input variable to the program

$$\eta = N\sigma/L \tag{10.1}$$

where N is the number of atoms on the line, σ is the diameter of each atom and L is the length of the line. A Monte Carlo move is made by choosing an atom at random or in order and giving it a small random displacement to the left or to the right. If the atom does not overlap with one of its neighbours, the move is accepted. If the atom overlaps the move is rejected and the old configuration is counted as the next step in the chain. This recipe generates a Markov chain whose limiting distribution is proportional to the Boltzmann factor of the fluid. Unweighted averages calculated over the states in the chain are equivalent to averages in the canonical ensemble. It is possible to calculate the radial distribution function and hence the pressure of the hard sphere fluid. The maximum possible displacement of an atom is adjusted so that about 50% of the trial moves are accepted. The radial distribution is calculated in the simulation by sorting all pair separations into a histogram [3]. Suppose we sort every tenth cycle, so that there are a total of τ_{run} sorts, and that a particular bin of the histogram, corresponding to the interval $(r, r + \delta r)$, contains $n_{his}(b)$ pairs. Then the average number of atoms whose distance from a given atom in the fluid lies in this interval, is

$$n(b) = n_{his}(b)/(N\tau_{run}). \tag{10.2}$$

The average number of atoms in same interval for the ideal gas at the same density is

$$n_{id}(b) = 2\rho\delta r. \tag{10.3}$$

The radial distribution function is

$$g(r + \frac{1}{2}\delta r) = n(b)/n_{id}(b). \tag{10.4}$$

The compression factor of the fluid can be calculated from the simulation by extrapolating $g(r)$ to contact,

$$PL/Nk_BT = 1 + (N\sigma/L)g(r \to \sigma^+). \tag{10.5}$$

This can be compared to the exact result which can be derived in the thermodynamic limit by factorizing the configurational partition function [20]

$$PL/Nk_BT = 1/(1-\eta). \tag{10.6}$$

10.3 Task

You begin by choosing a reasonable value of N and of η. We recommend you simulate about 50 atoms at a packing fraction of say 0.60. It is convenient to use a line of unit length and to calculate the appropriate value of σ. The program can be divided into a number of stages.

10.3.1 Input

We begin by setting up the atom positions. You should allow for two choices:

- Setting the 50 atoms at equal intervals on the line, with atom 1 at the origin;
- Reading a random, or aged configuration from the file `mc1ddat`.

The program should now ask you for a number of parameters:

- A run title (say 'test run one');
- The number of MC cycles (say 10000);
- The number of steps between output (say 100);
- The density (say 0.6);
- The interval for calculating $g(r)$ (say `DELR` = 0.02);
- The name of the input file for the configuration, which is also used for dumping the final configuration (say `mc1ddat`);
- A flag which tells the program whether you want to run from a disordered start or a lattice start (say 0 or 1).

You will need to use a random number generator, which you can start by calling the NAG library subroutine `G05CBF(0.0)`. The argument of this routine changes the seed of the random number generator.

10.3.2 Setup

- It is useful at this stage to zero the accumulator or histogram you will use to calculate the radial distribution function i.e. set `IRDF(I)=0` for `I=1, NRDF` (`NRDF=100`).
- You will need to choose a maximum displacement for the particles, say `MAXDIS = SIGMA/20.0`.
- You should convert the interval for the RDF calculation to units with $L = 1$ i.e. set `DELR = DELR * SIGMA`.
- Calculate the maximum distance for sorting for the RDF, i.e. set `RDFMAX = DELR * REAL(NRDF)`.
- Zero any other accumulators you may need.

10.3.3 Loop over cycles

In each cycle you will loop over all atoms, in order. You should try to move each atom one at a time.

```
RXINEW = RX(I) + (2.0 * G05CAF( DUMMY )-1.0) * MAXDIS
```

This generates a uniform random displacement between `-MAXDIS` and `+MAXDIS`. If `MAXDIS` is less than `SIGMA`, then the particle in the new position may overlap with its neighbours. We only need to check the neighbours to the left and right of the atom we have just moved. In checking for overlap make sure you apply the minimum image convention for the atom separation

```
RXIJ = RXIJ - ANINT( RXIJ )
```

and be careful when comparing distances that you only consider the magnitude not the sign of the interatomic separation. Remember overlaps occur if `RXIJ<SIGMA`. If an overlap occurs go back to the old configuration and use the old configuration as the next step in the chain.

10.3.4 Periodic operations

There are a number of operations which can be carried out periodically.

- Adjust the maximum displacement `MAXDIS` every ten cycles so that approximately half the configurations are accepted (optional).
- Output the following information every ten cycles:
  - the number of cycles;

  - the number of moves accepted;
  - the acceptance ratio;
  - the maximum displacement.

- Write the current configuration to the restart file, `mc1ddat`, every 500 cycles (optional).
- Calculate the radial distribution function every twenty cycles. This procedure involves looping over all distinct pairs of atoms and calculating the centre of mass separation using the minimum image convention. These separations are sorted into a histogram `IRDF` where the bins are of thickness `DELR`. The radial distribution function $g(r)$ is calculated from

```
GR(IR) = IRDF(IR)/REAL(N)/ACRDF/(DELR*DENS*2.0)
```

  where `ACRDF` is the number of sorts, `DENS` is N/L for $L = 1$, and the factor of 2 takes care of the fact that we are sorting atoms to the left and right of a given atom.

At the end of the simulation $g(r)$ as a function of r/σ should be tabulated and stored. The appropriate r/σ corresponds to the middle of the histogram bin [3]. At the end of the calculation you should plot $g(r)$ against r/σ . The compression factor of the fluid should be calculated.

10.4 Further work

- How is the compression factor affected by the length of the run?
- How is the compression factor affected by the histogram interval?
- How is the compression factor affected by using different seeds for the random number generator?
- How is the compression factor affected by changing the system size?
- How does the initial configuration affect the convergence of the run?
- Is the compression factor obtained from this program more or less accurate than the value obtained from the corresponding dynamic simulation (see section 11) ?

Figure 11.1: A one-dimensional system of $N = 5$ hard spheres in periodic boundaries (with two periodic images shown)

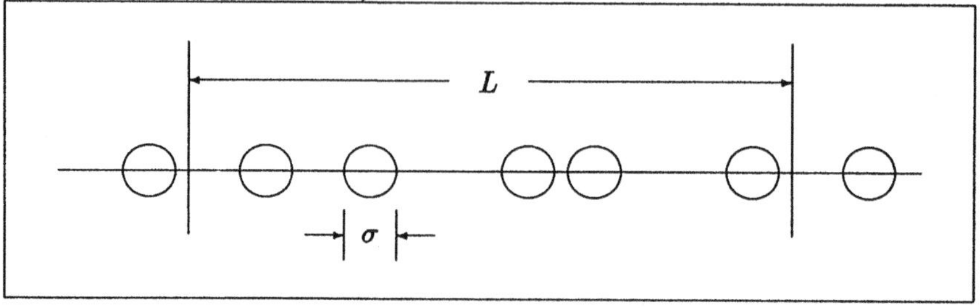

11 Hard sphere dynamics in one dimension

11.1 Summary

Write a molecular dynamics program for a one-dimensional system of hard spheres. Using a modest number of particles (e.g. $N = 50$), and choosing some suitable densities, run the program for a few thousand collisions and calculate the pressure. Compare with the exactly-known formula (valid in the limit $N \to \infty$).

11.2 Background

The earliest molecular dynamics simulations used the hard sphere model. A hard-sphere MD program differs from (say) a Lennard-Jones one, because the atoms move from collision to collision; the atomic velocities only change when atoms collide. The program addresses the questions:

- Which two atoms will collide next, and at what time?
- What are the post-collisional velocities?

The hard sphere model is very idealized, but a useful reference system in theories of liquid structure and dynamics [15].

In this exercise, hard sphere MD is illustrated using a one-dimensional example (see Figure 11.1). The system consists of a set of N hard spheres of diameter σ, with their centres on a line of length L (with $L > N\sigma$). (Equally well, we can consider N hard rods of length σ lying along this line.) The usual periodic boundary conditions are applied. The equation of state of this system is known, and it is particularly simple in the limit $N \to \infty$ [21]:

$$PL = NkT/(1 - \eta) \tag{11.1}$$

where $\eta = N\sigma/L$ is the packing fraction.

Each atom can only collide with its neighbour on either side. Suppose that at a given time $t = 0$, the centres of two neighbouring atoms i and j are at x_i, x_j and their velocities are v_i, v_j. Set $x_{ij} = x_i - x_j$ (with the minimum image convention applied) and $v_{ij} = v_i - v_j$. At time $t = 0$, we can assume that $|x_{ij}| \geq \sigma$, i. e. that the atoms are not already overlapping. If a collision is to occur at some future time, t_{ij}, then we must have

$$|x_{ij} + v_{ij}t_{ij}| = \sigma \tag{11.2}$$

or

$$t_{ij} = -(x_{ij} \pm \sigma)/v_{ij}. \tag{11.3}$$

Note that x_{ij} and v_{ij} must have opposite signs, otherwise the atoms are moving apart and we cannot obtain positive t_{ij}. If they are moving together, the smallest (positive) root of eqn (11.3) gives the collision time. So, we can answer the first question above. The simulation program should solve this equation for all neighbouring pairs and select the smallest positive t_{ij} which corresponds to the next collision. The whole system can then be advanced to this time, i. e. each x_i becomes $x_i + v_i t_{ij}$.

The outcome of a perfectly elastic collision can be calculated using the conservation laws. Assuming equal masses m, we can see immediately from momentum conservation, that if v_i changes by δv_{ij}, then v_j must change by $-\delta v_{ij}$. Conservation of energy (all kinetic here) then tells us how large δv_{ij} is. It is left as an exercise to show that $\delta v_{ij} = -v_{ij}$, i. e.

$$\begin{aligned} v_i' = v_i + \delta v_{ij} &= v_i - v_{ij} = v_j \\ v_j' = v_j - \delta v_{ij} &= v_j + v_{ij} = v_i \end{aligned} \tag{11.4}$$

where the primes denote post-collisional values. So the colliding particles just exchange velocities. After this, a search for the next collision takes place, and the whole process is repeated.

Lastly, we wish to calculate the pressure. The usual expression (in d dimensions) is [15]

$$PL^d = NkT + \frac{1}{d} < \sum_{i<j} x_{ij}.f_{ij} > \tag{11.5}$$

where the sum is over every distinct pair ij, and f_{ij} is the force acting between them. In more than one dimension, the $x_{ij}.f_{ij}$ term would be a scalar product of two vectors. The angle brackets represent an average. In our case, the forces act impulsively: each collision results in a momentum transfer $m\delta v_{ij}$. Bearing in mind that force is the time derivative of momentum, we can see that eqn (11.5) becomes

$$PL^d = NkT + \frac{m}{dt} \sum_{ij} x_{ij}\delta v_{ij}. \tag{11.6}$$

The sum is now over all collisions between pairs ij occurring during time t, the course of the simulation run.

11.3 Task

Write a molecular dynamics program for this system. You may find it useful to set the atomic diameter σ and the mass m to be unity. The initial positions can be set up on a regular lattice, but be sure to allow a short equilibration period if this is done. Make sure that initial velocities are picked from a proper Gaussian distribution (*why?*); adjust them to make the total momentum zero, and scale them, to make them consistent with $kT = 1$, for convenience.

For a modest system size ($N = 50$) and for packing fractions $\eta = N\sigma/L = 0.5, 0.7, 0.9$, run the program for a few thousand collisions. Compare the simulated pressure with that predicted by eqn (11.1) above.

11.4 Further work

- Are your results worse than expected? Try running the program several times, say for $N = 25$, making sure that different random numbers are used to generate the initial velocities each time. Does this suggest an explanation?

- For larger system sizes, it becomes important to cut down on the work involved in predicting future collisions. We should construct a list of future collision times, making sure to reduce these values whenever the configuration is advanced in time. Whenever two particles i and j collide, it should only be necessary to re-evaluate a few of the collision times in this list (*which ones?*); the rest are unaffected. Try to incorporate this improvement in your program.

- The MD results obtained here can be compared with MC results for the same system. This is the subject of another exercise: you might like to try it. The MC simulation is easier to code, but it is more tricky to obtain the pressure.

- Instead of periodic boundary conditions, it is possible to confine the system between hard walls. This would be a very simple model of a fluid in a pore. Try to adapt the simulation program in this way. Much is known of the statistical mechanics of this system, even without taking the limit $N \to \infty$ [22].

- The generalization of the one-dimensional hard-sphere program to three dimensions is quite simple. The major difference is that x_i and v_i become vectors, and so eqn (11.2) produces a quadratic equation in the time t_{ij} instead of a linear equation. The equation giving postcollisional velocities is also slightly modified: for smooth hard spheres the impulse is perpendicular to the colliding surfaces at the point of contact, and its magnitude is determined by the component of the relative velocity in that direction. Full details are given in standard references [3,23].

12 A Monte Carlo simulation of Lennard-Jones atoms in one dimension

12.1 Summary

Write a Monte Carlo simulation of Lennard Jones atoms in one dimension. You should write the program for a small number of atoms ($N = 50$). You will need to specify the density and the temperature of the fluid and you should calculate the internal energy and the pressure. This problem is an extension of that of section 10 and is most easily attempted after that program is written.

12.2 Background

The Monte Carlo simulation of continuous potentials has been described in a number of reviews [3,10,24,25]. You should simulate about 50 atoms on a line using the normal periodic boundary conditions. The potential between pairs of atoms is the Lennard-Jones potential with a cut-off at 2.5σ ,

$$\begin{aligned} v(r) &= 4\epsilon[(\sigma/r)^{12} - (\sigma/r)^{6}] \quad r \leq 2.5 \\ v(r) &= 0 \qquad\qquad\qquad\qquad r > 2.5 \end{aligned} \tag{12.1}$$

The structure of the program is essentially the same as that developed in section 10. The main difference is that the potential has an attractive as well as a repulsive component. The energy change during the trial move (δV) must be calculated explicitly. Our potential has a range of approximately three molecular diameters so that we have to develop a new technique for calculating the energy of a trial state i.e. it is no good just considering nearest neighbours. The simplest approach is to calculate the energy change arising from all the changes in the unique pairs distances during the trial move. If the overall change in energy is downhill (i.e. $\delta V \leq 0$) the move is immediately accepted. If the move is uphill (i.e. $\delta V > 0$) the move is accepted with a probability proportional to $\exp(-\beta\delta V)$. If the move is rejected then the old configuration becomes the next step in the Markov chain i.e. it is recounted in calculating the ensemble average [3].

12.3 Task

You might plan your program along the following lines.

12.3.1 Input

Read in the following variables:

- the run title;

- the number of cycles;
- the number of cycles between output lines;
- the number of steps between saving the configuration;
- the interval for updating the maximum displacement;
- the name of the file for dumping configurations;
- a flag to decide if you want a lattice start or a start from a file;

and the following in the usual reduced Lennard-Jones units:

- the density;
- the cutoff;
- the temperature.

The initial positions of the atoms on the lattice can now be set-up by reading them from the configuration file or placing the atoms on a lattice at equal intervals along the line.

12.3.2 Setup

Once the input data has been read by the program, the value of σ is calculated assuming the line is of unit length. Other lengths such as the cut-off are scaled to a line of unit length. It is also important to have a minimum length in this problem, say 0.7σ . Lennard-Jones atoms can overlap, but to avoid overflows in the calculation of the potential and its exponential, we shall say that any moves resulting in separations less than $r_{min} = 0.7\sigma$ will be rejected. You will also need to set a maximum displacement at say 0.1σ. Any accumulators are set to zero. At this point it is useful to calculate the total energy and virial of the system.

$$V = \sum_i \sum_{j>i} v(r_{ij}) \tag{12.2}$$

$$W = -\sum_i \sum_{j>i} r_{ij}(dv(r_{ij})/dr_{ij}) = \sum_i \sum_{j>i} w(r_{ij}) \tag{12.3}$$

These sums are for all pair separations less than 2.5σ. We can also calculate long-range corrections to these properties using a mean-field approximation (i.e. $g(r_{ij}) = 1.0$)

$$V_{lrc} = N\rho \int_{r_c}^{\infty} v(r_{ij}) dr_{ij} \tag{12.4}$$

$$W_{lrc} = N\rho \int_{r_c}^{\infty} w(r_{ij}) dr_{ij} \tag{12.5}$$

The virial is related to the pressure of the one-dimensional fluid

$$P = \frac{Nk_BT}{L} + \frac{< W >}{L}. \tag{12.6}$$

12.3.3 Main loop

The program now loops over cycles. Inside this loop we loop over all atoms on the line in order. Each atom is given a uniform random displacement left or right using a random number generator. If the move results in a significant overlap i.e. any $r_{ij} < r_{min}$, it is immediately rejected and the old configuration is recounted as the next step in the chain. If the move is not rejected on these grounds, the change in energy and virial for the move is calculated. Note that the long-range corrections are constant during these moves. If the change in energy is downhill (i.e. negative), the move is accepted and the total energy and virial of the line are updated. If the move is uphill (i.e. $\delta V > 0$ where $\delta V = V_{new} - V_{old}$), we calculate $\exp(-\beta\delta V)$ where $\beta = 1/k_BT$. We compare $\exp(-\beta\delta V)$ with a uniform random number in the range zero to one. If the exponential is greater than the random number we accept the move, if it is less we reject the move and recount the old configuration as the next step in the chain. This sequence of events requires some careful programming.

12.3.4 Periodic operations

In the main loop it is necessary to perform a number of periodic operations. The maximum displacement should be adjusted so that approximately half the attempted moves are accepted. Output, such as the number of cycles, the number of accepted moves, the energy and virial (including their long-range corrections), and the value of the maximum displacement should be written to the screen or to a file for inspection. The current configuration should be dumped to the configuration file to allow for a restart.

12.3.5 Winding up

At the end of the main loop you should use the accumulators to calculate the average configurational energy and pressure and the fluctuation in these properties. Throughout the simulation you have been incrementing the energy and virial of your initial configuration each time you accept an atom move. At this point you could recalculate the energy and virial of the complete line, using a sum over all pairs. This should agree with the value that you are carrying for the energy and virial to within the machine accuracy. It is a useful check that you are calculating all the interactions and the energy changes properly. In this exercise, we would like you to construct the program, and run it for a number of densities and temperatures, establishing the trends in the energy and pressure for your one dimensional fluid.

12.4 Further work

- Plot the internal energy of the fluid as a function of temperature for a number of fixed densities. Calculate the specific heat at constant volume by graphically differentiating the energy with respect to temperature. Plot C_v against T. Does this system exhibit a solid-liquid or a liquid-gas phase transition? You could also calculate C_v using the appropriate fluctuation formula in the canonical ensemble.

- You should consider how your results depend on the length of the run, and the seed of the random number generator.

- You can start the simulation from a lattice configuration or from a disordered configuration which you create at the end of your first run and update with subsequent runs. How does the starting configuration affect the convergence of the Markov chain? Is a 50% acceptance ratio for the trial moves the optimum value? How does the optimum acceptance ratio compare with the value you found in the Monte Carlo simulation of the hard spheres on a line (section 10)?

- This program is easily extended to three dimensions. How does the definition of the pressure depend on the dimensionality? How will the two long-range corrections change? If you rewrite the program for three dimensions you could compare your results for the energy and pressure with those from section 7.

13 Ising model simulations

13.1 Summary

This project involves using the Ising model to test various simulation techniques. The aim is to compare Monte Carlo, employing both asymmetric and symmetric transition probabilities, with a simple deterministic cellular automaton algorithm. There is the possibility also to look into multispin coding techniques for vector and parallel computers. This project would be suitable for an individual, or for a team, each individual pursuing a different aspect.

13.2 Background

The Ising model is described in a separate exercise (see section 9) but for completeness we give the details again here. The Ising model is one of the most fundamental models in statistical physics. At each site of a lattice is a spin s_i which may point up or down: $s_i = \pm 1$. There are interactions of the form

$$E_{ij} = -J s_i s_j \tag{13.1}$$

between nearest-neighbour spins i and j, where J (here assumed positive) is a coupling constant. This system can be thought of as a representing a ferromagnetic metal. We can add an external magnetic field (a term of the form $E_i = Hs_i$) if we wish, but here we consider the field-free case for simplicity. An isomorphism with the lattice gas model (each site either occupied or unoccupied) means that the system can also represent, in a highly idealized way, an atomic liquid or gas. The phase transitions in these types of model reflect those of real systems, thanks to universality.

The statistical mechanics of the one-dimensional Ising model, i. e. a chain of spins, can be worked out quite easily. There are no phase transitions, but it is a useful simulation 'test bed'. The two-dimensional Ising model, for infinite system size in zero field is also an exactly-solved problem. Here, there is a first-order phase transition between ordered and disordered states, below a critical temperature. Further details may be found in standard texts on statistical mechanics (for example [26]). So the simulated properties (for large enough systems) can be compared with known ones. Another approach is to study a system small enough that the statistical properties can be obtained by direct counting of states.

In this project we use small Ising systems to test out basic simulation methods. Useful background material on the simulation of spin systems can be found in the standard references [10,11,17]. The basic Monte Carlo method consists of repetitions of the following steps:

- select a spin (sequentially or at random);
- calculate a transition probability for flipping this spin;
- choose to flip the spin or not, according to this probability.

In the effort to find faster and faster algorithms, much interest has centred in recent years on the relative efficiencies of different ways of choosing the transition probabilities, on ways of coding the program so as to consider many spins at once, and on the rapid generation of good random number sequences. There have also been some investigations of deterministic methods (cellular automata) of generating configurations, which do not involve random numbers at all. By tricks such as these, impressive performance can be squeezed out of even small microcomputers, and truly awesome flip rates achieved on supercomputers. The above references provide excellent accounts of these developments.

In fact, most of the underlying ideas have been around for many years. The present project is very much in the spirit of the work of Friedberg and Cameron [27] who took a small Ising lattice, and used it to test the performance of their simulation program. Their paper describes the basic Monte Carlo method, covering many technical points such as the selection of spins for flipping, the division of the system into two independent sublattices, the basic multispin coding approach, the choice between different transition probabilities, the danger of being locked into a region of configuration space, and the analysis of the results for statistical errors. Some of these points will be treated below.

Figure 13.1: A one-dimensional configuration of spins with periodic boundary conditions

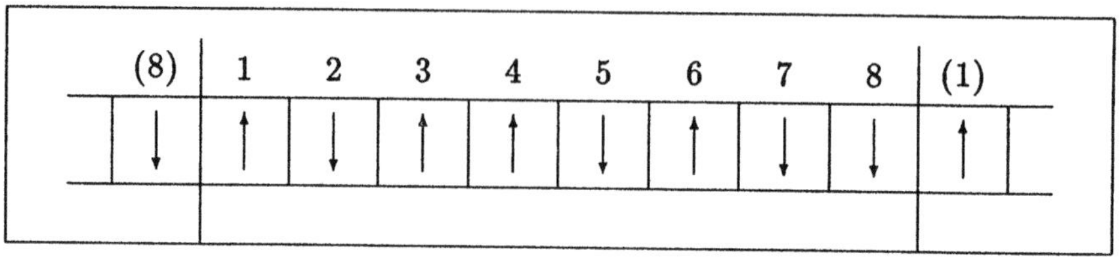

13.3 Monte Carlo simulation

As in section 9 we consider a one-dimensional system of 8 Ising spins, as shown in Figure 13.1. Each spin has two nearest neighbours, with interaction energies given by eqn (13.1). As usual, periodic boundary conditions apply.

The Metropolis formula [18] for the probability of accepting a spin flip with an associated energy change ΔE is

$$P(\Delta E) = \min(1, \exp(-\Delta E/kT)) \tag{13.2}$$

where T is the temperature and k Boltzmann's constant. In other words, if ΔE is negative (downhill), accept the flip unconditionally; if ΔE is positive (uphill), accept with probability $\exp(-\Delta E/kT)$. The alternative, symmetrical, Glauber formula [19] is

$$P(\Delta E) = \frac{\exp(-\Delta E/kT)}{1 + \exp(-\Delta E/kT)}. \tag{13.3}$$

In both cases, assuming that there is no bias in the way we *attempt* flips one way or another, both these prescriptions satisfy the detailed balance (microscopic reversibility) condition i. e. that $P(\Delta E)/P(-\Delta E) = \exp(-\Delta E/kT)$. This leads to proper canonical ensemble averages. Accepting a flip 'with a given probability' entails choosing a random number uniformly from a given range, typically (0,1).

Write two programs to simulate this Ising system: one using the Metropolis method and one employing the Glauber prescription. You may find it useful to work in reduced units, setting $J = 1$. For simplicity, you can select spins for flipping sequentially. Suggested input is as follows.

- The initial configuration (read in from a file, which you could prepare using the editor).
- The temperature, in reduced units, kT/J.
- The run length in Monte Carlo cycles (one cycle is one attempted flip per spin).

The NAG library provides random number generators: the subroutine `G05CCF` initializes the generator in a non-reproducible way, while the function `G05CAF ( DUMMY )` returns a random number in the range (0,1). Both `G05CAF` and its dummy argument `DUMMY` should be declared `DOUBLE PRECISION`. Suggested output at user-specified intervals:

- The total energy.
- The magnetization (i.e. number of up spins minus number of down spins).
- The ratio of flips accepted to flips attempted.

You might also like to print snapshots of the configurations. These programs will probably be fast enough to run interactively.

Run the programs (with your choice of temperature) to see what happens, to an initial configuration with the lowest possible energy, and to one with the highest possible energy, as shown in Figures 9.2, 9.3, under these two algorithms. What happens at very high temperatures and at very low temperatures?

13.4 Multispin coding

This simple one-dimensional system can also be used to illustrate the ideas behind multispin coding and the running of Monte Carlo simulations on parallel computers. We have been selecting spins sequentially for flipping; random selection is another way of going about things. Yet a third possibility is to look first at all the odd-numbered spins, flip them with the appropriate probabilities, and then consider all the even-numbered spins in a similar way. Because the interactions are restricted to nearest neighbours, it does not matter in what order we consider spins within each of these two sets: the calculations involved in flipping spins 1, 3, 5, and 7, for example, are independent of one another. These four spins could equally well be treated in parallel, and updated all at once. Then, attention could be focused on spins 2, 4, 6, 8, and so on. Try this updating scheme in your program, for the initial configurations discussed above. You should be able to see some potential pitfalls of this method in special cases, but for non-pathological starting conditions it should be just as valid as the other methods. In two (and also three) dimensions, it is possible to adopt a black-white checkerboard identification of two independent sublattices. This is the approach used by Friedberg and Cameron [27] and it has been described several times since (see e. g. Chapter 10 of ref. [10] and references therein).

13.5 Cellular automaton method

A cellular automaton (CA) uses completely deterministic rules for updating a configuration of (in this case) spins. For the Ising model on a lattice in zero applied field, as long as there are an even number of neighbours for each site, there is a simple rule that allows the system to evolve while conserving the energy. Consequently, this simulation probes the microcanonical rather than the canonical ensemble. Nonetheless,

Figure 13.2: Cellular automaton simulation

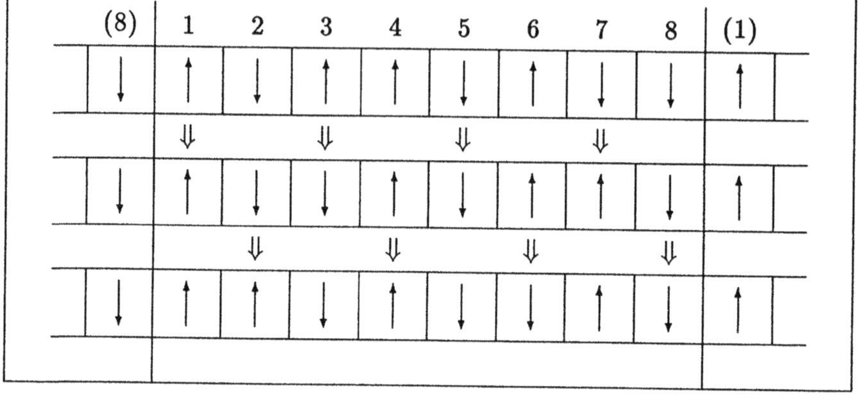

this is a potentially useful route to statistical mechanical properties. The motivation for introducing it is that, for such a simple model, the generation of the random numbers in conventional Monte Carlo can be the most time-consuming operation. A cellular automaton requires no random numbers, except to set up an initial configuration. This approach has been employed by Herrmann [28], and the original CA model is due to Pomeau and Vichniac [29,30]. The method is valid in any number of dimensions; here we take the one-dimensional example as a simple illustration.

The CA rule is based on the division into two sublattices, mentioned above, and the observation that if a spin is surrounded by equal numbers of up and down spins, then flipping it does not change the energy. The rule is simply that all such spins on a given sublattice should be flipped simultaneously. Then, the same procedure is applied to the other sublattice. This prescription is repeated many times in the course of a simulation. Consider the starting configuration of Figure 13.1, which is given again at the top of Figure 13.2. We apply the rule to the 'odd-numbered' sublattice first. Of these spins, only numbers 3 and 7 have one up and one down neighbour, so only these are flipped. This gives the second configuration in the Figure. Now we apply the rule to the 'even-numbered' sublattice. Of these, only numbers 2 and 6 qualify, so just these are flipped, giving the third configuration in Figure 13.2. Continue applying these flip rules for at least ten more steps. Examine the generated configurations. Do there appear to be any problems with the technique? It is easy to see that the lowest-energy and highest-energy configurations of Figures 9.2, 9.3, will not evolve at all under these rules; how about similar configurations with just one spin out of place? Possible failures of the ergodicity assumption are considered by Herrmann [28], but in higher-dimensional systems, and for total energies of interest (around the phase transition) it is not thought that this is a serious problem. You might like to write a simple simulation program for this system.

13.6 In search of more speed

If you tried out the exercises on Monte Carlo simulation by hand, you will have introduced some of the ideas that appear in fast Ising computer programs. For example, in MC simulations, the flip acceptance probabilities are exponential functions of the associated energy change (see eqns (13.2), (13.3)) However, you will have seen that it is not necessary to calculate them every time: you can draw up a table of the flip probabilities, and then look up these values for a given configuration of neighbouring spins. On a computer, this logical lookup operation is often faster than the alternative floating point arithmetic. There are examples of such coding for the Ising model, in three dimensions, in refs [11,17]. In the CA simulation, you applied a logical rule to determine the update sequence; this can be coded efficiently on a computer, as you can see in Herrmann's paper [28].

The division of a system into sublattices, each consisting of non-interacting spins, and the simultaneous updating of all the spins on a sublattice, is a natural approach on a parallel computer [31]. On serial machines it leads to multispin coding to improve the speed (see Chapter 1 of ref. [11] and references therein). For systems as simple as the Ising model, only one bit is required to represent the state of each spin. A single word of storage on the computer can therefore hold data for many spins (on different sublattices) simultaneously. If there exist in the computer language suitable bit-by-bit Boolean operations on these variables, then the MC updates can be carried out in parallel. Computer code for this type of MC simulation has been published [32].

You are invited to code the one-dimensional Ising model, and also the two-dimensional system on a square lattice, using any of the algorithms mentioned above. Use multi-spin coding if you like: you will need to investigate the language manuals to find out if suitable bit-by-bit operations exist. `FORTRAN` may not be the best language here: try `BASIC` (on a micro) or `C`, or even assembler. Also, you might like to try to program one of the parallel computers. On the DAP we have demonstration programs, using essentially conventional Monte Carlo methods, with fast random number generation [33] and also the cellular automaton approach. There is direct output to a video monitor, and this enables you to get a feel for the physics of the 2-dimensional Ising model. Try these demo programs out.

Returning to the work of Friedberg and Cameron [27], you might like to reproduce their calculations for a very small 4×4 lattice. Note how, in this case, the properties are calculated exactly by direct counting of the states. Note also that some of the states are known to be inaccessible according to the multispin update rule used, and so they are omitted from the calculation. You should consider whether sequential or random selection of spins in this simple two-dimensional case would suffer from the same problem.

14 The effects of small non-sphericities in the model atom-atom intermolecular potential on the predicted crystal structure of an A_2 molecu

14.1 Task

Most model intermolecular potentials assume the isotropic atom-atom model potential. However molecules are not superpositions of spherical charge distributions, and this practical shows that quite modest anisotropies in the effective shape of an atom can have a marked effect on the predicted crystal structure.

Consider an anisotropic atom-atom potential for a diatomic molecule of the form

$$U = \sum_{\substack{i=1,2 \\ j=3,4}} \frac{\epsilon}{(\alpha\rho_0 - 6)} \left[6\exp(-\alpha(R_{ij} - \rho(\Omega_{ij}))) - \frac{\alpha\rho_0^7}{(R_{ij} - \rho(\Omega_{ij}) + \rho_0)^6} \right], \tag{14.1}$$

where the minimum energy separation for an intermolecular pair of atoms $\rho(\Omega_{ij})$ depends on the relative orientation of the atoms. This can be defined by the unit vectors $\mathbf{z}_1$ and $\mathbf{z}_2$ along the intramolecular bonds, and a unit intermolecular atom-atom vector $\hat{\mathbf{R}}$. For this practical, consider small quadrupolar distortions of the atoms, as might represent the anisotropic effects of the equatorial lone pair density found in the halogens. Thus we have

$$\rho(\Omega_{ij}) = \rho_0 + \rho_{202}\tfrac{1}{2}(3\mathbf{z}_1.\hat{\mathbf{R}}^2 + 3\mathbf{z}_2.\hat{\mathbf{R}}^2 - 2). \tag{14.2}$$

The effects of variations in ρ_{202} on the optimum lattice parameters and lattice energies in different crystal structures will depend on the relative orientations of the nearest neighbour molecules in the lattice. For a diatomic molecule, two of the possible crystal structures which can be derived from the face-centred-cubic atomic structure are the Pa3 structure, when the atoms are stretched into molecules along the cube diagonals, and the R$\bar{3}$ structure, when the molecules lie along the three-fold symmetry axis. These structures have very different nearest neighbour orientations, and so the changes in the minimum lattice parameters with the change in the molecular shape will be different.

This effect can be studied using a crystal structure analysis program such as WMIN ([34]). You will need to write a subroutine to evaluate the anisotropic potential, which requires information about the bonded atoms in order to define the bond vectors. Study a hypothetical molecule with a bondlength= 1 length unit, $\epsilon = 1$ energy unit, and $\rho_0 = 2.0$ in both crystal structures. Consider variations $-0.2 \le \rho_{202} \le 0.2$. The details of the crystal structures are given in [35] which describes a fuller study of this kind. The results are used to rationalise the crystal structures adopted by the homonuclear diatomics.

15 Deriving a model potential for CS_2 by fitting to the low temperature crystal structure

15.1 Task

A useful method of testing and improving a model potential is to see whether it gives a reasonable prediction of the observed crystal structure, and if not, optimising the potential parameters using the structure. A successful prediction of the crystal structure does not imply that you have a potential which is accurate at all relative orientations, however, it is a good start to developing a potential suitable for a simulation of condensed phases.

We can examine the intermolecular potential for CS_2 by using a crystal structure analysis program which includes a parameter fitting option, such as WMIN ([34]). A reasonable starting point is the isotropic atom-atom Lennard-Jones 12-6 potential, developed as an effective pair potential for liquid CS_2, by Tildesley and Madden [36].

First predict the static crystal structure from this potential. Then use the fitting mode to optimise the potential parameters, and see whether the optimised potential gives a better predicted structure. You can then experiment with different functional forms for the model potential. Note that this fitting procedure is only sampling the potential in the relative orientations which are found in the experimental crystal structure, under the imposition of the observed space group symmetry (*Cmca*). The derivation of an accurate potential for CS_2 would be a major project, requiring a simultaneous fit to several properties, as can be appreciated from the discussion of various CS_2 potentials in the lecture course.

16 The Lennard-Jones fluid: a hard-sphere fluid in disguise?

16.1 Summary

The purpose of this exercise is for you to explore the state point dependence of the properties of the Lennard-Jones, LJ, fluid given by the `FORTRAN` program `ljeq`. You will attempt to predict its properties using formulae based on those of the hard-sphere fluid.

16.2 Background

The Lennard-Jones potential is:

$$\phi(r) = 4\epsilon((\sigma/r)^{12} - (\sigma/r)^6). \tag{16.1}$$

The thermodynamic properties of the LJ potential fluid come here from empirical parameterisations of simulation PVT data. Here we consider two expressions from the literature. The two equations of state used by program `ljeq` are those of Nicolas *et al.* [16] and Ree [37]. These will be referred to simply as $P(\rho, T)$. The program `ljeq` prints out the following thermodynamic quantities. The total energy per particle, E, which includes both kinetic and configurational terms is

$$E = 3k_BT/2 + \int_0^\rho \left(P - T \left(\frac{\partial P}{\partial T} \right)_\rho \right) d\rho/\rho^2 \tag{16.2}$$

The Helmholtz free energy in excess of the ideal gas term, is given by the formula [16]

$$A = \int_0^\rho (P/\rho - k_BT) d\rho/\rho \tag{16.3}$$

where k_B is the Boltzmann constant. This is performed analytically for the Ree equation of state but is evaluated numerically (in 100 equal ρ steps using Simpson's rule) in the case of the Nicolas *et al.* potential. The excess Gibbs free energy is given by

$$G = A + P/\rho \tag{16.4}$$

Remember that the "density", ρ is a *number* density. It denotes the number of particle centres to be found (on average) in a volume σ^3. The isothermal bulk modulus, K_0^T is [38]

$$\begin{aligned} K_0^T &= -V \left(\frac{\partial P}{\partial V} \right)_T \\ &= \rho \left(\frac{\partial P}{\partial \rho} \right)_T \end{aligned} \tag{16.5}$$

$$\left(\frac{\partial P}{\partial V} \right)_T = -\frac{1}{V} K_0^T . \tag{16.6}$$

The specific heat at constant volume, C_V, is

$$C_V = \left(\frac{\partial E}{\partial T} \right)_V . \tag{16.7}$$

$$\left(\frac{\partial V}{\partial T} \right)_P = - \left(\frac{\partial P}{\partial T} \right)_V / \left(\frac{\partial P}{\partial V} \right)_T . \tag{16.8}$$

Also,

$$\gamma = C_P/C_V = 1 - T \frac{\left(\frac{\partial P}{\partial T} \right)_V^2}{\left(\frac{\partial P}{\partial V} \right)_T C_V} \tag{16.9}$$

where C_P is the specific heat at constant pressure. Also the adiabatic bulk modulus K_0^s is given by

$$K_0^s = \gamma K_0^T , \tag{16.10}$$

and for the expansivity

$$\beta = \frac{1}{V}\left(\frac{\partial V}{\partial T}\right)_P. \tag{16.11}$$

16.3 The problem

The object of this exercise is to use the program `ljeq` to discover if one can consider the LJ fluid as a perturbed hard-sphere fluid. This is true if its properties are the same as or are deriveable from those of an equivalent hard-sphere fluid. The program `ljeq` prints out many physical properties of the LJ fluid.

16.3.1 Thermodynamic properties

Let the equivalent hard sphere diameter be σ_{HS}, given in units of σ. To a good approximation, this is only temperature dependent [39]

$$\sigma_{HS}/\sigma = 1.0217(1 - 0.0178/T^{1.256})/T^{1/12}. \tag{16.12}$$

The σ_{HS}/σ is printed out by `ljeq`. The LJ number density is $\rho = N\sigma^3/V$ i.e. the number of molecular centres in a volume σ^3. N is the number of particles in volume V. The reduced temperature $T \to k_B T/\epsilon$.

- Test the following hard-sphere based equations of state along the isotherms, $T = 1.5$ and 5.0.

  Van der Waals [40]:

  $$PV/Nk_BT = 1/(1 - \rho_{HS} b) + \rho_{HS} a/(k_B T). \tag{16.13}$$

  where $b = 2\pi\sigma_{HS}^3/3$, $\rho_{HS} = N\sigma_{HS}^3/V$ and a is a negative constant. Note that $\rho_{HS} = \rho(\sigma_{HS}/\sigma)^3$ and it is printed out by `ljeq`.

  Heyes and Woodcock [41]:

  $$PV/Nk_BT = b\rho_{HS}/(1 - \rho_{HS}/\rho_0)^2 + \rho_{HS} a/(k_B T). \tag{16.14}$$

  where $\rho_0 = 1.625\sigma_{HS}^3$. In this expression the first term on the right hand side is a simple but accurate representation of the hard-sphere equation of state. An equally valid choice would have been that of Carnahan-Starling [42]:

  $$PV/Nk_BT = \frac{1 + \eta + \eta^2 - \eta^3}{(1 - \eta)^3}. \tag{16.15}$$

  where $\eta = (\pi/6)\rho_{HS}$, the hard-sphere volume fraction.

16.3.2 Transport coefficients [39]:

For a hard-sphere fluid, the self-diffusion coefficients, D, shear viscosity, η and thermal conductivity, λ at finite density ρ divided by their zero density limiting values, D_0, η_0 and λ_0 are to a good approximation given by:

$$\begin{aligned} D/D_0 &= 1.271(2^{1/2}/\rho_{HS} - 1.384)/(2^{1/2}/\rho_{HS}), & (16.16)\\ \eta_0/\eta &= 0.2195(2^{1/2}/\rho_{HS} - 1.384), & (16.17)\\ \lambda_0/\lambda &= 0.1611(2^{1/2}/\rho_{HS} - 1.217). & (16.18) \end{aligned}$$

in the range $3/2 < 2^{1/2}/\rho_{HS} < 5/2$.

- Plot these transport coefficient ratios against ρ_{HS}^{-1} and see if you obtain a straight line as predicted above. `ljeq` prints out the values for the left-hand quantities. Be careful to substitute the 'correct' density to check all the above equations. Good luck!

16.4 Further tasks

- Run the FORTRAN programs `lj3d`, a three-dimensional LJ Molecular Dynamics program and `hs3d`, a three-dimensional hard-sphere Molecular Dynamics program and check the properties given by `ljeq`.

17 Time correlation functions

17.1 Summary

Calculate the velocity auto-correlation function (VACF) of a Lennard-Jones atom in a liquid, from its trajectory. Integrate the correlation function to obtain the diffusion coefficient. Calculate the root mean square displacement as a function of time and calculate the diffusion coefficient from the slope of the line at large times.

17.2 Background

The VACF is a function which describes how the velocity of an atom at a time, t_0, is correlated with its velocity at a time $t_0 + t$. At short times the velocities will be strongly correlated and at long times the correlation should fall to zero. The function is independent of the origin time, t_0, which we can set to zero. It is defined as

$$C_{vv}(t) = < \mathbf{v}(t).\mathbf{v}(0) > . \tag{17.1}$$

This average is over time origins in a molecular dynamics simulation or equivalently states in the microcanonical ensemble. Time correlation functions [15,43,44] are important in statistical mechanics because:

- they give a clear picture of the dynamics in a fluid;
- their time integrals may often be related to macroscopic transport coefficients;
- their Fourier transforms are often related to experimentally measurable spectra.

17.3 Task

In this problem you are given the velocities of a particular Lennard-Jones atom calculated in a molecular dynamics simulation. The velocities are in the normal reduced units, $v^* = v(m/\epsilon)^{1/2}$ where ϵ is the Lennard Jones well depth and m is the mass of the atom. The three components v_x^*, v_y^*, v_z^* are stored in the file **vacfdat** in the format `3F10.6`. Each line in the file corresponds to a particular time step and there are 5000 timesteps stored sequentially. The data is obtained from a simulation of Lennard-Jones atoms at $T^* = k_B T/\epsilon = 0.76$, $\rho^* = \rho\sigma^3 = 0.8442$. The velocities of atoms are written to the file at every step and the timestep is $\delta t^* = \delta t(\epsilon/m\sigma^2)^{1/2} = 0.005$ where σ is the Lennard-Jones diameter. A simple way to calculate the VACF is write it as a discretized time average.

$$C_{vv}(\tau) = (1/\tau_{max}) \sum_{\tau_0=1}^{\tau_{max}} \mathbf{v}(\tau_0).\mathbf{v}(\tau_0 + \tau) \tag{17.2}$$

In words, we average over τ_{max} time origins the dot product of $\mathbf{v}$ at time $\tau_0\delta t$ with $\mathbf{v}$ at a time $\tau\delta t$ later. Of course the value of $\tau + \tau_0$ must not exceed the total number of steps in the file. This average assumes that we want to use each successive point in the file as a time origin. This is probably inefficient since successive origins are highly correlated. We recommend that you try every 5th or 10th step as a time origin.

We recommend that you calculate C_{vv} for τ values from 0 to 200 or 300. You should experiment with the upper limit to see if you cover the whole interesting range of the function. You will need a simple method for calculating the time correlation function. A possible stategy is to read all data from the file **vacfdat** into three arrays `VX(10000)`, `VY(10000)`, `VZ(10000)` and then manipulate these arrays. There are more complicated methods for calculating the VACF [3], which require less storage, but these need not concern us in this exercise. When you have calculated your correlation function, you should normalize it by $C_{vv}(\tau = 0)$ so that it falls from one to zero and write this normalized VACF, $c_{vv}(t)$ to the file **vacfres**. Plot the correlation function against time in reduced units. The integral of the VACF can be related to the diffusion coefficient D

$$D = (1/3) \int_0^\infty dt\ C_{vv}(t) \tag{17.3}$$

or

$$D = (k_B T/m) \int_0^\infty dt\ c_{vv}(t) \tag{17.4}$$

Write a simple program to calculate D using the trapezoidal rule or Simpson rule. Convert D from reduced units to SI units for liquid argon (ϵ/k_B =123K , σ =0.334nm).

17.4 Further work

- How is the value of D affected by the choice of the upper limit used in numerical quadrature? At long times the integrand falls off algebraically as $t^{-d/2}$, where d is the dimension of the system. Can you use the functional form of the integrand at long times to correct your value of the diffusion coefficient?

- There is a file `vacf1dat` which contains the positions of atom 1 in exactly the same format as the velocities. (Note $r_x^* = r_x/\sigma$). The correlation of the mean-squared displacement of the centre of mass is also related to the diffusion coefficient at large times

$$2Dt = (1/3)\left\langle |\mathbf{r}_i(t) - \mathbf{r}_i(0)|^2 \right\rangle \tag{17.5}$$

Calculate the correlation function $< |\mathbf{r}_i(t) - \mathbf{r}_i(0)|^2 >$ and plot it against time. Periodic boundary conditions were not used in the simulation which created this data. Calculate the slope from the linear portion of the plot at long times. This should be equal to $6D$. Are the estimates D by both techniques in agreement?

18 Polymer reptation Monte Carlo

18.1 Summary

Write a Monte Carlo program to move, by a reptation algorithm, a flexible polymer molecule on a two-dimensional square lattice. Calculate the mean-squared end-to-end distance, and compare with the prediction of a simple random walk model.

18.2 Background

A simple approach to modelling flexible polymers is to place the atoms on the sites of a lattice, with site-site bonds joining the atoms of a given molecule. In Monte Carlo, we are allowed to make individual atomic moves from one site to another, as long as all the bonds remain intact. A simple approach to this is based on the 'slithering snake' or 'reptation' model [45] for the way real polymer chains move in a dense liquid. The head of the chain moves to a new position; the rest of the chain follows like a snake or lizard. This type of Monte Carlo move has been used in simulations by Wall and co-workers [46,47]. Taking all the atoms in a chain to be identical, it amounts to selecting a new head position, and deleting the tail atom at the other end of the chain. If the new head position turns out to be already occupied by another atom (in the same chain, in another chain, or belonging to a solvent molecule) then the move is rejected in the usual Monte Carlo way. Otherwise it is accepted. Either end of the chain can be chosen as the head, and it avoids problems of the chain becoming stuck, with the head buried in a cul-de-sac, if this choice is made randomly at each move.

Figure 18.1: A sequence of configurations for reptation Monte Carlo on a two-dimensional square lattice

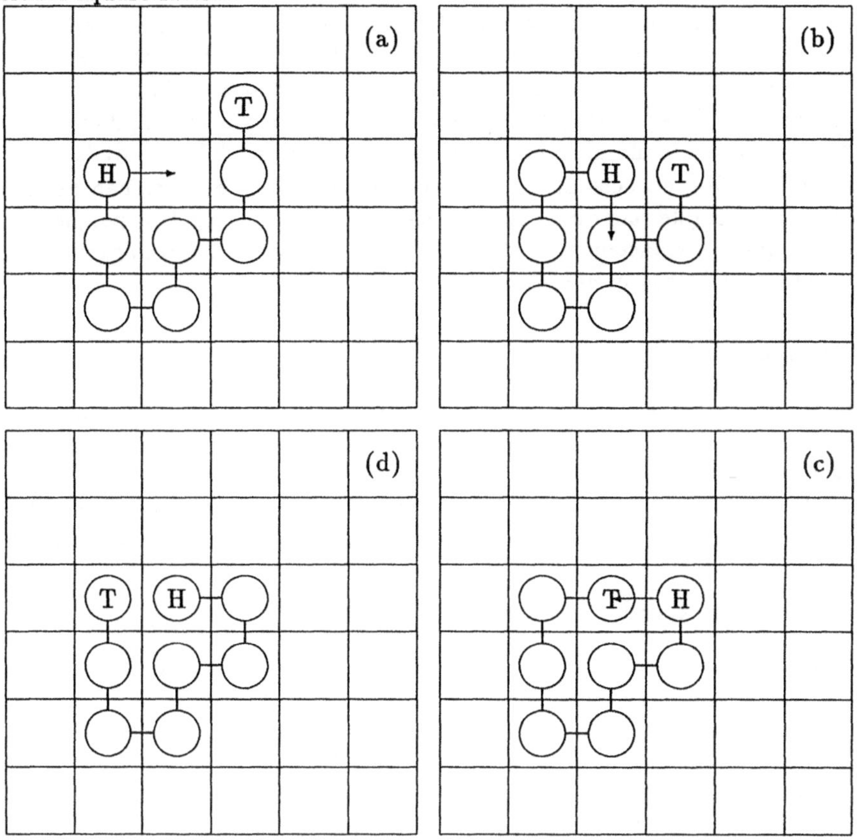

This method, and other MC techniques for polymers, are described in the standard references [3,11,12]. Here, we illustrate it with a typical sequence of steps as shown in the Figure. A single polymer chain of 8 atoms is simulated, on a square lattice. The initial configuration is shown in (a). A trial move of the chosen head atom is indicated by a vector. In this case, the new head site is vacant, and the move is accepted, giving configuration (b). The same head atom is chosen again, but this time the proposed move would lead to an overlap. Thus, the configuration is left unaltered, in (c). Now, in this example, the head and tail identities are switched as a result of a random choice. This time the trial head move is accepted: the new position will have been vacated by the tail atom when the move is completed. The result is configuration (d). This process is repeated until the simulation has run its course. In a real application, we might deal with hundreds of densely-packed polymer chains, and we would attempt to move each in turn.

A measure of the 'size' of a polymer molecule is the mean-square end-to-end distance $< \Delta r_N^2 > = < |\mathbf{r}_N - \mathbf{r}_1|^2 >$, where $\mathbf{r}_1$ is the position of atom 1 and $\mathbf{r}_N$ is the position of atom N in an N-atom chain. This can be calculated as a simulation average in the usual way. It is interesting to compare this quantity with the predictions of a simple random walk theory. In this, atom 1 is placed at an arbitrary point on the lattice, atom 2 is placed, with equal probability, at any of the neighbouring lattice sites, atom 3 placed randomly next to atom 2 and so on. The result is an N-step random walk, taking no account of the exclusion effects (i.e. sites can be multiply occupied). In a simple one-dimensional case, each site having two neighbours, the exact mean-square displacement is $< \Delta r_N^2 > = N$, taking the lattice spacing as unity. *What is the result in two dimensions?*

18.3 Task

Write a reptation Monte Carlo program for this system. Consider just one chain of about 8 atoms moving on a lattice. Choose the initial configuration in any way you like, making sure that no overlaps are allowed. In the simulation, move the head atom in any of the three 'non-backwards' directions. Measure the mean-square end-to-end distance as a simulation average. Compare with the simple random walk prediction.

18.4 Further work

- Repeat the exercise, but include the effects of a solvent by introducing obstacles: non-moving atoms that simply block one site each. The chain must wind its way in between the obstacles: the head atom can only move to vacant sites. Choose a density of obstacles by trial and error, and distribute them randomly, making sure again that there are no initial overlaps with chain atoms. Make sure that you don't exceed 50% occupation of sites by obstacles (*why?*). Compare the mean-square end-to-end distance with your first results.

- Repeat the original exercise, but start from a configuration in which every bond angle is a right angle (i. e. there are no straight-line sections) and only allow head atom moves to be at right angles to the existing bond (i. e. no 'backwards' *or* 'forwards' moves). Compare results.

- Consider trying out reptation Monte Carlo on a triangular two-dimensional lattice (every site has 6 neighbours) and on a three-dimensional cubic lattice.

- Try out this method for a dense system of many polymer chains, attempting to move each in turn.

19 Simulations of crystal surfaces

19.1 Summary

This project involves using the solid-on-solid model to simulate the surface of a crystal, using various Monte Carlo techniques. Special emphasis is placed on considering the requirements of microscopic reversibility, and the devising of kinetic Monte Carlo schemes. This project would be suitable for an individual, or for a team, each individual pursuing a different aspect.

19.2 Background

The computer simulation of crystal growth is a huge field of research (see [48] and references therein). Here we look at one of the simplest simulation models. The solid-on-solid (SOS) model can be thought of as a modification of the basic Ising or lattice gas model of statistical physics. It is specifically designed for the study of the interface between two phases [49,50]. As in the lattice gas model, each site of a simple cubic lattice (let us say) is either unoccupied or occupied by an atom. There are nearest-neighbour interactions between atoms. However, in addition, one direction (the z direction, say) is singled out. At sufficiently large, negative, values of z, all the lattice sites are taken to be occupied; at sufficiently large, positive, z, all the sites are unoccupied. For each value of x, y, there is just one value of z at which the changeover from occupied to unoccupied takes place. In other words, there are no 'bubbles' of one phase included in the other, and no 'overhangs'. The system is a set of columns of atoms, based on the 2-dimensional square x, y lattice: for each x, y lattice site, labelled i, let the height of the column be h_i. This set of heights specifies the state of the system.

The energy can be written in several ways, but we choose the simplest:

$$E = \epsilon \sum_{<ij>} |h_i - h_j|. \tag{19.1}$$

Here ϵ is an energy parameter. The sum is over distinct nearest neighbour sites in the square lattice (i.e. neighbouring columns). We count the interaction between each site i and each of its four neighbours j only once. The heights are measured from an arbitrary origin: note that shifting them all by the same amount does not change E. Thus, the absolute position of the surface is not relevant. However, when we consider driving the process of crystal growth, the rate of change of the average surface position $< h >$ will be of interest. Here the average $< ... >$ is taken over all the columns. The surface width W is important. We can define this as the root-mean-square deviation of h_i from the mean,

$$W = < (h - < h >)^2 >^{1/2} . \tag{19.2}$$

This system exhibits a phenomenon called surface 'roughening': above a critical temperature ($kT_R/\epsilon \approx 1.15$) the surface becomes rough and ill-defined, and the width diverges.

19.3 Metropolis Monte Carlo

It is relatively easy to write a Metropolis Monte Carlo program to simulate the SOS model. To change the state of the system, we must change the values of the h_i, typically by one unit at a time. A simple scheme is:

- select a column i (sequentially or at random);
- decide whether to try to increase or decrease h_i;
- calculate a transition probability for this change;
- choose to make the change or not, according to this probability.

For simplicity, let us choose to increase or decrease h_i with equal probability. Then, from eqn (19.1), it is easy to calculate the energy change ΔE associated with such a move. The Metropolis formula [18] for the probability of accepting the move is

$$P(\Delta E) = \min(1, \exp(-\Delta E/kT)) \tag{19.3}$$

where T is the temperature and k Boltzmann's constant. In other words, if ΔE is negative (downhill), accept the flip unconditionally; if ΔE is positive (uphill), accept with probability $\exp(-\Delta E/kT)$. As always, accepting a flip 'with a given probability' entails choosing a random number uniformly from a given range, typically (0,1).

Write a program to simulate a small (say 6×6 columns) SOS system using this method. Apply the usual periodic boundary conditions in the x and y directions. You may find it useful to work in reduced units, setting $\epsilon = 1$. For simplicity, you can select columns for moves in a sequential way. Suggested input is as follows.

- The initial configuration (read in from a file, which you could prepare using the editor).
- The temperature, in reduced units, kT/ϵ.
- The run length in Monte Carlo cycles (one cycle is one attempted move per column).

The NAG library provides random number generators: the subroutine `G05CCF` initializes the generator in a non-reproducible way, while the function `G05CAF ( DUMMY )` returns a random number in the range (0,1). Both `G05CAF` and its dummy argument `DUMMY` should be declared `DOUBLE PRECISION`. Suggested output at user-specified intervals:

- The total energy.
- The average height $< h >$ (just to check for drift).
- The interface width, defined by eqn (19.2).
- The ratio of moves accepted to moves attempted.

Run the program, starting from a flat surface, choosing temperatures above and below the approximate roughening temperature mentioned above.

19.4 Transition probabilities

It is instructive to consider the operation of the Metropolis method in more detail. Two kinds of move are contemplated. Increasing a column height $h_i \to h_i + 1$ means creating a new atom at the vacant 3-dimensional lattice site just above the top of the column. Depending upon the heights of neighbouring columns, this vacant lattice site might have from 0 to 4 neighbours in the lateral directions: call this number n. The energy change that goes into the Metropolis prescription depends on n. Referring to eqn (19.1), calculate this energy ΔE_n^+ for each n (it also depends on ϵ of course). The converse process, decreasing a column height $h_i \to h_i - 1$ corresponds to annihilating an atom from the top of a column. This atom might have from 0 to 4 lateral neighbours, depending on neigbouring column heights. Calculate ΔE_n^-, the associated energy change, for each n. You should have, of course, $\Delta E_n^+ = -\Delta E_n^-$, since one process is the reverse of the other. You should also note that, for a site with exactly two lateral neighbours, $\Delta E_2^+ = -\Delta E_2^- = 0$. These are called 'kink sites': at equilibrium they can be created or destroyed freely.

The transition probability $\Pi_n^\pm$ is a product of two terms. One is the underlying probability of selecting a certain trial move, $\alpha_n^\pm$, and we have been taking all these to be equal, for simplicity: $\alpha_n^+ = \alpha_n^- = \alpha$ for all n. The other is the Metropolis function, eqn (19.3), giving the probability of accepting such a move. In this case $P_n^\pm = \min(1, \exp(-\Delta E_n^\pm/kT))$. So we have $\Pi_n^\pm = \alpha_n^\pm P_n^\pm$ and (to account for the possibility that all the α's are *not* the same in general) microscopic reversibility applies to the Π functions. Here, this means

$$\frac{\Pi_n^+}{\Pi_n^-} = \exp(-\Delta E_n^+/kT) \qquad (19.4)$$

Now draw up a table of Boltzmann factors and transition probabilities for creation and annihilation moves, as functions of n. Your probabilities should satisfy eqn (19.4). At this point, you may wish to check your table with a tutor.

Having gone through this exercise, you probably realize that the heart of your Monte Carlo program can be speeded up. For each attempted move, creation or annihilation, you only require to calculate the appropriate number of lateral neighbours. The expensive exponentials can be pre-calculated, and looked up in a table. You may wish to make this change, and see if any speed improvement results.

19.5 Varying the prescription

Now we can try changing the basic Metropolis prescription. This is where things become interesting, because we can seek to make the Monte Carlo method model what actually happens in the real system, albeit in an idealized way.

Suppose that we wish to make the rates of selection and acceptance of creation moves independent of site, while the annihilation of atoms continues to depend on the binding energies with neighbours. This corresponds to a simple model of crystal growth: atoms arriving from the liquid add on to the growing surface irrespective of their environment, but to remove an atom it is necessary to break the bonds with its neighbours. We can arrange this in our Monte Carlo simulation, but must pay attention to microscopic reversibility. Thus we can set $P_n^+ = P^+$, a constant for all n. In fact, we can make this constant unity if we like, accepting every creation move that we attempt; we shall probably have to attempt such moves less often. Certainly, the underlying transition matrices for creation and annihilation will have to be different. We can set $\alpha_n^+ = \alpha^+$ and $\alpha_n^- = \alpha^-$ for all n, where $\alpha^+ \neq \alpha^-$. Finally, the annihilation move acceptance probabilities must be determined by the microscopic reversibility condition, eqn (19.4). Draw up a table, as you did in the last section, giving transition probabilities for both kinds of move, for each value of n, using this scheme. Check (for example) that the rates of kink site ($n = 2$) creation and annihilation are equal. If you like, adapt your simulation program along these lines. Consider one point: move acceptance probabilities cannot be greater than unity. Do you need to take any special measures to ensure this?

Suppose that we wish never to reject a move: at each site, we consider only increasing or decreasing h. Now the distinction between attempting moves and accepting moves is blurred. We select a site, and calculate two weights: one for creation and one for annihilation. Then, we use the weights, and a random number generator, to decide between these two possibilities. Consider more carefully how to construct this scheme, without violating microscopic reversibility. Note that there will be two separate sets of energy changes involved in the two possibilities: the creation and annihilation moves being considered are *not* the reverse of one another.

19.6 Kinetic Monte Carlo

In kinetic Monte Carlo, we drive crystal growth (or the reverse process) by scaling all the creation probabilities with respect to the annihilation probabilities. Now the rates of kink site ($n = 2$) creation and annihilation are not equal; instead they are related by

$$\Pi_2^+ = \exp(\Delta\mu/kT)\Pi_2^- \qquad (19.5)$$

where $\Delta\mu$ represents a chemical potential difference (zero at equilibrium). The rest of the probabilities are shifted in the same way. Incorporate this modification in your program. Now the crystal will grow (if $\Delta\mu > 0$), and you can measure the growth rate (by monitoring $< h >$) as a function of T and $\Delta\mu$.

19.7 Further work

This project can be taken in several directions. Much useful information on crystal growth simulations can be found in Muller-Krumbhaar's review [48], and the references therein. Here are a few possibilities.

- Study the growth of screw dislocations. This requires some modification of the boundary conditions, and you might like to consider how to change your program to handle this.
- Start the simulation from an irregular surface, or one with a specific defect or feature, and compare the evolution with that from a flat surface.
- Use different underlying lattices, to model the growth rates of different faces of a crystal.

Finally, demonstration programs for the SOS model are available on the DAP: try running them to see some of the features of this model.

20 Irreversible aggregation by Brownian dynamics simulation

20.1 Summary

This is an introduction to Brownian Dynamics, BD. In this exercise you will run a simple 3-dimensional BD simulation program that generates a "suspension" of Lennard-Jones, LJ, macromolecules in a fluid. *You* have to include the pair radial distribution function in the program and monitor the *fractal dimension* of the developing aggregates as a function of time.

20.2 Background

The object of this exercise is for you to learn about a method to model macromolecules in solution using the technique of Brownian Dynamics, BD [51,52]. It is very similar to Molecular Dynamics, MD.

Molecular Dynamics in its simplest form, i.e. for a system at *equilibrium*, is simply a numerical solution of Newton's equations of motion,

$$\ddot{\mathbf{R}}(t) = \mathbf{F}(t)/m. \tag{20.1}$$

where $\mathbf{R}$ is the position of an arbitrary molecule chosen out of N, m is the mass of the molecule (for simplicity, they all have the same mass here) and $\mathbf{F}$ is the net force on the molecule. For the x component of the force, $F_x(t)$ at time t then,

$$F_{xi} = -\sum_{i\neq j}^{N} \left(\frac{\partial\phi(r)}{\partial r}\right)_{r=r_{ij}} \frac{r_{xij}}{r_{ij}}, \tag{20.2}$$

where $\mathbf{r}_{ij} = \mathbf{R}_i - \mathbf{R}_j$. The subscript, i, denotes the molecule index and $\mathbf{R}_i$ is the position of molecule, i. In this work we will assume pairwise additive interactions of the form $\phi(r)$. The Lennard-Jones potential, [53]

$$\phi(r) = 4\epsilon((\sigma/r)^{12} - (\sigma/r)^6), \tag{20.3}$$

is a popular choice, which we will use here.

If we are interested in suspensions of particles in a liquid then the *details* of the comparatively fast movements of the 'solvent' molecules are not going to have a bearing on the movements of the large molecule, except in some kind of 'average' sense. This 'mean-field' approximation we can treat as follows. For particles $\sim 1\mu m$ in diameter then the Langevin equation is a reasonable governing equation for this situation,

$$\ddot{\mathbf{R}}(t) = \mathbf{F}(t)/m + \mathbf{S}(t)/m - \beta\dot{\mathbf{R}}(t), \tag{20.4}$$

where $\mathbf{S}(t)$ is an 'instantaneous' random force intended to approximate the resultant effect of the 'collisions' of the many millions of host fluid molecules on the macromolecule of interest. When we say 'instantaneous' here we mean 'instantaneous' on the time scale of the macromolecules (i.e. a time span in which they hardly move) whereas this can be a very long time indeed for the host fluid molecules. (Very long compared to the lifetime of their time correlation functions.) The force $\mathbf{S}(t)$ is the net result of the buffetings of the host fluid molecules upon the macromolecule during this small time for the macromolecule but long time for the host molecule. The solvent has another effect which is incorporated in the last term on the right hand side of eqn (20.4). This is a Stokes drag term, representing the resistance of the host medium to the 'flow' of the macromolecule. The Stokes friction coefficient, β, is defined as,

$$\beta = 3\pi\eta_s\sigma/m \tag{20.5}$$

where η_s is the viscosity of the pure solvent.

A finite difference algorithm is used to evolve the particles according to eqn (20.4). In fact, the particles should move continuously through space with time, t, given any preset equations of motion governing the particles. This is worth bearing in mind! There are N LJ molecules in the BD cell. The positions of the molecules, $\mathbf{R}$, in `FORTRAN` program `ljbd`, are updated in time steps of duration, h, using a *leapfrog* algorithm, [53], of the form,

$$\mathbf{R}_x(t+h) = \mathbf{R}_x(t) + \Delta\mathbf{R}_x(t), \tag{20.6}$$

for the x-component of R_x where we have, [53]

$$\Delta\mathbf{R}_x(t) = (\mathbf{F}_x(t) + S_x(h))h/\beta m, \tag{20.7}$$

where h is the time step and

$$< S_x^2(h) >= (2k_BTm\beta/h). \tag{20.8}$$

Now some technical details. The BD simulations are performed on a cubic unit cell of volume V . LJ reduced units are used throughout the program, i.e. $k_BT/\epsilon \rightarrow T$, and number density, $\rho = N\sigma^3/V$. Time is in $\sigma(m/\epsilon)^{1/2}$. The program prints out $\tau_p = \beta^{-1}$, the time for relaxation of macromolecular momentum,

$$\tau_p = m/3\pi\sigma\eta_s, \tag{20.9}$$

and also, τ_r, the time for significant structural evolution, i.e. movements of the macromolecules $\sim \sigma$,

$$\tau_r = 3\pi\sigma^3\eta_s/4k_BT. \tag{20.10}$$

20.3 The problem

- I would like you to modify program `ljbd` to include the calculation of the pair radial distribution function, $g(r)$,

  $$g(r) = n(r)/(4\pi\rho r^2 dr). \tag{20.11}$$

  where $n(r)$ is the average number of particles found in the annulus $r - dr/2 < r < r + dr/2$ about an arbitrary particle.

- Take $T = 1.0$ and $\rho = 0.6$ for the state point. Also, *in the first instance*, $N = 32$ and run each simulation several times from different starting configurations to assess the statistical noise. Let me know if you run into problems so I can give you some help!

- Once you have written the program and equilibrated your dispersion at the above state point, reset the state point to, $T = 0.2$ and $\rho = 0.1$ to look at irreversible aggregation. At this low temperature and density the LJ particles will "stick-together" more or less on contact. The small clusters will grow into larger clusters in some fashion.

  Why not calculate the *fractal dimension* D_f of the aggregates?

  $$g(r) \sim r^{D_f - d}, \tag{20.12}$$

  for $r \rightarrow \infty$, where d is the dimension of the space (= 3 here). The clusters formed should have a fractal dimension D_f of 2.5 if they form by particle-cluster aggregation and should be 1.78 if they form by cluster-cluster aggregation [54]. Good luck!

21 Diffusion-limited aggregation

21.1 Summary

Write a program to grow a crystal in two dimensions using diffusion limited aggregation. The simulation is performed on a two-dimensional square lattice. Calculate the fractal dimension of the crystal.

21.2 Background

The word fractal is derived from the latin word *fractus* which means irregular or fragmented. It was used by B. Mandelbrot [55] to describe a class of objects where the dimension can be, and often is, non-integer. We are used to Euclidean objects such as the sphere where the dimension is 3. What we mean by this is that the volume (V) of the object scales as the third power of a linear dimension (L) such as the diameter. However not all objects are as well behaved as this. The Sierpinski sponge [56], shown in figure 21.1, is constructed by dividing the face of the cube into nine smaller squares. The central square on each face, and the plug of material behind, is removed through to the opposite face. The construction is repeated on each of the eight remaining smaller squares and so on ad infinitum. The picture shows this construction taken to the third level. For this sponge $V(L)$ scales as L^D, where D the fractal dimension is 2.7268. The sponge illustrates another important feature of fractal objects which is their self similarity, i.e. when the object is studied at a microscopic level it has the same structure or pattern as the object viewed at a macroscopic level. These fractal structures are not merely mathematical nightmares, since approximations to them are common in nature. An area where the fractal concept is useful is that of aggregation or cluster growth, e.g. in the electrodeposition of zinc metal leaves on a surface [57].

One interesting model of cluster growth is diffusion limited aggregation [58]. A dynamic simulation of this model might proceed by placing a seed particle on a square lattice. A second particle is fired at the seed from a lattice point close to the firing circle (FC) and it executes a random walk on the lattice. If it explores a lattice point adjacent to the seed particle, it sticks and forms part of the growing cluster. If it crosses an escape circle (EC), some distance outside the firing circle, the walk is terminated and the particle is started again from a random lattice point on the the firing circle. A second and subsequent particle sticks if, after firing, it becomes adjacent to any particle already in the cluster. As the cluster grows the firing circle and the escape circle have to be moved back so that during the simulation FC is approximately five lattice spacings from the outside edge of the cluster. A typical run might comprise 3000 to 5000 shots. The geometry of the simulation is shown in figure 21.2.

21.3 Task

Write a program to grow a cluster using the DLA mechanism on a two-dimensional square lattice. You will need to choose a lattice size for your problem, and positions for the firing and escape circles. Draw out the cluster at intervals of 1000 shots. It is interesting to colour or mark shots in the intervals 1-1000, 1001-2000, etc, in different way, so that in your final picture we can establish where particular particles finish within the cluster as a function of the firing time. Calculate the fractal dimension of the patch by calculating the number of particles $N(r)$ inside a circle of radius r

Figure 21.1: The Sierpinski sponge[56]

Figure 21.2: The geometry of the DLA simulation

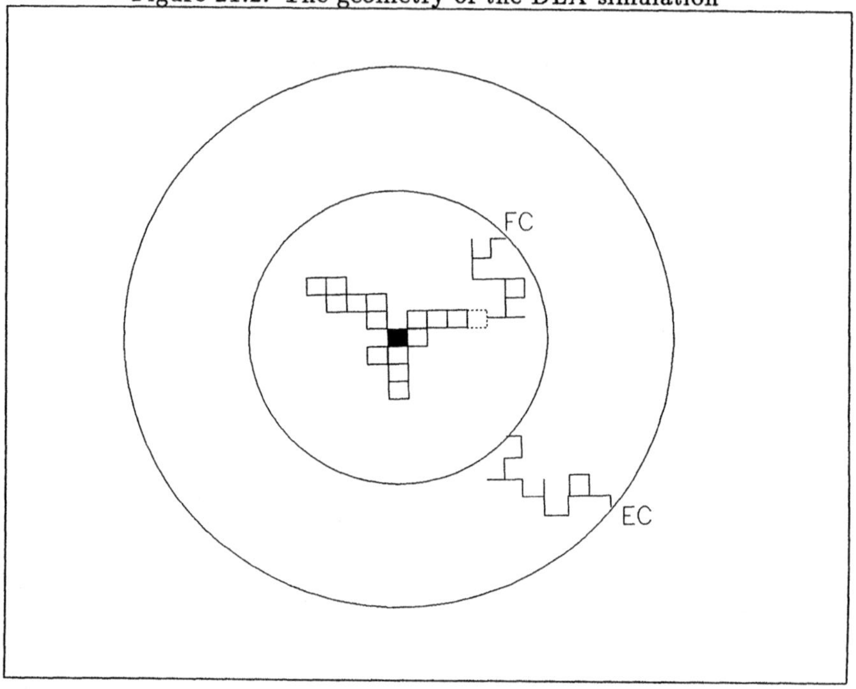

centred at the seed. A log-log plot of $N(r)$ against r should contain a linear portion where the slope is the fractal or Hausdorff dimension.

21.4 Further work

- Is the cluster really fractal, i.e. is it self similar on all length scales? Can any objects in nature be completely fractal in this respect?
- Repeat the exercise, using a triangular lattice in two dimensions. Does the shape of the lattice affect the fractal dimension of the cluster?
- Change the sticking probability of particles, i.e. allow particles to stick with a probability of 0.5. You will need to make decisions about what to do with particles which do not stick to the lattice at the first attempt. How does this change the fractal dimension of the patch?
- Design an algorithm which allows you to fire particles in straight lines to the cluster. This simulation will model crytal growth from a two-dimensional gas rather than a two-dimensional liquid. What is the fractal dimension of this patch?.

22 Shear flow in 2D Lennard-Jones fluid by non-equilibrium molecular dynamics

22.1 Summary

This task is an introduction to Non-Equilibrium Molecular Dynamics, NEMD. In this exercise you will work with a simple MD simulation program that generates a 2-dimensional Lennard-Jones, LJ, fluid. You will modify it to introduce shear flow (here, "plane couette flow") and thereby calculate the shear viscosity.

22.2 Background

The object of this exercise is to get some practice at writing a simple Non-Equilibrium Molecular Dynamics, NEMD, simulation program.

First, the essentials of the *equilibrium* program you will be working with are given below. A finite difference algorithm is used to evolve the particles according to slightly modified Newton's equations of motion. In fact, the particles should move continuously through space with time, t, given any preset equations of motion governing the particles. This is worth bearing in mind! There are N LJ molecules in the MD cell. The positions of the molecules, $\mathbf{R}$, in `FORTRAN` program `lj2d`, are updated in time steps of duration, h, using the *leapfrog* formulation of the Verlet algorithm, [59]

$$R_x(t+h) = R_x(t) + \Delta R_x(t), \tag{22.1}$$

and

$$R_y(t+h) = R_y(t) + \Delta R_y(t), \tag{22.2}$$

where at constant total energy we have, [59]

$$\Delta R_x(t) = \Delta R_x(t-h) + F_x(t)h^2/m, \tag{22.3}$$

$$\Delta R_y(t) = \Delta R_y(t-h) + F_y(t)h^2/m, \tag{22.4}$$

where $F_x(t)$ and $F_y(t)$ are the forces on a molecule at time t and m is the mass of the molecule (for simplicity, they all have the same mass here).

$$F_{xi} = -\frac{1}{m}\sum_{i\neq j}^{N}\left(\frac{\partial\phi(r)}{\partial r}\right)_{r=r_{ij}}\frac{r_{xij}}{r_{ij}}, \tag{22.5}$$

where $\mathbf{r}_{ij} = \mathbf{R}_i - \mathbf{R}_j$. The subscript, i, denotes the molecule index and $\mathbf{R}_i$ is the position of molecule, i. At constant temperature, T, (which is essential for sheared fluids because they are continuously trying to warm up!) we can maintain the temperature constant using the so-called *velocity* rescaling method as first derived by Woodcock, [60]

$$\Delta R_x(t) = \Delta R_x(t-h) \times K(t-h) + F_x(t)h^2/m, \tag{22.6}$$

$$\Delta R_y(t) = \Delta R_y(t-h) \times K(t-h) + F_y(t)h^2/m. \tag{22.7}$$

The constant, $K(t-h)$, is a scaling factor, which under shearing conditions, will on average be slightly less than unity.

$$K(t-h) = ((2N-3)k_BT/2E_k(t-h))^{1/2}, \tag{22.8}$$

where $E_k(t-h)$ is the kinetic energy of the MD cell of particles,

$$E_k(t-h) = \frac{1}{2}\sum_{i=1}^{N} m\dot{\mathbf{R}}(t-h)^2, \tag{22.9}$$

where to an adequate approximation,

$$\dot{\mathbf{R}} = \Delta\mathbf{R}(t-h)/h. \tag{22.10}$$

The MD simulations use particles interacting via the Lennard-Jones potential,

$$\phi(r) = 4\epsilon((\sigma/r)^{12} - (\sigma/r)^6). \tag{22.11}$$

The MD simulations are performed on a square unit cell of area A. LJ reduced units are used throughout the program, i.e. $k_BT/\epsilon \rightarrow T$, and number density, $\rho = N\sigma^3/V$.

Time is in $\sigma(m/\epsilon)^{1/2}$. Consequently, *your* shear rate is in $(\epsilon/m)^{1/2}/\sigma$, shear viscosity is in $(m\epsilon)^{1/2}/\sigma^2$ and stress is in $\epsilon\sigma^{-2}$.

The shear velocity profile, $\dot{\gamma}$, is to be introduced in the fluid using isokinetic SLLOD equations of motion [61,62]. The peculiar or thermal velocity is denoted by $\tilde{v}_\alpha$,

$$\dot{R}_x = v_x = \tilde{v}_x + \dot{\gamma} R_y, \tag{22.12}$$
$$\dot{R}_y = v_y = \tilde{v}_y, \tag{22.13}$$
$$\frac{d\tilde{v}_x}{dt} = F_x/m - \dot{\gamma}\tilde{v}_y, \tag{22.14}$$
$$\frac{d\tilde{v}_y}{dt} = F_y/m. \tag{22.15}$$

22.3 The problem

- I would like you to modify program `lj2d` to include a shear flow field. In addition to changing the program to include the extra terms in eqns (22.12)-(22.15), you will need to make the periodic boundaries of the cell compatible with the macroscopic shear velocity profile. We often use so-called *Lees-Edwards* periodic boundary conditions to achieve this [63]. These are succinctly,

$$\dot{R}^i_x = \dot{R}_x + n_i L_y \dot{\gamma}, \tag{22.16}$$
$$\dot{R}^i_y = \dot{R}_y, \tag{22.17}$$

where the superscript i refers to the image of the particle whose position in the 'real' MD cell is $\mathbf{R}$. The value of n_i (the image index) in the y-direction ranges from $-\infty < n_i < \infty$. The real MD cell corresponds to $n_i = 0$. The side length of the MD cell in the y direction is L_y. The position displacements are

$$R^i_x = \{R_x + n_i L_y \dot{\gamma} t\}, \tag{22.18}$$
$$R^i_y = R_y, \tag{22.19}$$

where t is the time duration of the simulation since the application of the shear velocity field. The notation $\{...\}$ is shorthand for: apply periodic boundary conditions so that R^i_x falls within the same limits as that of R_x (i.e. $0 < R^i_x < L_x$, where L_x is the sidelength of the cell in the x direction.). If the 'real' particle moves out of the cell, its image i enters at the position,

$$R_x \rightarrow \{R_x + n_i L_y \dot{\gamma} t\}, \tag{22.20}$$
$$R^i_y \rightarrow R_y + n_i L_y \tag{22.21}$$

Let me know if you run into problems so I can give you some help! Take $T = 0.5$ and $\rho = 0.75$ for the state point (it is close to the 2D LJ triple point [63,64]). Also, in the first instance set, $N = 50$ and run each simulation several times to assess the statistical noise.

- Once you have written the program to generate the positions of the particles at fixed $\dot{\gamma}$, then calculate the viscosity, η, from

$$P_{xy} = \frac{1}{A}\left(\sum_{i=1}^{N} m_i \tilde{v}_{xi}\tilde{v}_{yi} - \sum_{i=1}^{N-1}\sum_{j>i}^{N}(r_{xij}r_{yij}/r_{ij})\frac{d\phi(r_{ij})}{dr}\right), \tag{22.22}$$

where r_{xij} is the x component of $\mathbf{r}_{ij}$ and $A = (N/\rho)$, the area of the MD cell.

$$\eta = -P_{xy}/\dot{\gamma}. \tag{22.23}$$

23 The Fourier transform and applications to molecular dynamics

23.1 Introduction

The Fourier transform is one of the most important mathematical transforms. In molecular dynamics simulations it has many uses, some of which will be introduced here. It must be said that a brief introduction like this cannot hope to encompass more than a minute part of the subject, but what we cover will provide an outline of some of the commonest applications in MD. Students not acquainted with Fourier transforms should take time to read the following notes in preparation for the computational exercises to follow. We will deal with the following aspects.

- The definition of the Fourier transform
- Some mathematical properties of Fourier transforms
- Some examples of Fourier transforms
- The discrete Fourier transform

For further details refer to [65,66,67,68,69,70,71].

23.2 The Fourier Transform

We shall be concerned exclusively with the complex Fourier transform, since this is the most general form. There are few handicaps to this approach. Mathematically, we can write the Fourier transform $H(f)$ of a function $h(t)$ as:

$$H(f) = \int_{-\infty}^{\infty} h(t)\exp(-i2\pi ft)dt \tag{23.1}$$

and its so called inverse transform as:

$$h(t) = \int_{-\infty}^{\infty} H(f)\exp(i2\pi ft)df \tag{23.2}$$

The variables t and f are said to be *conjugate* variables in the time and frequency domains respectively. In this document we will refer almost exclusively to these variables, but they are not the only conjugate variables that commonly occur. Other examples are time, t, with angular frequency $\omega = 2\pi f$:

$$H(\omega) = \int_{-\infty}^{\infty} h(t) \exp(-i\omega t) dt \tag{23.3}$$

$$h(t) = \frac{1}{2\pi} \int_{-\infty}^{\infty} H(\omega) \exp(i\omega t) d\omega \tag{23.4}$$

and the position vector, $\mathbf{r}$ with the reciprocal space vector, $\mathbf{k}$:

$$H(\mathbf{k}) = \int_{-\infty}^{\infty} h(\mathbf{r}) \exp(-i\mathbf{r}.\mathbf{k}) d\mathbf{r} \tag{23.5}$$

$$h(\mathbf{r}) = \left\{\frac{1}{2\pi}\right\}^3 \int_{-\infty}^{\infty} H(\mathbf{k}) \exp(i\mathbf{r}.\mathbf{k}) d\mathbf{k} \tag{23.6}$$

The occurence of a Fourier transform and its inverse is dependent on certain mathematical conditions. They are intuitively reasonable but require rigorous proof. Briefly, they assume that the integrals (areas) of the functions h and H over the range concerned exist and are are bounded, and that the functions take a mean value at any discontinuity. These are known as the Dirichlet conditions. It will be important to remember these in subsequent sections. Any good textbook on the subject will discuss these more fully.

23.3 Some Properties of the Fourier Transform

In order to use the Fourier transform as an effective mathematical tool, it is essential to know some of the basic properties. We present here some of these properties, which will be useful when tackling the exercises. They will be presented without proof, but students should attempt to prove them to their own satisfaction at their leisure. (Note: In what follows the Fourier transform of a function $h(t)$ is indicated by $H(f) = FT(h(t))$ and the inverse Fourier transform by $h(t) = FT^{-1}(H(f))$.)

23.3.1 Even, Odd and Ordinary Functions

A function $f(x)$ of a variable x is *even* if $f(-x) = f(x)$ and *odd* if $f(-x) = -f(x)$. Ordinary functions do not have either of these properties. Using these definitions it can be shown that:

- An ordinary function is a sum of an even and an odd function.
- If $h(t)$ is real and even then $H(f)$ is real and even.
- If $h(t)$ is real and odd then $H(f)$ is imaginary and odd.

- If $h(t)$ is real and ordinary then $H(f)$ is complex with an even real part and odd imaginary part.
- If $h(t)$ is imaginary and even then $H(f)$ is imaginary and even.
- If $h(t)$ is imaginary and odd then $H(f)$ is real and odd.
- If $h(t)$ is imaginary and ordinary then $H(f)$ is complex with an odd real part and even imaginary part.
- If $h(t)$ is complex and even then $H(f)$ is complex and even.
- If $h(t)$ is complex and odd then $H(f)$ is complex and odd.
- If $h(t)$ is complex and with an even real part and odd imaginary part then $H(f)$ is real.
- If $h(t)$ is complex and with an odd real part and even imaginary part then $H(f)$ is imaginary.

23.3.2 Shift Theorems

Basically these show that a change in origin in one domain results in the appearance of a phase factor in the other domain. i.e.

- If $FT(h(t)) = H(f)$ then $FT(h(t-t_o)) = H(f)\ \exp(-i2\pi f t_o)$.
- If $FT^{-1}(H(f)) = h(t)$ then $FT^{-1}(H(f-f_o)) = h(t)\ \exp(i2\pi f_o t)$.

23.3.3 Scaling Theorems

These theorems show that a change of scale in one domain results in the inverse scaling in the other. i.e.

- If $FT(h(t)) = H(f)$ then $FT(h(\alpha t)) = \alpha^{-1}H(f/\alpha)$.
- If $FT^{-1}(H(f)) = h(t)$ then $FT^{-1}(H(\beta f)) = \beta^{-1}h(t/\beta)$

23.3.4 Parseval's Theorem

This important theorem provides a way of expressing the integral of the square of a function (say $h(t)$) in the time domain as an integral involving its Fourier transform (i.e. $H(f)$) in the frequency domain:

$$\int_{-\infty}^{\infty} h(t)^2 dt = \int_{-\infty}^{\infty} H^*(f)H(f)df$$

In the case where $h(t)$ can be expressed as a sum of harmonic frequencies, this offers a very simple way of evaluating the l.h.s. integral, since the r.h.s. is then a simple arithmetic sum. It is useful in evaluating the power in a fluctuating function (wave-form) described by $h(t)$. Parseval's theorem is a special case of a theorem concerning the product of two functions in the time domain.

23.3.5 The Convolution Theorem

Two of the most important properties of Fourier transforms concern the convolution integral and the correlation integral (below). A convolution integral has the form:

$$c(t) = \int h(t')g(t-t')dt'$$

Which is a convolution of the functions $h(t)$ and $g(t)$. It is permissible for these two functions to be the same. The convolution theorem says:

- If $c(t)$ is the convolution of $h(t)$ and $g(t)$ and these three functions have Fourier transforms $C(f)$, $H(f)$ and $G(f)$ respectively, then

$$C(f) = H(f)G(f).$$

This powerful theorem shows how a difficult integral problem in one domain becomes much easier in the other domain. This has important computational advantages in certain problems.

23.3.6 The Correlation Theorem

Correlation integrals are very common in molecular dynamics. They form the basis for the analysis of time dependent phenomena. A correlation integral has the form:

$$c(t) = \int h(t')g(t+t')dt'$$

Which is a correlation of the functions $h(t)$ and $g(t)$. If these two functions are the same, it is said to be an *auto*-correlation function. If they are different functions, it is called a *cross*-correlation function.

The correlation theorem says:

- If $c(t)$ is the correlation function of $h(t)$ and $g(t)$ and these three functions have Fourier transforms $C(f)$, $H(f)$ and $G(f)$ respectively, then

$$C(f) = H^*(f)G(f),$$

  where $H^*(f)$ is the complex conjugate of $H(f)$.

This is a theorem of immense power, both in theoretical studies and computational algorithms.

23.3.7 The Wiener-Khintchine Theorem

This theorem has application in the analysis of time dependent correlation functions, where it sheds light on the frequency dependence of such phenomena as energy dissipation etc. It says that:

- If $FT(c(t)) = C(f)$, where $c(t)$ is a time dependent correlation function, then:

$$C(f) = 2\int_0^\infty c(t)\cos(2\pi ft)dt$$

- and conversely:

$$c(t) = 2\int_0^\infty C(f)\cos(2\pi ft)df$$

This follows quite simply if it is assumed that the correlation function is symmetric about $t = 0$.

23.3.8 Derivatives of a Function

This provides the relationship between the derivative of a function and its Fourier transform. It is important because it helps in the solution of certain kinds of differential equation. Let $h(t)$ be a function of time and $H(f)$ its Fourier transform.

- If $h'(t) = \frac{d}{dt}h(t)$ is the derivative of the function $h(t)$, then:

$$FT(h'(t)) = i2\pi fH(f).$$

- If $h''(t) = \frac{d^2}{dt^2}h(t)$ is the second derivative of the function $h(t)$, then:

$$FT(h''(t)) = -(2\pi f)^2H(f).$$

Higher order derivatives have similar expressions.

23.3.9 The Dirac Delta Function

The Dirac delta function (which is not strictly a function at all), is a useful device for demonstrating certain properties of Fourier transforms and constructing formulae. It is sometimes referred to as the *impulse function.* It has the following properties:

- $\delta(t - t_o) = 0$ if $t \neq t_o$
- $\int_{-\infty}^{\infty} \delta(t - t_o)dt = 1$
- $\int_{-\infty}^{\infty} h(t)\delta(t - t_o)dt = h(t_o)$

where $h(t)$ is an ordinary function. The function $\delta(t)$ may be regarded as a rectangular function of vanishingly small width, but finite area (i.e. as a kind of sharp "pulse" function), though this is not a rigorous definition.

23.4 Some Example Fourier Transforms

We present here some examples of Fourier transformed functions. Any decent text-book on the subject will provide many more. The ones given have some relevance to the computational exercises outlined later.

23.4.1 Trigonometric Functions

$$FT(\cos(2\pi f_o t)) = \frac{1}{2}(\delta(f + f_o) + \delta(f - f_o))$$

$$FT(\sin(2\pi f_o t)) = \frac{i}{2}(\delta(f + f_o) - \delta(f - f_o))$$

The delta functions indicate that the continuous trigonometric functions in the time domain transform to single numbers in the frequency domain. This is an obvious example of the Fourier transform "projecting out" the frequency dependence of a function. Notice that the transform admits both positive and negative frequencies.

23.4.2 Gaussian Function

The gaussian function is common in all branches of physics, and molecular dynamics is no exception.
If

$$g(t) = (\alpha/\pi)^{1/2} \exp(-\alpha t^2)$$

then:

$$G(f) = \exp(-(\pi f)^2/\alpha)$$

Notice the transformed function is also a gaussian.

23.4.3 Exponential Function

A function decaying exponentially from a fixed value at given time origin (assumed to be $t = 0$ here), is a common physical occurrence i.e.

$$h(t) = A\exp(-\alpha t), \text{ when } t > 0$$

$$h(t) = A/2, \text{ when } t = 0$$

$$h(t) = 0, \text{ when } t < 0)$$

(Notice the definition of the function when $t = 0$, to comply with the Dirichlet conditions.) The Fourier transform is then:

$$H(f) = A(\alpha^2 + (2\pi f)^2)^{1/2} \exp(i \tan^{-1}\{-2\pi f/\alpha\})$$

Notice that the transform is *complex*. On the other hand, the double sided exponential:

$$h(t) = A\exp(-\alpha|t|), \text{ with } -\infty < t < \infty$$

has the Fourier transform:

$$H(f) = 2A\alpha/(\alpha^2 + (2\pi f)^2)$$

which is real.

23.4.4 Rectangular Window Function

This function is actually very common, although it is not often realised, particularly by novice users of the discrete Fourier transform. It is defined by:

$$w(t) = A, \text{ when } |t| < T/2$$

$$w(t) = A/2, \text{ when } |t| = T/2$$

$$w(t) = 0, \text{ when } |t| > T/2$$

Where T is the width of the window in the time domain, and A is the height. Notice the second equation, which is necessary to define the discontinuity at $|t| = T/2$ so that it satisfies the Dirichlet conditions. We then have:

$$W(f) = AT\sin(\pi Tf)/(\pi Tf)$$

It is important to note that the Fourier transform is not an impulse function (δ-function) in the frequency domain.

23.4.5 The Blackman-Harris Window Function

This function is used as an alternative to the rectangular window function described above. It is one of many alternative functions giving better resolution in harmonic analysis. It is defined by:

$$w(t) = \sum_{j=0}^{3} a_j \cos(2\pi jt/T)$$

where T is the width of the time window. The coefficients $a_0, \ldots, a_3$ have the values $0.35875, -0.48829, 0.14128$ and -0.01168 respectively. Its Fourier transform is easy to obtain, but it is the convolution of this with the transform of the rectangular window function that is important. (It is designed to produce a good replica of the δ function in the frequency domain).

23.4.6 The Dirac Delta Function

This follows easily from the definition of the delta function given above:

$$FT(A\delta(t)) = A$$

This shows that a spike at the origin in the time domain will give rise to a "plateau" in the frequency domain. If an unexpected plateau appears in any of your Fourier transforms, suspect this as the cause.

23.4.7 The Sampling Function

A continuous function in the time domain may be sampled by being multiplied by the following function, which samples at regular intervals Δt:

$$s(t) = \sum_{j=-\infty}^{\infty} \delta(t - j\Delta t)$$

and its Fourier transform is:

$$S(f) = \frac{1}{\Delta t} \sum_{j=-\infty}^{\infty} \delta(f - j/\Delta t)$$

This transform is central to the construction of the discrete Fourier transform.

23.5 The Discrete Fourier Transform

To convert the Fourier transform from its integral representation to a discrete representation amenable to digital processing, substantial modifications of the original time function are necessary. These modifications result in subtle changes in the properties of the transform that affect the accuracy and interpretation of the result. It is instructive to examine these modifications in turn to see their effects and, where necessary remedy them. The principal modifications required are sampling and windowing in the time domain and sampling in the frequency domain.

23.5.1 Sampling in the Time Domain

The continuous function of time $h(t)$ may be converted to a discrete representation by multiplying it by the sampling function $s(t)$ described above. The sampling function consists of an infinite train of regularly spaced impulse functions of unit area and infinitesimal width separated by the time interval Δt. Multiplication by this function produces a set of data points equally spaced in the time domain at intervals of Δt, and weighted by the value of the function $h(t)$ at the corresponding abscissa. Thus we have, in the time domain:

$$h(t)s(t) = \sum_{n=-\infty}^{+\infty} h(n\Delta t)\delta(t - n\Delta t)$$

However the effect of this multiplication on the Fourier transform is more complicated; it results in a *convolution* of the functions $H(f)$ and $S(f)$, which are the Fourier transforms of $h(t)$ and $s(t)$ respectively. We saw above that S(f) is itself a regular sequence of impulses separated by the frequency interval $1/\Delta t$ in the frequency domain. The convolution of this function with $H(f)$ results in a *continuous* function of frequency consisting of periodic replications of the function $H(f)$ centred on the locations of the impulse functions (i.e. the basic function $H(f)$ is repeated at intervals of $1/\Delta t$). We may write this as:

$$FT(h(t)s(t)) = \int_{-\infty}^{+\infty} S(f')H(f-f')df' = \frac{1}{\Delta t}\sum_{n=-\infty}^{+\infty} H(f-\frac{n}{\Delta t})$$

The periodic replication of $H(f)$ is important for a number of reasons. Firstly it is clearly an artefact of the sampling of the function $h(t)$, as it is not present in the true Fourier transform. Secondly, it is clearly possible for the replicated functions to overlap in the frequency domain, and when they do it constitutes an error in the Fourier transform known as *aliasing.* If however the original function $h(t)$ is band limited (i.e. does not possess frequency components higher than a critical frequency f_c say), then it is possible to avoid the problem of aliasing by choosing a sampling interval of $1/(2f_c)$. This is known as Nyquist sampling. In general however, the frequency components are not band limited, and Δt must be chosen to be as small as is practical, and thereby widen the gap between the impulse functions of $S(f)$ in the frequency domain.

23.5.2 Windowing in the Time Domain

The infinite set of data points produced by sampling the function $h(t)$ must be reduced to permit digital processing. This is done by multiplying the sampled function $h(t)s(t)$ by a rectangular window function $w(t)$ similar to that described in section 23.4.4 above and with the form:

$$w(t) = 1, \text{ when } -\Delta t/2 < t < T - \Delta t/2$$

$$w(t) = 1/2, \text{ when } t = \Delta t/2 \text{ or } t = T - \Delta t/2$$

$$w(t) = 0, \text{ otherwise}$$

Where $T = N\Delta t$. Notice that the left and right ends of the window are chosen so as not to coincide with any data points. The effect of this window in the time domain is as expected:

$$(h(t)s(t))w(t) = \sum_{n=0}^{N-1} h(n\Delta t)\delta(t-n\Delta t)$$

Which represents N sampled data points only. Once again we must consider the effect of this procedure on the Fourier transform, and again it is a convolution. The Fourier transform of the window function is of the form $\sin(\pi T f)/(\pi f)$ (times a phase factor), which has a sharp principal peak and smaller, decaying side peaks. When convoluted with the replicated functions $H(f)$ these side peaks introduce small ripples into the continuous function. These ripples diminish as the width of the window function (i.e. the parameter T) increases, but are implicitly always present and give rise to an error in the Fourier transform known as *leakage*. We will encounter leakage later, but for now note the importance of taking as wide a window as is possible, to reduce the magnitude of this effect.

23.5.3 Sampling in the Frequency Domain

So far, sampling and windowing has discretised the function $h(t)$ in the time domain. The Fourier transform however is still a continuous function (if replicated and with ripples). It is now discretised by multiplying with a sampling function in the frequency domain. Since the function in the frequency domain has a periodicity of $1/\Delta t$ and there are N sampled data points in the time domain, it is sensible to sample at intervals of $1/(N\Delta t)$ to give N data points in the frequency domain also. The sampling function is therefore:

$$W'(f) = \sum_{k=-\infty}^{+\infty} \delta(f - k/(N\Delta t))$$

In the time domain, this multiplication is equivalent to a convolution, which means that the sampling in the frequency domain is essentially accomplished by convoluting the inverse Fourier transform of $W'(f)$ with the sampled and windowed function $h(t)$ (c.f. sampling in the time domain and its effect in the frequency domain). In practice this is equivalent to a periodic replication of the time domain data, with a periodicity T. (Note: aliasing in the time domain does not arise because the first and last data points have been carefully chosen not to correspond to periodic images).

23.5.4 The Final Expression

The result of applying all of these operations is to produce the discrete version of the Fourier transform:

$$H(n\Delta f) = \Delta t \sum_{k=0}^{N-1} h(k\Delta t) \exp(-i2\pi nk/N) \tag{23.7}$$

where $n = 0, \ldots, N-1$ and $\Delta f = 1/(N\Delta t)$. The inverse discrete Fourier transform is similarly produced:

$$h(k\Delta t) = \Delta f \sum_{n=0}^{N-1} H(n\Delta f) \exp(i2\pi nk/N) \tag{23.8}$$

It should be noted that these sums resemble the appropriate Fourier integral in each case, but in a form based on the rectangular sum approximation. (Warning: FFT routines available on most computing systems do not incorporate the normalisation constants Δt and Δf; users must therefore supply these values themselves.) The simple outcome eqn (23.8) is what leads novice users to believe that nothing much has happened to the Fourier transform in this process, but this account should convince the reader that this assumption is untrue. It is very important to realise that the discrete Fourier transform regards the time and frequency domain functions as being periodic in their respective domains, the first with periodicity T, given by the window width, and the latter with periodicity $1/\Delta t$, where Δt is the time domain sampling interval.

24 Harmonic analysis

24.1 Summary

The use of the Fourier transform in harmonic analysis is perhaps its best known application. In essence it is used to "project out" the underlying frequencies in a given function or signal. In this exercise, the student will apply the discrete Fourier transform to an harmonic analysis of a simple function. The exercise will show the problems that arise from an ill-considered application of the transform and reveal the usefulness of window functions. The object of the exercise is to familiarise the student with the behaviour of the discrete Fourier transform under controlled conditions. It is better to unravel the difficulties in cases where the results are known, than waste time mis-managing real experimental data!

Students should use the 1D Fourier transform routine in the program library for this exercise. Note that this routine requires the processed arrays to be of length 2^N, where N is an integer.

24.2 Background

In harmonic analysis we are primarily concerned with the frequency components of a "signal" of some kind. In molecular dynamics for instance, we might analyse the fluctuations in the values of some property or other, to determine the timescales of the fluctuations and their power spectrum. In principle the signal may be of infinite duration, in which case the harmonic analysis in integral form is given by:

$$H(f) = \lim_{T\to\infty} \frac{1}{T} \int_{-\infty}^{\infty} h(t) \exp(-i2\pi f t) dt$$

but in practice, the signal is necessarily of finite duration and is sampled discretely. In this case the discrete form is employed:

$$H(n\Delta f) = \frac{\Delta t}{T}\sum_{k=0}^{N-1} h(k\Delta t)\exp(-i2\pi nk/N)$$

where T represents the time window over which the signal is sampled. (Note that since $T = N\Delta t$ the normalisation factor is just $1/N$.)

The harmonic analysis generally yields a *complex* function of frequency, but this merely means that both the amplitude and phase information are present in the transform. We can separate these two parts quite simply: recall that any complex number can be written as the product of a real number and an imaginary power of e:

$$(a + ib) = Ae^{i\theta}$$

where $A = (a^2 + b^2)^{1/2}$ and $\theta = \tan^{-1}(b/a)$. So in harmonic analysis the amplitude of a frequency component is obtained by multiplying the $H(n\Delta f)$ by its complex conjugate and taking the square root of the result. The phase is obtained from the inverse tangent of the ratio of the imaginary and real parts of $H(n\Delta f)$. We may also combine the positive and negative frequency components if we wish, to obtain a real result.

The major difficulty with the harmonic analysis of an unknown signal is the problem of "leakage", whereby alien frequencies are introduced into the resultant spectrum by the discrete Fourier transform itself. It is more likely to be a problem than "aliasing", which is more evident in short duration signals, which necessarily contain very high frequencies. (This fact derives from the "uncertainty principle", in which a signal that is located precisely in the time domain (i.e. of short duration) is widely dispersed in the frequency domain (i.e. has a broad spread of frequencies).)

24.3 Task

Consider a cosine function with a periodicity P of 8 seconds i.e.:

$$h(t) = \cos(2\pi t/P)$$

Suppose we have sampled this function at regular intervals Δt of 1 second over a time period T of 32 seconds (i.e. $T = 4P$). Construct an array of these data and use a discrete Fourier transform routine to obtain the Fourier transform. Does the result compare with the analytical result in Section 23.4.1? If it does not, do not proceed until you are completely satisfied with the results!

The second task is a repetion of the first, but with a small difference. Now set the period P of the cosine function to be 9.14, but retain the sampling interval of 1 second and the sampling period T of 32 seconds. What does the discrete Fourier transform produce in this case? Plot the function over the complete time interval

T and try to account for the observed result in terms of what you know about the discrete Fourier transform. Refer back to the previous sections if necessary!

Repeat the exercise again, but before performing the Fourier transform, multiply the function by the Blackman-Harris window function (section 23.4.5). Observe the result. Are you able to account for this result? If not, plot out the array of the cosine function multiplied by the Blackman Harris window function. Does this provide any clues?

Repeat the exercise for the sine function, using the same periodicities. You may wish to try an alternative window function, due to Hanning:

$$w(t) = \frac{1}{2}(1 - \cos(2\pi t/T)$$

Is it better or worse than the Blackman-Harris function?

Finally, consider the function:

$$h(t) \;=\; 10.0\cos(2\pi t/9.6) \;+\; 0.1\;\cos(2\pi t/6.15)$$

and suppose it is sampled regularly 128 times over a 100 second period. Try using the discrete Fourier transform to project out the two inherent frequencies; firstly using the raw FFT routine; secondly using the Hanning window function and finally using the Blackman-Harris window function. These results should convince you of the importance of a good window function in harmonic analysis.

24.4 Further Work

Use the discrete Fourier transform to obtain acceptable Fourier transforms of the following functions (there is no need to use window functions here).

- $y(t) = (\alpha/\pi)^{1/2}\exp(-\alpha t^2)$ with $\alpha = 0.5$ and $-10 < t < 10$
- $y(t) = \exp(-\alpha|t|)$ with $\alpha = 2.5$ and $-5 < t < 5$
- $y(t) = \exp(-\alpha t)$ with $\alpha = 3.0$ and $0 < t < 10$ (assume $y(t) = 0$ when $t < 0$).
- $y(t) = \cos(5t)\exp(-t^2)$ with $0 < t < 2$ (Note: $y(t)$ is even).

You should compare your results with those obtained analytically (see section 23.4).

25 Correlation functions

25.1 Summary

The object of this exercise is to investigate methods for evaluating time-dependent correlation functions. Two methods are considered; the conventional "moving origin" method and a method based on the fast Fourier transform. The student should acquire some idea of the relative merits of these methods and gain some practice at coding them in a simple application. Students should use the 1D Fourier transform routine in the program library for this exercise. Note that this routine requires the processed arrays to be of length 2^N, where N is an integer.

25.2 Background

In molecular dynamics simulations it is frequently of interest to evaluate correlation integrals of the form:

$$c(t) = \lim_{T\to\infty} \frac{1}{T}\int_0^T u(\tau)h(t+\tau)d\tau$$

where $u(t)$ and $h(t)$ are both functions of time. (They may be the same function if required.) Such functions reveal a cause-and-effect relationship between the functions $u(t)$ and $h(t)$, by which a change in $u(t)$ manifests a change in $h(t)$ at some time later.

In molecular dynamics the correlation integral is approximated by the discrete equation:

$$c(k\Delta t) = \left\langle \frac{\Delta t}{(N-k)} \sum_{n=0}^{N-k-1} u(n\Delta t)h((n+k)\Delta t) \right\rangle$$

where $k = 0, \ldots, N-1$ and $N\Delta t$ is a time interval over which the the correlation function is required. (Note: the angular brackets indicate an *ensemble average*, which in this context means an average over many time origins.)

This direct method however is not necessarily the fastest way to proceed. The correlation theorem (section 23.3.6) suggests that the business of correlation is simpler in the frequency domain. It is also well-known that the discrete Fourier transform (often called the *fast* Fourier transform or FFT) can Fourier transform any given function extremely quickly. A more efficient scheme therefore suggests itself:

- Fourier transform $u(t)$ and $h(t)$; obtain $U^*(f)$ and $H(f)$, (the asterisk indicates the complex conjugate).
- Multiply $U^*(f)$ and $H(f)$ together, as required by the correlation theorem. Obtain $C(f)$.
- Inverse Fourier transform $C(f)$ to obtain $c(t)$.

Using the discrete Fourier transform, the correlation function may be written as:

$$c(k\Delta t) = \left\langle \frac{\Delta t}{(N-k)(2N)} \sum_{n=0}^{2N-1} U^*\left(\frac{n}{2N\Delta t}\right) H\left(\frac{n}{2N\Delta t}\right) \exp\left(\frac{2\pi ink}{2N}\right) \right\rangle$$

where $k = 0, \ldots, N-1$ as before. Notice that the sum ranges over $2N$ and not N as might be assumed at first. In practice, it means that an array of N zeros is appended to the arrays U^* and H. This is done to correct an overlap or end effect which arises because we necessarily must sample both U^* and H over some finite time period (T say) which results in a meaningful correlation function of period $2T$. If we choose a time window of T in the discrete Fourier transform, the transform will incorrectly assume a periodicity of T for all the functions it encounters, including the correlation function. In consequence, it will overlap the correlation function with itself, leading to spurious results. Doubling the time interval to $2T$ removes this error.

25.3 Task

Write a FORTRAN routine to calculate the correlation function of two variables $u(n\Delta t)$ and $h(n\Delta t)$, where $n = 1, \ldots, 10000$. Take the correlation out to $100\Delta t$. Use the traditional method, with a new origin taken every $10\Delta t$. Use the routine in conjunction with a Lennard-Jones molecular dynamics program from the library and calculate the velocity autocorrelation function.

Next write a second routine, incorporating a Fast Fourier transform routine and calculate the same correlation function as above. Compare the two previous methods in terms of accuracy and speed. Try to find ways of improving the performance of both of these methods (e.g. can you vectorise either of them?) and try to decide which of them you would prefer in general applications.

25.4 Further Work

Suppose you have a dataset containing the accumulated results of a molecular dynamics simulation. Each data record contains M of real numbers, each of which specifies the value of a variable in the simulation at the same timestep. There are N data records corresponding to each timestep in the simulation. Design a program that uses the fast Fourier transform routine to calculate any desired auto- or cross-correlation function or functions as selected by the user.

26 Particle density calculations

26.1 Summary

The calculation of the Fourier transform of the particle density has elements which arise in several important areas of computer simulation. At the simplest level it provides a description of the order in a system (i.e. the melting factor), and at a more advanced level it provides the route to the dynamic structure factor. It also has much in common with electrostatic calculations using the Ewald sum. The student should gain experience of coding these elements efficiently and some insight into versatility of the three dimensional Fourier transform.

26.2 Background

One of the most powerful insights into the structure of liquids provided by molecular dynamics is the calculation of the dynamics structure factor $S(\mathbf{k},\omega)$. Among other uses it is central to the theory of neutron scattering. A principal part of the calculation of this function is the *spatial* Fourier transform of the particle density.

The particle density is defined by:

$$\rho(\mathbf{r},t) = \sum_{i=1}^{N} \delta(\mathbf{r} - \mathbf{r}_i(t))$$

and its spatial Fourier transform is:

$$\rho(\mathbf{k},t) = \sum_{i=1}^{N} \exp(-i\mathbf{k}.\mathbf{r}_i(t))$$

where $\mathbf{k}$ is a reciprocal space vector, which for a cubic MD cell has the form:

$$\mathbf{k}^{\dagger} = \frac{2\pi}{L}(l,m,n)$$

where l,m,n are integers and L is the width of the MD cell. (Note: there is an infinite set of functions $\rho(\mathbf{k},t)$, though only the lowest orders of $\mathbf{k}$ are normally of interest.)

We may calculate a correlation function for each $\rho(\mathbf{k},t)$ which is known as the intermediate scattering function $F(\mathbf{k},t)$:

$$F(\mathbf{k},t) = \langle \rho(\mathbf{k},t)\rho^*(\mathbf{k},0)\rangle$$

This function may be calculated using a sum over origins. It is normally scaled so that $F(\mathbf{k},0) = 1$.

Finally, to obtain the dynamic structure factor, the intermediate scattering function is Fourier transformed in the time domain:

$$S(\mathbf{k},\omega) = \int_{-\infty}^{\infty} F(\mathbf{k},t)\exp(-i\omega t)dt$$

The central importance of the Fourier transform in these calculations is manifest!

26.3 Task

Write a FORTRAN code that can be used to calculate the Fourier transform of the particle density, for a complete set of $\mathbf{k}$ vectors up to l,m,n values of (say) 5, suitable for processing the data provided by a typical atomic MD program every timestep. (Assume a cubic MD cell.) Test the routine by using it to analyse the periodicity in perfect crystals with face centred cubic, simple cubic and body centered cubic lattices. Find out if it possible to use the same routine with hexagonal lattices.

Can you adapt the routine to provide some method of indicating when a molecular dynamics simulation has melted?

Develop the code further to enable the calculation of the intermediate scattering function and the dynamic structure factor. On the basis of what you know about the correlation theorem (and assuming you have not yet exploited this) can you devise a way of proceeding *directly* to the dynamic structure factor, without calculating the intermediate scattering function first? (Unfortunately, because of the heavy amount of computing this kind of calculation entails, it will not be possible to test these developments fully.)

26.4 Further Work

Suppose that instead of point particles we had a system consisting of electrostatic charges represented by gaussian charge density distibutions:

$$\rho_i(\mathbf{r}) = q_i(\alpha^2/\pi)^{3/2}\exp(-\alpha^2 r^2)$$

where q_i is the net charge on the i'th particle, and α is a parameter expressing the "width" of the gaussian function (note it is the same for all charges). Can you adapt the calculation of the Fourier transform of the particle density to calculate the Fourier transform of the charge density? What is the relationship between this calculation and the Ewald summation method for calculating the electrostatic contribution to the configuration energy?

27 Quantum simulations

27.1 Summary

Most dynamical simulations performed to date have used classical mechanics. Recently however much work has concentrated on quantum mechanical simulations, and it is likely that such simulations will grow in importance in the future. In this exercise the student will gain some experience of using a fairly simple quantum mechanical simulation method that is based on the discrete Fourier Transform. Its application to simple one-dimensional systems will provide some insight into the difficulties that arise in quantum simulations, not only from the radically different approach that must be adopted, but from rationalising the results obtained.

27.2 Background

It is well known that the discrete Fourier transform may be used to solve differential equations. One particular application that has aroused much interest recently is its use in solving the time dependent Schrödinger equation:

$$\left\{\frac{-\hbar^2}{2m}\nabla^2 + V(x,y,z)\right\}\psi(x,y,z,t) = i\hbar\frac{\partial}{\partial t}\psi(x,y,z,t)$$

We will consider the application to one dimensional, single particle problems only.

The one dimensional Schrödinger equation can be easily discretized: the functions $\psi(x,t)$ and $V(x)$ are simply multiplied by the sampling function (section 23.4.7) to give:

$$\psi(x,t) \Rightarrow \sum_{n=-\infty}^{\infty} \psi(n\Delta x,t)\delta(x - n\Delta x)$$

$$V(x) \Rightarrow \sum_{n=-\infty}^{\infty} V(n\Delta x)\delta(x - n\Delta x)$$

Both of these sampled functions are nominally of infinite extent, but we may truncate them in regions where $\psi(x,t)$ makes a negligible contibution to the probability density.

The use made of the discrete Fourier transform in this method is to calculate the kinetic energy of the wavefunction as discrete points from the second derivative of the wavefunction. In principle this could be done by a simple second difference method:

$$\frac{\partial^2}{\partial x^2}\psi(n\Delta x,t) = \Delta x^{-2}\{\psi((n+1)\Delta x,t) - 2\psi(n\Delta x,t) + \psi((n-1)\Delta x,t)\}$$

but this has been found to accumulate errors and give rise to instability in the motion of the wavefunction. The discrete Fourier transform method relies on the principles outlined in section 23.3.8 and is much more accurate than the second difference method. We simply Fourier transform the wavefunction and multiply the result by the square of the reciprocal space vector and then inverse Fourier transform the result to obtain the required second derivative. The price paid for the improved accuracy however, is that the system becomes pseudo-periodic. This is of course less of a problem if the physical system is genuinely periodic.

Propagation of the wavefunction forward in time is done using the leapfrog algorithm:

$$\psi(n\Delta x,(k+1)\Delta t) = \psi(n\Delta x,(k-1)\Delta t) + 2\Delta t\frac{\partial}{\partial t}\psi(n\Delta x,k\Delta t)$$

where:

$$\frac{\partial}{\partial t}\psi(n\Delta x,k\Delta t) = -i\hbar^{-1}\left\{\frac{-\hbar^2}{2m}\psi''(n\Delta x,k\Delta t) + V(n\Delta x)\psi(n\Delta x,k\Delta t)\right\}$$

according to the time dependent Schrödinger equation. (ψ'' is the second derivative of ψ).

The algorithm for solving the time dependent Schrödinger equation is therefore:

1. Discretize the starting wavefunction (at $t = 0$) and the potential function.
2. Calculate the wavefunction at timestep $-\Delta t$ using Runge-Kutta or alternative method.
3. Calculate the second derivative of the wavefunction at each point.
4. Calculate the product $V(n\Delta x)\psi(n\Delta x)$ for all points.
5. Propagate the wavefunction for one timestep according to the Schrödinger equation.
6. Renormalise wavefunction (if necessary).

Steps 3 to 6 are repeated for however many time steps are required. Note that the renormalisation in step 6 provides an additional measure of the stability of the algorithm (besides energy conservation). If the algorithm becomes unstable, the normalisation condition becomes seriously violated at each timestep.

27.3 Task

Take a copy of the program MDKOS from the library and examine it closely to familiarise yourself with the coding. Then run it with the following potential functions:

- A constant potential $V = 0$.
- A parabolic potential $V = \frac{1}{2}k(x - x_0)^2$ (find suitable values of k).
- A step potential: $V(x) = 0$; $(x < L/2)$, $V(x) = h$; $(x > L/2)$, where L is the width of the simulation cell, h is the barrier height.
- Any other potential that might interest you (e.g. double well potential).

Your objectives in each case should be:

- To cast the problem with the correct physical dimensions to ensure stability of the algorithm and an absence of artefacts.
- To rationalise the behaviour of the wavefunction in terms of quantum mechanics. (i.e. do you believe the results?)

27.4 Further Work

Extension of this method to more than one particle or dimension is difficult. However, it is possible to change the above method to deal with two interacting particles in one dimension without undue difficulty. Recognising that the two-particle one-dimensional system has only two degrees of freedom and the fact that the centre of mass of the system is not affected by the inter-particle interaction, can you work out how the adaptation can be made?

28 Percolation cluster statistics of 1D fluids

28.1 Summary

In this exercise you will write a simple simulation program to produce a 1-dimensional fluid. You will make use of it by investigating the distribution of clusters generated by the program. This leads to a probing of the long range connectivity of the fluid. If the fluid is "infinitely connected", it is said to *percolate*. The cluster statistics in the vicinity of the percolation threshold will be investigated by you, to test universal scaling relationships.

28.2 Background

The object of this exercise is to get some practice at writing a simple simulation program and applying it to a topical area of research.

In recent years there has been an interest in cluster distributions and cluster properties within simple fluids [72]. I would like you to look for clusters that extend over all space. (These are important in the "real world" because they determine the macroscopic behaviour of a number of weakly aggregating systems; the rigidity of gels and conduction through random networks, for example. Long-range connectivity in the largely-randomly positioned particles has a determining effect on the properties of these materials [73].) For this purpose, a cluster is defined as a group of molecules interconnected by an arbitrary coordination-distance. Above a certain density of particles in an infinite system there is a finite probability that clusters will form spanning all space. This is called a *percolation transition or threshold.* The density at which this occurs and the statistics and properties of the clusters in this region of the phase diagram are interesting because there are universal exponents that characterise them for non-interacting particles thrown randomly onto the sites of a lattice [74]. In lattice systems "universality" refers to the independence of the value of the exponents to the number of nearest neighbours or lattice coordination number. In calculations on off-lattice or continuum noninteracting disks [75] and spheres [76] the exponents are the same as those from the lattice simulations. Recent studies on continuum interacting fluids generated by Monte Carlo simulation suggest that the values of these exponents are the same for these assemblies, as well as being independent of the nature and range of the interaction potential. The fluids studied so far by MC simulation have been hard-spheres [77], square-wells [78] and adhesive-spheres [79].

Generally in percolation studies, a set of particles is considered to be part of the same cluster if each member is separated from at least one of the others by a distance $\leq \sigma_s$, which is arbitrary but is usually $\sim \sigma$, the core diameter of the particle. A percolating cluster is a special cluster having infinite extent. Within the framework of the periodically repeating cells of your simulations, a sufficient criterion for percolation is for a particle and its image to belong to the same cluster. One parameter to vary is the ratio σ_s/σ and see what effect it has on the percolation threshold and on the exponents. As this ratio diminishes to unity, we term this the soft-core to hard-core transition.

To maintain consistency with accepted notation, the density is given the symbol, p, here, rather than, ρ, as is usual for the reduced density of simple fluids. For an infinite number of molecules there is a well-defined density, p_c, at which there is a finite probability of finding a percolating cluster [80]. Above and below the percolation density many finite sized clusters exist. (As you will discover this transition is smeared for finite periodic systems.) These occupy the holes in the percolating cluster above p_c. The simulation methods employed in this exercise (e.g. [81]) will make use of these clusters. The distribution of different sized clusters is characterised by the cluster number distribution function, n_s, [82,83,84] which is the time-average number of clusters containing s particles, N_s divided by N i.e. $n_s = N_s/N$. This is consistent with the definition used in other continuum works [76,82]. Note however that the cluster number definition used in lattice studies is pN_s/N [74]. For finite periodic systems there is an upper bound on s, i.e. $1 \leq s \leq N$.

The behaviour of lattice systems *close to* p_c is described in terms of the following critical exponents. In the discussion below we will use the value for the exponents taken from random population of 1D lattices. The summations involving n_s will also only involve non-percolating clusters. Let $\xi(p)$ be the "correlation" lengthscale of the largest cluster, then [80]

$$\xi(p) \propto \mid p - p_c \mid^{-\nu} . \tag{28.1}$$

As $\nu = 1.0$ [80] then ξ diverges at p_c, approaching from either side and is typical of critical behaviour. In one dimension $p_c = 1$, the close packing density, and therefore it is only possible to probe the system for $p < p_c$. In the finite systems studied here the percolation transition is smeared out over a density range. As density increases from zero the fraction of configurations that generate a percolating cluster, P, increases from zero to unity. The larger the number of particles in the cell, N, then the narrower is the density range for this transition. The density at which $P = 0.5$ shows the smallest N dependence. The density at which $P = 0.5$ is best taken as the percolation threshold for that (N, σ_s) combination. As N increases we find

$$p_c(N \to \infty) = p_c(N) - aL^{-1/\nu} \tag{28.2}$$

where L is the sidelength of the simulation cell and a is a positive constant. The next function we will consider is the zero'th moment of n_s.

$$\Omega = \sum_s p(n_s(p) - n_s(p_c)) \propto \mid p - p_c \mid^{2-\alpha} . \tag{28.3}$$

In 1D $\alpha = 1.0$ [80]. Here and for all the sums below we interested in the "singular" or nonanalytic part of the sum over all cluster sizes [80]. For each sum we must subtract off the analytic background. This is so we isolate that part of the sum dominated by the representative (i.e. largest) cluster in the vicinity of the percolation threshold. For completeness, there is a quantity called the percolation probability, P_∞, which is the fraction of molecules found in the percolating cluster. It only applies for $p \geq p_c$ [79,80]. In arbitrary dimension,

$$P_\infty \propto (p - p_c)^\beta. \tag{28.4}$$

In one dimension $\beta = 0.0$ as p cannot exceed p_c [80]. In lattice studies this corresponds to the fraction of occupied sites belonging to the infinite percolating network. The susceptibility, χ, is

$$\chi = \sum_s{}' s^2 n_s(p) \propto \mid p - p_c \mid^{-\gamma} \tag{28.5}$$

where $\gamma = 1.0$ in one dimension [80]. The "′" denotes the omission of the largest cluster at each sample configuration. The largest cluster discovered each time step can either be a percolating cluster, should one (or more) exist, or the largest non-percolating cluster (should there not be a percolating cluster). In polymer science the *susceptibility*, χ, corresponds to the *mass average molecular mass* whereas $\sum_s' s n_s(p)$ corresponds to the *number average molecular mass*. Close to p_c,

$$\chi = c_- \mid p - p_c \mid^{-\gamma} \tag{28.6}$$

for $p \leq p_c$.

The functions Ω, P_∞ and χ require finite-scaling corrections for these small periodic systems, as discussed in the Appendix. At the percolation threshold [85],

$$n_s(p_c) \propto s^{-\tau}, \tag{28.7}$$

$$n_s(p) = n_s(p_c) f(z), z \equiv (p - p_c) s^\sigma; s \rightarrow \infty, p \rightarrow p_c \tag{28.8}$$

where $f(z)$ is a universal function and $\sigma = 1.0$ in one dimension. In random lattice percolation the critical exponents are interrelated by scaling laws,

$$\tau = 2 + 1/\delta, \sigma = 1/(\nu + \beta), 2 - \alpha = \gamma + 2\beta = \beta(\delta + 1) = d\nu \tag{28.9}$$

where d is the dimensionality ($d = 1$, here) [80,85]. $\tau = 2$ in one dimension.

28.3 The problem

- I would like you to make a contribution to our understanding of percolation by investigating the percolation cluster statistics of some simple 1-dimensional fluids.
  - Hard-sphere fluid by random parking Monte Carlo
  - Hard-sphere fluid by Molecular Dynamics
  - Hard-sphere fluid by Metropolis Monte Carlo
  - Lennard-Jones fluid by Molecular Dynamics

- Lennard-Jones fluid by Metropolis Monte Carlo
- Soft-sphere fluid by Molecular Dynamics
- Soft-sphere fluid by Metropolis Monte Carlo

Choose one of these and let me know so I can give you some help! A typical simulation program can be separated into the following parts.

- At the beginning there is a section for reading in control parameters (usually from a separate dataset or file). These might include, for example in a MD program, the number of time steps to be calculated and the magnitude of the time step.
- Then the program has to decide upon the starting coordinates of the molecules (and velocities for MD). The molecules could start from a lattice and be given Maxwell-Boltzmann random velocities, or alternatively the molecular configuration and velocities could be read in from a dataset coming from a previous calculation. Note that at some stage you will have to start the molecules off from a lattice!
- Now we arrive at the most time consuming part of the program. This is where we move the particles, either with time in MD or randomly in MC. In MD the force on every particle is evaluated by going through every pair interaction. In MC one usually evaluates the potential energy of a single particle moved one at a time. The *periodic boundary conditions* enter this part of the code. During this stage the expressions for the quantities in statistical mechanical ensemble averages are evaluated.
- Now we move the particles, all together in MD or (usually) one at a time in MC, for a single pass through the previous section. Again periodic boundary conditions need to be taken into account.
- Once the desired number of passes has been completed, the calculation is closed down with the saving of molecular coordinates (and velocities for MD) onto a dataset, ready for future use in a resumed simulation. Other quantities, such as ensemble averages can also be carried over to the next simulation segment.

(If you want to save some time, use the `FORTRAN` program `lj1d` , which is a 1 dimensional Lennard-Jones Molecular Dynamics fluid program. Modify it to build in a percolation cluster search.) Have a look at the paper by Geiger and Stanley [82] who used Molecular Dynamics computer simulations to evaluate the percolation exponents of model liquid water. Take $T = 1.0$ and $N = 50, 100, 200....$ and run each simulation several times to assess the statistical noise.

- Once you have written the program to generate the positions of the particles, calculate $p_c(N)$ for $\sigma_s/\sigma = 1.1$ and 1.5 by establishing the ρ at which $P = 0.5$. A percolating cluster in 1D is easy to find! If two particle centres are within σ_s they are in a cluster together. If a third particle centre is within σ_s of either of these two, then it also is in the same cluster, i.e. we now have a triplet. This procedure is continued. If a particle and its image are also in the same cluster, then within the repeating framework of periodic boundary conditions, all of space is part of a percolating cluster (i.e. we can 'hop' from one particle to the next and span all available space).
- Determine χ from eqn (28.5) and plot it versus $p - p_c(N)$ close to $p_c(N)$. Also test the 'laws' (28.7)-(28.9).

28.4 Appendix: Finite-scaling hypothesis

The *finite-scaling hypothesis* supposes that physical properties are homogeneous functions of the critical-coupling parameter, $\epsilon = | (p - p_c(L))/p_c(L) |$ and the length-scale of the system, L. First let us recall the notion of homogeneous functions in the case of two variables. If $f(x, y)$ is a homogeneous function of two variables it obeys the following for an arbitrary constant, λ,

$$f(x,y) = \lambda^c f(\lambda^a x, \lambda^b y) \tag{28.10}$$

e.g. if $f(x, y) = x^2y^3$, then $c = -2a - 3b$. By choosing $\lambda^b = y^{-1}$ then $f(x, y)$ can be rewritten as,

$$f(x,y) = y^{-c/b} f(y^{-a/b}x, 1). \tag{28.11}$$

Returning to two cluster averages discussed in the text, $P_\infty = f(\epsilon, L)$ and $\chi = g(\epsilon, L)$ can be written in the homogeneous forms,

$$P_\infty = L^{-\beta/\nu} f(L^{1/\nu}\epsilon, 1) \tag{28.12}$$

where $c = \beta$, $a = -1$ and $b = \nu$, and

$$\chi = L^{\gamma/\nu} g(L^{1/\nu}\epsilon, 1) \tag{28.13}$$

where the quantity $L^{1/\nu}\epsilon \propto (L/\xi)^{1/\nu}$ from eqn (28.1) is therefore essentially a ratio of two lengthscales. In the limit, $L \to \infty$ and $\xi \ll L$ then $P_\infty \sim \epsilon^\beta$ and $\chi \sim \epsilon^{-\gamma}$. This indicates that $f(L^{1/\nu}, 1) \to (L^{1/\nu}\epsilon)^\beta$ and $g(L^{1/\nu}\epsilon, 1) \to (L^{1/\nu}\epsilon)^{-\gamma}$ in this limit. In the other limit, $L^{1/\nu}\epsilon \to 0, \xi \gg L$ then f and g must reduce to L-dependent constants.

29 Quaternions and constraints

29.1 Summary

This project compares two techniques for the simulation of small rigid molecules: the use of quaternion parameters to describe molecular rotation, and the introduction of constraints into the atomic equations of motion. It is best attempted by a small team, divided into two subgroups, one to study each technique.

29.2 Background

Most molecules are legitimately regarded as being composed of atoms, held together by covalent bonds which are strong compared with the intermolecular interactions. While many interesting molecules are highly flexible, and none are truly rigid, it is considered a reasonable approximation in some cases to treat all the internal bond lengths and angles as being fixed. In this limit, the molecule becomes a rigid body. In classical mechanics we may treat it in either of two ways (and here it may be useful to refer to a standard text on Classical Mechanics, such as Goldstein [86]). We may separate the centre-of-mass translation from rotation about the centre of mass, and solve the associated equations of motion. The orientation of a molecule is represented by Euler angles [87,88] or, to avoid certain difficulties in the equation of motion, by quaternions [89]. Alternatively, we may solve the atomic equations of motion, imposing holonomic constraints upon the intramolecular separations, to fix the internal degrees of freedom. This gives rise to the SHAKE and RATTLE approaches [90,91]. Constraint dynamics is the more widely-applicable approach, since it can handle the case of flexible as well as rigid molecules. However, in the rigid limit, there seems to be little to choose between the two approaches. In comparing them, it is important to use similar underlying algorithms to integrate the equations of motion [92]. Historically, this has not always been the case: quaternions have been used in conjunction with Gear predictor-corrector methods, while constraints grew up with the Verlet algorithm (compare e.g. [36] with [93]). Here, we use similar low-order Verlet-like algorithms in both cases. Much of the background material can be found in the original references cited above, and in standard texts [3]. Here we present a summary only.

29.2.1 Quaternion algorithm

The rotational equations of motion of a rigid body may be treated as two first-order differential equations. One equation links the torque τ acting on the body with the time derivative of angular momentum $\mathbf{l}$, and hence (via the moment of inertia) with the rate of change of the angular velocities

$$\dot{\mathbf{l}}^s = \tau^s. \tag{29.1}$$

The superscript s indicates that these vectors and time derivatives are measured in a space-fixed axis system. The second differential equation shows how the angular velocities determine the rate of change of the parameters characterizing the orientation of the body. These in turn specify the rotation matrix $\mathbf{A}$, used to convert from space-fixed to body-fixed axes thus:

$$\mathbf{e}^b = \mathbf{A}.\mathbf{e}^s. \tag{29.2}$$

The body-fixed frame is most conveniently chosen to diagonalize the inertia tensor, giving (for nonlinear molecules) three principal moments of inertia

$$I_{xx} = \sum_a m_a(y_a^2 + z_a^2) = \sum_a m_a(|\mathbf{r}_a|^2 - x_a^2) \tag{29.3}$$

and similarly for I_{yy}, I_{zz}. The sum is over the atoms a in a molecule, m_a is the atomic mass, and coordinates are measured from the molecular centre of mass. Then, off-diagonal elements being zero, angular velocity and angular momentum are linked by

$$l_x^b = I_{xx}.\omega_x^b \tag{29.4}$$

and similarly for y and z.

The conventional orientational parameters are the three Euler angles; for reasons explained elsewhere [88,89] these can only be used at the expense of some difficulties in integrating the equations of motion. Instead, Evans [89] proposed the use of four quaternion parameters. The quaternions, for which we use the shorthand $\mathbf{Q} = (q_0, q_1, q_2, q_3)$, are linked by a normalization condition $q_0^2 + q_1^2 + q_2^2 + q_3^2 = 1$. They can be related to the Euler angles, but we only need to know that the rotation matrix takes the form

$$\mathbf{A} = \begin{pmatrix} q_0^2 + q_1^2 - q_2^2 - q_3^2 & 2(q_1q_2 + q_0q_3) & 2(q_1q_3 - q_0q_2) \\ 2(q_1q_2 - q_0q_3) & q_0^2 - q_1^2 + q_2^2 - q_3^2 & 2(q_2q_3 + q_0q_1) \\ 2(q_1q_3 + q_0q_2) & 2(q_2q_3 - q_0q_1) & q_0^2 - q_1^2 - q_2^2 + q_3^2 \end{pmatrix}. \tag{29.5}$$

Note that the reference position, when body-fixed and space-fixed axes coincide, and $\mathbf{A}$ is the unit matrix, corresponds to $\mathbf{Q} = (1, 0, 0, 0)$. The time derivatives of these variables are related to the angular velocity components, say in body-fixed axes:

$$\begin{pmatrix} \dot{q}_0 \\ \dot{q}_1 \\ \dot{q}_2 \\ \dot{q}_3 \end{pmatrix} = \frac{1}{2} \begin{pmatrix} q_0 & -q_1 & -q_2 & -q_3 \\ q_1 & q_0 & -q_3 & q_2 \\ q_2 & q_3 & q_0 & -q_1 \\ q_3 & -q_2 & q_1 & q_0 \end{pmatrix} \begin{pmatrix} 0 \\ \omega_x^b \\ \omega_y^b \\ \omega_z^b \end{pmatrix}. \tag{29.6}$$

In most quaternion simulations, following Evans, these equations are fed into a Gear predictor-corrector algorithm, together with the equations linking torques with angular momentum (and hence angular velocity). Here, though, we follow Fincham [92], and adopt a leapfrog Verlet approach. For the translational motion we store centre-of-mass positions $\mathbf{R}(t)$ and half-step velocities $\mathbf{V}(t - \frac{1}{2}\delta t)$. For the rotation, we store quaternions $\mathbf{Q}(t)$ and half-step angular momenta in space-fixed axes, $\mathbf{l}^s(t - \frac{1}{2}\delta t)$. From the current positions and orientations, we calculate the total force on each molecule, giving the centre-of-mass acceleration $\mathbf{A}(t)$, and the torque $\tau^s(t)$. The translational leapfrog algorithm is

$$\begin{aligned} \mathbf{V}(t + \frac{1}{2}\delta t) &= \mathbf{V}(t - \frac{1}{2}\delta t) + \delta t \mathbf{A}(t) \\ \mathbf{R}(t + \delta t) &= \mathbf{R}(t) + \delta t \mathbf{V}(t + \frac{1}{2}\delta t). \end{aligned} \tag{29.7}$$

The rotational leapfrog algorithm is very similar:

$$\begin{aligned} \mathbf{l}^s(t+\frac{1}{2}\delta t) &= \mathbf{l}^s(t-\frac{1}{2}\delta t) + \delta t \tau^s(t) \\ \mathbf{Q}(t+\delta t) &= \mathbf{Q}(t) + \delta t \dot{\mathbf{Q}}(t+\frac{1}{2}\delta t). \end{aligned} \tag{29.8}$$

The only catch is in calculating $\dot{\mathbf{Q}}(t+\frac{1}{2}\delta t)$ via eqn (29.6): we need $\mathbf{Q}(t+\frac{1}{2}\delta t)$, on the right of this equation. These same quantities are also needed in the rotation matrix used in the first step of the conversion $\mathbf{l}^s(t+\frac{1}{2}\delta t) \rightarrow \mathbf{l}^b(t+\frac{1}{2}\delta t) \rightarrow \omega^b(t+\frac{1}{2}\delta t)$ (the second step uses the principal moments of inertia). To obtain $\mathbf{Q}(t+\frac{1}{2}\delta t)$ we undertake a preliminary advance of just half a step:

$$\begin{aligned} \mathbf{l}^s(t) &= \mathbf{l}^s(t-\frac{1}{2}\delta t) + \frac{1}{2}\delta t \tau^s(t) \\ \mathbf{Q}(t+\frac{1}{2}\delta t) &= \mathbf{Q}(t) + \frac{1}{2}\delta t \dot{\mathbf{Q}}(t). \end{aligned} \tag{29.9}$$

Again, it is straightforward to convert $\mathbf{l}^s(t) \rightarrow \omega^b(t)$, and we already have $\mathbf{Q}(t)$, so $\dot{\mathbf{Q}}(t)$ is easily obtained from eqn (29.6). After calculating these auxiliary quantities (which do not need to be stored) the main algorithm equations (29.8) can be implemented.

29.2.2 Constraint algorithm

Here we summarize the SHAKE algorithm for constraint dynamics [90] Atomic dynamics is handled using the Verlet method. We store current atomic positions $\mathbf{r}_a(t)$ and values from the previous step $\mathbf{r}_a(t-\delta t)$. We calculate interatomic forces, and hence the atomic accelerations $\mathbf{a}_a(t)$. These do not include the internal bond constraint forces which we denote $\mathbf{g}_a(t)$. However, we can see how these will enter the algorithm. We write this:

$$\begin{aligned} \mathbf{r}_a(t+\delta t) &= 2\mathbf{r}_a(t) - \mathbf{r}_a(t-\delta t) + \delta t^2 \mathbf{a}_a(t) + \delta t^2 \mathbf{g}_a(t)/m_a \\ &= \tilde{\mathbf{r}}_a(t+\delta t) + \delta t^2 \mathbf{g}_a(t)/m_a \end{aligned} \tag{29.10}$$

Here m_a is the atomic mass and the tilde denotes the new positions which *would* have been reached in the absence of the constraint forces. The latter are in fact approximations to the true constraint forces (since the integration algorithm itself is only approximate). They are determined by the requirement that the new positions satisfy the bond length constraints. Each such constraint force is taken to be directed along a bond at time t, so it takes the form $\mathbf{g}_a = \lambda_{ab}\mathbf{r}_{ab} + \lambda_{ac}\mathbf{r}_{ac} + \ldots$ where the λs are undetermined multipliers. We can insert these expressions into the above eqns (29.10). For each bond constraint, say on the distance r_{ab}, we subtract the corresponding pair of equations to give $\mathbf{r}_{ab}(t+\delta t)$; then, setting $|\mathbf{r}_{ab}(t+\delta t)|^2$ equal to the square of the desired bond length, we obtain a quadratic equation involving the

λs. In fact we obtain a system of equations: the same number as there are multipliers. These may be solved by an iterative method [90]. When the constraint forces have been calculated, they are used in eqns (29.10) to update positions. The whole process is then repeated step by step.

29.3 Task

Consider a single triatomic molecule, with equal mass atoms (for simplicity), and three equal-length interatomic bonds. Write MD programs for this molecule, in the absence of external forces, using the quaternion leapfrog algorithm and the SHAKE constraint algorithm, as described briefly above. For the quaternion case, you will need to work out the moments of inertia: take the three-fold symmetry axis to point in the z-direction, and note that I_{xx} and I_{yy} will be identical to each other. For SHAKE, begin by writing down explicitly the updating equations involving the undetermined multipliers (see [3] p94). In both cases, you will need to choose initial conditions (angular velocities or positions at the previous time step) corresponding to rigid rotation of the molecule about the centre of mass.

Choosing the same starting conditions, compare the performance of the two algorithms as the time step is increased. It will be worthwhile to monitor the kinetic energy: it should not change with time. Then, include a simple harmonic pair potential, attracting each atom to the coordinate origin. Now there will be forces and torques, but the total energy (kinetic plus potential) should be conserved.

30 The chemical potential of liquid butane

30.1 Summary

This projects uses equilibrium configurations from a molecular dynamics simulation of liquid butane to calculate the chemical potential using the Widom particle insertion method. This project might be tackled by a small team.

30.2 Background

The calculation of mechanical properties, such as the configurational energy or virial, is straightforward using either the Metropolis Monte Carlo method, or molecular dynamics. The calculation of *statistical* properties such as the free energy, entropy, or chemical potential is trickier. In principle, there is no reason why the free energy cannot be calculated as a straightforward ensemble average. In practice, the statistical error is high because the techniques do not sample the appropriate regions of phase space. A variety of methods have been proposed to facilitate the calculation of these *statistical* properties. These include thermodynamic integration, umbrella sampling, and the probability ratio method. A brief review of these techniques can be found in [3]. The most straightforward method of calculating the chemical potential is the particle insertion method originally proposed by Widom [94]. In this technique a fictitious molecule is inserted into the fluid, without affecting the trajectories of the real molecules. The configurational energy between this test particle and the real molecules is calculated and the long-range correction from molecules beyond the cut-off is included. If V_{test} is the test particle energy then the residual chemical potential μ_r is given by

$$\mu_r = -k_B T \ln < \exp(-\beta V_{test}) > \tag{30.1}$$

in the canonical ensemble [95,96].

A set of configurations is provided from a simulation of liquid butane. The simulation of 108 molecules is at 291K and a density of 0.583 gm l^{-3}. A cutoff of half the boxlength was used for the intermolecular interactions. The potential consists of three components.

- CH_2 and CH_3 groups are treated as single identical sites. There is a site-site potential between united CH_2 and CH_3 atoms on different molecules. The parameters are ϵ/k_B =72K and σ =0.3923nm. The C-C bondlength is 0.155nm.
- There is a bond-bending potential

  $$V(\theta) = \frac{1}{2} k_\theta (\theta - < \theta >)^2 \tag{30.2}$$

  where k_θ =520 kJmol^{-1}, and $< \theta >$=109.47 degrees.
- The torsional potential is given by

  $$V(\phi) = \sum_{i=1}^{5} c_i (\cos\phi)^i \tag{30.3}$$

  where the sum is over five terms $c_0 = 1116$K, $c_1 = 1462$K, $c_2 = -1578$K, $c_3 = -368$K, $c_4 = 3156$K, $c_5 = -3788$K.

You will find 1000 configuration in the file butandat. Each configuration contains the coordinates of 532 atoms from configurations separated by timesteps of 2×10^{-15}s. They were written using a formatted write statement of the form

```
WRITE(*,'(12F10.3)')
:            RX1, RY1, RZ1, RX2, RY2, RZ2, RX3, RY3, RZ3, RX4, RY4, RZ4
```

where the integer refers to a particular atom in a molecule.

30.3 Task

You are invited to write an analysis routine to calculate the chemical potential of butane by particle insertion. We recommend that you use a lattice of at least $7 \times 7 \times 7$ test particles to analyse each configuration on the tape. Some points to consider are

- How will you orient the test molecules?
- How will you handle the three internal degrees of freedom? You could treat butane as mixture of its conformers.
- How will you handle the long-range corrections?
- How will you check for convergence of the results?
- Is a lattice of test particles the best approach?
- Does the Widom formula depend on the ensemble you are considering? How important is this point?

30.4 Further work

- Can you remove some of the internal degrees of freedom without significantly changing the results?
- There is an alternative to the straight particle insertion which calculates the distribution of test particle energies $F(V)$ and the distribution of real particle energies $G(V)$. The ratio of these two distributions can be used to calculate the residual chemical potential [97]. You will need to study this method and decide whether it can be applied to your problem.

31 Molecular dynamics on the Distributed Array Processor

31.1 Summary

This project is suitable for an individual, or a small team, but requires some investment of time and effort, since the programming is non-standard. Consult with the author before attempting it. The programs involved are stored on the VAXstation which acts as host to the DAP.

Examine, and improve, programs which have been provided to carry out the molecular dynamics of a two-dimensional Lennard-Jones system, using a fast, massively-parallel computer, the Distributed Array Processor. Consider various strategies for efficient simulation on this type of architecture, and compare with techniques for coarse-grained parallel computers.

31.2 Background

To simulate ever larger and more complicated systems, we must make the maximum use of novel computer architectures as they are developed [98]. Considerable experience has been gained in the writing of vectorized (or easily-vectorizable) code for supercomputers and attached processors, like CRAY, CYBER, FPS, which rely on the pipeline approach. The introduction of parallelism into computer design is seen as the next significant step towards more computer power [99]. The current challenge is to write simulation programs for parallel architecture machines. As can be seen in recent review articles [31,100] the first steps have already been taken along this road.

Parallel computers can be broadly divided into two categories. Coarse-grained machines are composed of a relatively small number of processors, often connected together in a fairly flexible way. Examples are the transputer-based machines, and large conglomerates such as the LCAP system at IBM. This structure allows quite sophisticated programming methods to be used: in principle, each processor can be performing a separate task, giving the possibility of full MIMD (multiple-instruction, multiple-datastream) operation. Of course, the programs may, of necessity, be quite complicated, and processor synchronization for transfer of information is a major headache. The writing of molecular dynamics code for this kind of machine is the subject of a separate project. Fine-grained, massively parallel, machines consist of large numbers of processors, linked in a regular array, with fast inter-processor connections. Examples are the Connection Machine and the Distributed Array Processor (DAP). These machines are well-suited for SIMD (single-instruction, multiple-datastream) operation: all the processors are essentially doing the same thing. The programs, and programming language, required to do this can be relatively simple. A limiting factor, however, is the fixed machine architecture: a problem can either be mapped onto it, or it cannot. The current project is concerned with the writing of a molecular dynamics program for a machine of this kind. Several different avenues of approach will be explored.

31.3 `FORTRAN` on the DAP

The DAP is programmed in an extended `FORTRAN` language called `FORTRAN Plus`. A manual is provided near the machine for reference, but here we review one or two salient points. The DAP is actually a 32×32 square array of processors, each with its own data store. 32×32 'matrix' objects can be manipulated in single statements. They are declared rather like doubly-dimensioned `FORTRAN` arrays, except that the actual dimensions are left blank:

```
REAL*4 RX(,), RY(,), VX(,), VY(,), FX(,), FY(,)
```

These objects can be manipulated very simply. The following code carries out the updating associated with the velocity Verlet algorithm. `DT` (the timestep) and `DT2` are ordinary 'scalar' variables.

```
      DT2 = DT / 2.0

C     ** VELOCITY VERLET PART 1 **

      VX = VX + FX * DT2
      VY = VY + FY * DT2
      RX = RX + VX * DT
      RY = RY + VY * DT

C     ** FORCE EVALUATION **

      CALL FORCE

C     ** VELOCITY VERLET PART 2 **

      VX = VX + FX * DT2
      VY = VY + FY * DT2
```

All 1024 values are updated at once. It is possible to refer to individual values, by the usual double index notation, e.g. RX(I,J), with I and J between 1 and 32. It is equally valid to use a single index, e.g. RX(I), with I between 1 and 1024, treating the 'matrix' as a 'long vector'. By far the most useful feature of the language is the use of LOGICAL variables as 'masks'. To illustrate this, the following piece of code serves to put atoms back inside the simulation box (with coordinates between 0 and BOX) when they stray outside:

```
      RX(RX.GT.BOX) = RX - BOX
      RY(RY.GT.BOX) = RY - BOX
      RX(RX.LT.0.0) = RX + BOX
      RY(RY.LT.0.0) = RY + BOX
```

An expression like RX.GT.BOX evaluates to give a LOGICAL MATRIX which is put in the place of the indices on the left of the assignment. Only in positions where this MATRIX is TRUE will the assignment be made; the other values will be left unchanged. LOGICAL variables, each occupying one bit on the DAP, are treated very efficiently indeed. This masking often takes the place of IF statements, when code for a scalar machine is parallelized.

VECTORS of length 32 are also allowed: they are declared as objects thus:

```
      REAL*4 AVEC()
```

Figure 31.1: The SUMR and SUMC functions

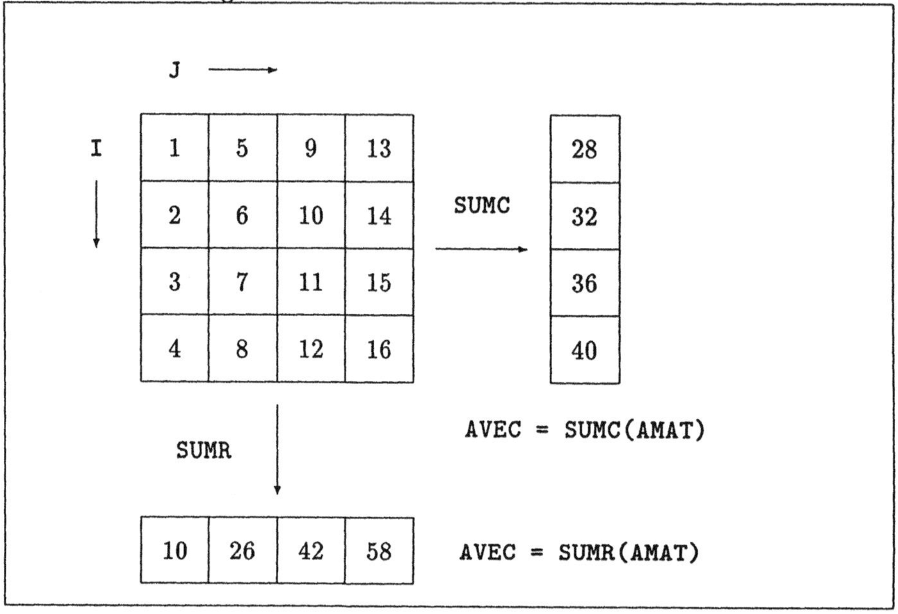

and treated in one go by statements like the above. A vector can be extracted from a matrix. The statement

```
AVEC = AMAT(I,)
```

sets AVEC equal to the Ith row of MATRIX AMAT while

```
AVEC = AMAT(,J)
```

sets it equal to the Jth column.

A large number of routines for handling vectors and matrices are provided, and described in the manuals. We shall only describe one or two here. For reasons of space, the figures are drawn for a 4×4 DAP, but the meaning should be clear. It is possible to convert a MATRIX AMAT(,), to give a VECTOR AVEC() by 'SUMming the Rows' with the SUMR function, or by 'SUMming the Columns' with the SUMC function, as shown in Figure 31.1. Note the usual FORTRAN convention for indices I, J: in the figure the value 3 is stored in AMAT(3,1). It is sometimes useful to expand a vector to make a MATrix of identical Columns using MATC or a MATrix of identical Rows using MATR. This is shown in Figure 31.2. A matrix can be filled up with a single number using the MAT function, e.g. AMAT = MAT(3.5) or simply by a statement like AMAT = 3.5.

Figure 31.2: The MATR and MATC functions

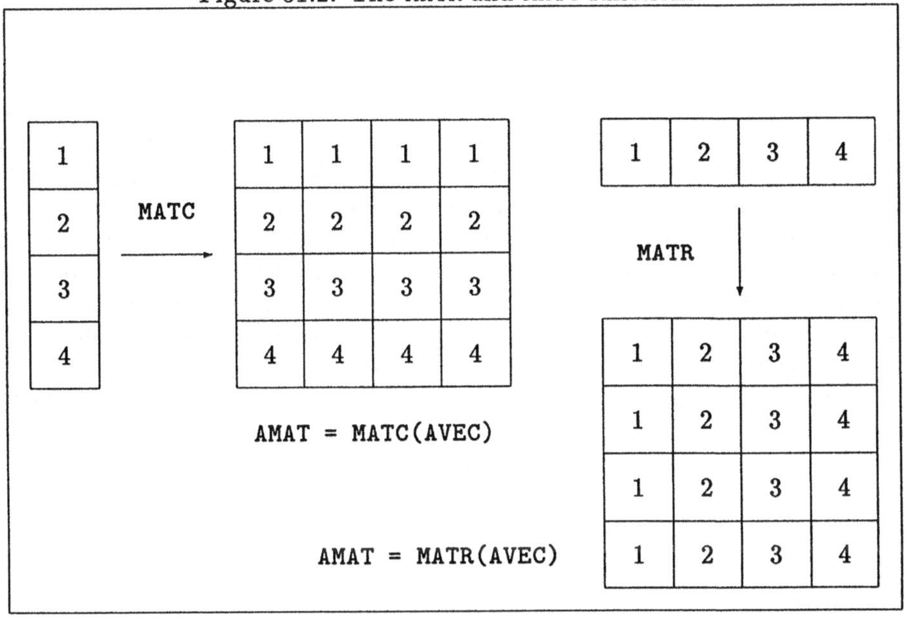

31.4 Simple force evaluation

You are now ready to see a program using a simple force evaluation routine written in FORTRAN Plus. Look at the files MD.FOR and MD.DFP. The former is the main program; it is written in standard FORTRAN, to run on the host machine (here a VAXstation) and it handles the data input and output, and governs the overall course of the simulation. It calls three ENTRY SUBROUTINES in MD.DFP: MD_START to transfer the initial configuration into the DAP and set up the graphics, MD_RUN to carry out the run itself, and MD_STOP to clean up afterwards. The data conversion routines and graphics routines are stored elsewhere; they need not concern us. Our attention is on the MD_RUN routine, and within it the routine MD_STEP which actually carries out one molecular dynamics step. In this routine can be seen the Velocity Verlet algorithm given above, complete with implementation of periodic boundary conditions, a call to a routine KINENG which evaluates the total kinetic energy of the system, and the all-important FORCE routine. This is the most expensive routine: here is where we must make most use of parallelism in one form or another.

To begin with, we adopt the simplest approach: the evaluation of interactions between all pairs of atoms. This 'all-pairs' algorithm is considered by Fincham [31]. Atoms are considered 32 at a time; for convenience in this example the number of atoms is assumed to be a multiple of 32, but this restriction can easily be removed. All the interactions between two such sets of atoms are placed in 32×32 MATRIX variables, using the MATC and MATR functions discussed above. In the program we use statements like

```
RXIJ = MATC ( RXI ) - MATR ( RXJ )
```

Suppose that there are 256 atoms: we envisage the complete matrix of 256×256 pair interactions as being split into $8 \times 8 = 64$ blocks, each block being 32×32. Because of Newton's third law, we only need to consider blocks along the diagonal, and in the upper triangle, of this big matrix. For the blocks on the diagonal, where we are considering interactions within a given set of 32 atoms we are doing some double-counting of pairs, so the maximum efficiency of the machine is not quite being used. A more serious point is that interactions for atoms outside the potential cutoff are simply set to zero, using a logical mask; this will not save any time, so clearly this method will not be very efficient for short-range potentials.

Try running this program for various numbers of atoms, choosing reasonable densities and time-steps. Note that the energy conservation will not be especially impressive, as we have written the program using 32-bit REAL*4 variables. You can modify the program to use higher precision: REAL*n, n=4,5,6,7,8 are all possible on the DAP. Consult an advisor if you want to follow this up.

31.5 The Brode-Ahrlichs algorithm

We now describe a more elegant all-pairs algorithm due to Brode and Ahrlichs [101]. This method was originally devised for vector processors in which long pipelines are an advantage, but it is easily programmable on a transputer multiprocessor in a systolic loop configuration and, as we shall see, on the DAP. The essential feature is that data is shifted, in a synchronized fashion, from each processor to the next, around a cyclic chain.

Shifting operations are an important part of the FORTRAN Plus language on the DAP. In the current application, we consider storing the coordinates of 1024 atoms in MATRIX variables, and treat them simply as long vectors, ignoring the two-dimensional connectivity of the DAP. Data is cyclically shifted from one processor to the next by the functions SHLC (SHift to the Left with Cyclic boundaries) and SHRC (SHift to the Right with Cyclic boundaries). This is illustrated in Figure 31.3. (Analogous functions with planar boundaries exist: SHLP shifts data to the left, disposing of the leftmost data items and feeding in zeroes from the right.) These functions have an optional second argument, to determine how many places to the left or right the data will be shifted. The simplest case of a single shift may be coded more simply, by an indexing convention, if the appropriate boundary conditions are predefined by a GEOMETRY(CYCLIC) or GEOMETRY(PLANAR) declaration. In cyclic boundaries, AMAT(+), SHLC(AMAT), and SHLC(AMAT,1) are all equivalent; similarly AMAT(-), SHRC(AMAT), and SHRC(AMAT,1) mean the same.

Now we consider the Brode-Ahlrichs method. We work with two copies of the atomic coordinates. One set of coordinates, stored in MATRIX variables RXI(,), RYI(,),

Figure 31.3: Long-vector shifting functions

1	2	3	4	5	6	7	8	9	10	11	12	13	14	15	16

↓ SHLC

2	3	4	5	6	7	8	9	10	11	12	13	14	15	16	1

```
AMAT = SHLC(AMAT)
```

Figure 31.4: Brode-Ahlrichs algorithm

Step																	
1	1	2	3	4	5	6	7	8	9	10	11	12	13	14	15	16	RXI
	2	3	4	5	6	7	8	9	10	11	12	13	14	15	16	1	RXJ
2	1	2	3	4	5	6	7	8	9	10	11	12	13	14	15	16	RXI
	3	4	5	6	7	8	9	10	11	12	13	14	15	16	1	2	RXJ
3	1	2	3	4	5	6	7	8	9	10	11	12	13	14	15	16	RXI
	4	5	6	7	8	9	10	11	12	13	14	15	16	1	2	3	RXJ

let's say, remains static. The second copy, `RXJ(,)`, `RYJ(,)`, is progressively shifted in a cyclic way. This is illustrated in Figure 31.4. We concentrate on the x-coordinates here, but the y-coordinates are treated in exactly the same way. Also, we work with two copies of the atomic forces, `FXI(,)`, `FYI(,)` and `FXJ(,)`, `FYJ(,)`. These are set to zero at the start of the force routine. We shift the second, '`J`' copy around with the corresponding coordinates. At step 1, the atomic indices differ by 1 (modulo 1024). By subtracting the two matrices `RXI-RXJ` we obtain a set of differences in x-coordinates for all N pairs of this kind; similarly for the y-coordinates. Periodic boundary corrections are applied, and the corresponding pair forces can all be worked out in parallel, and put in `MATRIX` variables `FXIJ`, `FYIJ`. Then, the force acting on each member of the pair can be added in to the appropriate accumulators:

```
FXI = FXI + FXIJ
FYI = FYI + FYIJ
FXJ = FXJ - FXIJ
FYJ = FYJ - FYIJ
```

After this we shift all the '`J`' variables and move on to step 2, and so on. At each step we are calculating a new set of N pair interactions. When shall we stop? There are $\frac{1}{2}N(N-1)$ pair interactions in all, and we are computing them N at a time, so we will be done after $\frac{1}{2}(N-1)$ steps. Clearly, things are simple if N is odd. *You should consider carefully the minor complication resulting from N being even.*

At the end, we simply sum the 'I' and 'J' forces, to give the total force acting on each atom:

```
FX = FXI + FXJ
FY = FYI + FYJ
```

Write a force routine for the two-dimensional Lennard-Jones system based on this approach, run it and check for energy conservation and speed.

The Brode-Ahlrichs algorithm is a neat approach to the 'all-pairs' force evaluation, but we still have not saved any time through the introduction of a potential cutoff. However, it is possible to make a simple modification that partially achieves this. The coordinates, together with associated velocities, can be sorted before entering the force routine. Suppose we sort into order of x-coordinate. Then, as we go through the steps of the Brode-Ahlrichs method, we examine the x-separations as they are calculated. Initially, the x-separations will be very small, but they will all grow as the two copies of the configuration become more out-of-registry with each other. As soon as *all* the x-separations exceed the cutoff, we can stop work: there will be no more pairs within range. Sorting is done efficiently by a standard parallel algorithm, provided for you in the routine `SORT` in the file `MD_SORT.DFP`. (This routine assumes the `COMMON` block structure in `MD.DFP`.) The `FORTRAN Plus` logical function `ALL` can be used as in the statement

```
IF ( ALL ( RXIJ .GT. RCUT ) ) GOTO 1000
```

to carry out the test.

Other systolic loop simulation algorithms, some of which may be suitable for the DAP, are described by Fincham [31]. They have descriptive names like 'pass the parcel' and 'tractor tread'. You might like to investigate these further.

31.6 Mesh algorithms

So far, we have exploited the algorithmic parallelism of the calculation of forces between all pairs: different pair forces can be computed independently, and in parallel. It is clear that we must use geometrical parallelism, i.e. the use of different regions of space in parallel, if we are to exploit the short-ranged nature of typical interatomic potentials. In the last section we did this, in a crude way, by sorting the x-coordinates, thus mapping one dimension of space onto the one-dimensional long-vectors used to store the configuration.

For our two-dimensional system, a two-dimensional mapping would enable us to use our knowledge of the cutoff to cut down on unnecessary work. Two approaches seem promising, and they are both mentioned in Fincham's article [31]. The first is to divide our two-dimensional space into a regular array of cells, and assign each atom to a cell. Each cell is quite small, so that no two atoms could have their centres in the same cell, because of the repulsive part of the interaction potential. Each cell, then, is either unoccupied or singly occupied. This technique is quite common on scalar machines [102], and it can be vectorized or parallelized without too much difficulty, although some indirect addressing is involved. On the DAP, we would allocate one processor to each cell. With 1024 cells, we could handle a system of several hundred atoms, and it is not difficult to extend the idea to use multiples of 1024 cells. It is simplest to consider a square lattice of cells, but hexagonal cells would be more efficient (*you might like to consider why later*). Then, interactions are considered cell-by-cell rather than atom-by-atom. The program can easily access neighbouring cells, since they have indices close to each other. We only have to consider neighbouring cells out to a certain distance, because of the cutoff, and so we can save time. To offset this advantage, a certain amount of needless work will be done in computing functions for empty cells. The fraction of cells that are empty will depend on the density: the method will work best for dense systems.

A related method has been proposed by Boris [103], again with a vector computer in mind. The atoms are sorted in two dimensions, such that `RX(I,J)` is less than `RX(I+1,J)` for all `I` while `RY(I,J)` is less than `RY(I,J+1)` for all `J`. This corresponds to a kind of distorted square lattice, with exactly one atom allocated to each cell. Thus, on the DAP, we would typically consider a 1024-atom system. By this means, no time is wasted on computing non-existent interactions, since all the cells are full. On the other hand, the irregular nature of the underlying division of space means that interactions would have to be computed 'further out' than in the case of a regular grid, to ensure that nothing within the cutoff is missed. As the atoms move around, of course, the two-dimensional list must be updated, but this can be done by a parallel method.

On the DAP, for either of these methods, the searching of neighbouring cells can be done by shifting the data in `MATRIX` variables. All interactions between atoms in 1024 cells and atoms in all the adjacent cells in a particular direction, for example, can be calculated at once.

Two-dimensional shifts (in North, South, East or West directions) can be applied with automatic inclusion of cyclic boundary conditions, or with planar boundaries. Only the former case concerns us here; in the latter case, data is shifted off one edge of the matrix, and disappears, while zeroes are shifted in on the opposite side. These functions are illustrated in Figure 31.5. Again an optional argument can be used to specify the number of places to shift in the chosen direction. Also, given that (say) cyclic boundaries have been preselected by a `GEOMETRY CYCLIC` declaration, a condensed notation is allowed: `AMAT(-, )`, `SHSC(AMAT)`, and `SHSC(AMAT,1)` are all equivalent. Similarly `AMAT( ,+)`, `SHWC(AMAT)`, and `SHWC(AMAT,1)` are all equivalent. Shifts in both directions (e.g. `AMAT(+,-)`) are also allowed. These functions can be

Figure 31.5: Two-dimensional shifting functions

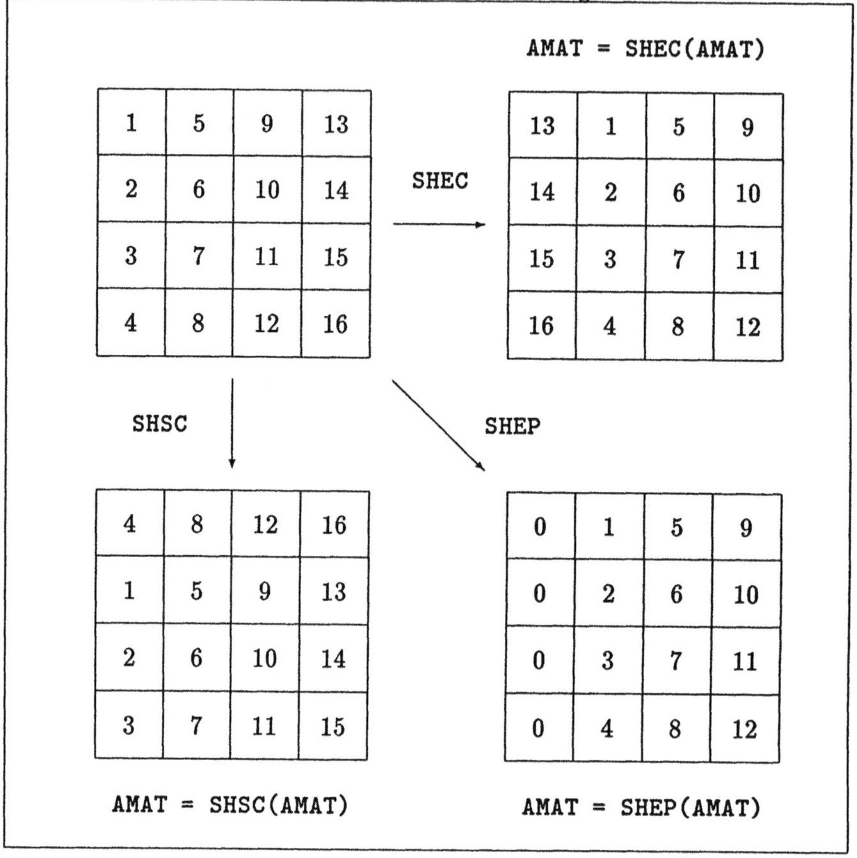

used in statements like

```
RXIJ = RX - SHSC(RX,ISHIFT)
```

to calculate atomic separations.

Adopting one of the methods described above, write a molecular dynamics force routine for the 2-dimensional Lennard-Jones system, to work fairly efficiently with a short potential cutoff.

References

[1] Maitland, G. C., Rigby, M., Smith, E. B., and Wakeham, W. A., (1981) *Intermolecular Forces* (Clarendon Press Oxford).

[2] Gray, C. G., and Gubbins, K. E., (1984) *The Theory of Molecular Fluids. 1. Fundamentals* (Clarendon Press, Oxford)

[3] Allen, M. P., and Tildesley, D. J., (1987) *Computer Simulation of Liquids* (Oxford University Press).

[4] Price, S. L., Stone, A. J., and Alderton, M., (1984) Molec. Phys. **52** 987.

[5] Price, S. L., and Stone, A. J., (1987), J. Chem. Phys. **86** 2859.

[6] Williams, D. E., and Cox, S. R., (1984) Acta Cryst. **B40** 404.

[7] Jones, J. E., and Ingham, A. E., (1925) Proc. Roy. Soc. Lond. **A107** 636.

[8] Ashcroft, N. W., and Mermin, N. D., (1976) *Solid State Physics* (Holt-Saunders).

[9] Anastasiou, N., and Fincham, D., (1982) Comput. Phys. Comm. **25** 159.

[10] Binder, K., (1986) *Monte Carlo Methods in Statistical Physics*, Topics in Current Physics **7** (2nd edition, Springer, Berlin).

[11] Binder, K., (1987) *Applications of the Monte Carlo Method in Statistical Physics*, Topics in Current Physics **36** (2nd edition, Springer, Berlin).

[12] Gould, H., and Tobochnik, J., (1988) *An Introduction to Computer Simulation Methods. Applications to Physical Systems* (Addison Wesley).

[13] Rahman, A., (1964) Phys. Rev. **136A** 405.

[14] Verlet, L., (1967) Phys. Rev. **159** 98.

[15] Hansen, J. P., and McDonald, I. R., (1986) *Theory of Simple Liquids* (2nd edition, Academic Press).

[16] Nicolas, J. J., Gubbins, K. E., Streett, W. T., and Tildesley, D. J., (1979) Molec. Phys. **37** 1429.

[17] Heermann, D. W., (1986) *Computer Simulation Methods in Theoretical Physics* (Springer, Berlin).

[18] Metropolis, N., Rosenbluth, A. W., Rosenbluth, M. N., Teller, A. H., and Teller, E., (1953) J. Chem. Phys. **21** 1087.

[19] Glauber, R. J., (1963) J. Math. Phys. **4**, 294.

[20] Thompson, C. J., (1972) *Mathematical Statistical Mechanics* (McMillan New York), p81-83.

[21] Tonks, L., (1936) Phys. Rev. **50** 955.

[22] Robledo, A., and Rowlinson, J. S., (1986) Molec. Phys. **58** 711.

[23] Erpenbeck, J. J., and Wood, W. W., (1977) in *Statistical Mechanics B*, Modern Theoretical Chemistry **6**, 1 (edited by B. J. Berne, Plenum).

[24] Valleau, J. P., and Whittington, S. G., (1977) *A guide to Monte Carlo for statistical mechanics. 1. Highways.* In *Statistical Mechanics A. Modern Theoretical Chemistry* (ed. B.J. Berne), Vol 5, 137-168 (Plenum Press, New York).

[25] Ciccotti, G., Frenkel, D., and McDonald, I. R., (1987) *Simulation of liquids and solids. Molecular dynamics and Monte Carlo methods in statistical mechanics* (North Holland).

[26] Huang, K., (1963) *Statistical Mechanics* (Wiley).

[27] Friedberg, R., and Cameron, J. E., (1970) J. Chem. Phys. **52** 6049.

[28] Herrmann, H. J., (1986) J. Stat. Phys. **45** 145.

[29] Pomeau, Y., (1984) J. Phys. **A17** L415.

[30] Vichniac, G., (1984) Physica **D10** 96.

[31] Fincham, D., (1987) Molec. Simulation **1** 1.

[32] Zorn, R., Herrmann, H. J., and Rebbi, C., (1981) Comput. Phys. Commun. **23** 337.

[33] The MC demo program on the DAP was written by consultants for Active Memory Technology working with researchers at Edinburgh University. The underlying work is described by Reddaway, S. F., Scott, D. M., and Smith, K. A., (1985) Comput. Phys. Commun. **37** 351.

[34] Busing, W. R., (1981) *WMIN, a computer program to model molecules and crystals in terms of potential energy functions*, (Oak Ridge National Laboratory Report ORNL-5747).

[35] Price, S. L., (1987) Molec. Phys. **62** 45.

[36] Tildesley D. J., and Madden, P. A., (1981) Molec. Phys. **42** 1137.

[37] Ree, F. H., (1980) J. Chem. Phys. **73** 5401.

[38] Heyes, D. M., (1984) J. Chem. Soc. Faraday Trans. II **80** 1363.

[39] Hammonds, K. D., and Heyes, D. M., (1988) J. Chem. Soc. Faraday Trans. II **84** 705.

[40] Atkins, P. W., (1987) *Physical Chemistry* (Oxford University Press).

[41] Heyes, D. M. and Woodcock, L. V. (1986) Molec. Phys. **59** 1369.

[42] Carnahan, N. F., and Starling, K. E., (1969) J. Chem. Phys. **51** 635.

[43] McQuarrie, D. A., (1976) *Statistical Mechanics* (Harper and Row New York).

[44] Chandler, D., (1987) *Introduction to Modern Statistical Mechanics* (Oxford University Press, New York).

[45] de Gennes, P., (1971) J. Chem. Phys. **55** 572.

[46] Wall, F. T., and Mandel, F., (1975) J. Chem. Phys. **63** 4592.

[47] Wall, F. T., Chin, J. C., and Mandel, F., (1977) J. Chem. Phys. **66** 3066.

[48] Muller-Krumbhaar, H., (1986) in ref [10], p261.

[49] Gilmer, G., (1980) Science **208** 355.

[50] Weeks, J., and Gilmer, G., (1979) Adv. Chem. Phys. **40** 157.

[51] Ermak, D. L., and McCammon, J. A., (1978) J. Chem. Phys **69** 1352.

[52] Allen, M. P., (1982) Molec. Phys. **47** 599.

[53] Heyes, D. M., (1988) J. Non-Newt. Fluid Mech. **27** 47.

[54] Jullien, R., (1987) Contemp. Phys. **28** 477.

[55] Mandelbrot, B. B., (1977) *Fractals: form, chance, and dimension* (W.H. Freeman).

[56] Pfeifer, P., and Avnir, D., (1983) J. Chem. Phys. **79** 3558.

[57] Meakin, P., (1983) Phys. Rev. **A27** 1495.

[58] Witten, T. A., and Sanders, L. M., (1983) Phys. Rev. **B27** 5686.

[59] Fincham, D. and Heyes, D. M., (1985) Adv. Chem. Phys. **LXII** 493.

[60] Woodcock, L. V., (1971) Chem. Phys. Lett. **10** 257.

[61] Evans, D. J., and Morriss, G. P., (1984) Phys. Rev. **A30** 1528, equations (5.1)-(5.6).

[62] Heyes, D. M., (1986) Molec. Phys. **57** 1265.

[63] Koch, S. W., Desai R. C., and Abraham, F. F., (1983) Phys. Rev. **A27** 2152.

[64] Udink, C., and Frenkel, D., (1987) Phys. Rev. **B35** 6933.

[65] Brigham, E. O., (1974) *The Fast Fourier Transform* (Prentice Hall N.J.).

[66] Stuart, R. D., (1982) *An Introduction to Fourier Analysis* (Science Paperbacks, Chapman and Hall, London).

[67] Champeney, D. C., (1985) *Fourier Transforms in Physics* (Student Monographs in Physics, Adam Hilger Ltd. Bristol).

[68] Chatfield, C., (1975) *The Analysis of Time Series: Theory and Practice* (Chapman and Hall, London).

[69] Gopal, E. S. R., (1974) *Statistical Mechanics and Properties of Matter* (Ellis Harwood Ltd. Chichester, UK and John Wiley and Sons).

[70] Harris, F. J., (1978) Proc. I.E.E.E. **66** 51.

[71] Kosloff D., and Kosloff, R., (1983) J. Comput. Phys. **52** 35.

[72] Honeycutt, J.D., and Andersen, H.C., (1987) J. Phys. Chem. **91** 4950.

[73] Herrmann, H.J., (1986) Phys. Rep. **136**, 153.

[74] Stauffer, D., (1986) *On Growth and Form* (ed H.E. Stanley and N. Ostwesky (Martinus Nijhoff) p 79; also Stauffer, D., (1985) *Introduction to Percolation Theory*, (Taylor & Francis, London).

[75] Gawlinski, E.T., and Stanley, H.E., (1981) J. Phys. **A14** L291.

[76] Balberg, I., and Binenbaum, N., (1985) Phys. Rev. **A31** 1222.

[77] Balberg, I. and Binenbaum N., (1987) Phys. Rev. **A35**, 5174.

[78] Bug, A. L. R., Safran, S. A., Grest, G. S., and Webman, I., (1985) Phys. Rev. Lett. **55** 1896.

[79] Seaton, N. A., and Glandt, E. D., (1987) J. Chem. Phys. **86** 4668.

[80] Stauffer, D., (1979), Phys. Rep. **54** 1.

[81] Heyes, D. M., (1987) J. Chem. Soc. Faraday Trans. II **83** 1985.

[82] Geiger, A., and Stanley, H. E., (1982), Phys. Rev. Lett. **49** 1895.

[83] Holian, B. L., and Grady, E. G., (1988) Phys. Rev. Lett. **60** 1355.

[84] Balberg, I., (1988) Phys. Rev. **B37** 2391.

[85] Lu, J. P., and Birman, J. L., (1987), J. Stat. Phys.} **46** 1057.

[86] Goldstein, H., (1980) *Classical Mechanics* (2nd edition, Addison-Wesley, Reading MA).

[87] Rahman, A., and Stillinger, F. H., (1971) J. Chem. Phys. **55** 3336.

[88] Barojas, J., Levesque, D., and Quentrec, B., Phys. Rev. **A7** 1092.

[89] Evans, D. J., (1977) Molec. Phys. **34** 317; Evans, D. J.. and Murad, S., (1977) Molec. Phys. **34** 327.

[90] Ryckaert, J. P., Ciccotti, G., and Berendsen, H. J. C., (1977) J. Comput. Phys. **23** 327.

[91] Andersen, H. C., (1983) J. Comput. Phys. **52** 24.

[92] Fincham, D., (1981) CCP5 Newsletter **2**, 6.

[93] Ciccotti, G., Ferrario, M., and Ryckaert, J. P., (1982) Molec. Phys. **47** 1253.

[94] Widom, B., (1963) J. Chem. Phys. **39** 2802.

[95] Romano, S., and Singer, K., (1979) Molec. Phys. **37** 1765.

[96] Powles, J. G., Evans, W. A. B., and Quirke, N., (1982) Molec. Phys. **47** 1347.

[97] Fincham, D., Quirke, N., and Tildesley, D. J., (1986) J. Chem. Phys. **84** 4535.

[98] Abraham, F. F., (1985) Adv. Phys. **35** 1.

[99] Hockney R. W., and Jesshope, C. R., (1981) *Parallel Computers* (Adam Hilger, Bristol).

[100] Rapaport, D. C., Comput. Phys. Rep. (to be published).

[101] Brode, S., and Ahrlichs, R., (1986) Comput. Phys. Commun. **42** 51.

[102] Quentrec, B., and Brot, C., (1975) J. Comput. Phys. **13** 430.

[103] Boris, J., (1986) J. Comput. Phys. **66** 1.

SUBJECT INDEX

GPSR Compliance

The European Union's (EU) General Product Safety Regulation (GPSR) is a set of rules that requires consumer products to be safe and our obligations to ensure this.

If you have any concerns about our products, you can contact us on ProductSafety@springernature.com

In case Publisher is established outside the EU, the EU authorized representative is:

Springer Nature Customer Service Center GmbH
Europaplatz 3
69115 Heidelberg, Germany

Batch number: 10816188

Printed by Printforce, the Netherlands